D1538242

SCHAUM'S OUTLINE OF

Theory and Problems of
BEGINNING CHEMISTRY

Third Edition

David E. Goldberg, Ph.D.

Professor of Chemistry
Brooklyn College
City University of New York

Schaum's Outline Series

McGRAW-HILL
New York   Chicago   San Francisco
Lisbon   London   Madrid   Mexico City   Milan
New Delhi   San Juan   Seoul   Singapore   Sydney   Toronto

The McGraw·Hill Companies

DAVID E. GOLDBERG received his Ph.D. from The Pennsylvania State University and joined the faculty of Brooklyn College in 1959. His primary interests are chemical and computer science education. He is the author or coauthor of 15 books and over 35 journal articles, as well as numerous booklets for student use. His books have been translated into seven foreign languages.

Schaum's Outline of Theory and Problems of
BEGINNING CHEMISTRY

Copyright © 2005, 1999, 1991 by The McGraw-Hill Companies, Inc. All rights reserved. Printed in the United States of America. Except as permitted under the United States Copyright Act of 1976, no part of this publication may be reproduced or distributed in any form or by any means, or stored in a data base or retrieval system, without the prior written permission of the publisher.

1 2 3 4 5 6 7 8 9 0   VFM VFM   0 9 8 7 6 5 4

ISBN 0-07-144780-6

This book is printed on recycled, acid-free paper containing a minimum of 50% recycled, de-inked fiber.

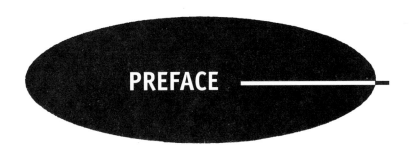

PREFACE

This book is designed to help students do well in their first chemistry course, especially those who have little or no chemistry background. It can be used effectively in a course preparatory to a general college chemistry course as well as in a course in chemistry for liberal arts students. It should also provide additional assistance to students in the first semester of a chemistry course for nurses and others in the allied health fields. It will prove to be of value in a high school chemistry course and in a general chemistry course for majors.

The book aims to help the student develop both problem-solving skills and skill in precise reading and interpreting scientific problems and questions. Analogies to everyday life introduce certain types of problems to make the underlying principles less abstract. Many of the problems were devised to clarify particular points often confused by beginning students. To ensure mastery, the book often presents problems in parts, then asks the same question as an entity, to see if the student can do the parts without the aid of the fragmented question. It provides some figures that have proved helpful to a generation of students.

The author gratefully acknowledges the help of the editors at McGraw-Hill.

DAVID E. GOLDBERG

TO THE STUDENT

This book is designed to help you understand chemistry fundamentals. Learning chemistry requires that you master chemical terminology and be able to perform calculations with ease. Toward these ends, many of the examples and problems are formulated to alert you to questions that sound different but are actually the same (Problem 3.16 for example) or questions that are different but sound very similar (Problems 5.13 and 7.25, for example). You should not attempt to memorize the solutions to the problems. (There is enough to memorize, without that.) Instead, you must try to understand the concepts involved. Your instructor and texts usually teach generalities (e.g., Atoms of all main group elements except noble gases have the number of outermost electrons equal to their group number.), but the instructor asks specific questions on exams (e.g., How many outermost electrons are there in a phosphorus atom?) You must not only know the principle, but also in what situations it applies.

You must practice by working many problems, because in addition to the principles, you must get accustomed to the many details involved in solving problems correctly. The key to success in chemistry is working very many problems! To get the most from this book, use a 5×8 card to cover up the solutions while you are doing the problems. Do not look at the answer first. It is easy to convince yourself that you know how to do a problem by looking at the answer, but generating the answer yourself, as you must do on exams, is not the same. After you have finished, compare your result with the answer given. If the method differs, it does not mean that your method is necessarily incorrect. If your answer is the same, your method is probably correct. Otherwise, try to understand what the difference is, and where you made a mistake, if you did so.

Some of the problems given after the text are very short and/or very easy (Problems 5.12 and 5.14, for example). They are designed to emphasize a particular point. After you get the correct answer, ask yourself why such a question was asked. Many other problems give analogies to everyday life, to help you understand a chemical principle (Problems 2.13 with 2.14, 4.6, 5.15 with 5.16, 7.13 through 7.16 and 10.41, for example).

Make sure you understand the chemical meaning of the terms presented throughout the semester. For example, "significant figures" means something very different in chemical calculations than in economic discussions. Special terms used for the first time in this book will be *italicized*. Whenever you encounter such a term, use it repeatedly until you thoroughly understand its meaning. If necessary, use the Glossary to find the meanings of unfamiliar terms.

Always use the proper units with measurable quantities. It makes quite a bit of difference if your pet is 4 in. tall or 4 ft tall! After Chapter 2, always use the proper number of significant figures in your calculations. Do yourself a favor and use the same symbols and abbreviations for chemical quantities that are used in the text. If you use a different symbol, you might become confused later when that symbol is used for a different quantity.

Some of the problems are stated in parts. After you do the problem by solving the various parts, see if you would know how to solve the same problem if only the last part were asked.

The conversion figure on page 348 shows all the conversions presented in the book. As you proceed, add the current conversions from the figure to your solution techniques.

CONTENTS

CHAPTER 1

Basic Concepts

1.1. INTRODUCTION

Chemistry is the study of matter and energy and the interactions between them. In this chapter, we learn about the elements, which are the building blocks of every type of matter in the universe, the measurement of matter (and energy) as mass, the properties by which the types of matter can be identified, and a basic classification of matter. The symbols used to represent the elements are also presented, and an arrangement of the elements into classes having similar properties, called a *periodic table*, is introduced. The periodic table is invaluable to the chemist for many types of classification and understanding.

Scientists have gathered so much data that they must have some way of organizing information in a useful form. Toward that end, scientific laws, hypotheses, and theories are used. These forms of generalization are introduced in Sec. 1.7.

1.2. THE ELEMENTS

An *element* is a substance that cannot be broken down into simpler substances by ordinary means. A few more than 100 elements and the many combinations of these elements—compounds or mixtures—account for all the materials of the world. Exploration of the moon has provided direct evidence that the earth's satellite is composed of the same elements as those on earth. Indirect evidence, in the form of light received from the sun and stars, confirms the fact that the same elements make up the entire universe. Before it was discovered on the earth, helium (from the Greek *helios*, meaning "sun") was discovered in the sun by the characteristic light it emits.

It is apparent from the wide variety of different materials in the world that there are a great many ways to combine elements. Changing one combination of elements to another is the chief interest of the chemist. It has long been of interest to know the composition of the crust of the earth, the oceans, and the atmosphere, since these are the only sources of raw materials for all the products that humans require. More recently, however, attention has focused on the problem of what to do with the products humans have used and no longer desire. Although elements can change combinations, they cannot be created or destroyed (except in nuclear reactions). The iron in a piece of scrap steel might rust and be changed in form and appearance, but the quantity of iron has not changed. Since there is a limited supply of available iron and since there is a limited capacity to dump unwanted wastes, recycling such materials is extremely important.

The elements occur in widely varying quantities on the earth. The 10 most abundant elements make up 98% of the mass of the crust of the earth. Many elements occur only in traces, and a few elements are synthetic. Fortunately for humanity, the elements are not distributed uniformly throughout the earth. The distinct properties of the different elements cause them to be concentrated more or less, making them more available as raw materials. For example, sodium and chlorine form salt, which is concentrated in beds by being dissolved in bodies of water that later dry up. Other natural processes are responsible for the distribution of the elements that now exists on the earth. It is interesting to note that different conditions on the moon—for example, the lack of water and air on the surface—might well cause a different sort of distribution of elements on the earth's satellite.

1.3. MATTER AND ENERGY

Chemistry focusses on the study of matter, including its composition, its properties, its structure, the changes that it undergoes, and the laws governing those changes. *Matter* is anything that has mass and occupies space. Any material object, no matter how large or small, is composed of matter. In contrast, light, heat, and sound are forms of energy. *Energy* is the ability to produce change. Whenever a change of any kind occurs, energy is involved; and whenever any form of energy is changed to another form, it is evidence that a change of some kind is occurring or has occurred.

The concept of mass is central to the discussion of matter and energy. The *mass* of an object depends on the quantity of matter in the object. The more mass the object has, the more it weighs, the harder it is to set into motion, and the harder it is to change the object's velocity once it is in motion.

Matter and energy are now known to be somewhat interconvertible. The quantity of energy producible from a quantity of matter, or vice versa, is given by Einstein's famous equation

$$E = mc^2$$

where E is the energy, m is the mass of the matter that is converted to energy, and c^2 is a constant—the square of the velocity of light. The constant c^2 is so large,

$$90\,000\,000\,000\,000\,000 \text{ meters}^2/\text{second}^2 \qquad \text{or} \qquad 34\,600\,000\,000 \text{ miles}^2/\text{second}^2$$

that tremendous quantities of energy are associated with conversions of minute quantities of matter to energy. The quantity of mass accounted for by the energy contained in a material object is so small that it is not measurable. Hence, the mass of an object is very nearly identical to the quantity of matter in the object. Particles of energy have very small masses despite having no matter whatsoever; that is, all the mass of a particle of light is associated with its energy. Even for the most energetic of light particles, the mass is small. The quantity of mass in any material body corresponding to the energy of the body is so small that it was not even conceived of until Einstein published his theory of relativity in 1905. Thereafter, it had only theoretical significance until World War II, when it was discovered how radioactive processes could be used to transform very small quantities of matter into very large quantities of energy, from which resulted the atomic and hydrogen bombs. Peaceful uses of atomic energy have developed since that time, including the production of the greater part of the electric power in many countries.

The mass of an object is directly associated with its weight. The *weight* of a body is the pull on the body by the nearest celestial body. On earth, the weight of a body is the pull of the earth on the body, but on the moon, the weight corresponds to the pull of the moon on the body. The weight of a body is directly proportional to its mass and also depends on the distance of the body from the center of the earth or moon or whatever celestial body the object is near. In contrast, the mass of an object is independent of its position. At any given location, for example, on the surface of the earth, the weight of an object is directly proportional to its mass.

When astronauts walk on the moon, they must take care to adjust to the lower gravity on the moon. Their masses are the same no matter where they are, but their weights are about one-sixth as much on the moon as on the earth because the moon is so much lighter than the earth. A given push, which would cause an astronaut to jump 1 ft high on the earth, would cause her or him to jump 6 ft on the moon. Since weight and mass are directly proportional on the surface of the earth, chemists have often used the terms interchangeably. The custom formerly was to use the term *weight*, but modern practice tends to use the term *mass* to describe quantities of matter. In this text, the term *mass* is used, but other chemistry texts might use the term *weight*, and the student must be aware that some instructors still prefer the latter.

The study of chemistry is concerned with the changes that matter undergoes, and therefore chemistry is also concerned with energy. Energy occurs in many forms—heat, light, sound, chemical energy, mechanical energy, electrical energy, and nuclear energy. In general, it is possible to convert each of these forms of energy to others.

Except for reactions in which the quantity of matter is changed, as in nuclear reactions, the **law of conservation of energy** is obeyed:

> Energy can be neither created nor destroyed (in the absence of nuclear reactions).

In fact, many chemical reactions are carried out for the sole purpose of converting energy to a desired form. For example, in the burning of fuels in homes, chemical energy is converted to heat; in the burning of fuels in automobiles, chemical energy is converted to energy of motion. Reactions occurring in batteries produce electrical energy from the chemical energy stored in the chemicals from which the batteries are constructed.

1.4. PROPERTIES

Every substance (Sec. 1.5) has certain characteristics that distinguish it from other substances and that may be used to establish that two specimens are the same substance or different substances. Those characteristics that serve to distinguish and identify a specimen of matter are called the *properties* of the substance. For example, water may be distinguished easily from iron or gold, and—although this may appear to be more difficult—iron may readily be distinguished from gold by means of the different properties of the metals.

EXAMPLE 1.1.   Suggest three ways in which a piece of iron can be distinguished from a piece of gold.

Ans.   Among other differences,

1. Iron, but not gold, will be attracted by a magnet.
2. If a piece of iron is left in humid air, it will rust. Under the same conditions, gold will undergo no appreciable change.
3. If a piece of iron and a piece of gold have exactly the same volume, the iron will have a lower mass than the gold.

Physical Properties

The properties related to the *state* (gas, liquid, or solid) or appearance of a sample are called *physical properties*. Some commonly known physical properties are density (Sec. 2.6), state at room temperature, color, hardness, melting point, and boiling point. The physical properties of a sample can usually be determined without changing its composition. Many physical properties can be measured and described in numerical terms, and comparison of such properties is often the best way to distinguish one substance from another.

Chemical Properties

A chemical reaction is a change in which at least one substance (Sec. 1.5) changes its composition and its set of properties. The characteristic ways in which a substance undergoes chemical reaction or fails to undergo chemical reaction are called its *chemical properties*. Examples of chemical properties are flammability, rust resistance, reactivity, and biodegradability. Many other examples of chemical properties will be presented in this book. Of the properties of iron listed in Example 1.1, only rusting is a chemical property. Rusting involves a change in composition (from iron to an iron oxide). The other properties listed do not involve any change in composition of the iron; they are physical properties.

1.5. CLASSIFICATION OF MATTER

To study the vast variety of materials that exist in the universe, the study must be made in a systematic manner. Therefore, matter is classified according to several different schemes. Matter may be classified as organic or inorganic. It is *organic* if it is a compound of carbon and hydrogen. (A more rigorous definition of organic must wait until Chap. 18.) Otherwise, it is *inorganic*. Another such scheme uses the composition of matter as a basis for classification; other schemes are based on chemical properties of the various classes. For example, substances may be classified as acids, bases, or salts (Chap. 8). Each scheme is useful, allowing the study of a vast variety of materials in terms of a given class.

In the method of classification of matter based on composition, a given specimen of material is regarded as either a pure substance or a mixture. An outline of this classification scheme is shown in Table 1-1. The term *pure substance* (or merely *substance*) refers to a material all parts of which have the same composition and that has a definite and unique set of properties. In contrast, a *mixture* consists of two or more substances and has a somewhat arbitrary composition. The properties of a mixture are not unique, but depend on its composition. The properties of a mixture tend to reflect the properties of the substances of which it is composed; that is, if the composition is changed a little, the properties will change a little.

**Table 1-1   Classification of Matter
Based on Composition**

Substances
   Elements
   Compounds
Mixtures
   Homogeneous mixtures (solutions)
   Heterogeneous mixtures (mixtures)

Substances

There are two kinds of substances—elements and compounds. *Elements* are substances that cannot be broken down into simpler substances by ordinary chemical means. Elements cannot be made by the combination of simpler substances. There are slightly more than 100 elements, and every material object in the universe consists of one or more of these elements. Familiar substances that are elements include carbon, aluminum, iron, copper, gold, oxygen, and hydrogen.

Compounds are substances consisting of two or more elements chemically combined in definite proportions by mass to give a material having a definite set of properties different from that of any of its constituent elements. For example, the compound water consists of 88.8% oxygen and 11.2% hydrogen by mass. The physical and chemical properties of water are distinctly different from those of both hydrogen and oxygen. For example, water is a liquid at room temperature and pressure, while the elements of which it is composed are gases under these same conditions. Chemically, water does not burn; hydrogen may burn explosively in oxygen (or air). Any sample of pure water, regardless of its source, has the same composition and the same properties.

There are millions of known compounds, and thousands of new ones are discovered or synthesized each year. Despite such a vast number of compounds, it is possible for the chemist to know certain properties of each one, because compounds can be classified according to their composition and structure, and groups of compounds in each class have some properties in common. For example, organic compounds are generally combustible in excess oxygen, yielding carbon dioxide and water. So any compound that contains carbon and hydrogen may be predicted by the chemist to be combustible in oxygen.

$$\text{Organic compound} + \text{oxygen} \longrightarrow \text{carbon dioxide} + \text{water} + \text{possible other products}$$

Mixtures

There are two kinds of mixtures—homogeneous mixtures and heterogeneous mixures. *Homogeneous mixtures* are also called *solutions*, and *heterogeneous mixtures* are sometimes simply called mixtures. In heterogeneous mixtures, it is possible to see differences in the sample merely by looking, although a microscope might be required. In contrast, homogeneous mixtures look the same throughout the sample, even under the best optical microscope.

EXAMPLE 1.2.   A teaspoon of salt is added to a cup of warm water. White crystals are seen at the bottom of the cup. Is the mixture homogeneous or heterogeneous? Then the mixture is stirred until the salt crystals disappear. Is the mixture now homogeneous or heterogeneous?

Ans.   Before stirring, the mixture is heterogeneous; after stirring, the mixture is homogeneous—a solution.

Distinguishing a Mixture from a Compound

Let us imagine an experiment to distinguish a mixture from a compound. Powdered sulfur is yellow and it dissolves in carbon disulfide, but it is not attracted by a magnet. Iron filings are black and are attracted by a magnet, but do not dissolve in carbon disulfide. You can mix iron filings and powdered sulfur in any ratio and get a yellowish-black mixture—the more sulfur that is present, the yellower the mixture will be. If you put the mixture in a test tube and hold a magnet alongside the test tube just above the mixture, the iron filings will be attracted, but the sulfur will not. If you pour enough (colorless) carbon disulfide on the mixture, stir, and then pour off the resulting yellow liquid, the sulfur dissolves but the iron does not. The mixture of iron filings and powdered sulfur is described as a mixture because the properties of the combination are still the properties of its components.

If you mix sulfur and iron filings in a certain proportion and then heat the mixture, you can see a red glow spread through the mixture. After it cools, the black solid lump that is produced—even if crushed into a powder—does not dissolve in carbon disulfide and is not attracted by a magnet. The material has a new set of properties; it is a compound, called iron(II) sulfide. It has a definite composition; and if, for example, you had mixed more iron with the sulfur originally, some iron(II) sulfide and some leftover iron would have resulted. The extra iron would not have become part of the compound.

1.6.   REPRESENTATION OF ELEMENTS

Each element has an internationally accepted *symbol* to represent it. A list of the names and symbols of the elements is found on page 349 of this book. Note that symbols for the elements are for the most part merely abbreviations of their names, consisting of either one or two letters. The first letter of the symbol is always written as a capital letter; the second letter, if any, is always written as a lowercase (small) letter. The symbols of a few elements do not suggest their English names, but are derived from the Latin or German names of the elements. The 10 elements whose names do not begin with the same letter as their symbols are listed in Table 1-2. For convenience, on page 349 of this book, these elements are listed twice—once alphabetically by name and again under the letter that is the first letter of their symbol. It is important to memorize the names and symbols of the most common elements. To facilitate this task, the most familiar elements are listed in Table 1-3. The elements with symbols in **bold** type should be learned first.

Table 1-2   Symbols and Names with Different Initials

Symbol	Name	Symbol	Name
Ag	Silver	Na	Sodium
Au	Gold	Pb	Lead
Fe	Iron	Sb	Antimony
Hg	Mercury	Sn	Tin
K	Potassium	W	Tungsten

The Periodic Table

A convenient way of displaying the elements is in the form of a periodic table, such as is shown on page 350 of this book. The basis for the arrangement of elements in the periodic table will be discussed at length in Chaps. 3 and 4. For the present, the periodic table is regarded as a convenient source of general information about the elements. It will be used repeatedly throughout the book. One example of its use, shown in Fig. 1-1, is to classify the elements as metals or nonmetals. All the elements except hydrogen that lie to the left of the stepped line drawn on the periodic table, starting to the left of B and descending stepwise to a point between Po and At, are metals. The other elements are nonmetals. It is readily seen that the majority of elements are metals.

Table 1-3   Important Elements Whose Names and Symbols Should Be Known

1 H																	2 He
3 Li	4 Be											5 B	6 C	7 N	8 O	9 F	10 Ne
11 Na	12 Mg											13 Al	14 Si	15 P	16 S	17 Cl	18 Ar
19 K	20 Ca	21 Sc	22 Ti	23 V	24 Cr	25 Mn	26 Fe	27 Co	28 Ni	29 Cu	30 Zn	31 Ga	32 Ge	33 As	34 Se	35 Br	36 Kr
37 Rb	38 Sr								46 Pd	47 Ag	48 Cd		50 Sn	51 Sb	52 Te	53 I	54 Xe
55 Cs	56 Ba				74 W				78 Pt	79 Au	80 Hg		82 Pb	83 Bi			86 Rn

	92 U	

The smallest particle of an element that retains the composition of the element is called an *atom*. Details of the nature of atoms are given in Chaps. 3 and 4. The symbol of an element is used to stand for one atom of the element as well as for the element itself.

1.7.   LAWS, HYPOTHESES, AND THEORIES

In chemistry, as in all sciences, it is necessary to express ideas in terms having very precise meanings. These meanings are often unlike the meanings of the same words in nonscientific usage. For example, the meaning of the word *property* as used in chemistry can be quite different from its meaning in ordinary conversation. Also, in chemical terminology, a concept may be represented by abbreviations, such as symbols or formulas, or by some mathematical expression. Together with precisely defined terms, such symbols and mathematical expressions constitute a language of chemistry. This language must be learned well. As an aid to recognition of special terms, when such terms are used for the first time in this book, they will be *italicized*.

A statement that generalizes a quantity of experimentally observable phenomena is called a scientific *law*. For example, if a person drops a pencil, it falls downward. This result is predicted by the law of gravity. A generalization that attempts to explain why certain experimental results occur is called a *hypothesis*. When a hypothesis is accepted as true by the scientific community, it is then called a *theory*. One of the most important scientific laws is the law of conservation of mass: During any process (chemical reaction, physical change, or

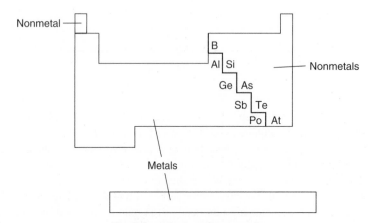

Fig. 1-1.   Metals and nonmetals

even a nuclear reaction) mass is neither created nor destroyed. Because of the close approximation that the mass of an object is the quantity of matter it contains (excluding the mass corresponding to its energy) the law of conservation of mass can be approximated by the law of conservation of matter: During an ordinary chemical reaction, matter can be neither created nor destroyed.

EXAMPLE 1.3. When a piece of iron is left in moist air, its surface gradually turns brown and the object gains mass. Explain this phenomenon.

Ans.   The brown material is an iron oxide, rust, formed by a reaction of the iron with the oxygen in the air.

$$\text{Iron} + \text{oxygen} \longrightarrow \text{an iron oxide}$$

The increase in mass is just the mass of the combined oxygen. When a long burns, the ash (which remains) is much lighter than the original log, but this is not a contradiction of the law of conservation of matter. In addition to the log, which consists mostly of compounds containing carbon, hydrogen, and oxygen, oxygen from the air is consumed by the reaction. In addition to the ash, carbon dioxide and water vapor are produced by the reaction.

$$\text{Log} + \text{oxygen} \longrightarrow \text{ash} + \text{carbon dioxide} + \text{water vapor}$$

The total mass of the ash plus the carbon dioxide and the water vapor is equal to the total mass of the log plus the oxygen. As always, the law of conservation of matter is obeyed as precisely as chemists can measure. The law of conservation of mass is fundamental to the understanding of chemical reactions. Other laws related to the behavior of matter are equally important, and learning how to apply these laws correctly is a necessary goal of the study of chemistry.

Solved Problems

1.1.   Are elements heterogeneous or homogeneous?

　　　　Ans.   Homogeneous. They look alike throughout the sample because they are alike throughout the sample.

1.2.   How can you tell if the word *mixture* means any mixture or a heterogeneous mixture?

　　　　Ans.   You can tell from the context. For example, if a problem asks if a sample is a solution or a mixture, the word *mixture* means heterogeneous mixture. If it asks whether the sample is a compound or a mixture, it means any kind of mixture. Such usage occurs in ordinary English as well as in technical usage. For example, the word *day* has two meanings—one is a subdivision of the other. "How many hours are there in a day? What is the opposite of night?"

1.3.   Are compounds heterogeneous or homogeneous?

　　　　Ans.   Homogeneous. They look alike throughout the sample because they are alike throughout the sample. Since there is only one substance present, even if it is a combination of elements, it must be alike throughout.

1.4.   A generality states that all compounds containing both carbon and hydrogen burn. Do octane and propane burn? (Each contains only carbon and hydrogen.)

　　　　Ans.   Yes, both burn. It is easier to learn that all organic compounds burn than to learn a list of millions of organic compounds that burn. On an examination, however, a question will probably specify one particular organic compound. You must learn a generality and be able to respond to a specific example of it.

1.5.   Sodium is a very reactive metallic element; for example, it liberates hydrogen gas when treated with water. Chlorine is a yellow-green, choking gas, used in World War I as a poison gas. Contrast these properties with those of the compound of sodium and chlorine—sodium chloride—known as table salt.

　　　　Ans.   Salt does not react with water to liberate hydrogen, is not reactive, and is not poisonous. It is a white solid and not a silvery metal or a green gas. In short, it has its own set of properties; it is a compound.

1.6. TNT is a compound of carbon, nitrogen, hydrogen, and oxygen. Carbon occurs is two common forms—graphite (the material in "lead pencils") and diamond. Oxygen and nitrogen comprise over 98% of the atmosphere. Hydrogen is an element that reacts explosively with oxygen. Which of the properties of the elements determines the properties of TNT?

Ans.   The properties of the elements do not matter. The properties of the compound are quite independent of those of the elements. A compound has its own distinctive set of properties. TNT is most noted for its explosiveness.

1.7. What properties of stainless steel make it more desirable for many purposes than ordinary steel?

Ans.   Its resistance to rusting and corrosion.

1.8. What properties of DDT make it useful? What properties make it undesirable?

Ans.   DDT's toxicity to insects is its useful property; its toxicity to humans, birds, and other animals makes it undesirable. It is stable, that is, nonbiodegradable (does not decompose spontaneously to simpler substances in the environment). This property makes its use as an insecticide more difficult.

1.9. Name an object or an instrument that changes:

(*a*)   chemical energy to heat   (*d*)   electrical energy to light

(*b*)   chemical energy to electrical energy   (*e*)   motion to electrical energy

(*c*)   electrical energy to chemical energy   (*f*)   electrical energy to motion

   Ans.   (*a*)   gas stove   (*d*)   lightbulb

           (*b*)   battery   (*e*)   generator or alternator

           (*c*)   rechargeable battery   (*f*)   electric motor

1.10. A sample contains 88.8% oxygen and 11.2% hydrogen by mass, and is gaseous and explosive at room temperature. (*a*) Is the sample a compound or a mixture? (*b*) After the sample explodes and cools, it is a liquid. Is the sample now a compound or a mixture? (*c*) Would it be easier to change the percentage of oxygen before or after the explosion?

Ans.   (*a*)   The sample is a mixture. (The compound of hydrogen and oxygen with this composition—water—is a liquid under these conditions.)

      (*b*)   It is a compound.

      (*c*)   Before the explosion. It is easy to add hydrogen or oxygen to the gaseous mixture, but you cannot change the composition of water.

1.11. Name one exception to the statement that nonmetals lie to the right of the stepped line in the periodic table (page 350).

Ans.   Hydrogen

1.12. Calculate the ratio of the number of metals to the number of nonmetals in the periodic table (page 350).

Ans.   There are 109 elements whose symbols are presented, of which 22 are nonmetals and 87 are metals, so the ratio is 3.95 metals per nonmetal.

1.13. Give the symbol for each of the following elements: (*a*) iron, (*b*) copper, (*c*) carbon, (*d*) sodium, (*e*) silver, (*f*) aluminum.

Ans.   (*a*) Fe   (*b*) Cu   (*c*) C   (*d*) Na   (*e*) Ag   (*f*) Al

1.14. Name each of the following elements: (*a*) K, (*b*) P, (*c*) Cl, (*d*) H, (*e*) O.

Ans.   (*a*) Potassium   (*b*) Phosphorus   (*c*) Chlorine   (*d*) Hydrogen   (*e*) Oxygen

1.15. Distinguish between a theory and a law.

    Ans.   A law tells what happens under a given set of circumstances, while a theory attempts to explain *why* that behavior occurs.

1.16. Distinguish clearly between (*a*) mass and matter and (*b*) mass and weight.

    Ans.   (*a*) *Matter* is any kind of material. The mass of an object depends mainly on the matter that it contains. It is affected only very slightly by the energy in it.

              (*b*) *Weight* is the attraction of the earth on an object. It depends on the mass of the object and its distance to the center of the earth.

CHAPTER 2

Mathematical Methods in Chemistry

2.1. INTRODUCTION

Physical sciences, and chemistry in particular, are quantitative. Not only must chemists describe things qualitatively, but also they must measure them quantitatively and compute numeric results from the measurements. The factor-label method is introduced in Sec. 2.2 to aid students in deciding how to do certain calculations. The metric system (Sec. 2.3) is a system of units designed to make the calculation of measured quantities as easy as possible. Exponential notation (Sec. 2.4) is designed to enable scientists to work with numbers that range from incredibly huge to unbelievably tiny. The scientist must report the results of the measurements so that any reader will have an appreciation of how precisely the measurements were made. This reporting is done by using the proper number of significant figures (Sec. 2.5). Density calculations are introduced in Sec. 2.6 to enable the student to use all the techniques described thus far. Temperature scales are presented in Sec. 2.7.

The units of each measurement are as important as the numeric value, and must always be stated with the number. Moreover, we will use the units to help us in our calculations (Sec. 2.2).

2.2. FACTOR-LABEL METHOD

The units of a measurement are an integral part of the measurement. In many ways, they may be treated as algebraic quantities, like *x* and *y* in mathematical equations. *You must always state the units when making measurements and calculations.*

The units are very helpful in suggesting a good approach for solving many problems. For example, by considering units in a problem, you can easily decide whether to multiply or divide two quantities to arrive at the answer. The *factor-label method*, also called *dimensional analysis* or the *factor-unit method*, may be used for quantities that are directly proportional to one another. (When one quantity goes up, the other does so in a similar manner. For example, when the number of dimes in a piggy bank goes up, so does the amount in dollars.) Over 75% of the problems in general chemistry can be solved with the factor-label method. Let us look at an example to introduce the factor-label method.

How many dimes are there in 9.40 dollars? We know that

$$10 \text{ dimes} = 1 \text{ dollar} \qquad \text{or} \qquad 1 \text{ dime} = 0.1 \text{ dollar}$$

We may divide both sides of the first of these equations by 10 dimes or by 1 dollar, yielding

$$\frac{10 \text{ dimes}}{10 \text{ dimes}} = \frac{1 \text{ dollar}}{10 \text{ dimes}} \qquad \text{or} \qquad \frac{10 \text{ dimes}}{1 \text{ dollar}} = \frac{1 \text{ dollar}}{1 \text{ dollar}}$$

Since the numerator and denominator (top and bottom) of the fraction on the left side of the first equation are the same, the ratio is equal to 1. The ratio 1 dollar/10 dimes is therefore equal to 1. By analogous argument, the first ratio of the equation to the right is also equal to 1. That being the case, we can multiply any quantity by either ratio without changing the value of that quantity, because multiplying by 1 does not change the value of anything. We call each ratio a *factor*; the units are the *labels*.

We can use the equation 1 dime = 0.1 dollar to arrive at the following equivalent equations:

$$\frac{1 \text{ dime}}{1 \text{ dime}} = \frac{0.1 \text{ dollar}}{1 \text{ dime}} \qquad\qquad \frac{1 \text{ dime}}{0.1 \text{ dollar}} = \frac{0.1 \text{ dollar}}{0.1 \text{ dollar}}$$

Students often like to use the equations above to avoid using decimal fractions, but these might be more useful later (Sec. 2.3).

To use the factor-label method, start with the quantity given (not a rate or ratio). Multiply that quantity by a factor, or more than one factor, until an answer with the desired units is obtained.

Back to the problem:

$$9.40 \text{ dollars}\left(\frac{10 \text{ dimes}}{1 \text{ dollar}}\right) = \qquad \text{or} \qquad 9.40 \text{ dollars}\left(\frac{1 \text{ dime}}{0.1 \text{ dollar}}\right) =$$

The dollar in the denominator cancels the dollars in the quantity given (the unit, not the number). It does not matter if the units are singular (dollar) or plural (dollars). We multiply by the number in the numerator of the ratio and divide by the number in the denominator. That gives us

$$9.40 \text{ \cancel{dollars}}\left(\frac{10 \text{ dimes}}{1 \text{ \cancel{dollar}}}\right) = 94 \text{ dimes} \qquad \text{or} \qquad 9.40 \text{ \cancel{dollars}}\left(\frac{1 \text{ dime}}{0.1 \text{ \cancel{dollar}}}\right) = 94 \text{ dimes}$$

EXAMPLE 2.1.   How many dollars are there in 220 dimes?

Ans.
$$220 \text{ \cancel{dimes}}\left(\frac{1 \text{ dollar}}{10 \text{ \cancel{dimes}}}\right) = 22.00 \text{ dollars}$$

In this case, the unit dimes canceled. Suppose we had multiplied by the original ratio:

$$220 \text{ dimes}\left(\frac{10 \text{ dimes}}{1 \text{ dollar}}\right) = \frac{2200 \text{ dimes}^2}{\text{dollar}}$$

Indeed, this expression has the same value, but the units are unfamiliar and the answer is useless.

More than one factor might be required in a single problem. The steps can be done one at a time, but it is more efficient to do them all at once.

EXAMPLE 2.2.   Calculate the number of seconds in 2.50 h.

Ans.   We can first calculate the number of minutes in 2.50 h.

$$2.50 \text{ h}\left(\frac{60 \text{ min}}{1 \text{ h}}\right) = 150 \text{ min}$$

Then we can change the minutes to seconds:

$$150 \text{ min}\left(\frac{60 \text{ s}}{1 \text{ min}}\right) = 9000 \text{ s}$$

Better, however, is to do both multiplications in the same step:

$$2.50\,\text{h}\left(\frac{60\,\text{min}}{1\,\text{h}}\right)\left(\frac{60\,\text{s}}{1\,\text{min}}\right) = 9000\,\text{s}$$

EXAMPLE 2.3.   Calculate the speed in feet per second of a jogger running 7.50 miles per hour (mi/h).

Ans.

$$\frac{7.50\,\text{mi}}{1\,\text{h}}\left(\frac{5280\,\text{ft}}{1\,\text{mi}}\right) = \frac{39\,600\,\text{ft}}{1\,\text{h}}$$

$$\frac{39\,600\,\text{ft}}{1\,\text{h}}\left(\frac{1\,\text{h}}{60\,\text{min}}\right) = \frac{660\,\text{ft}}{1\,\text{min}}$$

$$\frac{660\,\text{ft}}{1\,\text{min}}\left(\frac{1\,\text{min}}{60\,\text{s}}\right) = \frac{11.0\,\text{ft}}{1\,\text{s}}$$

Alternatively,

$$\frac{7.50\,\text{mi}}{1\,\text{h}}\left(\frac{5280\,\text{ft}}{1\,\text{mi}}\right)\left(\frac{1\,\text{h}}{60\,\text{min}}\right)\left(\frac{1\,\text{min}}{60\,\text{s}}\right) = \frac{11.0\,\text{ft}}{1\,\text{s}}$$

It is usually more reassuring, at least at the beginning, to do such a problem one step at a time. But if you look at the combined solution, you can see that it is easier to do the whole thing at once. With an electronic calculator, we need to press the equals $\boxed{=}$ key only once, and not round until the final answer (Sec. 2.5).

We will expand our use of the factor-label method in later sections.

2.3.   METRIC SYSTEM

Scientists measure many different quantities—length, volume, mass (weight), electric current, voltage, resistance, temperature, pressure, force, magnetic field intensity, radioactivity, and many others. The *metric system* and its recent extension, *Système International d'Unités* (SI), were devised to make measurements and calculations as simple as possible. In this section, length, area, volume, and mass will be introduced. Temperature will be introduced in Sec. 2.7 and used extensively in Chap. 12. The quantities to be discussed here are presented in Table 2-1. Their units, abbreviations of the quantities and units, and the legal standards for the quantities are also included.

Table 2-1   Metric Units for Basic Quantities

Quantity	Abbreviation	Fundamental Unit	Abbreviation of Unit	Standard	Comment
Length or distance	*l*	meter	m	meter	
	d				
Area	*A*	meter^2	m^2	meter^2	
Volume	*V*	meter^3	m^3	meter^3	SI unit
		or liter	L		older metric unit
					1 m^3 = 1000 L
Mass	*m*	gram	g	kilogram	1 kg = 1000 g

Length (Distance)

The unit of length, or distance, is the *meter*. Originally conceived of as one ten-millionth of the distance from the north pole to the equator through Paris, the meter is more accurately defined as the distance between two scratches on a platinum-iridium bar kept in Paris. The U.S. standard is the distance between two scratches on a similar bar kept at the National Institute of Standards and Technology. The meter is about 10% greater than the yard—39.37 in. to be more precise.

Larger and smaller distances may be measured with units formed by the addition of prefixes to the word *meter*. The important metric prefixes are listed in Table 2-2. The most commonly used prefixes are *kilo*, *milli*, and *centi*. The prefix kilo means 1000 times the fundamental unit, no matter to which fundamental unit it is attached. For example, 1 kilodollar is 1000 dollars. The prefix milli indicates one-thousandth of the fundamental unit. Thus, 1 millimeter is 0.001 meter; 1 mm = 0.001 m. The prefix centi means one-hundredth. A centidollar is one cent; the name for this unit of money comes from the same source as the metric prefix.

Table 2-2   Metric Prefixes

Prefix	Abbreviation	Meaning	Example
giga	G	1 000 000 000	1 Gm = 1 000 000 000 m
mega	M	1 000 000	1 Mm = 1 000 000 m
kilo	**k**	**1000**	**1 km = 1000 m**
deci	d	0.1	1 dm = 0.1 m
centi	**c**	**0.01**	**1 cm = 0.01 m**
milli	**m**	**0.001**	**1 mm = 0.001 m**
micro	μ	0.000 001	1 μm = 0.000 001 m
nano	n	1×10^{-9}	1 nm = 1×10^{-9} m
pico	p	1×10^{-12}	1 pm = 1×10^{-12} m

EXAMPLE 2.4.   Since meter is abbreviated m (Table 2-1) and milli is abbreviated m (Table 2-2), how can you tell the difference?

Ans.   Since milli is a prefix, it must always precede a quantity. If m is used without another letter, or if the m follows another letter, then m stands for the unit *meter*. If m precedes another letter, m stands for the prefix *milli*.

The metric system was designed to make calculations easier than using the English system in the following ways:

Metric	**English**
Subdivisions of all dimensions have the same prefixes with the same meanings and the same abbreviations.	There are different names for subdivisions.
Subdivisions all differ by powers of 10.	Subdivisions differ by arbitrary factors, rarely powers of 10.
There are no duplicate names with different meanings.	The same names often have different meanings.
The abbreviations are generally easily recognizable.	The abbreviations are often hard to recognize (e.g., lb for pound and oz for ounce).

Beginning students sometimes regard the metric system as difficult because it is new to them and because they think they must learn English-metric conversion factors (Table 2-3). Engineers do have to work in both systems in the United States, but scientists generally do not work in the English system at all. Once you familiarize yourself with the metric system, it is much easier to work with than the English system is.

Table 2-3   Some English-Metric Conversions

	Metric	**English**
Length	1 meter	39.37 inches
	2.54 centimeters	1 inch
Volume	1 liter	1.06 U.S. quarts
Mass	1 kilogram	2.2045 pounds (avoirdupois)
	28.35 grams	1 ounce

EXAMPLE 2.5.   (*a*) How many feet (ft) are there in 1.450 miles (mi)? (*b*) How many meters (m) are there in 1.450 kilometers (km)?

Ans.   (*a*) $1.450 \text{ mi} \left(\dfrac{5280 \text{ ft}}{1 \text{ mi}} \right) = 7656 \text{ ft}$

(*b*) $1.450 \text{ km} \left(\dfrac{1000 \text{ m}}{1 \text{ km}} \right) = 1450 \text{ m}$

You can do the calculation of part (*b*) in your head (merely move the decimal point in 1.450 three places to the right). The calculation of part (*a*) requires a calculator or pencil and paper.

Instructors often require English-metric conversions for two purposes: to familiarize the student with the relative sizes of the metric units in terms of the more familiar English units, and for practice in conversions (Sec. 2.2). Once you really get into the course, the number of English-metric conversions that you do is very small.

One of the main advantages of the metric system is that the same prefixes are used with all quantities, and the prefixes always have the same meanings.

EXAMPLE 2.6.   The unit of electric current is the ampere. What is the meaning of 1 milliampere?

Ans.   1 milliampere $= 0.001$ ampere   1 mA $= 0.001$ A

Even if you do not recognize the quantity, the prefix always has the same meaning.

EXAMPLE 2.7.   How many centimeters are there in 5.000 m?

Ans.   Each meter is 100 centimeters (cm); 5.000 m is 500.0 cm.

Area

The extent of a surface is called its *area*. The area of a rectangle (or a square, which is a rectangle with all sides equal) is its length times its width.

$$A = l \times w$$

The dimensions of area are thus the product of the dimensions of two distances. The area of a state or country is usually reported in square miles or square kilometers, for example. If you buy interior paint, you can expect a gallon of paint to cover about 400 ft^2. These units are stated aloud "square feet," but are usually written ft^2. The exponent (the superscript number) means that the unit is multiplied that number of times, just as it does with a number. For example, ft^2 means ft $\times$ ft.

EXAMPLE 2.8.   State aloud the area of Rhode Island, 1214 mi^2.

Ans.   "Twelve hundred fourteen square miles."

EXAMPLE 2.9.   A certain square is 3.0 m on each side. What is its area?

Ans. $A = l^2 = (3.0 \text{ m})^2 = 9.0 \text{ m}^2$

Note the difference between "3 meters, squared" and "3 square meters."

$$(3 \text{ m})^2 \quad \text{and} \quad 3 \text{ m}^2$$

The former means that the coefficient (3) is also squared; the latter does not.

EXAMPLE 2.10.   A rectangle having an area of 22.0 m^2 is 4.00 m wide. How long is it?

Ans.
$$A = l \times w$$

$$22.0 \text{ m}^2 = l(4.00 \text{ m})$$

$$l = 5.50 \text{ m}$$

Note that the length has a unit of distance (meter).

EXAMPLE 2.11.   What happens to the area of a square when the length of each side is doubled?

Ans.   Let l = original length of side; then l^2 = original area; $2l$ = new length of side; so $(2l)^2$ = new area. The area has increased from l^2 to $4l^2$; it has increased by a factor of 4. (See Problem 2.24.)

Volume

The SI unit of volume is the cubic meter, m^3. Just as area is derived from length, so can volume be derived from length. Volume is length $\times$ length $\times$ length. Volume also can be regarded as area $\times$ length. The cubic meter is a rather large unit; a cement mixer usually can carry between 2 and 3 m^3 of cement. Smaller units are dm^3, cm^3, and mm^3; the first two of these are reasonable sizes to be useful in the laboratory.

The older version of the metric system uses the *liter* as the unit of volume. It is defined as 1 dm^3. Chemists often use the liter in preference to m^3 because it is about the magnitude of the quantities with which they deal. The student has to know both units and the relationship between them.

Often it is necessary to multiply by a factor raised to a power. Consider the problem of changing 5.00 m^3 to cubic centimeters:

$$5.00 \text{ m}^3 \left(\frac{100 \text{ cm}}{1 \text{ m}} \right)$$

If we multiply by the ratio of 100 cm to 1 m, we will still be left with m^2 (and cm) in our answer. We must multiply by (100 cm/m) three times;

$$5.00 \text{ m}^3 \left(\frac{100 \text{ cm}}{1 \text{ m}} \right)^3 = 5\,000\,000 \text{ cm}^3$$

$$\left(\frac{100 \text{ cm}}{1 \text{ m}} \right)^3 \quad \text{means} \quad \left(\frac{100 \text{ cm}}{1 \text{ m}} \right)\left(\frac{100 \text{ cm}}{1 \text{ m}} \right)\left(\frac{100 \text{ cm}}{1 \text{ m}} \right)$$

and includes 100^3 cm^3 in the numerator and 1 m^3 in the denominator.

EXAMPLE 2.12.   How many liters are there in 1 m^3?

Ans.
$$1 \text{ m}^3 \left(\frac{10 \text{ dm}}{1 \text{ m}} \right)^3 \left(\frac{1 \text{ L}}{1 \text{ dm}^3} \right) = 1000 \text{ L}$$

One cubic meter is 1000 L. The liter can have prefixes just as any other unit can. Thus 1 mL is 0.001 L, and 1 kL is 1000 L = 1 m^3.

Mass

Mass is a measure of the quantity of material in a sample. We can measure that mass by its weight—the attraction of the sample to the earth or by its inertia—the resistance to change in its motion. Since weight and mass are directly proportional as long as we stay on the surface of the earth, chemists sometimes use these terms interchangeably. (Physicists do not do that.)

The unit of mass is the *gram*. [Since 1 g is *a very small mass*, the legal standard of mass in the United States is the kilogram. A **standard** for a type of measurement is an easily measured quantity that is chosen for comparison with all units and subunits. Usually, the basic unit is chosen as the standard, but the kilogram (not the gram) is chosen as the standard of mass because it is easier to measure precisely. Since there can be only one standard for a given type of measurement, and the cubic meter has been chosen for volume, the liter is not a standard of volume.]

It is very important that you get used to writing the proper abbreviations for units and the proper symbols at the beginning of your study of chemistry, so you do not get mixed up later.

EXAMPLE 2.13. What is the difference between mg and Mg, two units of mass?

Ans. Lowercase m stands for *milli* (Table 2-2), and 1 mg is 0.001 g. Capital M stands for *mega*, and 1 Mg is 1 000 000 g. It is obviously important not to confuse the capital M and lowercase m in such cases.

2.4. EXPONENTIAL NUMBERS

The numbers that scientists use range from enormous to extremely tiny. The distances between the stars are literally astronomical—the star nearest to the sun is 23 500 000 000 000 mi from it. As another example, the number of atoms of calcium in 40.0 g of calcium is 602 000 000 000 000 000 000 000, or 602 thousand billion billion. The diameter of one calcium atom is about 0.000 000 02 cm. To report and work with such large and small numbers, scientists use *exponential notation*. A typical number written in exponential notation looks as follows:

$$4.13 \times 10^3, \text{interpreted as} \quad \underbrace{4.13}_{\text{Coefficient}} \quad \times \quad \underbrace{\underbrace{10}_{\text{Base}} \underbrace{^3}_{\text{Exponent}}}_{\text{Exponential part}}$$

The coefficient is merely a decimal number written in the ordinary way. That coefficient is multiplied by the exponential part, made up of the base (10) and the exponent. (Ten is the only base that will be used in numbers in exponential form in the general chemistry course.) The exponent tells how many times the coefficient is multiplied by the base.

$$4.13 \times 10^3 = 4.13 \times 10 \times 10 \times 10 = 4130$$

Since the exponent is 3, the coefficient is multiplied by three 10s.

EXAMPLE 2.14. What is the value of 10^5?

Ans. When an exponential is written without an explicit coefficient, a coefficient of 1 is implied:

$$1 \times 10 \times 10 \times 10 \times 10 \times 10 = 100\ 000$$

There are five 10s multiplying the implied 1.

EXAMPLE 2.15. What is the value of 10^1? of 10^0?

Ans. $10^1 = 10$. There is one 10 multiplying the implied coefficient of 1.
$10^0 = 1$. There are no 10s multiplying the implied coefficient of 1.

EXAMPLE 2.16. Write 2.0×10^4 in decimal form.

Ans. $$2.0 \times 10 \times 10 \times 10 \times 10 = 20\ 000$$

When scientists write numbers in exponential form, they prefer to write them so that the coefficient has one and only one digit to the left of the decimal point, and that digit is not zero. That notation is called *standard exponential form*, or *scientific notation*.

EXAMPLE 2.17. Write 455 000 in standard exponential form.

Ans. $455\,000 = 4.55 \times 10 \times 10 \times 10 \times 10 \times 10 = 4.55 \times 10^5$

The number of 10s is the number of places in 455 000 that the decimal point must be moved to the left to get one (nonzero) digit to the left of the decimal point.

Using an Electronic Calculator

If you use an electronic calculator with exponential capability, note that there is a special key (labeled $\boxed{\text{EE}}$ or $\boxed{\text{EXP}}$) on the calculator which means "times 10 to the power." If you wish to enter 5×10^3, push $\boxed{5}$, then the special key, then $\boxed{3}$. Do *not* push $\boxed{5}$, then the multiply key, then $\boxed{1}$, then $\boxed{0}$, then the special key, then $\boxed{3}$. If you do so, your value will be 10 times too large. See the Appendix.

EXAMPLE 2.18. List the keystrokes on an electronic calculator which are necessary to do the following calculation:

$$6.5 \times 10^3 + 4 \times 10^2 =$$

Ans.

$$\boxed{6}\ \boxed{.}\ \boxed{5}\ \boxed{\text{EXP}}\ \boxed{3}\ \boxed{+}\ \boxed{4}\ \boxed{\text{EXP}}\ \boxed{2}\ \boxed{=}$$

On the electronic calculator, to change the sign of a number, you use the $\boxed{+/-}$ key, not the $\boxed{-}$ (minus) key. The $\boxed{+/-}$ key can be used to change the sign of a coefficient or an exponent, depending on when it is pressed. If it is pressed *after* the $\boxed{\text{EE}}$ or $\boxed{\text{EXP}}$ key, it works on the exponent rather than the coefficient.

EXAMPLE 2.19. List the keystrokes on an electronic calculator which are necessary to do the following calculation:

$$(7.7 \times 10^{-5})/(-4.5 \times 10^6) =$$

Ans.

$$\boxed{7}\ \boxed{.}\ \boxed{7}\ \boxed{\text{EXP}}\ \boxed{5}\ \boxed{+/-}\ \boxed{\div}\ \boxed{4}\ \boxed{.}\ \boxed{5}\ \boxed{+/-}\ \boxed{\text{EXP}}\ \boxed{6}\ \boxed{=}$$

The answer is -1.7×10^{-11}.

2.5. SIGNIFICANT DIGITS

No matter how accurate the measuring device you use, you can make measurements only to a certain degree of accuracy. For example, would you attempt to measure the length of your shoe (a distance) with an automobile odometer (mileage indicator)? The mileage indicator has tenths of miles (or kilometers) as its smallest scale division, and you can estimate to the nearest hundredth of a mile, or something like 50 ft, but that would be useless to measure the length of a shoe. No matter how you tried, or how many measurements you made with the odometer, you could not measure such a small distance. In contrast, could you measure the distance from the Empire State Building in New York City to the Washington Monument with a 10-cm ruler? You might at first think that it would take a long time, but that it would be possible. However, it would take so many separate measurements, each having some inaccuracy in it, that the final result would be about as bad as measuring the shoe size with the odometer. The conclusion that you should draw from this discussion is that you should use the proper measuring device for each measurement, and that no matter how hard you try, each measuring device has a certain limit to its accuracy, and you cannot measure more accurately with it. In general, you should estimate each measurement to one-tenth the smallest scale division of the instrument that you are using.

Precision is the closeness of a set of measurements to each other; accuracy is the closeness of the average of a set of measurements to the true value. Scientists report the precision for their measurements by using a certain number of digits. They report all the digits they know for certain plus one extra digit which is an estimate. *Significant figures* or *significant digits* are digits used to report the precision of a measurement. (Note the difference between the use of the word *significant* here and in everyday use, where it indicates "meaningful.") For example, consider the rectangular block pictured in Fig. 2-1. The ruler at the top of the block is divided into centimeters. You can estimate the length of a block to the nearest tenth of a centimeter (millimeter), but you cannot estimate the number of micrometers or even tenths of a millimeter, no matter how much you try. You should report the length of the block as 5.4 cm. Using the ruler at the bottom of the block, which has divisions in tenths of centimeters (millimeters), allows you to see for certain that the block is more than 5.4 cm but less than 5.5 cm. You can estimate it as 5.43 cm. Using the extra digit when you report the value allows the person who reads the result to determine that you used the more accurate ruler to make this latter measurement.

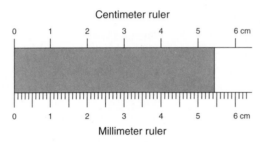

Fig. 2-1.   Accuracy of measurement

EXAMPLE 2.20.   Which of the two rulers shown in Fig. 2-1 was used to make each of the following measurements? (*a*) 2.75 cm, (*b*) 1.3 cm, (*c*) 5.11 cm, (*d*) 4.2 cm, and (*e*) 0.90 cm.

Ans.   The measurements reported in (*a*), (*c*), and (*e*) can easily be seen to have two decimal places. Since they are reported to the nearest hundredth of a centimeter, they must have been made by the more accurate ruler, the millimeter ruler. The measurements reported in (*b*) and (*d*) were made with the centimeter ruler at the top. In part (*e*), the 0 at the end shows that this measurement was made with the more accurate ruler. Here the distance was measured as more nearly 0.90 cm than 0.89 or 0.91 cm. Thus, the results are estimated to the nearest hundredth of a centimeter, but that value just happens to have a 0 as the estimated digit.

Zeros as Significant Digits

Suppose that we want to report the measurement 4.95 cm in terms of meters. Is our measurement any more or less precise? No, changing to another set of units does not increase or decrease the precision of the measurement. Therefore, we must use the same number of significant digits to report the result. How do we change a number of centimeters to meters?

$$4.95 \text{ cm} \left(\frac{1 \text{ m}}{100 \text{ cm}} \right) = 0.0495 \text{ m}$$

The zeros in 0.0495 m do not indicate anything about the precision with which the measurement was made; they are not significant. (They are important, however.) In a properly reported number, all nonzero digits are significant. Zeros are significant only when they help to indicate the precision of the measurement. The following rules are used to determine when zeros are significant in a properly reported number:

1. All zeros to the left of the first nonzero digit are nonsignificant. The zeros in 0.018 and 007 are not significant (except perhaps to James Bond).

2. All zeros between significant digits are significant. The 0 in 4.03 is significant.

3. All zeros to the right of the decimal point and to the right of the last nonzero digit are significant. The zeros in 7.000 and 6.0 are significant.

4. Zeros to the right of the last nonzero digit in a number with no decimal places are uncertain; they may or may not be significant. The zeros in 500 and 8 000 000 are uncertain. They may be present merely to indicate the magnitude of the number (i.e., to locate the decimal point), or they may also indicate something about the precision of measurement. [*Note*: Some elementary texts use an overbar to denote the last significant 0 in such numbers ($1\overline{0}0$). Other texts use a decimal point at the end of an integer, as in 100., to signify that the zeros are significant. However, these practices are not carried into most regular general chemistry texts or into the chemical literature.] A way to avoid the ambiguity is given in Example 2.23.

EXAMPLE 2.21. Underline the significant zeros in each of the following measurements, all in kilograms: (*a*) 7.00, (*b*) 0.7070, (*c*) 0.0077, and (*d*) 70.0.

Ans.   (*a*) 7.00, (*b*) 0.7070, (*c*) 0.0077, and (*d*) 70.0. In (*a*), the zeros are to the right of the last nonzero digit and to the right of the decimal point (rule 3), so they are significant. In (*b*), the leading 0 is not significant (rule 1), the middle 0 is significant because it lies between two significant 7s (rule 2), and the last 0 is significant because it is to the right of the last nonzero digit and the decimal point (rule 3). In (*c*), the zeros are to the left of the first nonzero digit (rule 1), and so are not significant. In (*d*), the last 0 is significant (rule 3), and the middle 0 is significant because it lies between the significant digits 7 and 0 (rule 2).

EXAMPLE 2.22. (*a*) How many significant digits are there in 1.60 cm? (*b*) How many decimal places are there in that number?

Ans.   (*a*) 3 and (*b*) 2.   Note the difference in these questions!

EXAMPLE 2.23. How many significant zeros are there in the number 8 000 000?

Ans.   The number of significant digits cannot be determined unless more information is given. If there are 8 million people living in New York City and one person moves out, how many are left? The 8 million people is an estimate, indicating a number nearer to 8 million than to 7 million or 9 million people. If one person moves, the number of people is still nearer to 8 million than to 7 or 9 million, and the population is still properly reported as 8 million.

   If you win a lottery and the state deposits $8 000 000 to your account, when you withdraw $1, your balance will be $7 999 999. The precision of the bank is much greater than that of the census takers, especially since the census takers update their data only once every 10 years.

   To be sure that you know how many significant digits there are in such a number, you can report the number in standard exponential notation because all digits in the coefficient of a number in standard exponential form are significant. The population of New York City would be 8×10^6 people, and the bank account would be $8.000\,000 \times 10^6$ dollars.

EXAMPLE 2.24. Change the following numbers of meters to millimeters. Explain the problem of zeros at the end of a whole number, and how the problem can be solved. (*a*) 7.3 m, (*b*) 7.30 m, and (*c*) 7.300 m.

Ans.   (*a*) $7.3 \text{ m} \left(\dfrac{1000 \text{ mm}}{1 \text{ m}} \right) = 7300 \text{ mm}$ (two significant digits)

       (*b*) $7.30 \text{ m} \left(\dfrac{1000 \text{ mm}}{1 \text{ m}} \right) = 7300 \text{ mm}$ (three significant digits)

       (*c*) $7.300 \text{ m} \left(\dfrac{1000 \text{ mm}}{1 \text{ m}} \right) = 7300 \text{ mm}$ (four significant digits)

       The magnitudes of the answers are the same, just as the magnitudes of the original values are the same. The numbers of millimeters all look the same, but since we know where the values came from, we know how many significant digits each contains. We can solve the problem by using standard exponential form: (*a*) 7.3×10^3 mm, (*b*) 7.30×10^3 mm, (*c*) 7.300×10^3 mm.

EXAMPLE 2.25. Change each of the following measurements to meters. (*a*) 3456 mm, (*b*) 345.6 mm, (*c*) 34.56 mm, and (*d*) 3.456 mm.

Ans.   (*a*) 3.456 m   (*b*) 0.3456 m   (*c*) 0.03456 m   (*d*) 0.003456 m
       All the given values have four significant digits, so each answer also has to have four significant digits (and they do since the leading zeros are not significant.)

Significant Digits in Calculations

   Special note on significant figures: Electronic calculators do not keep track of significant figures at all. The answers they yield very often have fewer or more digits than the number justified by the measurements. You must keep track of the significant figures and decide how to report the answer.

Addition and Subtraction

We must report the results of our calculations to the proper number of significant digits. We almost always use our measurements to calculate other quantities, and the results of the calculations must indicate to the reader the limit of precision with which the actual measurements were made. The rules for significant digits as the result of additions or subtractions with measured quantities are as follows.

We may keep digits only as far to the right as the uncertain digit in the least accurate measurement. For example, suppose you measured a block with the millimeter ruler (Fig. 2-1) as 5.71 cm and another block with the centimeter ruler as 3.2 cm. What is the length of the two blocks together?

$$\begin{array}{r} 3.2 \ \text{cm} \\ +5.71 \ \text{cm} \\ \hline 8.91 \ \text{cm} \rightarrow 8.9 \ \text{cm} \end{array}$$

Since the 2 in the 3.2-cm measurement is uncertain, the 9 in the result is also uncertain. To report 8.91 cm would indicate that we knew the 8.9 for sure and that the 1 was uncertain. Since this is more precise than our measurements justify, we must *round* our reported result to 8.9 cm. That result says that we are unsure of the 9 and certain of the 8.

The rule for addition or subtraction can be stated as follows: Keep digits in the answer only as far to the right as the measurement in which there are digits least far to the right.

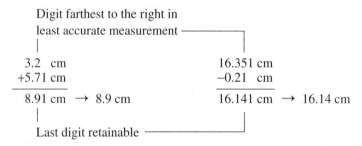

It is not the *number* of significant digits, but their *positions* which determine the number of digits in the answer in addition or subtraction. For example, in the following problems the numbers of significant digits change despite the final digit being retained in each case:

$$\begin{array}{r} 98.7 \ \text{cm} \\ +42.1 \ \text{cm} \\ \hline 140.8 \ \text{cm} \end{array} \qquad \begin{array}{r} 24.3 \ \text{cm} \\ -21.5 \ \text{cm} \\ \hline 2.8 \ \text{cm} \end{array}$$

Rounding

We have seen that we must sometimes reduce the number of digits in our calculated result to indicate the precision of the measurements that were made. To reduce the number, we round digits other than integer digits, using the following rules.

If the first digit which we are to drop is less than 5, we drop the digits without changing the last digit retained.

$$7.437 \rightarrow 7.4$$

If the first digit to be dropped is equal to or greater than 5, the last digit retained is increased by 1:

$$7.46 \rightarrow 7.5 \qquad 7.96 \ \ \rightarrow 8.0 \qquad \text{(Increasing the last digit retained caused a carry.)}$$
$$7.56 \rightarrow 7.6 \qquad 7.5501 \rightarrow 7.6$$

A slightly more sophisticated method may be used if the first digit to be dropped is a 5 and there are no digits or only zeros after the 5. Change the last digit remaining only if it is odd to the next higher even digit. The following numbers are rounded to one decimal place:

$$7.550 \rightarrow 7.6 \qquad 7.55 \rightarrow 7.6$$
$$7.450 \rightarrow 7.4 \qquad 7.45 \rightarrow 7.4$$

Use this method only if your instructor or text does.

For rounding digits in any whole-number place, use the same rules, except that instead of actually dropping digits, replace them with (nonsignificant) zeros. For example, to round 6718 to two significant digits

$$6718 \rightarrow 6700 \qquad \text{(two significant digits)}$$

EXAMPLE 2.26. Round the following numbers to two significant digits each: (*a*) 0.0654, (*b*) 65.4, and (*c*) 654.

Ans.   (*a*) 0.065   (*b*) 65   (*c*) 650
         [Do not merely drop the 4. (*b*) and (*c*) obviously cannot be the same.]

Multiplication and Division

In multiplication and division, different rules apply than apply to addition and subtraction. It is the *number* of significant digits in each of the values given, rather than their positions, which governs the number of significant digits in the answer. In multiplication and division, the answer retains as many significant digits as there are in the value with the *fewest* significant digits.

EXAMPLE 2.27. Perform each of the following operations to the proper number of significant digits:
(*a*) 1.75 cm × 4.041 cm, (*b*) 2.00 g/3.00 cm^3, and (*c*) 6.39 g/2.13 cm^3.

Ans.   (*a*)   1.75 cm × 4.041 cm = 7.07 cm^2. There are three significant digits in the first factor and four in the second. The answer can retain only three significant digits, equal to the smaller number of significant digits in the factors.

         (*b*)   2.00 g/3.00 cm^3 = 0.667 g/cm^3. There are three significant digits, equal to the number of significant digits in each number. Note that the number of *decimal places* is different in the answer, but in multiplication or division, the number of decimal places is immaterial.

         (*c*)   6.39 g/2.13 cm^3 = 3.00 g/cm^3. Since there are three significant digits in each number, there should be three significant digits in the answer. In this case, we had to *add zeros*, not round, to get the proper number of significant digits.

When we multiply or divide a measurement by a defined number, rather than by another measurement, we may retain in the answer the number of significant digits that occur in the measurements. For example, if we multiply a number of meters by 1000 mm/m, we may retain the number of significant digits in the number of millimeters that we had in the number of meters. The 1000 is a defined number, not a measurement, and it can be regarded as having as many significant digits as needed for any purpose.

EXAMPLE 2.28. How many significant digits should be retained in the answer when we calculate the number of centimeters in 6.137 m?

Ans.
$$6.137 \text{ m}\left(\frac{100 \text{ cm}}{1 \text{ m}}\right) = 613.7 \text{ cm}$$

The number of significant digits in the answer is 4, equal to the number in the 6.137-m measurement. The (100 cm/m) is a definition and does not limit the number of significant digits in the answer.

2.6. DENSITY

Density is a useful property with which to identify substances. *Density*, symbolized *d*, is defined as mass per unit volume:

$$\text{Density} = \frac{\text{mass}}{\text{volume}}$$

or in symbols,

$$d = \frac{m}{V}$$

Since it is a quantitative property, it is often more useful for identification than a qualitative property such as color or smell. Moreover, density determines whether an object will float in a given liquid. If the object is less dense than the liquid, it will float. It is also useful to discuss density here for practice with the factor-label method of solving problems, and as such, it is often emphasized on early quizzes and examinations.

Density is a ratio—the number of grams per milliliter, for example. In this regard, it is similar to speed.

The word *per* means *divided by*. To get a speed in miles per hour, divide the number of miles by the number of hours.

Say the following speed aloud: 50 miles/hour.

"Fifty miles per hour." (The division symbol is read as the word *per*.)

Distinguish carefully between density and mass, which are often confused in everyday conversation.

EXAMPLE 2.29.   Which weighs more, a pound of bricks or a pound of feathers?

Ans.   Since a pound of each is specified, neither weighs more. But everyone knows that bricks are heavier than feathers. The confusion stems from the fact that *heavy* is defined in the dictionary as either "having great mass" or "having high density." *Per unit volume*, bricks weigh more than feathers; that is, bricks are denser.

It is relative densities that determine whether an object will float in a liquid. If the object is less dense, it will float (unless it dissolves, of course).

EXAMPLE 2.30.   Which has a greater mass, a large wooden desk or a metal sewing needle? Which one will float in water?

Ans.   The desk has a greater mass. (You can pick up the needle with one finger, but not the desk.) Since the desk is so much larger (greater volume), it displaces more than its own mass of water. Its density is less than that of water, and it will float despite its greater mass. The needle is so small that it does not displace its own mass of water, and thus it sinks.

In doing numerical density problems, you may always use the equation $d = m/V$ or the same equation rearranged into the form $V = m/d$ or $m = dV$. You are often given two of these quantities and asked to find the third. You will use the equation $d = m/V$ if you are given mass and volume; but if you are given density and either of the others, you probably should use the factor-label method. That way, you need not manipulate the equation and then substitute; you can solve immediately. You need not memorize any density value except that of water— approximately 1.00 g/mL throughout its liquid range.

EXAMPLE 2.31.   Calculate the density of a 4.00-L body which has a mass of 7.50 kg.

Ans.
$$d = \frac{m}{V} = \frac{(7.50 \text{ kg})}{(4.00 \text{ L})} = 1.88 \text{ kg/L}$$

EXAMPLE 2.32.   What is the mass of 10.00 mL of gold, which has a density of 19.3 g/mL?

Ans.   Using the factor-label method, we find

$$10.00 \text{ mL}\left(\frac{19.3 \text{ g}}{1 \text{ mL}}\right) = 193 \text{ g} \quad \text{(The density is given to three significant digits.)}$$

With the equation:

$$d = \frac{m}{V}$$

$$m = Vd = (10.00 \text{ mL})\left(\frac{19.3 \text{ g}}{1 \text{ mL}}\right) = 193 \text{ g}$$

EXAMPLE 2.33.   What is the volume of 72.4 g of lead? (Density = 11.3 g/mL.)

Ans.
$$V = \frac{m}{d} = \frac{72.4 \text{ g}}{11.3 \text{ g/mL}} = 6.41 \text{ mL}$$

or
$$72.4 \text{ g}\left(\frac{1 \text{ mL}}{11.3 \text{ g}}\right) = 6.41 \text{ mL}$$

Use of the equation requires manipulation of the basic equation followed by substitution. Dividing by a ratio of units also may cause confusion. The factor-label method gives the same result by merely relying on the units. It also allows combination of the solution with other conversions.

EXAMPLE 2.34.   What mass of sulfuric acid is there in 44.4 mL of solution of density 1.85 g/mL which contains 96.0% sulfuric acid by mass?

Ans.   Note that the *solution* has the density 1.85 g/mL and that every 100.0 g of it contains 96.0 g of sulfuric acid.

$$44.4 \text{ mL solution}\left(\frac{1.85 \text{ g solution}}{1 \text{ mL solution}}\right)\left(\frac{96.0 \text{ g acid}}{100.0 \text{ g solution}}\right) = 78.9 \text{ g acid}$$

2.7.   TEMPERATURE SCALES

Scientists worldwide (and everyone else outside the United States) use the Celsius temperature scale, in which the freezing point of pure water is defined as 0°C and the normal boiling point of pure water is defined as 100°C. The *normal boiling point* is the boiling point at 1.00-atm pressure (Chap. 12). The Fahrenheit temperature scale is used principally in the United States. It has the freezing point defined as 32.0°F and the normal boiling point defined as 212°F. A comparison of the Celsius and Fahrenheit temperature scales is presented in Fig. 2-2. The temperature differences between the freezing point and normal boiling point on the two scales are 180°F and 100°C, respectively. To convert Fahrenheit temperatures to Celsius temperatures, subtract 32.0°F and then multiply the result by $\frac{100}{180}$ or $\frac{5}{9}$.

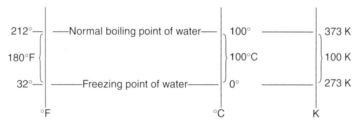

Fig. 2-2.   Comparison of Celsius, Fahrenheit, and Kelvin temperature scales

EXAMPLE 2.35.   Change 98.6°F to degrees Celsius.

Ans. $$°C = (98.6°F - 32.0°F)\tfrac{5}{9} = 37.0°C$$

To convert from Celsius to Fahrenheit, multiply by $\frac{9}{5}$, then add 32.0°F.

EXAMPLE 2.36.   Convert 20.0°C to degrees Fahrenheit.

Ans. $$(20.0°C)\tfrac{9}{5} + 32.0°F = 68.0°F$$

Kelvin Temperature Scale

The Kelvin temperature scale is defined to have the freezing point of pure water as 273.15 K and the normal boiling point of pure water as 373.15 K. The unit of the Kelvin temperature scale is the *kelvin*. (We do not use "degrees" with kelvins.) Thus, its temperatures are essentially 273° higher than the same temperatures on the Celsius scale. To convert from degrees Celsius to kelvins, merely add 273° to the Celsius temperature. To convert in the opposite direction, subtract 273° from the Kelvin temperature to get the Celsius equivalent. (See Fig. 2-2.)

EXAMPLE 2.37.   Convert 100°C and −27°C to kelvins.

Ans. $$100°C + 273° = 373 \text{ K}$$

$$-27°C + 273° = 246 \text{ K}$$

EXAMPLE 2.38. Change 171 K and 422 K to degrees Celsius.

Ans.
$$171 \text{ K} - 273° = -102°\text{C}$$

$$422 \text{ K} - 273° = 149°\text{C}$$

Note that a *change in temperature* in kelvins is the same as the equivalent change in temperature in degrees Celsius.

EXAMPLE 2.39. Convert 0°C and 30°C to kelvins. Calculate the change in temperature from 0°C to 30°C on both temperature scales.

Ans.
$$0°\text{C} + 273° = 273 \text{ K}$$

$$30°\text{C} + 273° = 303 \text{ K}$$

The temperature difference on the two scales is

$$
\begin{array}{cc}
30°\text{C} & 303 \text{ K} \\
\underline{-0°\text{C}} & \underline{-273 \text{ K}} \\
30°\text{C} & 30 \text{ K}
\end{array}
$$

Solved Problems

FACTOR-LABEL METHOD

2.1.   (*a*) Write the reciprocal for the following factor label: 4.0 mi/h. (*b*) Which of these—the reciprocal or the original factor label—is multiplied to change miles to hours?

   Ans.   (*a*) 1.0 h/4.0 mi   (*b*) The reciprocal:

$$10.0 \text{ mi} \left(\frac{1.0 \text{ h}}{4.0 \text{ mi}} \right) = 2.5 \text{ h}$$

2.2.   (*a*) Write the reciprocal for the following factor label: 5.00 dollars/pound. (*b*) Which of these—the reciprocal or the original factor label—is multiplied to change pounds to dollars?

   Ans.   (*a*) 1 pound/5.00 dollars   (*b*) The original:

$$4.6 \text{ pounds} \left(\frac{5.00 \text{ dollars}}{1 \text{ pound}} \right) = 23 \text{ dollars}$$

2.3.   Calculate the number of hours one must work at 12.00 dollars/h to earn 150.00 dollars.

   Ans.
$$150.00 \text{ dollars} \left(\frac{1 \text{ h}}{12.00 \text{ dollars}} \right) = 12.5 \text{ h}$$

2.4.   Calculate the number of cents in 4.58 dollars. (*a*) Do the calculation by converting first the dollars to dimes and then the dimes to cents. (*b*) Repeat with a direct calculation.

   Ans.   (*a*) $4.58 \text{ dollars} \left(\dfrac{10 \text{ dimes}}{1 \text{ dollar}} \right) = 45.8 \text{ dimes}$

   $45.8 \text{ dimes} \left(\dfrac{10 \text{ cents}}{1 \text{ dime}} \right) = 458 \text{ cents}$

   (*b*) $4.58 \text{ dollars} \left(\dfrac{10 \text{ dimes}}{1 \text{ dollar}} \right) \left(\dfrac{10 \text{ cents}}{1 \text{ dime}} \right) = 458 \text{ cents}$

   or $4.58 \text{ dollars} \left(\dfrac{100 \text{ cents}}{1 \text{ dollar}} \right) = 458 \text{ cents}$

2.5.   Percentages can be used as factors. The percentage of something is the number of parts of that thing per 100 parts total. Whatever unit(s) is (are) used for the item in question is also used for the total. If a certain ring is 64% gold, write six factors that can be used to solve problems with this information.

Ans.

64 oz gold	36 oz other metal	36 oz other metal
100 oz total	100 oz total	64 oz gold

or the reciprocals of these:

100 oz total	100 oz total	64 oz gold
64 oz gold	36 oz other metal	36 oz other metal

2.6. Calculate the number of children in a second-grade class of 18 girls in which there are 60% girls.

Ans.
$$18 \text{ girls}\left(\frac{100 \text{ children}}{60 \text{ girls}}\right) = 30 \text{ children}$$

2.7. If pistachio nuts cost \$6.00 per pound. (*a*) How many pounds of nuts can be bought for \$21.00? (*b*) How much does 2.43 pounds of nuts cost?

Ans.   (*a*) $21.00 \text{ dollars}\left(\dfrac{1 \text{ pound}}{6.00 \text{ dollars}}\right) = 3.50 \text{ pounds}$

         (*b*) $2.43 \text{ pounds}\left(\dfrac{6.00 \text{ dollars}}{1 \text{ pound}}\right) = 14.58 \text{ dollars}$

2.8. If 44% of a certain class is female, how large is the class if there are 77 females?

Ans.
$$77 \text{ females}\left(\frac{100 \text{ people}}{44 \text{ females}}\right) = 175 \text{ people}$$

2.9. If income tax for a certain student is 14%, how much gross pay does that student need to earn to be able to keep \$425? (Assume no other deductions.)

Ans.   If 14% goes to tax, $100\% - 14\% = 86\%$ is kept.

$$425 \text{ dollars kept}\left(\frac{100 \text{ dollars gross}}{86 \text{ dollars kept}}\right) = 494.19 \text{ dollars gross}$$

2.10. Commercial sulfuric acid solution is 96.0% H_2SO_4 in water. How many grams of H_2SO_4 are there in 175 g of the commercial solution?

Ans.
$$175 \text{ g solution}\left(\frac{96.0 \text{ g } H_2SO_4}{100 \text{ g solution}}\right) = 168 \text{ g } H_2SO_4$$

METRIC SYSTEM

2.11. (*a*) Is there a unit "ounce" that means a weight and another that means a volume? (*b*) Is there a metric unit that means both weight and volume? (*c*) What is the difference in weight between an ounce of gold and an ounce of lead?

    Ans.   (*a*)   Yes, 16 fluid oz = 1 pint, 16 oz avoirdupois = 1 pound, and 12 oz Troy = 1 pound Troy.

            (*b*)   No.

            (*c*)   Gold is measured in Troy ounces, and lead is measured in avoirdupois ounces.

2.12. Is there a difference between a U.S. gallon and an imperial gallon?

    Ans.   They are two different units despite both having the word *gallon* in their names. The U.S. gallon is equal to 3.7853 L; the imperial gallon is equal to 4.5460 L.

2.13. There are nickels in one container and pennies in another. Each container holds the same amount of money. (*a*) Which container holds more coins? (*b*) In which container is each coin more valuable?

    Ans.   The more valuable the coin, the fewer are required to make a certain amount of money. The nickels are each more valuable, so there are more pennies.

2.14. A certain mass is measured in kilograms and again in grams. Which measurement involves (*a*) the larger number? (*b*) the larger unit?

    Ans.   The larger the unit, the fewer are required to measure a given distance. There are more grams because the kilogram is larger. Compare this to Problem 2.13.

2.15. (a) Would 0.22 m^3 of water fill a thimble, a gallon jug, or an average bathtub? (b) 1.0 cm^3?

    Ans.   (*a*) A bathtub   (*b*) A thimble

2.16. What two instances are described in the text in which a standard for a type of measurement is different from the unit with no prefix?

    Ans.   1.   The kilogram is the standard of mass, both in SI and legally by act of Congress in the United States.

                2.   The liter is a unit of volume in the (older version of the) metric system, but the cubic meter is the standard.

2.17. Give the basic SI or metric unit and its abbreviation for each of the following types of measurements: (*a*) mass, (*b*) volume, and (*c*) distance.

    Ans.   (*a*) gram (g)   (*b*) cubic meter (m^3) or liter (L)   (*c*) meter (m)

2.18. Make the following conversions:

    (*a*)   Change 46 cm to meters   (*c*)   Change 1.5 L to milliliters

    (*b*)   Change 0.28 kg to grams   (*d*)   Change 1.23 kg to milligrams

    Ans.   (*a*) $46 \text{ cm}\left(\dfrac{1 \text{ m}}{100 \text{ cm}}\right) = 0.46 \text{ m}$ (*c*) $1.5 \text{ L}\left(\dfrac{1000 \text{ mL}}{1 \text{ L}}\right) = 1500 \text{ mL}$

                 (*b*) $0.28 \text{ kg}\left(\dfrac{1000 \text{ g}}{1 \text{ kg}}\right) = 280 \text{ g}$ (*d*) $1.23 \text{ kg}\left(\dfrac{1000 \text{ g}}{1 \text{ kg}}\right)\left(\dfrac{1000 \text{ mg}}{1 \text{ g}}\right) = 1.23 \times 10^6 \text{ mg}$

2.19. (a) How many 1 cm^3 blocks can be placed on the bottom left edge of a 1 m^3 box? (b) How many such rows of these blocks can be placed on the bottom of the box? (c) How many such layers fit within the box? (d) How many 1 cm^3 can fit into 1 m^3? (e) Cube the following equation: $100 \text{ cm} = 1 \text{ m}$.

    Ans.   (*a*) 100 blocks   (*b*) 100 rows   (*c*) 100 layers   (*d*) 1 million (10^6) cubic centimeters equal 1 cubic meter ($100 \times 100 \times 100 = 10^6$)   (*e*) $(100 \text{ cm})^3 = (1 \text{ m})^3$, $(100)^3 \text{ (cm)}^3 = (1)^3 \text{ (m)}^3$, $1 \times 10^6 \text{ cm}^3 = 1 \text{ m}^3$

2.20. What is wrong with the following problems?

    (*a*)   Change 13 cm to kilograms

    (*b*)   Change 2.0 cm^2 to mm

    (*c*)   Change 4.4 mL to cm^2

    Ans.   (*a*)   You cannot convert a distance to a mass.

                 (*b*)   You cannot convert an area to a distance.

                 (*c*)   You cannot convert a volume to an area.

2.21. (a) Change 0.0250 mi^2 to square yards. (1 mi = 1760 yd). (b) Change 0.0250 km^2 to square meters.

    Ans.   (*a*) $0.0250 \text{ mi}^2\left(\dfrac{1760 \text{ yd}}{1 \text{ mi}}\right)^2 = 77\,400 \text{ yd}^2$

                 (*b*) $0.0250 \text{ km}^2\left(\dfrac{1000 \text{ m}}{1 \text{ km}}\right)^2 = 25\,000 \text{ m}^2$

2.22. (*a*) Calculate the number of centimeters in 4.58 km. Do the calculation by converting first the kilometers to meters and then the meters to centimeters. (*b*) Repeat with a direct calculation. (*c*) Compare this problem with Problem 2.4.

Ans. (*a*) $4.58 \text{ km} \left(\dfrac{1000 \text{ m}}{1 \text{ km}} \right) = 4580 \text{ m}$

$4580 \text{ m} \left(\dfrac{100 \text{ cm}}{1 \text{ m}} \right) = 458\,000 \text{ cm}$

(*b*) $4.58 \text{ km} \left(\dfrac{1000 \text{ m}}{1 \text{ km}} \right) \left(\dfrac{100 \text{ cm}}{1 \text{ m}} \right) = 458\,000 \text{ cm}$

(*c*) The methods are the same.

2.23. Change (*a*) 2.00 mg to kilograms. (*b*) 2.00 mm to kilometers. (*c*) 2.00 cm to kilometers.

Ans. (*a*) $2.00 \text{ mg} \left(\dfrac{0.001 \text{ g}}{1 \text{ mg}} \right) \left(\dfrac{1 \text{ kg}}{1000 \text{ g}} \right) = 2.00 \times 10^{-6} \text{ kg}$

(*b*) $2.00 \text{ mm} \left(\dfrac{0.001 \text{ m}}{1 \text{ mm}} \right) \left(\dfrac{1 \text{ km}}{1000 \text{ m}} \right) = 2.00 \times 10^{-6} \text{ km}$

(*c*) $2.00 \text{ cm} \left(\dfrac{0.01 \text{ m}}{1 \text{ cm}} \right) \left(\dfrac{1 \text{ km}}{1000 \text{ m}} \right) = 2.00 \times 10^{-5} \text{ km}$

Note the identical processes in all three parts and the similar answers in the first two.

2.24. Draw a square 2 cm × 2 cm. In the upper left corner of that square, draw another square 1 cm × 1 cm. How many of the small squares will fit into the large one?

Ans. Four small squares will fit; that is, $4 \text{ cm}^2 = (2 \text{ cm})^2$. (See Fig. 2-3).

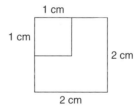

Fig. 2-3.

2.25. (*a*) What is the ratio in areas of a square 1 cm on each side and a square 1 mm on each side? (*b*) What is the ratio in volumes of a cube 1 cm on each side and a cube 1 mm on each side?

Ans. (*a*) The ratio of centimeters to millimeters is 10 : 1; the ratio of their squares is therefore 100 : 1.

(*b*) The ratio of the cubes is 1000 : 1.

2.26. Explain in terms of units why area × distance yields volume.

Ans. $\text{m}^2 \times \text{m} = \text{m}^3$

2.27. Give the volume corresponding to each of the following numbers of liters in terms of a dimension involving length cubed: (*a*) 2000 L, (*b*) 2 L, (*c*) 0.002 L, and (*d*) 2×10^{-6} L.

Ans. (*a*) 2 m^3 (*b*) 2 dm^3 (*c*) 2 cm^3 (*d*) 2 mm^3

2.28. Give the volume in liters of each of the following: (*a*) 3 m³, (*b*) 3 dm³, (*c*) 3 cm³, and (*d*) 3 mm³.

Ans. (*a*) 3000 L (*b*) 3 L (*c*) 0.003 L (*d*) 3×10^{-6} L

2.29. Make the following conversions:

(a) Change 6.3 cm^2 to m^2

(b) Change 13 cm^3 to m^3

(c) Change 9.2 L to cm^3

(d) Change 9.6 dm^3 to L

(e) Change 1.7 m^3 to kL

(f) Change 3.3 m^2 to cm^2

(g) Change 219 mL to cm^3

Ans.   (a) $6.3 \text{ cm}^2 \left(\dfrac{1 \text{ m}}{100 \text{ cm}} \right)^2 = 0.000\,63 \text{ m}^2$

(e) $1.7 \text{ m}^3 \left(\dfrac{1 \text{ kL}}{1 \text{ m}^3} \right) = 1.7 \text{ kL}$

(b) $13 \text{ cm}^3 \left(\dfrac{1 \text{ m}}{100 \text{ cm}} \right)^3 = 0.000\,013 \text{ m}^3$

(f) $3.3 \text{ m}^2 \left(\dfrac{100 \text{ cm}}{1 \text{ m}} \right)^2 = 33\,000 \text{ cm}^2$

(c) $9.2 \text{ L} \left(\dfrac{1000 \text{ cm}^3}{1 \text{ L}} \right) = 9200 \text{ cm}^3$

(g) $219 \text{ mL} \left(\dfrac{1 \text{ cm}^3}{1 \text{ mL}} \right) = 219 \text{ cm}^3$

(d) $9.6 \text{ dm}^3 \left(\dfrac{1 \text{ L}}{1 \text{ dm}^3} \right) = 9.6 \text{ L}$

2.30. Give the volume corresponding to each of the following in terms of a dimension involving length cubed: (a) 1 kL, (b) 2 L, (c) 3 mL, and (d) 4 μL.

Ans.   (a) 1 m^3 (b) 2 dm^3 (c) 3 cm^3 (d) 4 mm^3

EXPONENTIAL NUMBERS

2.31. In the number 5.0×10^3, identify (a) the coefficient, (b) the exponent, (c) the base, and (d) the exponential part.

Ans.   (a) 5.0   (b) 3   (c) 10   (d) 10^3

2.32. What is the meaning of each of the following: (a) ten squared, (b) ten cubed, (c) ten to the fourth power?

Ans.   (a) 10^2 (b) 10^3 (c) 10^4

2.33. Which of the following numbers are in standard exponential form?

(a) 11.0×10^{-3}

(b) 6.7×10^0

(c) 3×10^7

(d) 0.77×10^7

(e) 1.23×10^{-4}

(f) 0.12×10^5

(g) 0.1×10^5

(h) 4.4×10^3

(i) 1.0×10^4

Ans.   (b), (c), (e), (h), and (i)

2.34. Show by writing out the explicit meaning of the exponential parts that $(3.0 \times 10^3)^2 = 9.0 \times 10^6$.

Ans.
$$(3.0 \times 10^3)^2 = (3.0 \times 10^3)(3.0 \times 10^3)$$
$$= 3.0 \times 3.0 \times (10 \times 10 \times 10) \times (10 \times 10 \times 10)$$
$$= 9.0 \times 10^6$$

2.35. Simplify:

(a) $\dfrac{1}{10^{-3}}$

(b) $\dfrac{1}{10^6}$

(c) $\dfrac{10^{-4}}{10^{-4}}$

Ans.   (a) 10^3 (b) 10^{-6} (c) 10$^0 = 1$

2.36. Perform the following calculations:

(a) $(2.20 \times 10^{10})(3.50 \times 10^4)$ (e) $(2.20 \times 10^{-10})/(3.50 \times 10^{-4})$

(b) $(2.20 \times 10^{10}) + (3.50 \times 10^4)$ (f) $(2.20 \times 10^{-10}) - (3.50 \times 10^{-4})$

(c) $(2.20 \times 10^{10})/(3.50 \times 10^4)$ (g) $(2.20 \times 10^{21}) + (3.5 \times 10^{19})$

(d) $(2.20 \times 10^{-10})/(3.50 \times 10^4)$ (h) $(2.20 \times 10^{10})(3.50 \times 10^{10})$

Ans.   (a) 7.70×10^{14} (c) 6.29×10^5 (e) 6.29×10^{-7} (g) 2.24×10^{21}

(b) 2.20×10^{10} (d) 6.29×10^{-15} (f) -3.50×10^{-4} (h) 7.70×10^{20}

2.37. Convert the following numbers to decimal form: (a) 5.6×10^2, (b) 7.07×10^{-2}, and (c) 6.11×10^0.

Ans.   (a) 560   (b) 0.0707   (c) 6.11

2.38. Perform the following calculations:

(a) $1.70 \times 10^6 + 9.7 \times 10^5$ (d) $6.02 \times 10^{23} - 1.3 \times 10^{22}$

(b) $1.11 \times 10^{-10} + 4.28 \times 10^{-9}$ (e) $1.40 \times 10^{-11} - 2.30 \times 10^{-9}$

(c) $7.7 \times 10^3 + 4.05 \times 10^3$

Ans.   (a) 2.67×10^6

(b) 4.39×10^{-9}

(c) 1.18×10^4

(d) 5.89×10^{23}

(e) -2.29×10^{-9}

2.39. Perform the following calculations:

(a) $(4.30 \times 10^8)(1.60 \times 10^7)$ (d) $(1.65 \times 10^{-19})/(1.10 \times 10^{-22})$

(b) $(3.45 \times 10^{-3})(2.00 \times 10^{-1})$ (e) $(6.02 \times 10^{23})/(6.02 \times 10^{22})$

(c) $(4.13 \times 10^1)/(7.1 \times 10^4)$

Ans.   (a) 6.88×10^{15}

(b) 6.90×10^{-4}

(c) 5.8×10^{-4}

(d) 1.50×10^3

(e) $1.00 \times 10^1 = 10.0$

2.40. Perform the following calculations:

(a) $(1.25 \times 10^6 + 2.10 \times 10^5)/(3.11 \times 10^{-3})$

(b) $(7.44 \times 10^{-4} + 2.46 \times 10^{-3})(6.16 \times 10^4 + 2.5 \times 10^3)$

(c) $(7.7 \times 10^9 + 2.3 \times 10^9) - (2.7 \times 10^8)$

(d) $(6.02 \times 10^{23} + 2.0 \times 10^{21}) + (4.20 \times 10^{22})$

(e) $(2.2 \times 10^{-9} + 1.1 \times 10^{-10}) - (3 \times 10^{-3})$

Ans.   (a) 4.69×10^8

(b) 2.05×10^2

(c) 9.7×10^9

(d) 6.46×10^{23}

(e) -3×10^{-3}

2.41. (*a*) Change 2.67 mi^2 to square feet. (*b*) Change 2.67 km^2 to square meters.

Ans.   (*a*) $2.67 \text{ mi}^2 \left(\dfrac{5280 \text{ ft}}{1 \text{ mi}}\right)^2 = 7.44 \times 10^7 \text{ ft}^2$ (*b*) $2.67 \text{ km}^2 \left(\dfrac{1000 \text{ m}}{1 \text{ km}}\right)^2 = 2.67 \times 10^6 \text{ m}^2$

2.42. Calculate the number of cubic meters in $6.4 \times 10^5 \text{ mm}^3$.

Ans. $6.4 \times 10^5 \text{ mm}^3 \left(\dfrac{1 \text{ m}}{1000 \text{ mm}}\right)^3 = 6.4 \times 10^{-4} \text{ m}^3$

2.43. Which is bigger, 4.0 L or 4000 mm^3?

Ans.   4.0 L is bigger. It is equal to 4000 cm^3 or 4 000 000 mm^3.

2.44. Change 7.3 g/cm^3 to kilograms per cubic meter.

Ans. $\dfrac{7.3 \text{ g}}{\text{cm}^3} \left(\dfrac{100 \text{ cm}}{1 \text{ m}}\right)^3 \left(\dfrac{1 \text{ kg}}{1000 \text{ g}}\right) = 7.3 \times 10^3 \text{ kg/m}^3$

2.45. Given that 1 L = 1 dm^3, calculate the number of liters in (*a*) 1 m^3, (*b*) 1 cm^3, and (*c*) 1 mm^3.

Ans.   (*a*) $1 \text{ m}^3 \left(\dfrac{10 \text{ dm}}{1 \text{ m}}\right)^3 \left(\dfrac{1 \text{ L}}{1 \text{ dm}^3}\right) = 1000 \text{ L}$ (*c*) $1 \text{ mm}^3 \left(\dfrac{1 \text{ dm}}{100 \text{ mm}}\right)^3 \left(\dfrac{1 \text{ L}}{1 \text{ dm}^3}\right) = 1 \times 10^{-6} \text{ L}$

         (*b*) $1 \text{ cm}^3 \left(\dfrac{1 \text{ dm}}{10 \text{ cm}}\right)^3 \left(\dfrac{1 \text{ L}}{1 \text{ dm}^3}\right) = 0.001 \text{ L}$

SIGNIFICANT DIGITS

2.46. What is the difference between the number of significant digits and the number of decimal places in a measurement?

Ans.   The number of significant digits is the number of digits that reflect the precision of the measurement. The number of decimal places is the number of digits after the decimal point. The two have little to do with each other, so do not get them confused.

2.47. Underline the significant digits in each of the following measurements. If a digit is uncertain, place a question mark below it.
(*a*) 2.03 m, (*b*) 47.0 m, (*c*) 51.05 m, (*d*) 9.00 × 10^3 m, (*e*) 6.100 m, (*f*) 0.0400 m, (*g*) 1.110 m, (*h*) 30.0 m, (*i*) 400 m

Ans.   (*a*)   2.03 m   rule 2   (*f*)   0.0400 m   rules 1 and 3

         (*b*)   47.0 m   rule 3   (*g*)   1.110 m   rule 3

         (*c*)   51.05 m   rule 2   (*h*)   30.0 m   rules 2 and 3

         (*d*)   9.00 × 10^3 m   rule 3   (*i*)   400   rule 4

         (*e*)   6.100 m   rule 3

2.48. To the correct number of significant digits, calculate the number of

(*a*) millimeters in 4.00 m   (*c*) millimeters in 13.57 cm

(*b*) kilograms in 49 g   (*d*) centimeters in 0.0040 m

Ans.   (*a*)   4.00 × 10^3 mm (three significant digits)   (*c*)   135.7 mm (four significant digits)

         (*b*)   0.049 kg (two significant digits)   (*d*)   0.40 cm (two significant digits)

2.49. In which of the following problems are the numbers of significant digits retained?

(a) $(65.4\ \text{cm}^2)/(60.5\ \text{cm})$ (c) $(65.4\ \text{cm}^2)(60.5\ \text{cm})$

(b) $(65.4\ \text{cm}) + (60.5\ \text{cm})$ (d) $(65.4\ \text{cm}) - (60.5\ \text{cm})$

    Ans.   (a) and (c). [In (b) the number increases to 4, in (d) it decreases to 2.]

2.50. Calculate the answer to each of the following expressions to the correct number of significant digits:

(a) $6.06\ \text{cm} \times 1.8\ \text{cm}$ (e) $29.0\ \text{cm} + 0.004\ \text{cm}$

(b) $6.06\ \text{cm} - 1.8\ \text{cm}$ (f) $9.44\ \text{cm} + 4.66\ \text{cm} - 0.02\ \text{cm}$

(c) $(6.06\ \text{g})/(1.8\ \text{cm}^3)$ (g) $4.77\ \text{cm} + 1.506\ \text{cm}$

(d) $1.234\ \text{cm} + 5.6\ \text{cm} - 0.078\ \text{cm}$

    Ans.   (a) $11\ \text{cm}^2$ (c) $3.4\ \text{g/cm}^3$ (e) $29.0\ \text{cm}$ (g) $6.28\ \text{cm}$

             (b) $4.3\ \text{cm}$ (d) $6.8\ \text{cm}$ (f) $14.08\ \text{cm}$

2.51. Calculate the answer to each of the following expressions to the correct number of significant digits:

(a) $(1.66\ \text{cm} + 7.3\ \text{cm}) \times 4.00\ \text{cm}$ (e) $1.11 \times 10^5\ \text{cm} \times 3.17\ \text{cm} \times 1.00 \times 10^3\ \text{cm}$

(b) $2.29\ \text{cm} \times (1.7\ \text{cm} - 4.213\ \text{cm})$ (f) $5.7 \times 10^{-3}\ \text{cm} - 7.8 \times 10^{-3}\ \text{cm}$

(c) $(1.91\ \text{cm} \times 4.8\ \text{cm}) - 5.71\ \text{cm}^2$ (g) $(2.170\ \text{cm} \times 7.11\ \text{cm}) + (5.820\ \text{cm}^2 - 7.16\ \text{cm}^2)$

(d) $7130\ \text{cm} - 2.13\ \text{cm}$

    Ans.   (a) $36\ \text{cm}^2$ (The sum has only two significant digits.)   (b) $-5.7\ \text{cm}^2$ (c) $3.5\ \text{cm}^2$ (d) $7128\ \text{cm}$

             (e) $3.52 \times 10^8\ \text{cm}^3$ (f) $-2.1 \times 10^{-3}\ \text{cm}$ (g) $14.1\ \text{cm}^2$

2.52. Perform the following calculations to the proper number of significant digits. All measurements are in centimeters.

(a) $1.20 \times 10^8 - 9.8 \times 10^6$ (d) $6.66 \times 10^7 - 6.02 \times 10^6$

(b) $6.66 \times 10^{-3} - 4.78 \times 10^{-2}$ (e) $2.6 \times 10^{-19} - 1.2 \times 10^{-17}$

(c) $7.7 \times 10^9 - 3.7 \times 10^8$

    Ans.   (a) $1.10 \times 10^8\ \text{cm}$ (d) $6.06 \times 10^7\ \text{cm}$

             (b) $-4.11 \times 10^{-2}\ \text{cm}$ (e) $-1.2 \times 10^{-17}\ \text{cm}$

             (c) $7.3 \times 10^9\ \text{cm}$

2.53. Calculate each answer to the correct number of significant digits: (a) $(3.69\ \text{g})/(1.23\ \text{mL})$, (b) $(369\ \text{g})/(1.23\ \text{L})$, and (c) $(987\ \text{g})/(1.5\ \text{L})$.

    Ans.   (a) $3.00\ \text{g/mL}$ (b) $3.00 \times 10^2\ \text{g/L}$ (c) $660\ \text{g/L} = 0.66\ \text{kg/L}$.

             Three significant digits are required in (a) and in (b); two significant digits are required in (c).

2.54. Explain why every zero in the coefficient is significant in a number properly expressed in standard exponential notation.

    Ans.   No zeros are needed for the sole purpose of determining the magnitude of the number. (The exponential part of the number does that.) The only other reason a zero would be present is if it were significant.

2.55. To how many decimal places should you report the reading of (a) a 50-mL buret (graduated in 0.1 mL) and (b) an analytic balance (calibrated to 0.001 g)?

    Ans.   (a) 2   (b) 4

             (One estimated digit in each.)

2.56. Round each of the following calculated results to the nearest integer (whole number): (*a*) 1.545 cm, (*b*) 5.500 cm, (*c*) 5.5 cm, (*d*) 4.501 cm, (*e*) 45.95 cm, and (*f*) 4.500 cm.

Ans.   (*a*) 2 cm   (*b*) 6 cm   (*c*) 6 cm   (*d*) 5 cm   (*e*) 46 cm   (*f*) 5 cm (or 4 cm with the sophisticated system).

2.57. Add, expressing the answer in the proper number of significant figures:

(*a*)   2000.0 cm + 0.02 cm

(*b*) 2.0000×10^3 cm + 2×10^{-2} cm

Ans.   (*a*)

$$\begin{array}{r} 2000.0 \ \text{cm} \\ 0.02 \ \text{cm} \\ \hline 2000.02 \ \text{cm} \rightarrow 2000.0 \ \text{cm} \end{array}$$

The fraction added is so small relative to the first number that it does not make any difference within the number of significant figures allowed.

(*b*)   This part is exactly the same as part (*a*). The same numbers, expressed in scientific notation with the same number of significant digits, are used. Be aware that when a small number is added to a much larger one, the small number may not affect the value of the larger within the precision of its measurement.

2.58. Determine the answers to the following expressions to the correct number of significant digits:

(*a*)   18.00 mL(2.50 g/mL)

(*b*)   0.207 m − 9.2 mm

(*c*)   167 cm − 15.3 cm

(*d*) $\dfrac{9.00 \text{ g}}{1.50 \text{ mL}}$

Ans.   (*a*)   18.00 mL(2.50 g/mL) = 45.0 g   (three significant digits)

(*b*)   0.207 m − 9.2 mm = 0.207 m − 0.0092 m = 0.198 m   (three decimal places)

(*c*)   167 cm − 15.3 cm = 151.7 cm → 152 cm

(*d*) $\dfrac{9.00 \text{ g}}{1.50 \text{ mL}}$ = 6.00 g/mL   (three significant digits)

2.59. The radius of a circle is 3.00 cm. (*a*) What is the diameter of the circle? (*d* = 2*r*) (*b*) What is the area of the circle? ($A = \pi r^2$)

Ans.   (*a*)   The diameter is twice the radius:

$$d = 2r = 2(3.00 \text{ cm}) = 6.00 \text{ cm}$$

The 2 here is a definition, not a measurement, and does not limit the accuracy of the answer.

(*b*) $A = \pi r^2 = (3.141\ 59)(3.00 \text{ cm})^2 = 28.3 \text{ cm}^2$

We can use the value of π to as many significant digits as we wish. It is not a measurement.

2.60. Perform the following calculations to the proper number of significant digits:

(*a*) $(1.20 \times 10^8 \text{ g} + 9.7 \times 10^7 \text{ g})/(3.04 \times 10^5 \text{ L})$

(*b*) $(1.66 \times 10^{-3} \text{ m} + 6.78 \times 10^{-2} \text{ m})(4.1 \times 10^4 \text{ m} + 1.3 \times 10^3 \text{ m})$

(*c*) $(7.7 \times 10^9 \text{ m} + 3.3 \times 10^9 \text{ m}) - (9.4 \times 10^4 \text{ m})$

(*d*) $(6.02 \times 10^{23} \text{ atoms} + 6.02 \times 10^{21} \text{ atoms}) - (3.95 \times 10^{22} \text{ atoms})$

(*e*) $1 \text{ cm} - (1.6 \times 10^{-6} \text{ cm} + 1.1 \times 10^{-7} \text{ cm})$

Ans.   (*a*) 7.14×10^2 g/L (The sum has three significant digits.)

(*b*) 2.9×10^3 m^2 (The second sum limits the answer to two significant digits.)

(*c*) 1.10×10^{10} m (The sum has three significant digits, and the number subtracted is too small to affect the sum at all.)

(*d*) 5.69×10^{23} atoms

(*e*)   1 cm (The sum is too small to affect 1 cm at all.)

DENSITY

2.61. What mass in grams is present in 45.0 mL of sulfuric acid solution, which has a density of 1.86 g/mL?

> *Ans.* Start with the 45.0 mL, not with the ratio of grams to milliliters. Multiply that value by the ratio to get the desired answer:
>
> $$45.0 \text{ mL}\left(\frac{1.86 \text{ g}}{1.00 \text{ mL}}\right) = 83.7 \text{ g}$$

2.62. What is the density of gold in kilograms per liter (kg/L) if it is 19.3 g/mL?

> *Ans.*
> $$\frac{19.3 \text{ g}}{\text{mL}}\left(\frac{1 \text{ kg}}{1000 \text{ g}}\right)\left(\frac{1000 \text{ mL}}{1 \text{ L}}\right) = \frac{19.3 \text{ kg}}{1 \text{ L}}$$

2.63. Calculate the density of gold (19.3 kg/L) in kilograms per cubic meter (kg/m^3).

> *Ans.*
> $$\frac{19.3 \text{ kg}}{1 \text{ L}}\left(\frac{1000 \text{ L}}{1 \text{ m}^3}\right) = \frac{19\,300 \text{ kg}}{1 \text{ m}^3}$$

2.64. Calculate the density of a solution of which 15.33 g occupies 8.2 mL.

> *Ans.*
> $$d = \frac{m}{V} = \frac{(15.33 \text{ g})}{(8.2 \text{ mL})} = 1.9 \text{ g/mL}$$

2.65. The density of air is about 1.25 g/L. The density of water is about 1.00 g/mL. Which is denser?

> *Ans.* Water is denser, 1000 g/L. Be sure that when you compare two densities, you use the same units for both.

2.66. The density of lead is 11.34 g/mL. (*a*) What volume is occupied by 175 g of lead? (*b*) What is the mass of 175 mL of lead?

> *Ans.* (*a*) $175 \text{ g}\left(\dfrac{1 \text{ mL}}{11.34 \text{ g}}\right) = 15.4 \text{ mL}$ (*b*) $175 \text{ mL}\left(\dfrac{11.34 \text{ g}}{1 \text{ mL}}\right) = 1980 \text{ g} = 1.98 \text{ kg}$

2.67. The density of gold is 19.3 g/mL, that of lead is 11.34 g/mL, and that of liquid mercury is 13.6 g/mL. Will a cube of solid gold float in liquid mercury? Will a cube of lead?

> *Ans.* Lead will float in mercury, because it is less dense than mercury, but gold will not float because it is denser.

2.68. Convert a density of 2.7 g/mL to (*a*) kg/L and (*b*) kg/m^3.

> *Ans.* (*a*) $\dfrac{2.7 \text{ g}}{\text{mL}}\left(\dfrac{1 \text{ kg}}{10^3 \text{ g}}\right)\left(\dfrac{10^3 \text{ mL}}{1 \text{ L}}\right) = \dfrac{2.7 \text{ kg}}{\text{L}}$ (*b*) $\dfrac{2.7 \text{ kg}}{\text{L}}\left(\dfrac{10^3 \text{ L}}{1 \text{ m}^3}\right) = \dfrac{2.7 \times 10^3 \text{ kg}}{\text{m}^3}$

2.69. Calcuate the density of a board 2.00 m long, 2.5 cm thick, and 0.50 dm wide which has a mass of 2.25 kg. Will the board float in water (density 1.00 g/mL)?

> *Ans.* The volume of the board is length × width × thickness
>
> $$2.00 \text{ m}\left(\frac{100 \text{ cm}}{1 \text{ m}}\right) = 200 \text{ cm long} \quad 0.50 \text{ dm}\left(\frac{10 \text{ cm}}{1 \text{ dm}}\right) = 5.0 \text{ cm wide}$$
>
> $$V = 200 \text{ cm} \times 5.0 \text{ cm} \times 2.5 \text{ cm} = 2.5 \times 10^3 \text{ cm}^3 = 2.5 \text{ L}$$
>
> The density is the mass divided by the volume:
>
> $$(2.25 \text{ kg})/(2.5 \text{ L}) = 0.90 \text{ kg/L}$$
>
> The board will float in water, since it is less dense. The density of water is 1.00 kg/L.

TEMPERATURE SCALES

2.70. Convert $-40°F$ to degrees Celsius.

Ans. $$°C = (°F - 32°)\tfrac{5}{9} = (-40° - 32°)\tfrac{5}{9} = -40°C$$

2.71. Change the following temperatures in degrees Celsius to kelvins: (*a*) 15°C, (*b*) −15°C, and (*c*) −265°C.

Ans. (*a*) $15°C + 273° = 288$ K

   (*b*) $-15°C + 273° = 258$ K

   (*c*) $-265°C + 273° = 8$ K

2.72. Change the following temperatures in kelvins to degrees Celsius: (*a*) 330 K, (*b*) 282 K, and (*c*) 302 K.

Ans.   (*a*) 57°C   (*b*) 9°C   (*c*) 29°C

Supplementary Problems

2.73. Change 42.30 km/h to meters per second.

Ans. $$\frac{42.30 \text{ km}}{\text{h}}\left(\frac{1000 \text{ m}}{1 \text{ km}}\right)\left(\frac{1 \text{ h}}{60 \text{ min}}\right)\left(\frac{1 \text{ min}}{60 \text{ s}}\right) = 11.75 \text{ m/s}$$

2.74. Match the value on the first line below with each corresponding value on the second line. (More than one value might correspond for each.)

1 μL   1 kL   1 L   1 mL
1 dm^3 1 cm^3 1 mm^3 1 m^3 10^3 L 10^{-3} L 10^0 L 10^{-6} L

Ans. $1 \text{ μL} = 1 \text{ mm}^3 = 10^{-6}$ L $1 \text{ L} = 1 \text{ dm}^3 = 10^0$ L

   $1 \text{ kL} = 1 \text{ m}^3 = 10^3$ L $1 \text{ mL} = 1 \text{ cm}^3 = 10^{-3}$ L

2.75. What exponential number best represents each of the following metric prefixes? (*a*) deci, (*b*) nano, (*c*) centi, (*d*) micro, (*e*) milli, and (*f*) kilo

Ans.   (*a*) 10^{-1} (*b*) 10^{-9} (*c*) 10^{-2} (*d*) 10^{-6} (*e*) 10^{-3} (*f*) 10^3

2.76. A certain beaker has a mass of 141.2 g. When 13.0 mL of an organic liquid is placed in the beaker, the total mass is 150.5 g. Calculate the density of the liquid to the proper number of significant digits.

Ans.   The mass of the liquid is 150.5 g − 141.2 g = 9.3 g.
   Its density is (9.3 g)/(13.0 mL) = 0.72 g/mL.

2.77. A beaker plus contents has a combined mass of 130.3 g. The beaker has a mass of 129.1 g. If the density of the contents is 2.04 g/mL, what is the volume of the contents to the correct number of significant digits?

Ans. $$(130.3 \text{ g} - 129.1 \text{ g})\left(\frac{1 \text{ mL}}{2.04 \text{ g}}\right) = 0.59 \text{ mL}$$

   There are only two significant digits since the subtraction yields an answer with two significant digits. That answer limits the final volume to two significant digits.

2.78. Make the following English-metric conversions:

(*a*)   Change 4.22 in. to cm   (*d*)   Change 21.7 lb to grams

(*b*)   Change 2.35 kg to pounds   (*e*)   Change 60.0 mi/h to meters per second

(*c*)   Change 6.13 L to fluid ounces   (*f*)   Change 3.50 in.^{2} to cm^2

(g) Change 242.4 mL to quarts (U.S.)   (i) Change 6.49 kg to ounces

(h) Change 13.71 ft to meters   (j) Change 433 km to miles

Ans.   (a) $4.22 \text{ in.} \left(\dfrac{2.54 \text{ cm}}{1 \text{ in.}} \right) = 10.7 \text{ cm}$

(b) $2.35 \text{ kg} \left(\dfrac{2.20 \text{ lb}}{1 \text{ kg}} \right) = 5.17 \text{ lb}$

(c) $6.13 \text{ L} \left(\dfrac{1.06 \text{ qt}}{1 \text{ L}} \right) \left(\dfrac{32 \text{ oz}}{1 \text{ qt}} \right) = 208 \text{ oz}$

(d) $21.7 \text{ lb} \left(\dfrac{1.0 \text{ kg}}{2.2 \text{ lb}} \right) \left(\dfrac{1000 \text{ g}}{1 \text{ kg}} \right) = 9860 \text{ g} = 9.86 \times 10^3 \text{ g}$

(e) $\dfrac{60.0 \text{ mi}}{\text{h}} \left(\dfrac{1760 \text{ yd}}{1 \text{ mi}} \right) \left(\dfrac{36 \text{ in.}}{1 \text{ yd}} \right) \left(\dfrac{1 \text{ m}}{39.37 \text{ in.}} \right) \left(\dfrac{1 \text{ h}}{3600 \text{ s}} \right) = \dfrac{26.8 \text{ m}}{\text{s}}$

(f) $3.50 \text{ in.}^2 \left(\dfrac{2.54 \text{ cm}}{1 \text{ in.}} \right)^2 = 22.6 \text{ cm}^2$

(g) $242.4 \text{ mL} \left(\dfrac{1 \text{ L}}{1000 \text{ mL}} \right) \left(\dfrac{1.06 \text{ qt}}{1 \text{ L}} \right) = 0.257 \text{ qt}$

(h) $13.71 \text{ ft} \left(\dfrac{12 \text{ in.}}{1 \text{ ft}} \right) \left(\dfrac{2.54 \text{ cm}}{1 \text{ in.}} \right) \left(\dfrac{1 \text{ m}}{100 \text{ cm}} \right) = 4.18 \text{ m}$

(i) $6.49 \text{ kg} \left(\dfrac{2.2 \text{ lb}}{1 \text{ kg}} \right) \left(\dfrac{16 \text{ oz}}{1 \text{ lb}} \right) = 228 \text{ oz}$

(j) $433 \text{ km} \left(\dfrac{0.621 \text{ mi}}{1 \text{ km}} \right) = 269 \text{ mi}$

2.79. Round the atomic masses (found in the periodic table) of the first 20 elements to three significant digits each.

Ans.

H	1.01	C	12.0	Na	23.0	S	32.1
He	4.00	N	14.0	Mg	24.3	Cl	35.5
Li	6.94	O	16.0	Al	27.0	Ar	39.9
Be	9.01	F	19.0	Si	28.1	K	39.1
B	10.8	Ne	20.2	P	31.0	Ca	40.1

2.80. What is wrong with the following problem? Change 125 g to milliliters.

Ans.   You cannot change a mass to a volume (unless you have a value for density).

2.81. What is the volume of 125 g of water?

Ans.   The one density that you must know is that of water, about 1.00 g/mL.

$$125 \text{ g} \left(\dfrac{1 \text{ mL}}{1.00 \text{ g}} \right) = 125 \text{ mL}$$

2.82. Express, in standard exponential notation, the number of

(a) grams in 7.65 ng   (e) milliliters in 3.21 L

(b) meters in 8.76 dm   (f) grams in 4.32 kg

(c) kilograms in 9.87 mg   (g) centimeters in 5.43 m

(d) liters in 6.54 mL

Ans.   (a) $7.65 \times 10^{-9} \text{ g}$ (e) $3.21 \times 10^3 \text{ mL}$

(b) $8.76 \times 10^{-1} \text{ m}$ (f) $4.32 \times 10^3 \text{ g}$

(c) $9.87 \times 10^{-6} \text{ kg}$ (g) $5.43 \times 10^2 \text{ cm}$

(d) $6.54 \times 10^{-3} \text{ L}$

2.83. Express, in standard exponential notation, the number of:

(a) liters in 1.0 m^3 (d) cubic meters in 1.0 mL   (g) cubic meters in 1.0 L

(b) liters in 1.0 mL   (e) cubic centimeters in 1.0 L   (h) cubic meters in 1.0 cm^3

(c) kiloliters in 1.0 m^3 (f) milliliters in 1.0 m^3 (i) cubic meters in 1.0 mm^3

Ans.   (a) 1.0×10^3 L   (d) 1.0×10^{-6} m^3 (g) 1.0×10^{-3} m^3

(b) 1.0×10^{-3} L   (e) 1.0×10^3 cm^3 (h) 1.0×10^{-6} m^3

(c)   1.0 kL   (f) 1.0×10^6 mL   (i) 1.0×10^{-9} m^3

2.84. The density of an alloy (a mixture) containing 60.0% gold plus other metals is 14.2 g/cm^3. If gold costs $600.00 per Troy ounce, what is the value of the gold in a 14.6-cm^3 piece of jewelry? (1 Troy ounce $= 31.103$ g)

Ans. $14.6 \text{ cm}^3 \left(\dfrac{14.2 \text{ g alloy}}{1 \text{ cm}^3} \right) \left(\dfrac{60.0 \text{ g gold}}{100.0 \text{ g alloy}} \right) \left(\dfrac{1 \text{ oz gold}}{31.1 \text{ g gold}} \right) \left(\dfrac{600.00 \text{ dollars}}{1 \text{ oz gold}} \right) = 2400 \text{ dollars}$

2.85. A solution with a density of 1.15 g/mL contains 33.0 g of sugar per 100 mL of solution. What mass of solution will contain 166 g of sugar? What mass of water is required to make this solution?

Ans. $166 \text{ g sugar} \left(\dfrac{100 \text{ mL solution}}{33.0 \text{ g sugar}} \right) \left(\dfrac{1.15 \text{ g solution}}{1 \text{ mL solution}} \right) = 578 \text{ g solution}$

$578 \text{ g solution} - 166 \text{ g sugar} = 412 \text{ g water}$

2.86. A gold alloy chain has a density of 14.6 g/cm^3. The cross-sectional area of the chain is 5.00 mm^2. What length of chain has a mass of 9.70 g?

Ans. $V = 9.70 \text{ g} \left(\dfrac{1 \text{ cm}^3}{14.6 \text{ g}} \right) \left(\dfrac{10 \text{ mm}}{1 \text{ cm}} \right)^3 = 664 \text{ mm}^3$

$\text{Length} = \dfrac{V}{A} = \dfrac{664 \text{ mm}^3}{5.00 \text{ mm}^2} = 133 \text{ mm} = 13.3 \text{ cm}$

2.87. How much does the chain of the prior problem cost if the gold alloy costs $500/Troy ounce? (1 Troy ounce $= 31.1$ g)

Ans. $9.70 \text{ g} \left(\dfrac{1 \text{ oz}}{31.1 \text{ g}} \right) \left(\dfrac{\$500.00}{1 \text{ oz}} \right) = \$156$

2.88. Calculate the density of sulfuric acid, 1.86 g/mL, in kilograms per liter.

Ans. $\dfrac{1.86 \text{ g}}{\text{mL}} \left(\dfrac{1 \text{ kg}}{1000 \text{ g}} \right) \left(\dfrac{1000 \text{ mL}}{1 \text{ L}} \right) = \dfrac{1.86 \text{ kg}}{\text{L}}$

2.89. An ashtray is made of 55.0 g of glass and 45.0 g of iron. The density of the glass is 2.50 g/mL, and that of the iron is 7.86 g/mL. What is the density of the ashtray?

Ans.   The density of the ashtray is its mass divided by its volume. Its mass is 100.0 g. Its volume is the total of the volumes of the glass and iron:

$$55.0 \text{ g glass} \left(\dfrac{1 \text{ mL}}{2.50 \text{ g}} \right) = 22.0 \text{ mL glass}$$

$$45.0 \text{ g iron} \left(\dfrac{1 \text{ mL}}{7.86 \text{ g}} \right) = 5.73 \text{ mL iron}$$

The total volume is 27.7 mL, and the density is

$$d = \dfrac{m}{V} = \dfrac{(100.0 \text{ g})}{(27.7 \text{ mL})} = 3.61 \text{ g/mL}$$

2.90. Calculate the mass of a gold sphere (density $= 19.3 \text{ g/cm}^3$) (a) of radius 3.00 cm and (b) of diameter 3.00 cm. $V = \frac{4}{3}\pi r^3$

Ans.   (a) $V = \frac{4}{3}\pi(3.00 \text{ cm})^3 = 113 \text{ cm}^3$ (b) $r = \frac{d}{2} = 1.50 \text{ cm}$

   $m = 113 \text{ cm}^3\left(\frac{19.3 \text{ g}}{1 \text{ cm}^3}\right) = 2180 \text{ g} = 2.18 \text{ kg}$ $V = \frac{4}{3}\pi(1.50 \text{ cm})^3 = 14.1 \text{ cm}^3$

   $m = 14.1 \text{ cm}^3\left(\frac{19.3 \text{ g}}{1 \text{ cm}^3}\right) = 272 \text{ g}$

2.91. How far can Mary drive her car on $50.00 if the car goes 25.4 mi/gal and gasoline costs $1.899 per gallon?

Ans. $50.00 \text{ dollars}\left(\frac{1 \text{ gal}}{1.899 \text{ dollars}}\right)\left(\frac{25.4 \text{ mi}}{1 \text{ gal}}\right) = 669 \text{ mi}$

2.92. (a) What is the mass of 18.6 mL of mercury? (Density $= 13.6 \text{ g/mL}$) (b) What is the volume of 18.6 g of mercury?

Ans.   (a) $18.6 \text{ mL}\left(\frac{13.6 \text{ g}}{1.00 \text{ mL}}\right) = 253 \text{ g}$

   (b) $\frac{18.6 \text{ g}}{13.6 \text{ g/mL}} = 1.37 \text{ mL}$

   (a) When we multiplied the mass per milliliter by the volume, the units mL canceled, and we were left with grams, which is the unit of mass we wanted. (b) When we divided the mass by the mass per milliliter, the units grams canceled, and we were left with mL, which is the unit of volume we wanted. If we had multiplied in part (b) or divided in part (a), the units would not have canceled, and we would have obtained a result that was not appropriate to the problem. The units helped us decide whether to multiply or divide.
   Instead of dividing by a ratio we can do the same thing by inverting the ratio and multiplying by it.

2.93. Calculate the density of 150.0-g rectangular block with length 15.00 cm, width 2.50 cm, and thickness 1.50 cm.

Ans. $\frac{150.0 \text{ g}}{(15.00 \text{ cm})(2.50 \text{ cm})(1.50 \text{ cm})} = 2.67 \text{ g/cm}^3$

2.94. Using the data of Problem 2.67, determine whether a cube of metal 1.50 cm on each side is mercury, gold, or lead. The mass of the cube is 65.2 g.

Ans.   The volume of a cube is equal to its length, cubed, or $(1.50 \text{ cm})^3 = 3.375 \text{ cm}^3$.

   $d = \frac{m}{V} = \frac{(65.2 \text{ g})}{(3.375 \text{ cm}^3)} = 19.3 \text{ g/cm}^3$

   The metal is gold.

2.95. Commercial sulfuric acid solution is 96.0% H_2SO_4 in water and has a density of 1.86 g/mL. How many grams of H_2SO_4 are there in 37.1 mL of the commercial solution?

Ans. $37.1 \text{ mL solution}\left(\frac{1.86 \text{ g solution}}{1 \text{ mL solution}}\right)\left(\frac{96.0 \text{ g } H_2SO_4}{100 \text{ g solution}}\right) = 66.2 \text{ g } H_2SO_4$

   Quantity given   Density of solution   Percent

2.96. What mass of H_2SO_4 is present in 41.7 g of solution which is 96.0% H_2SO_4 in water?

Ans. $41.7 \text{ g solution}\left(\frac{96.0 \text{ g } H_2SO_4}{100 \text{ g solution}}\right) = 40.0 \text{ g } H_2SO_4$

Atoms and Atomic Masses

3.1. INTRODUCTION

In this chapter we will discuss

1. Dalton's postulates regarding the existence of the atom and the laws on which those postulates are based

2. Atomic mass—its uses and limitations

3. The structure of the atom

4. The existence of isotopes

5. The periodic table, which for now is presented only enough to introduce the concepts of periodic groups or families and the numbers of electrons in the outermost electron shells

3.2. ATOMIC THEORY

In 1804, John Dalton proposed the existence of atoms. He not only postulated that atoms exist, as had ancient Greek philosophers, but he also attributed to the atom certain properties. His postulates were as follows:

1. Elements are composed of indivisible particles, called atoms.

2. All atoms of a given element have the same mass, and the mass of an atom of a given element is different from the mass of an atom of any other element.

3. When elements combine to form a given compound, the atoms of one element combine with those of the other element(s) in a definite ratio to form molecules. Atoms are not destroyed in this process.

4. Atoms of two or more elements may combine in different ratios to form different compounds.

5. The most common ratio of atoms is 1:1, and where more than one compound of two or more elements exists, the most stable is the one with 1:1 ratio of atoms. (This postulate is incorrect.)

Dalton's postulates stimulated great activity among chemists, who sought to prove or disprove them. The fifth postulate was very quickly shown to be incorrect, and the first three have had to be modified in light of later knowledge; however, the first four postulates were close enough to the truth to lay the foundations for a basic understanding of mass relationships in chemical compounds and chemical reactions.

Dalton's postulates can be used to explain three quantitative laws that had been developed shortly before he proposed his theory.

The *Law of Conservation of Mass* states that mass is neither created nor destroyed in a chemical reaction or physical change.

The *Law of Definite Proportions* states that every chemical compound is made up of elements in a definite ratio by mass.

The *Law of Multiple Proportions* states that when two or more different compounds are formed from the same elements, the ratio of masses of each element in the compounds for a given mass of any other element is a small whole number.

EXAMPLE 3.1.   Calculate the mass of sodium chloride formed by the complete reaction of 10.0 g of sodium with 15.4 g of chlorine. What law allows this calculation?

Ans.   The Law of Conservation of Mass:

$$10.0 \text{ g} + 15.4 \text{ g} = 25.4 \text{ g total}$$

EXAMPLE 3.2.   Calculate the mass of oxygen that will combine with 2.00 g of magnesium if 0.660 g of oxygen reacts with 1.00 g of magnesium. What law allows this calculation?

Ans.   The Law of Definite Proportions states that twice as much oxygen will react with twice as much magnesium to yield the same ratio of masses:

$$0.660 \text{ g oxygen} \left(\frac{2.00 \text{ g magnesium}}{1.00 \text{ g magnesium}} \right) = 1.32 \text{ g oxygen}$$

EXAMPLE 3.3.   Show that the following data illustrate the Law of Multiple Proportions:

	Element 1	Element 1
Compound 1	1.00 g	1.33 g
Compound 2	1.00 g	2.66 g

Ans.   For a fixed mass of element 1 in the two compounds (1.00 g), the ratio of masses of the other element is a small integral ratio: (2.66 g)/(1.33 g) = 2.00.

Dalton argued that these laws are entirely reasonable if the elements are composed of atoms. For example, the reason that mass is neither gained nor lost in a chemical reaction is that the atoms in the reaction merely change partners with one another; they do not appear or disappear. The definite proportions of compounds stem from the fact that the compounds comprise a definite ratio of atoms (postulate 3), each with a definite mass (postulate 2). The law of multiple proportions is due to the fact that different numbers of atoms of one element can react with a given number of atoms of a second element (postulate 4), and since atoms must combine in whole-number ratios, the ratio of masses must also be in whole numbers.

3.3. ATOMIC MASSES

Once Dalton's hypotheses had been proposed, the next logical step was to determine the *relative* masses of the atoms of the elements. Since there was no way at that time to determine the mass of an individual atom, the relative masses were the best information available. That is, one could tell that an atom of one element had a mass twice as great as an atom of a different element (or 15/4 times as much, or 17.3 times as much, etc.). How could even these relative masses be determined? They could be determined by taking equal (large) numbers of atoms of two elements and by determining the ratio of masses of these collections of atoms.

For example, a large number of sodium atoms have a total mass of 23.0 g, and an equal number of chlorine atoms have a total mass of 35.5 g. Since the number of atoms of each kind is equal, the ratio of masses of one sodium atom to one chlorine atom is 23.0 to 35.5. How can one be sure that there are equal numbers of sodium and chlorine atoms? One ensures equal numbers by using a compound of sodium and chlorine in which there are equal numbers of atoms of the two elements (i.e., sodium chloride, common table salt).

A great deal of difficulty was encountered at first, because Dalton's fifth postulate gave an incorrect ratio of numbers of atoms in many cases. Such a large number of incorrect results were obtained that it soon became apparent that the fifth postulate was not correct. It was not until some 50 years later than an experimental method was devised to determine the atomic ratios in compounds, at which time the scale of relative atomic masses was determined in almost the present form. These relative masses are called *atomic masses*, or sometimes *atomic weights*.

Atomic masses are so small that an appropriate unit was developed to report them—an *atomic mass unit* (amu). 1 amu $= 1.66 \times 10^{-24}$ g. The atomic mass of the lightest element, hydrogen, was originally taken to be 1 amu. The modern values of the atomic masses are based on the most common kind of carbon atom, called "carbon-12" and written ^{12}C, as the standard. The mass of ^{12}C is measured in the modern *mass spectrometer*, and ^{12}C is *defined* to have an atomic mass of exactly 12 amu. On this scale hydrogen has an atomic mass of 1.008 amu.

The atomic mass of an element is the relative mass of an average atom of the element compared with ^{12}C, which by definition has a mass of exactly 12 amu. Thus, since a sulfur atom has a mass 8/3 times that of a carbon atom, the atomic mass of sulfur is

$$12 \text{ amu} \times \frac{8}{3} = 32 \text{ amu}$$

A complete list of the modern values of the atomic masses of the elements is given in the Table of Elements, page 349, and in the periodic table, page 350.

3.4. ATOMIC STRUCTURE

From 50 to 100 years after Dalton proposed his theory, various discoveries were made that show that the atom is not indivisible, but really is composed of parts. Natural radioactivity and the interaction of electricity with matter are two different types of evidence for this subatomic structure. The most important subatomic particles are listed in Table 3-1, along with their most important properties. The protons and neutrons are found in a very tiny *nucleus* (plural, *nuclei*). The electrons are found outside the nucleus. The information in Table 3-1 must be memorized, but only the whole number part of the masses (1 amu, 1 amu, 0 amu) must be remembered.

Table 3-1   Subatomic Particles

	Charge (e)	Mass (amu)	Location
Proton	+1	1.00728	In nucleus
Neutron	0	1.00894	In nucleus
Electron	−1	0.0005414	Outside nucleus

There are two types of electric charges that occur in nature—positive and negative. Charges of these two types are opposite one another, and cancel the effect of the other. Bodies with opposite charge types attract one another; those with the same charge type repel one another. If a body has equal numbers of charges of the two types, it has no net charge and is said to be *neutral*. The charge on the electron is a fundamental unit of electric charge (equal to 1.6×10^{-19} Coulomb) and is given the symbol e.

EXAMPLE 3.4.   Using the data of Table 3-1, find the charge on a nucleus that contains (*a*) 8 protons and 8 neutrons and (*b*) 8 protons and 10 neutrons.

Ans.   (*a*) 8(+1) + 8(0) = +8   (*b*) 8(+1) + 10(0) = +8

Both nuclei have the same charge. Although the nuclei have different numbers of neutrons, the neutrons have no charges, so they do not affect the charge on the nucleus.

EXAMPLE 3.5.   To the nearest integer, calculate the mass (in amu) of a nucleus that contains (*a*) 17 protons and 18 neutrons and (*b*) 17 protons and 20 neutrons.

Ans.   (*a*) 17(1 amu) + 18(1 amu) = 35 amu

(*b*) 17(1 amu) + 20(1 amu) = 37 amu

The nuclei differ in mass (but not in charge).

Uncombined atoms as a whole are electrically neutral.

EXAMPLE 3.6.   Refer to Table 3-1 and deduce which two of the types of subatomic particles in an uncombined atom occur in equal numbers.

Ans.   The number of positive charges must equal the number of negative charges, since the atom has a net charge of zero. The number of positive charges, as shown in the table, is equal to the number of protons. The number of negative charges, also from the table, is equal to the number of electrons. Therefore, in an uncombined atom, the number of protons must equal the number of electrons.

The number of protons in the nucleus determines the chemical properties of the element. That number is called the *atomic number* of the element. Atomic number is symbolized Z. Each element has a different atomic number. An element may be identified by giving its name or its atomic number. Atomic numbers may be specified by use of a subscript before the symbol of the element. For example, carbon may be designated $_6$C. The subscript is really unnecessary, since all carbon atoms have atomic number 6 and all atoms with atomic number 6 are carbon atoms, but it is sometimes useful to include it. Atomic numbers are listed in the periodic table, page 350, and in the Table of Elements, page 349.

EXAMPLE 3.7.   (*a*) What is the charge on a magnesium nucleus? (*b*) What is the charge on a magnesium atom?

Ans.   (*a*) +12, equal to the atomic number of magnesium (from the Table of Elements, p. 349). (*b*) 0 (all uncombined atoms have a net charge of 0). Note that these questions sound very much alike, but are very different. You must read questions in chemistry very carefully.

3.5.   ISOTOPES

Atoms having the same number of protons but different numbers of neutrons are called *isotopes* of one another. The number of neutrons does not affect the chemical properties of the atoms appreciably, so all isotopes of a given element have essentially the same chemical properties. Different isotopes have different masses (contrary to Dalton's second postulate) and different nuclear properties, however.

The sum of the number of protons and the number of neutrons in the isotope is called the *mass number* of the isotope. Mass number is symbolized A. Isotopes are usually distinguished from one another by their mass numbers, given as a superscript before the chemical symbol for the element. Carbon-12 is an isotope of carbon with a symbol ^{12}C.

EXAMPLE 3.8.   (*a*) What is the sum of the number of protons and the number of neutrons in ^{12}C? (*b*) What is the number of protons in ^{12}C? (*c*) What is the number of neutrons in ^{12}C?

Ans.   (*a*) 12, its mass number. (*b*) 6, its atomic number, given in the periodic table. (*c*) 12 − 6 = 6. Note that the mass numbers for most elements are *not* given in the periodic table.

EXAMPLE 3.9.   Choose the integer quantities from the following list: (*a*) atomic number, (*b*) atomic mass, and (*c*) mass number.

Ans.   Atomic number and mass number are integer quantities; atomic mass is not in general equal to an integer.

EXAMPLE 3.10.   Choose the quantities that appear in the periodic table from the following list: (*a*) atomic number, (*b*) mass number, and (*c*) atomic mass.

Ans.   Atomic number and atomic mass appear in the periodic table. The mass numbers of only those few elements that do not occur naturally appear there in parentheses.

The *atomic mass* of an element is the weighted average of the masses of the individual isotopes of the element. (Thus, unless an element consists of only one stable isotope, the atomic mass is not equal to the mass of any atom.)

EXAMPLE 3.11. Naturally occurring copper consists of 69.17% ^{63}Cu and 30.83% ^{65}Cu. The mass of ^{63}Cu is 62.939 598 amu, and the mass of ^{65}Cu is 64.927 793 amu. What is the atomic mass of copper?

Ans.   The *weighted average* is the sum of the mass of each isotope times its fraction present:

$$62.939\ 598 \text{ amu}\left(\frac{69.17}{100}\right) + 64.927\ 793 \text{ amu}\left(\frac{30.83}{100}\right) = 63.55 \text{ amu}$$

3.6. PERIODIC TABLE

The *periodic table* is an extremely useful tabulation of the elements. It is constructed so that each vertical column contains elements that are chemically similar. The elements in the columns are called *groups*, or families. (Elements in some groups can be very similar to one another. Elements in other groups are less similar. For example, the elements of the first group resemble one another more than the elements of the fourth group from the end, headed by N.) Each row in the table is called a *period* (Fig. 3-1).

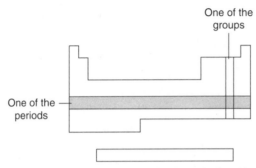

Fig. 3-1.   A period and a group in the periodic table

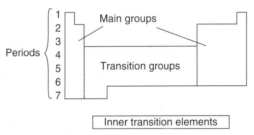

Fig. 3-2.   Periods and element types in the periodic table

There are three distinct areas of the periodic table—the *main group elements*, the *transition group elements*, and the *inner transition group elements* (Fig. 3-2). We will focus our attention at first on the main group elements, whose properties are easiest to learn and to understand.

The periods and the groups are identified differently. The periods are labeled from 1 to 7. Some reference is made to period numbers. The groups are referred to extensively by number. Unfortunately, the groups have been labeled in three different ways (Table 3-2):

1. *Classical*: Main groups are labeled IA through VIIA plus 0. Transition groups are labeled IB through VIII (although not in that order).

2. *Amended*: Main groups and transition groups are labeled IA through VIII and then IB through VIIB plus 0.

3. *Modern*: Groups are labeled with Arabic numerals from 1 through 18.

Although the modern designation is seemingly simpler, it does not point up some of the relationships that the older designations do. In this book, the classical system will be followed, with the modern number sometimes included in parentheses. The designations are tabulated in Table 3-2. You should check to see which designation your instructor or text uses, and use that designation yourself.

Table 3-2   Periodic Group Designations

Classical	IA	IIA	IIIB	. . .	VIII*	IB	IIB	IIIA	. . .	VIIA	0
Amended	IA	IIA	IIIA	. . .	VIII*	IB	IIB	IIIB	. . .	VIIB	0
Modern	1	2	3	. . .	8 9 10	11	12	13	. . .	17	18

*Note that three columns are collected as one group (VIII) in the two older designations.

EXAMPLE 3.12.   (a) What is the relationship between the *numbers* of the first eight columns in all three designations? (b) of the last eight columns?

Ans.   (a) They are all the same, 1 through 8. (b) The modern designation is 10 higher for each of these groups except the last, where it is 18 higher.

Several important groups are given names. Group IA (1) metals (not including hydrogen) are called the *alkali metals*. Group IIA (2) elements are known as the *alkaline earth metals*. Group VIIA (17) elements are called the *halogens*. Group IB (11) metals are known as the *coinage metals*. Group 0 (18) elements are known as the *noble gases*. These names lessen the need for using group numbers and thereby lessen the confusion from the different systems.

The electrons in atoms are arranged in shells. (A more detailed account of electronic structure will be presented in Chap. 4.) The maximum number of electrons that can occupy any shell n is given by

$$\text{Maximum number in shell } n = 2n^2$$

Since there are only a few more than 100 electrons total in even the biggest atoms, it can easily be seen that the shells numbered 5 or higher never get filled with electrons. Another important limitation is that the outermost shell, called the *valence shell*, can never have more than eight electrons in it. The number of electrons in the valence shell is a periodic property.

Shell number	1	2	3	4	5	6	7
Maximum number of electrons	2	8	18	32	50	72	98
Maximum number of outermost shell	2	8	8	8	8	8	8

EXAMPLE 3.13.   (a) How many electrons would fit in the first seven shells of an atom if the shells filled to their capacity in numeric order? (b) Why does this not happen?

Ans.   (a) A total of 280 $(2 + 8 + 18 + 32 + 50 + 72 + 98)$ electrons could be held in this fictional atom.   (b) Filling one atom with 280 electrons does not happen because the shells do not fill to capacity before the next ones start to fill and because even the biggest atom has only a few more than 100 electrons total.

EXAMPLE 3.14.   (a) What is the maximum number of electrons in the third shell of an atom in which there are electrons in the fourth shell? (b) What is the maximum number of electrons in the third shell of an atom in which there are no electrons in the fourth shell?

Ans.   (a) Eighteen is the maximum number in the third shell.   (b) Eight is the maximum number if there are no electrons in higher shells. (There can be no electrons in higher shells if there are none in the fourth shell.)

EXAMPLE 3.15.   What is the maximum number of electrons in the first shell when it is the outermost shell? when it is not the outermost shell?

Ans.   The maximum number is 2. It does not matter if it is the outermost shell or not, 2 is the maximum number of electrons in the first shell.

EXAMPLE 3.16. Arrange the 12 electrons of magnesium into shells.

Ans.   2   8   2

The first two electrons fill the first shell, and the next eight fill the second shell. That leaves two electrons in the third shell.

The number of outermost electrons is crucial to the chemical bonding of the atom. (See Chap. 5.) For main group elements, the number of outermost electrons is equal to the classical group number, except that it is 2 for helium and 8 for the other noble gases. (It is equal to the modern group number minus 10 except for helium and the first two groups.)

Electron dot diagrams use the chemical symbol to represent the nucleus plus the inner electrons and a dot to represent each valence electron. Such diagrams will be extremely useful in Chap. 5.

EXAMPLE 3.17. Write the electron dot symbols for elements Na through Ar.

Ans.   Na·   Mg: ·Al: ·Si: ·P: :S: :Cl: :Ar:

Solved Problems

ATOMIC THEORY

3.1. If 10.0 g of sodium and 20.0 g of chlorine are mixed, they react to form 25.4 g of sodium chloride. Calculate the mass of chlorine that does not react.

Ans. The Law of Conservation of Mass allows us to determine the mass of unreacted chlorine.

$$10.0\text{ g} + 20.0\text{ g} = 25.4\text{ g} + x$$

$$x = 4.6\text{ g}$$

3.2. A sample of a compound contains 1.00 g of carbon and 1.33 g of oxygen. How much oxygen would be combined with 7.00 g of carbon in another sample of the same compound? What law allows you to do this calculation?

Ans. Since it is the same compound, it must have the same ratio of masses of its elements, according to the Law of Definite Proportions. Thus there is $7.00\text{ g C}\left(\frac{1.33\text{ g O}}{1.00\text{ g C}}\right) = 9.31\text{ g O}$ in the second sample.

3.3. Determine in each part whether the two samples are samples of the same compound or not. What law allows this calculation?

	Sample 1		Sample 2	
	Element 1	**Element 2**	**Element 1**	**Element 2**
(a)	6.11 g	8.34 g	7.48 g	10.2 g
(b)	43.1 g	88.3 g	12.7 g	17.3 g

Ans. If they are the same compound, they will have the same ratio of masses according to the Law of Definite Proportions.

(a) $\dfrac{6.11\text{ g element 1}}{8.34\text{ g element 2}} = 0.733$ $\dfrac{7.48\text{ g element 1}}{10.2\text{ g element 2}} = 0.733$

These are samples of the same compound.

(b) $\dfrac{43.1\text{ g element 1}}{88.3\text{ g element 2}} = 0.488$ $\dfrac{12.7\text{ g element 1}}{17.3\text{ g element 2}} = 0.734$

These are not samples of the same compound.

3.4. Which part of Problem 3.3 illustrates the Law of Multiple Proportions?

Ans.   Part (*b*).

$$\frac{0.734}{0.488} = \frac{1.50}{1.00} = \frac{3}{2}$$

3.5. Why was Dalton's contribution different from that of the ancient Greeks who postulated the existence of atoms?

Ans.   Dalton based his postulates on experimental evidence; the Greeks did not. Also, he postulated that atoms of different elements are different.

3.6. Could anyone have done Dalton's work 100 years earlier than Dalton did?

Ans.   No; the laws on which his theory was based had not been established until shortly before his work.

3.7. If 20 kg of coal burns, producing 100 g of ashes, how is the Law of Conservation of Mass obeyed?

Ans.   The coal plus oxygen has a certain mass. The ashes plus the carbon dioxide (and a few other compounds that are produced) must have a combined mass that totals the same as the combined mass of the coal and oxygen. The law does not state that the total mass before and after the reaction must be the mass of the solids only.

3.8. A sample of water purified from an iceberg in the Arctic Ocean contains 11.2% hydrogen by mass. (*a*) What is the percent of hydrogen by mass in water prepared in the lab from hydrogen and oxygen? (*b*) How can you predict that percentage?

Ans.   (*a*) 11.2%   (*b*) The law of definite proportions states that all water, no matter its source, contains the same percentage of its constituent elements.

3.9. Chemists do not use the Law of Multiple Proportions in their everyday work. Why was this law introduced in this book?

Ans.   It is important in the history of chemistry, but mostly it was introduced to show that Dalton's atomic theory was based on experimental data.

3.10. What happens to a postulate of a scientific theory if it leads to incorrect results?

Ans.   It is changed or abandoned.

ATOMIC MASSES

3.11. The average man in a certain class weighs 190 lb. The average woman in that class weighs 135 lb. The male/female ratio is 3:2. What is the average weight of a person in the class?

Ans.   Three of every five people are male. Thus:

$$\left(\frac{3}{5}\right)(190 \text{ lb}) + \left(\frac{2}{5}\right)(135 \text{ lb}) = 168 \text{ lb}$$

3.12. (*a*) Determine the total weight of all the men and the total weight of all the women in your chemistry class (by using an anonymous questionnaire) (*b*) Attempt to determine the ratio of the average weight of each man to that of each woman without counting the numbers of each. (*c*) Attempt the same determination at a dated party.

Ans.   (*b*)   You cannot determine the ratio of averages without knowing something about the numbers present.

(*c*)   At a dated party there is apt to be a 1:1 ratio of men to women; therefore the ratio of the averages is equal to the ratio of the total weights.

3.13. If 75.77% of naturally occurring chlorine atoms have a mass of 34.968 852 amu and 24.23% have a mass of 36.965 903 amu, calculate the atomic mass of chlorine.

Ans.

$$\left(\frac{75.77}{100}\right)(34.968\,852 \text{ amu}) + \left(\frac{24.23}{100}\right)(36.965\,903 \text{ amu}) = 35.45 \text{ amu}$$

ATOMIC STRUCTURE

3.14. If uncombined atoms as a whole are neutral, how can they be made up of charged particles?

Ans.   The number of positive charges and negative charges (protons and electrons) are equal, and the effect of the positive charges cancels the effect of the negative charges.

3.15. Explain, on the basis of the information in Table 3-1, (*a*) why the nucleus is positively charged, (*b*) why the nucleus contains most of the mass of the atom, (*c*) why electrons are attracted to the nucleus, and (*d*) why the atom may be considered mostly empty space.

Ans.   (*a*) The only positively charged particles, the protons, are located there.   (*b*) The protons and neutrons, both massive compared to the electrons, are located there.   (*c*) The nucleus is positively charged, and they are negatively charged.   (*d*) The nucleus is tiny compared to the atom as a whole, and the nucleus contains most of the mass of the atom. Thus, the remainder of the atom contains little mass and may be thought of as mostly empty space.

3.16. (*a*) How many protons are there in the nucleus of a neon atom? (*b*) How many protons are there in the neon atom?

Ans.   (*a*) and (*b*) 10 (all the protons are in the nucleus, so it is not necessary to specify the nucleus). Here are two questions that sound different, but really are the same. Again, you must read the questions carefully. You must understand the concepts and the terms; you must not merely memorize.

ISOTOPES

3.17. Complete the following table (for uncombined atoms):

	Symbol	Atomic Number	Mass Number	Number of Protons	Number of Electrons	Number of Neutrons
(*a*)	^{3}He					
(*b*)		8	18			
(*c*)				24		28
(*d*)			107		47	
(*e*)					29	36

Ans.

	Symbol	Atomic Number	Mass Number	Number of Protons	Number of Electrons	Number of Neutrons
(*a*)	^{3}He	2	3	2	2	1
(*b*)	^{18}O	8	18	8	8	10
(*c*)	^{52}Cr	24	52	24	24	28
(*d*)	^{107}Ag	47	107	47	47	60
(*e*)	^{65}Cu	29	65	29	29	36

(*a*) Atomic number is determined from the identity of the element and is equal to the number of protons and to the number of electrons. Its value is given in the periodic table. The mass number is given, and the number of neutrons is equal to the mass number minus the atomic number.   (*b*) The atomic number and the number of protons are always the same, and these are the same as the number of electrons in uncombined atoms. (*c*) The sum of the number of protons and neutrons equals the mass number. In general, if we use *Z* for the atomic number, *A* for the mass number, and *n* for the number of neutrons, we get the simple equation:

$$A = Z + n$$

Once we are given two of these, we may calculate the third. Also,

$$Z = \text{number of protons} = \text{number of electrons in uncombined atom}$$

3.18. What subscripts could be used in the symbols in the first column of the answer to Problem 3.17?

Ans.   The atomic numbers can be used as subscripts to the left of the chemical symbols.

3.19. (*a*) Which columns of the answer to Problem 3.17 are always the same as other columns for uncombined atoms? (*b*) Which column is the sum of two others? (*c*) Which column is the same as the superscripts of the symbols?

Ans.   (*a*)   Atomic number, number of protons, and number of electrons

           (*b*)   Mass number = number of protons + number of neutrons

           (*c*)   The superscripts in the first column are repeated in the column of mass numbers

3.20. The numbers of which subatomic particle are different in atoms of two isotopes?

Ans.   The number of neutrons differs.

3.21. Consider the answer to Problem 3.20 and the fact that ^{16}O and ^{18}O have very similar chemical properties. Which is more important in determining chemical properties, nuclear charge or nuclear mass?

Ans.   Since the charge is the same and the mass differs, the chemical properties are due to the charge.

3.22. What is the net charge on an atom that contains 8 protons, 8 neutrons, and 10 electrons?

Ans.   The charge is -2. There are 10 electrons, each with -1 charge for a total of -10, and there are 8 protons, each with a $+1$ charge, for a total of $+8$. It does not matter how many neutrons there are, since they have no charge. $+8 + (-10) = -2$. If the initial letter of the particle stands for the number of that particle, then

$$\text{Charge} = (+1)p + (-1)e + (0)n = (+1)(8) + (-1)(10) + (0)(8) = -2$$

3.23. Two uncombined atoms are isotopes of each other. One has 17 electrons. How many electrons does the other have?

Ans.   The second isotope also has 17 electrons. The only difference between isotopes is the number of neutrons.

3.24. What is the mass number of an atom with 8 protons, 8 neutrons, and 8 electrons?

Ans.   The mass number is the number of protons plus neutrons; the number of electrons is immaterial.

$$8 + 8 = 16$$

3.25. Complete the following table (for uncombined atoms):

	Symbol	Atomic Number	Mass Number	Number of Protons	Number of Electrons	Number of Neutrons
(*a*)	$^{78}_{34}$Se					
(*b*)		20	44			
(*c*)			14			8
(*d*)			50		24	
(*e*)			80	34		

Ans.

	Symbol	Atomic Number	Mass Number	Number of Protons	Number of Electrons	Number of Neutrons
(a)	$^{78}_{34}\text{Se}$	34	78	34	34	44
(b)	$^{44}_{20}\text{Ca}$	20	44	20	20	24
(c)	$^{14}_{6}\text{C}$	6	14	6	6	8
(d)	$^{50}_{24}\text{Cr}$	24	50	24	24	26
(e)	$^{80}_{34}\text{Se}$	34	80	34	34	46

PERIODIC TABLE

3.26. How many electrons are there in the outermost shell of (a) H and (b) He?

Ans.   (a) One. There is only one electron in the whole atom. The shell can hold two electrons, but in the free hydrogen atom, there is only one electron present.   (b) Two.

3.27. How are the 19 electrons of the potassium atom arranged in shells?

Ans.   2   8   8   1

The first shell of any atom holds a maximum of 2 electrons, and the second shell holds a maximum of 8. Thus, the first 2 electrons of potassium fill the first shell, and the next 8 fill the second shell. The outermost shell of any atom can hold at most 8 electrons. In potassium, there are 9 electrons left, which would fit into the third shell if it were not the outermost shell; however, if we put the 9 electrons into the third shell, it would be the outermost shell. Therefore, we put 8 of the remaining electrons in that shell. That leaves 1 electron in the fourth shell.

3.28. How many electrons are there in the outermost shell of each of the following elements? (a) Li, (b) Be, (c) Al, (d) Si, (e) N, (f) O, (g) Cl, and (h) Ar.

Ans.   (a) 1   (b) 2   (c) 3   (d) 4   (e) 5   (f) 6   (g) 7   (h) 8

3.29. Write the electron dot symbol for each atom in Problem 3.28.

Ans.   Li·   Be: :Al·   :Si·   :N·   :Ö: :Cl: :Är:

Supplementary Problems

3.30. Distinguish clearly between (a) neutron and nucleus, (b) mass number and atomic mass, (c) atomic number and mass number, (d) atomic number and atomic mass, and (e) atomic mass and atomic mass unit.

Ans.   (a) The nucleus is a distinct part of the atom. Neutrons are subatomic particles that, along with protons, are located in the nucleus.   (b) Mass number refers to individual isotopes. It is the sum of the numbers of protons and neutrons. Atomic mass refers to the naturally occurring mixture of isotopes, and is the relative mass of the average atom compared to ^{12}C.   (c) Atomic number equals the number of protons in the nucleus; mass number is defined in part (b).   (d) These terms are defined in parts (b) and (c).   (e) Atomic mass unit is the unit of atomic mass.

3.31. Can you guess from data in the periodic table what the most important isotopes of bromine are? With the additional information that ^{80}Br does not occur naturally, how can you amend that guess?

Ans.   The atomic mass of Br is about 80, so one might be tempted to guess that ^{80}Br is the most important isotope. Knowing that it is not, one then can guess that an about-equal mixture of ^{79}Br and ^{81}Br occurs, which is correct.

3.32. The ratio of atoms of Y to atoms of X in a certain compound is 1:3. The mass of X in a certain sample is 3.00 g, and the mass of Y in the sample is 4.00 g. What is the ratio of their atomic masses?

Ans. Let N = number of atoms of Y; then $3N$ = number of atoms of X. A_Y and A_X represent their atomic masses. So

$$\frac{N(A_Y)}{3N(A_X)} = \frac{4.00 \text{ g}}{3.00 \text{ g}}$$

The Ns cancel, and the ratio of atomic masses is $A_Y/A_X = 4/1$.

3.33. Two oxides of nitrogen have compositions as follows:

Compound 1	46.68% N	53.32% O
Compound 2	36.86% N	63.14% O

Show that these compounds obey the law of multiple proportions.

Ans. The ratio of the mass of one element in one compound to the mass of that element in the other compound—for a fixed mass of the other element—must be in a ratio of small whole numbers. Let us calculate the mass of oxygen in the two compounds per gram of nitrogen. That is, the fixed mass will be 1.000 g N.

$$1.000 \text{ g N}\left(\frac{53.32 \text{ g O}}{46.68 \text{ g N}}\right) = 1.142 \text{ g O} \qquad 1.000 \text{ g N}\left(\frac{63.14 \text{ g O}}{36.86 \text{ g N}}\right) = 1.713 \text{ g O}$$

Is the ratio of the masses of oxygen a ratio of small integers?

$$\frac{1.713}{1.142} = \frac{1.50}{1} = \frac{3}{2}$$

The ratio of 1.50:1 is equal to the ratio 3:2, and the law of multiple proportions is satisfied. Note that the ratio of masses of nitrogen to oxygen is not necessarily a ratio of small integers; the ratio of mass of oxygen in one compound to mass of oxygen in the other compound is what must be in the small integer ratio.

3.34. (*a*) Compare the number of seats in a baseball stadium during a midweek afternoon game to the number during a weekend doubleheader (when two games are played successively). (*b*) Compare the number of locations available for electrons in the second shell of a hydrogen atom and the second shell of a uranium atom.

Ans. (*a*) The number of seats is the same at both times. (The number *occupied* probably is different.)

(*b*) The number of locations is the same. In uranium, all 8 locations are filled, whereas in hydrogen they are empty.

3.35. Show that the following compounds obey the law of multiple proportions.

	% Cobalt	% Oxygen
Compound 1	71.06	28.94
Compound 2	78.65	21.35

Does the law require that the ratio (mass of cobalt)/(mass of oxygen) be integral for each of these compounds?

Ans. The ratio of mass of cobalt to mass of oxygen is not necessarily integral. Take a constant mass of cobalt, for example, 1.000 g Co. From the ratios given, there are

$$\text{Compound 1} \qquad 1.000 \text{ g Co}\left(\frac{28.94 \text{ g O}}{71.06 \text{ g Co}}\right) = 0.4073 \text{ g O}$$

$$\text{Compound 2} \qquad 1.000 \text{ g Co}\left(\frac{21.35 \text{ g O}}{78.65 \text{ g Co}}\right) = 0.2715 \text{ g O}$$

The ratio of masses of O (per 1.000 g Co) in the two compounds is

$$\frac{0.4073}{0.2715} = \frac{1.500}{1.000} = \frac{3}{2}$$

The ratio of masses of oxygen in the two compounds (for a given mass of Co) is the ratio of small integers, as required by the law of multiple proportions. The ratio of mass of cobalt to mass of oxygen is not integral.

3.36. Complete a table like that of Problem 3.25 for uncombined atoms having (*a*) symbol ^{235}U; (*b*) atomic number 12, 14 neutrons; (*c*) mass number 3, 1 electron; (*d*) 27 protons, 33 neutrons; (*e*) mass number 83, 47 neutrons; and (*f*) 6 electrons, 8 neutrons.

Ans.

Symbol	Atomic Number	Mass Number	Number of Protons	Number of Electrons	Number of Neutrons
(*a*) $^{235}_{92}U$	92	235	92	92	143
(*b*) $^{26}_{12}Mg$	12	26	12	12	14
(*c*) $^{3}_{1}H$	1	3	1	1	2
(*d*) $^{60}_{27}Co$	27	60	27	27	33
(*e*) $^{83}_{36}Kr$	36	83	36	36	47
(*f*) $^{14}_{6}C$	6	14	6	6	8

3.37. Does hydrogen peroxide, H_2O_2, have the same composition as water, H_2O? Does your answer violate the law of definite proportions? Explain briefly.

Ans. The compounds have different ratios of hydrogen to oxygen atoms and thus different mass ratios. The law of definite proportions applies to *each compound individually*, not to the two different compounds. Both H_2O and H_2O_2 follow the law of definite proportions (and together they also follow the law of multiple proportions).

3.38. Which of the following familiar metals are main group elements and which are transition metals? (*a*) Cu, (*b*) Ni, (*c*) Pb, (*d*) Au, (*e*) Sn, and (*f*) Al.

Ans. (*c*), (*e*), and (*f*) are main group elements.

3.39. In which section (main group, transition group, inner transition group) are the nonmetals found?

Ans. Main group.

3.40. In which atom is it more difficult for you to predict the number of valence electrons—Fe or F?

Ans. Fe (F is a main group element, with valence electrons equal to its classical group number.)

3.41. How are the 37 electrons of Rb arranged in shells?

Ans.   2   8   18   8   1

The first three shells hold maximums of 2, 8, and 18, respectively. The nine remaining electrons cannot all fit into the fourth shell, because that would then be the outermost shell. So eight go into the fourth shell, leaving one in the fifth shell.

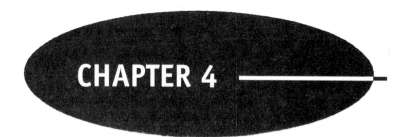

CHAPTER 4

Electronic Configuration of the Atom

4.1. INTRODUCTION

In Chap. 3 the elementary structure of the atom was introduced. The facts that protons, neutrons, and electrons are present in the atom and that electrons are arranged in shells allow us to explain isotopes (Chap. 3), chemical bonding (Chap. 5), and much more. However, with this simple theory, we still have not been able to deduce why the transition metal groups and inner transition metal groups arise, and many other important generalities. In this chapter we introduce a more detailed description of the electronic structure of the atom which begins to answer some of these more difficult questions.

The modern theory of the electronic structure of the atom is based on experimental observations of the interaction of electricity with matter, studies of electron beams (cathode rays), studies of radioactivity, studies of the distribution of the energy emitted by hot solids, and studies of the wavelengths of light emitted by incandescent gases. A complete discussion of the experimental evidence for the modern theory of atomic structure is beyond the scope of this book. In this chapter only the results of the theoretical treatment will be described. These results will have to be memorized as "rules of the game," but they will be used so extensively throughout the general chemistry course that the notation used will soon become familiar. In the rest of this course, the elementary theory presented in Chap. 3 will suffice. You should study only those parts of this chapter that are covered in your course.

4.2. BOHR THEORY

The first plausible theory of the electronic structure of the atom was proposed in 1914 by Niels Bohr (1885–1962), a Danish physicist. To explain the hydrogen spectrum (Fig. 4-1), he suggested that in each hydrogen atom, the electron revolves about the nucleus in one of several possible circular orbits, each having a definite radius corresponding to a definite energy for the electron. An electron in the orbit closest to the nucleus has the lowest energy. With the electron in that orbit, the atom is said to be in its lowest energy state, or *ground state*. If a discrete quantity of additional energy were absorbed by the atom in some manner, the electron might be able to move into another orbit having a higher energy. The hydrogen atom would then be in an *excited state*. An atom in the excited state will return to the ground state and give off its excess energy as light in the process.

In returning to the ground state, the energy may be emitted all at once, or it may be emitted in a stepwise manner (but not continuously) as the electron drops from a higher allowed orbit to allowed orbits of lower and lower energy. Since each orbit corresponds to a definite energy level, the energy of the light emitted will correspond to the definite differences in energy between levels. Therefore, the light emitted as the atom returns to its ground state will have a definite energy or a definite set of energies (Fig. 4-2). The discrete amounts of energy emitted or absorbed by an atom or molecule are called *quanta* (singular, *quantum*). A quantum of light energy is called a *photon*.

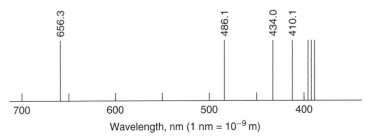

Fig. 4-1.   Visible spectrum of hydrogen

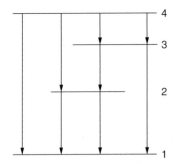

Fig. 4-2.   Possible return paths for electron in orbit 4

       $4 \rightarrow 1$ $4 \rightarrow 2 \rightarrow 1$ $4 \rightarrow 3 \rightarrow 2 \rightarrow 1$ $4 \rightarrow 3 \rightarrow 1$

       Only electron transitions down to the second orbit cause emission of visible light. Other transitions may involve infrared or ultraviolet light.

The wavelength of a photon—a quantum of light—is inversely proportional to the energy of the light, and when the light is observed through a spectroscope, lines of different colors, corresponding to different wavelengths, are seen. The origin of the visible portion of the hydrogen spectrum is shown schematically in Fig. 4-3.

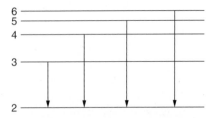

Fig. 4-3.   The origin of the visible spectrum of hydrogen (not drawn to scale)

Bohr's original idea of orbits of discrete radii has been greatly modified, but the concept that the electron in the hydrogen atom occupies definite energy levels still applies. It can be calculated that an electron in a higher energy level is located on average farther away from the nucleus than one in a lower energy level. It is customary to refer to the successive energy levels as electron *shells*. The terms *energy level* and *shell* are used interchangeably. The shells are sometimes designated by capital letters, with K denoting the lowest energy level, as follows:

Energy level	1	2	3	4	5	...
Shell notation	K	L	M	N	O	...

The electrons in atoms other than hydrogen also occupy various energy levels. With more than one electron in each atom, the question of how many electrons can occupy a given level becomes important. The maximum number of electrons that can occupy a given shell depends on the shell number. For example, in any atom, the first shell can hold a maximum of only 2 electrons, the second shell can hold a maximum of 8 electrons, the third shell can hold a maximum of 18 electrons, and so forth. The maximum number of electrons that can occupy any particular shell is $2n^2$, where n is the shell number.

EXAMPLE 4.1.   What is the maximum number of electrons that can occupy the N shell?

Ans.   The N shell in an atom corresponds to the fourth energy level ($n = 4$); hence the maximum number of electrons it can hold is

$$2n^2 = 2(4)^2 = 2 \times 16 = 32$$

4.3. QUANTUM NUMBERS

The modern theory of the electronic structure of the atom stems from a complex mathematical equation (called the Schrödinger equation), which is beyond the mathematical requirements for the general chemistry course. We therefore take the results of the solution of this equation as postulates. Solution of the equation yields three *quantum numbers*, with a fourth and final quantum number obtained from experimental results. The quantum numbers are named and have limitations as shown in Table 4-1. Each electron is specified in terms of its four quantum numbers that govern its energy, its orientation in space, and its possible interactions with other electrons. Thus, listing the values of the four quantum numbers describes the probable location of the electron, somewhat analogously to listing the section, row, seat, and date on a ticket to a rock concert. To learn to express the electronic structure of an atom, it is necessary to learn (1) the names, symbols, and permitted values of the quantum numbers and (2) the order of increasing energy of electrons as a function of their sets of quantum numbers.

Table 4-1   Quantum Numbers

Name	Symbol	Limitations
Principal quantum number	n	Any positive integer
Angular momentum quantum number or Azimuthal quantum number	l	$0, \ldots, n - 1$ in integer steps
Magnetic quantum number	m_l	$-l, \ldots, 0, \ldots, +l$ in integer steps
Spin quantum number	m_s	$-\frac{1}{2}, +\frac{1}{2}$

The principal quantum number of an electron is denoted n. It is the most important quantum number in determining the energy of the electron. In general, the higher the principal quantum number, the higher the energy of the electron. Electrons with higher principal quantum numbers are also apt to be farther away from the nucleus than electrons with lower principal quantum numbers. The values of n can be any positive integer: 1, 2, 3, 4, 5, 6, 7, ... The first seven principal quantum numbers are the only ones used for electrons in ground states of atoms.

The angular momentum quantum number is denoted l. It also affects the energy of the electron, but in general not as much as the principal quantum number does. In the absence of an electric or magnetic field around the atom, only n and l have any effect on the energy of the electron. The value of l can be 0 or any positive integer up to, but not including, the value of n for that electron.

The magnetic quantum number, denoted m_l, determines the orientation in space of the electron, but does not ordinarily affect the energy of the electron. Its values depend on the value of l for that electron, ranging from $-l$ through 0 to $+l$ in integral steps. Thus, for an electron with an l value of 3, the possible m_l values are $-3, -2, -1, 0, 1, 2,$ and 3.

The spin quantum number, denoted m_s, is related to the "spin" of the electron on its "axis." It ordinarily does not affect the energy of the electron. Its possible values are $-\frac{1}{2}$ and $+\frac{1}{2}$. The value of m_s does not depend on the value of any other quantum number.

The permitted values for the other quantum numbers when $n = 2$ are shown in Table 4-2. The following examples will illustrate the limitations on the values of the quantum numbers (Table 4-1).

Table 4-2   Permitted Values of Quantum Numbers When $n = 2$

n	2	2	2	2	2	2	2	2
l	0	0	1	1	1	1	1	1
m_l	0	0	-1	0	$+1$	-1	0	$+1$
m_s	$-\frac{1}{2}$	$+\frac{1}{2}$	$-\frac{1}{2}$	$-\frac{1}{2}$	$-\frac{1}{2}$	$+\frac{1}{2}$	$+\frac{1}{2}$	$+\frac{1}{2}$

Note: Each vertical set of four quantum numbers represents values for one electron.

EXAMPLE 4.2.   What are the first seven permitted values for n?

Ans.   1, 2, 3, 4, 5, 6, and 7.

EXAMPLE 4.3.   What values of l are permitted for an electron with principal quantum number $n = 3$?

Ans.   0, 1, and 2. (l can have integer values from 0 up to $n - 1$.)

EXAMPLE 4.4.   What values are permitted for m_l for an electron in which the l value is 2?

Ans. $-2, -1, 0, 1,$ and 2. (m_l can have integer values from $-l$ through 0 up to $+l$.)

EXAMPLE 4.5.   What values are permitted for m_s for an electron in which $n = 2, l = 1,$ and $m_l = 0$?

Ans. $-\frac{1}{2}$ and $+\frac{1}{2}$. (The values that the other quantum numbers have do not matter; the m_s value must be either $-\frac{1}{2}$ or $+\frac{1}{2}$.)

The values of the angular momentum quantum number are often given letter designations, so that when they are stated along with principal quantum numbers, less confusion results. The letter designations of importance in the ground states of atoms are the following:

l Value	Letter Designation
0	s
1	p
2	d
3	f

4.4.   QUANTUM NUMBERS AND ENERGIES OF ELECTRONS

The energy of the electrons in an atom is of paramount importance. The n and l quantum numbers determine the energy of each electron (apart from the effects of external electric and magnetic fields, which are most often not of interest in general chemistry courses). The energies of the electrons increase as the sum $n + l$ increases; the lower the value of $n + l$ for an electron in an atom, the lower is its energy. For two electrons with equal values of $n + l$, the one with the lower n value has lower energy. Thus, we can fill an atom with electrons starting with its lowest-energy electrons by starting with the electrons with the lowest sum $n + l$.

EXAMPLE 4.6. Arrange the electrons in the following list in order of increasing energy, lowest first:

	n	l	m_l	m_s
(a)	4	1	1	$-\frac{1}{2}$
(b)	2	1	-1	$\frac{1}{2}$

	n	l	m_l	m_s
(c)	3	2	-1	$\frac{1}{2}$
(d)	4	0	0	$-\frac{1}{2}$

Ans. Electron (b) has the lowest value of $n+l$ $\ (2+1=3)$, and so it is lowest in energy of the four electrons. Electron (d) has the next-lowest sum of $n+l$ $\ (4+0=4)$ and is next in energy (despite the fact that it does not have the next-lowest n value). Electrons (a) and (c) both have the same sum of $n+l$ $\ (4+1=3+2=5)$. Therefore, in this case, electron (c), the one with the lower n value, is lower in energy. Electron (a) is highest in energy.

The *Pauli exclusion principle* states that no two electrons in the same atom can have the same set of four quantum numbers. Along with the order of increasing energy, we can use this principle to deduce the order of filling of electron shells in atoms.

EXAMPLE 4.7. Use the Pauli principle and the $n+l$ rule to predict the sets of quantum numbers for the 13 electrons in the ground state of an aluminum atom.

Ans. We want all the electrons to have the lowest energy possible. The lowest value of $n+l$ will have the lowest n possible and the lowest l possible. The lowest n permitted is 1 (Table 4-1). With that value of n, the only value of l permitted is 0. With $l=0$, the value of m_s must be 0 $\ (-0\cdots +0)$. The value of m_s can be either $-\frac{1}{2}$ or $+\frac{1}{2}$. Thus, the first electron can have either

$$n=1, \qquad l=0, \qquad m_l=0, \qquad m_s=-\tfrac{1}{2} \qquad \text{or} \qquad n=1, \qquad l=0, \qquad m_l=0, \qquad m_s=+\tfrac{1}{2}$$

The second electron also can have $n=1, l=0$, and $m_l=0$. Its value of m_s can be either $+\frac{1}{2}$ or $-\frac{1}{2}$, but not the same as that for the first electron. If it were, this second electron would have the same set of four quantum numbers that the first electron has, which is not permitted by the Pauli principle. If we were to try to give the third electron the same values for the first three quantum numbers, we would be stuck when we came to assign the m_s value. Both $+\frac{1}{2}$ and $-\frac{1}{2}$ have already been used, and we would have a duplicate set of quantum numbers for two electrons, which is not permitted. We cannot use any other values for l or m_l with the value of $n=1$, and so the third electron must have the next-higher n value, $n=2$. The l values could be 0 or 1, and since 0 will give a lower $n+l$ sum, we choose that value for the third electron. Again the value of m_l must be 0 since $l=0$, and m_s can have a value $-\frac{1}{2}$ (or $+\frac{1}{2}$). For the fourth electron, $n=2, l=0, m_l=0$, and $m_s=+\frac{1}{2}$ (or $-\frac{1}{2}$ if the third were $+\frac{1}{2}$). The fifth electron can have $n=2$ but not $l=0$, since all combinations of $n=2$ and $l=0$ have been used. Therefore, $n=2, l=1, m_l=-1$, and $m_s=-\frac{1}{2}$ are assigned. The rest of the electrons in the aluminum atom are assigned quantum numbers somewhat arbitrarily as shown in Table 4-3.

Table 4-3   Quantum Numbers of the Electrons of Aluminum

Electron	1	2	3	4	5	6	7	8	9	10	11	12	13
n	1	1	2	2	2	2	2	2	2	2	3	3	3
l	0	0	0	0	1	1	1	1	1	1	0	0	1
m_l	0	0	0	0	-1	0	1	-1	0	1	0	0	-1
m_s	$-\frac{1}{2}$	$+\frac{1}{2}$	$-\frac{1}{2}$	$+\frac{1}{2}$	$-\frac{1}{2}$	$-\frac{1}{2}$	$-\frac{1}{2}$	$+\frac{1}{2}$	$+\frac{1}{2}$	$+\frac{1}{2}$	$-\frac{1}{2}$	$+\frac{1}{2}$	$-\frac{1}{2}$

4.5.   SHELLS, SUBSHELLS, AND ORBITALS

Electrons having the same value of n in an atom are said to be in the same *shell*. Electrons having the same value of n and the same value of l in an atom are said to be in the same *subshell*. (Electrons having the same values of n, l and m_l in an atom are said to be in the same *orbital*.) Thus, the first two electrons of aluminum

(Table 4-3) are in the first shell and in the same subshell. The third and fourth electrons are in the same shell and subshell with each other. They are also in the same shell with the next six electrons (all have $n = 2$) but a different subshell ($l = 0$ rather than 1). With the letter designations of Sec. 4.3, the first two electrons of aluminum are in the $1s$ subshell, the next two electrons are in the $2s$ subshell, and the next six electrons are in the $2p$ subshell. The following two electrons occupy the $3s$ subshell, and the last electron is in the $3p$ subshell.

Since the possible numerical values of l depend on the value of n, the number of subshells within a given shell is determined by the value of n. The number of subshells within a given shell is merely the value of n, the shell number. Thus, the first shell has one subshell, the second shell has two subshells, and so forth. These facts are summarized in Table 4-4. Even the atoms with the most electrons do not have enough electrons to completely fill the highest shells shown. The subshells that hold electrons in the ground states of the biggest atoms are in **boldface**.

Table 4-4   Arrangement of Subshells in Electron Shells

Energy level n	Type of Subshell	Number of Subshells
1	s	1
2	s, p	2
3	s, p, d	3
4	s, p, d, f	4
5	s, p, d, f, g	5
6	s, p, d, f, g, h	6
7	s, p, d, f, g, h, i	7

EXAMPLE 4.8.   What are the values of n and l in each of the following subshells? (a) $2p$, (b) $3s$, (c) $5d$, and (d) $4f$.

Ans.   (a) $n = 2, l = 1$ (b) $n = 3, l = 0$ (c) $n = 5, l = 2$ (d) $n = 4, l = 3$

EXAMPLE 4.9.   Show that there can be only two electrons in any s subshell.

Ans.   For any given value of n, there can be a value of $l = 0$, corresponding to an s subshell. For $l = 0$ there can be only one possible m_l value: $m_l = 0$. Hence, n, l, and m_l are all specified for a given s subshell. Electrons can then have spin values of $m_s = +\frac{1}{2}$ or $m_s = -\frac{1}{2}$. Thus, every possible set of four quantum numbers is used, and there are no other possibilities in that subshell. Each of the two electrons has the first three quantum numbers in common and has a different value of m_s. The two electrons are said to be *paired*.

Depending on the permitted values of the magnetic quantum number m_l, each subshell is further broken down into units called *orbitals*. The number of orbitals per subshell depends on the type of subshell but not on the value of n. Each consists of a maximum of two electrons; hence, the maximum number of electrons that can occupy a given subshell is determined by the number of orbitals available. These relationships are presented in Table 4-5. The maximum number of electrons in any given energy level is thus determined by the subshells it contains. The first shell can contain 2 electrons; the second, 8 electrons; the third, 18 electrons; the fourth, 32 electrons; and so on.

Table 4-5   Occupancy of Subshells

Type of Subshell	Allowed Values of m_l	Number of Orbitals	Maximum Number of Electrons
s	0	1	2
p	$-1, 0, 1$	3	6
d	$-2, -1, 0, 1, 2$	5	10
f	$-3, -2, -1, 0, 1, 2, 3$	7	14

Suppose we want to write the electronic configuration of titanium (atomic number 22). We can rewrite the first 13 electrons that we wrote above for aluminum and then just keep going. As we added electrons, we filled the first shell of electrons first, then the second shell. When we are filling the third shell, we have to ask if the electrons with $n = 3$ and $l = 2$ will enter before the $n = 4$ and $l = 0$ electrons. Since $n + l$ for the former is 5 and that for the latter is 4, we must add the two electrons with $n = 4$ and $l = 0$ before the last 10 electrons with $n = 3$ and $l = 2$. In this discussion, the values of m_l and m_s tell us how many electrons can have the same set of n and l values, but do not matter as to which come first.

	14	15	16	17	18	19	20	21	22
n	3	3	3	3	3	4	4	3	3
l	1	1	1	1	1	0	0	2	2
m_l	0	+1	-1	0	+1	0	0	-2	-1
m_s	$-\frac{1}{2}$	$-\frac{1}{2}$	$+\frac{1}{2}$	$+\frac{1}{2}$	$+\frac{1}{2}$	$-\frac{1}{2}$	$+\frac{1}{2}$	$-\frac{1}{2}$	$-\frac{1}{2}$
$n + l$	4	4	4	4	4	4	4	5	5

Thus, an important development has occurred because of the $n + l$ rule. The fourth shell has started filling before the third shell has been completed. This is the origin of the transition series elements. Thus, titanium, atomic number 22, has two electrons in its $1s$ subshell, two electrons in its $2s$ subshell, six electrons in its $2p$ subshell, two electrons in its $3s$ subshell, six electrons in its $3p$ subshell, two electrons in its $4s$ subshell, and its last two electrons in the $3d$ subshell.

We note in the electronic configuration for electrons 13 through 20 for titanium that when the $(n + l)$ sum was 4 we added the $3p$ electrons before the $4s$ electrons. Since each of these groups has an $(n + l)$ sum of 4 [the $(n + l)$ values are the same] we add electrons having the lower n value first.

We conventionally use a more condensed notation for electronic configurations, with the subshell notation and a superscript to denote the number of electrons in that subshell. To write the detailed electronic configuration of any atom, showing how many electrons occupy each of the various subshells, one needs to know only the order of increasing energy of the subshells, given above, and the maximum number of electrons that will fit into each, given in Table 4-5. A convenient way to designate such a configuration is to write the shell and subshell designation, and add a superscript to denote the number of electrons occupying that subshell. For example, the electronic configuration of the titanium atom is written as follows:

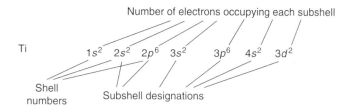

The shell number is represented by 1, 2, 3, ..., and the letters designate the subshells. The superscript numbers tell how many electrons occupy each subshell. Thus, in this example, there are two electrons in the $1s$ subshell, two in the $2s$ subshell, six in the $2p$ subshell, two in the $3s$ subshell, six in the $3p$ subshell, two in the $4s$ subshell, and two in the $3d$ subshell. (The $3d$ subshell can hold a *maximum* of 10 electrons, but in this atom this subshell is not filled.) The total number of electrons in the atom can easily be determined by adding the numbers in all the subshells, that is, by adding all the superscripts. For titanium, this sum is 22, equal to its atomic number.

EXAMPLE 4.10.   Write the electronic configuration of aluminum.

Ans. $1s^2 2s^2 2p^6 3s^2 3p^1$.

4.6. SHAPES OF ORBITALS

The *Heisenberg uncertainty principle* states that, since the energy of the electron is known precisely, its exact position cannot be known. It is possible to learn only the probable location of the electron in the vicinity of the atomic nucleus. The mathematical details of expressing the probability are quite complex, but it is possible to give an approximate description in terms of values of the quantum numbers n, l, and m_l. The shapes of the more important orbitals are shown in Fig. 4-4 for the case of the hydrogen atom. This figure shows that, in general, an electron in the $1s$ orbital is equally likely to be found in any direction about the nucleus. (The maximum probability is at a distance corresponding to the experimentally determined radius of the hydrogen atom). In contrast, in the case of an electron in a $2p$ orbital, there are three possible values of the quantum number m_l. There are three possible regions in which the electron is most likely to be found. It is customary to depict these orbitals as being located along the cartesian (x, y, and z) axes of a three-dimensional graph. Hence, the three probability distributions are labeled p_x, p_y, and p_z, respectively.

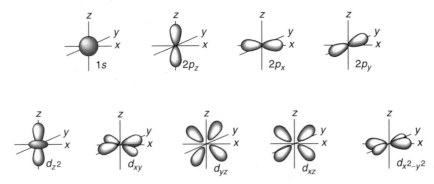

Only these nine orbital shapes need be memorized.

Fig. 4-4. Shapes of various orbitals

4.7. BUILDUP PRINCIPLE

With each successive increase in atomic number, a given atom has one more electron than the previous atom. Thus, it is possible to start with hydrogen and, adding one electron at a time, build up the electronic configuration of each of the other elements. In the buildup of electronic structures of the atoms of the elements, the last electron is added to the lowest-energy subshell possible. The relative order of the energies of all the important subshells in an atom is shown in Fig. 4-5. The energies of the various subshells are plotted along the vertical axis. The subshells are displaced left to right merely to avoid overcrowding. The order of increasing energy is as follows:

$$1s, 2s, 2p, 3s, 3p, 4s, 3d, 4p, 5s, 4d, 5p, 6s, 4f, 5d, 6p, 7s, 5f, 6d.$$

EXAMPLE 4.11.   Draw the electronic configuration of Ne and Mg on a figure like Fig. 4-5. Use an arrow pointing up to represent an electron with a given spin, and an arrow pointing down to represent an electron with the opposite spin.

Ans.   The drawings are shown in Fig. 4-6.

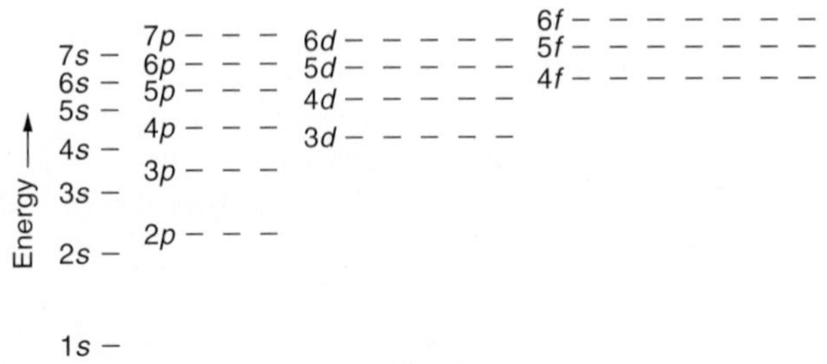

Fig. 4-5.   Energy level diagram for atoms containing more than one electron (not drawn to scale)

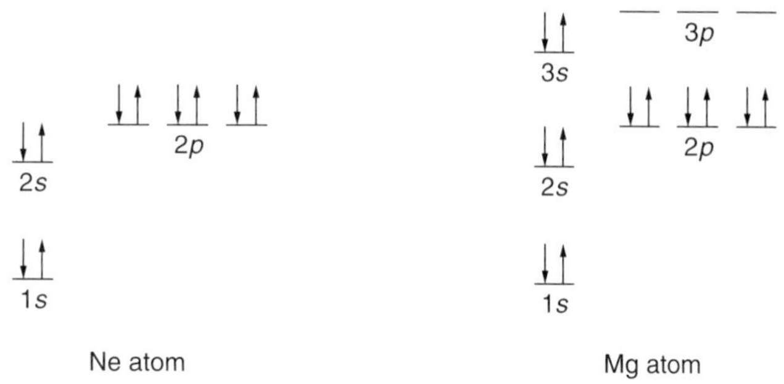

Fig. 4-6. Electronic configurations of Ne and Mg

The electronic configurations of the first 20 elements are given in Table 4-6. The buildup principle is crudely analogous to filling a vessel that has a layered base with marbles (see Fig. 4-7). The available spaces are filled from the bottom up.

In atoms with partially filled p, d, or f subshells, the electrons stay unpaired as much as possible. This effect is called *Hund's rule of maximum multiplicity*. Thus the configurations of the carbon and nitrogen atoms are as follows:

Table 4-6　Detailed Electronic Configuration of the First 20 Elements

H	$1s^1$	1	Na	$1s^22s^22p^63s^1$	11
He	$1s^2$	2	Mg	$1s^22s^22p^63s^2$	12
Li	$1s^22s^1$	3	Al	$1s^22s^22p^63s^23p^1$	13
Be	$1s^22s^2$	4	Si	$1s^22s^22p^63s^23p^2$	14
B	$1s^22s^22p^1$	5	P	$1s^22s^22p^63s^23p^3$	15
C	$1s^22s^22p^2$	6	S	$1s^22s^22p^63s^23p^4$	16
N	$1s^22s^22p^3$	7	Cl	$1s^22s^22p^63s^23p^5$	17
O	$1s^22s^22p^4$	8	Ar	$1s^22s^22p^63s^23p^6$	18
F	$1s^22s^22p^5$	9	K	$1s^22s^22p^63s^23p^64s^1$	19
Ne	$1s^22s^22p^6$	10	Ca	$1s^22s^22p^63s^23p^64s^2$	20

In turns out that fully filled or half-filled subshells have greater stability than subshells having some other numbers of electrons. One effect of this added stability is the fact that some elements do not follow the $n + l$ rule exactly. For example, copper would be expected to have a configuration

$$n + l \text{ configuration for Cu} \qquad 1s^22s^22p^63s^23p^64s^23d^9$$

$$\text{Actual configuration for Cu} \qquad 1s^22s^22p^63s^23p^64s^13d^{10}$$

The actual configuration has two subshells of enhanced stability (filled $3d$ and half-filled $4s$) in contrast to one subshell (filled $4s$) of the expected configuration. (There are also some elements whose configurations do not follow the $n + l$ rule and which are not so easily explained.)

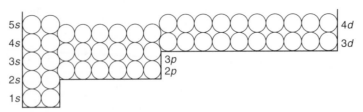

Fig. 4-7. Vessel with layered base

4.8. ELECTRONIC STRUCTURE AND THE PERIODIC TABLE

The arrangement of electrons in successive energy levels in the atom provides an explanation of the periodicity of the elements, as found in the periodic table. The charges on the nuclei of the atoms increase in a regular manner as the atomic number increases. Therefore, the number of electrons surrounding the nucleus increases also. The number and arrangement of the electrons in the outermost shell of an atom vary in a periodic manner (compare Table 4-6). For example, all the elements in Group IA—H, Li, Na, K, Rb, Cs, Fr—corresponding to the elements that begin a new row or period, have electronic configurations with a single electron in the outermost shell, specifically an s subshell.

H	$1s^1$
Li	$1s^2 2s^1$
Na	$1s^2 2s^2 2p^6 3s^1$
K	$1s^2 2s^2 2p^6 3s^2 3p^6 4s^1$
Rb	$1s^2 2s^2 2p^6 3s^2 3p^6 4s^2 3d^{10} 4p^6 5s^1$
Cs	$1s^2 2s^2 2p^6 3s^2 3p^6 4s^2 3d^{10} 4p^6 5s^2 4d^{10} 5p^6 6s^1$
Fr	$1s^2 2s^2 2p^6 3s^2 3p^6 4s^2 3d^{10} 4p^6 5s^2 4d^{10} 5p^6 6s^2 4f^{14} 5d^{10} 6p^6 7s^1$

The noble gases, located at the end of each period, have electronic configurations of the type $ns^2 np^6$, where n represents the number of the outermost shell. Also, n is the number of the period in the periodic table in which the element is found.

Since atoms of all elements in a given group of the periodic table have analogous arrangements of electrons in their outermost shells and different arrangements from elements of other groups, it is reasonable to conclude that the outermost electronic configuration of the atom is responsible for the chemical characteristics of the element. Elements with similar arrangements of electrons in their outer shells will have similar properties. For example, the formulas of their oxides will be of the same type. The electrons in the outermost shells of the atoms are referred to as *valence electrons*. The outermost shell is called the *valence shell*.

As the atomic numbers of the elements increase, the arrangements of electrons in successive energy levels vary in a periodic manner. As shown in Fig. 4-5, the energy of the 4s subshell is lower than that of the 3d subshell. Therefore, at atomic number 19, corresponding to the element potassium, the 19th electron is found in the 4s subshell rather than the 3d subshell. The fourth shell is started before the third shell is completely filled. At atomic number 20, calcium, a second electron completes the 4s subshell. Beginning with atomic number 21 and continuing through the next nine elements, successive electrons enter the 3d subshell. When the 3d subshell is complete, the following electrons occupy the 4p subshell through atomic number 36, krypton. In other words, for elements 21 through 30, the last electrons added are found in the 3d subshell rather than in the valence shell. The elements Sc through Zn are called *transition elements*, or d block elements. A second series of transition elements begins with yttrium, atomic number 39, and includes 10 elements. This series corresponds to the placement of 10 electrons in the 4d subshell.

The elements may be divided into types (Fig. 4-8), according to the position of the last electron added to those present in the preceding element. In the first type, the last electron added enters the valence shell. These elements are called the *main group elements*. In the second type, the last electron enters a d subshell in the next-to-last shell. These elements are the *transition elements*. The third type of elements has the last electron enter the f subshell in the $n-2$ shell—the second shell below the valence shell. These elements are the *inner transition elements*.

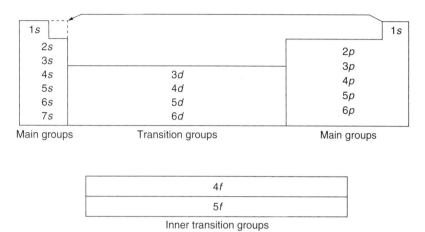

Fig. 4-8. Periodic table as an aid to assigning electronic configurations

An effective way to determine the detailed electronic configuration of any element is to use the periodic table to determine which subshell to fill next. Each s subshell holds a maximum of 2 electrons; each p subshell holds a maximum of 6 electrons; each d subshell holds a maximum of 10 electrons; and each f subshell holds a maximum of 14 electrons (Table 4-5). These numbers match the numbers of elements in a given period in the various blocks. To get the electronic configuration, start at hydrogen (atomic number $= 1$) and continue in order of atomic number, using the periodic table of Fig. 4-8.

EXAMPLE 4.12.   Using the periodic table, determine the detailed electronic configuration of magnesium.

Ans.   Starting at hydrogen, we put two electrons into the $1s$ subshell, then two more electrons into the $2s$ subshell. We continue (at atomic number 5) with the $2p$ subshell, and enter six electrons there, corresponding to the six elements (elements 5 to 10, inclusive) in that p block of the periodic table. We have two more electrons to put into the $3s$ subshell, which is next. Thus, we always start at hydrogen, and we end at the element required. The number of electrons that we add to each subshell is equal to the number of elements in the block of the periodic table. In this case, we added electrons from hydrogen to magnesium, following the atomic numbers in order, and we got a configuration $1s^2 2s^2 2p^6 3s^2$.

EXAMPLE 4.13.   Write the detailed electronic configurations for K, S, and Y.

Ans.
$$\begin{aligned}
&\text{K} && 1s^2 2s^2 2p^6 3s^2 3p^6 4s^1 \\
&\text{S} && 1s^2 2s^2 2p^6 3s^2 3p^4 \\
&\text{Y} && 1s^2 2s^2 2p^6 3s^2 3p^6 4s^2 3d^{10} 4p^6 5s^2 4d^1
\end{aligned}$$

In each case, the superscripts total to the atomic number of the element.

EXAMPLE 4.14.   Determine the detailed electronic configuration of Gd, atomic number 64.

Ans.   Gd $1s^2 2s^2 2p^6 3s^2 3p^6 4s^2 3d^{10} 4p^6 5s^2 4d^{10} 5p^6 6s^2 5d^1 4f^7$

We note that the $5d$ subshell started before the $4f$ subshell, but only one electron entered that shell before the $4f$ subshell started. Indeed, the periodic table predicts this correct configuration for Gd better than the $n+l$ rule or other common memory aids.

Instead of writing out the entire electronic configuration of an atom, especially an atom with many electrons, we sometimes abbreviate the configuration by using the configuration of the previous noble gas and represent the rest of the electrons explicitly. For example, the full configuration of cobalt can be given as

$$\text{Co}   \quad 1s^2 2s^2 2p^6 3s^2 3p^6 4s^2 3d^7$$

Alternatively, we can use the configuration of the previous noble gas, Ar, and add the extra electrons:

$$\text{Co} \qquad [\text{Ar}]\, 4s^2 3d^7$$

To determine the electronic configuration in this manner, start with the noble gas of the previous period and use the subshell notation from only the period of the required element. Thus, for Co, the notation for Ar (the previous noble gas) is included in the square brackets, and the $4s^2 3d^7$ is obtained across the fourth period. It is suggested that you do not use this notation until you have mastered the full notation. Also, on examinations, use the full notation unless the question or the instructor indicates that the shortened notation is acceptable.

Solved Problems

BOHR THEORY

4.1. Draw a picture of the electron jump corresponding to the third line in the visible emission spectrum of hydrogen according to the Bohr theory.

> *Ans.*   In hydrogen, only jumps to or from the second orbit are in the visible region of the spectrum. The electron falls from the fifth orbit to the second. The picture is shown as the third arrow of Fig. 4-3.

4.2. When electrons fall to lower energy levels, light is given off. What energy effect is expected when an electron jumps to a higher-energy orbit?

> *Ans.*   Absorption of energy is expected. The energy may be light energy (of the same energies as are given off in emission), or it may be heat or other types of energy.

4.3. What changes of orbit can the electron of hydrogen make in going from its fifth orbit to its ground state?

> *Ans.*   It can go from (*a*) $5 \rightarrow 4 \rightarrow 3 \rightarrow 2 \rightarrow 1$ (*b*) $5 \rightarrow 4 \rightarrow 3 \rightarrow 1$ (*c*) $5 \rightarrow 4 \rightarrow 2 \rightarrow 1$
> (*d*) $5 \rightarrow 4 \rightarrow 1$ (*e*) $5 \rightarrow 3 \rightarrow 2 \rightarrow 1$ (*f*) $5 \rightarrow 3 \rightarrow 1$ (*g*) $5 \rightarrow 2 \rightarrow 1$ (*h*) $5 \rightarrow 1$.

QUANTUM NUMBERS

4.4. What values are permitted for m_l in an electron in which the l value is 0?

> *Ans.*   0. (The value of m_l may range from -0 to $+0$; that is, it must be 0.)

4.5. Why must there be *two* electrons in an outer shell before the expansion of the next inner shell beyond eight electrons?

> *Ans.*   The ns orbital must be filled before electrons enter the $(n-1)d$ orbitals in the shell below.

4.6. (*a*) In a football stadium, can two tickets have the same set of section, row, seat, and date? How many of these must be different to have a legal situation? (*b*) In an atom, can two electrons have the same set of $n, l, m_l,$ and m_s? How many of these must be different to have a "legal" situation?

> *Ans.*   (*a*) At least one of the four must be different   (*b*) At least one of the four must be different

4.7. What are the possible values of m_l for an electron with $l = 4$?

> *Ans.* $-4, -3, -2, -1, 0, +1, +2, +3,$ and $+4$.

4.8. What are the permitted values of m_s for an electron with $n = 6, l = 4,$ and $m_l = -2$?

> *Ans.* $m_s = -\frac{1}{2}$ or $+\frac{1}{2}$, no matter what values the other quantum numbers have.

4.9. What is the maximum number of electrons that can occupy the first shell? The second shell?

   Ans.   The first shell can hold a maximum of two electrons. The second shell can hold a maximum of eight electrons.

4.10. What is the maximum number of electrons that can occupy the third shell?

   Ans. $n = 3$; hence, the maximum number of electrons the shell can hold is

$$2n^2 = 2(3)^2 = 2 \times 9 = 18$$

4.11. What is the maximum number of electrons that can occupy the third shell before the start of the fourth shell?

   Ans.   The maximum number of electrons in any outermost shell (except the first shell) is eight. The fourth shell starts before the d subshell of the third shell starts.

QUANTUM NUMBERS AND ENERGIES OF ELECTRONS

4.12. Arrange the electrons in the following list in order of increasing energy, lowest first:

	n	l	m_l	m_s			n	l	m_l	m_s
(a)	4	2	-1	$\frac{1}{2}$		(c)	4	1	1	$-\frac{1}{2}$
(b)	5	0	0	$-\frac{1}{2}$		(d)	4	1	-1	$\frac{1}{2}$

   Ans.   Electrons (c) and (d) have the lowest value of $n + l$ ($4 + 1 = 5$) and the lowest n, and so they are tied for lowest in energy of the four electrons. Electron (b) also has an equal sum of $n + l$ ($5 + 0 = 5$), but its n value is higher than those of electrons (c) and (d). Electron (b) is therefore next in energy (despite the fact that it has the highest n value). Electron (a) has the highest sum of $n + l$ ($4 + 2 = 6$) and is highest in energy.

4.13. If the n and l quantum numbers are the ones affecting the energy of the electron, why are the m_l and m_s quantum numbers important?

   Ans.   Their permitted values tell us how many electrons there can be with that given energy. That is, they tell us how many electrons are allowed in a given subshell.

4.14. How many electrons are permitted in the fifth shell? Explain why no atom in its ground state has that many electrons in that shell.

   Ans.   For $n = 5$, there could be as many $2(5)^2 = 50$ electrons. However, there are only a few more than 100 electrons in even the biggest atom. By the time you put 2 electrons in the first shell, 8 in the second, 18 in the third, and 32 in the fourth, you have already accounted for 60 electrons. Moreover, the fifth shell cannot completely fill until several overlying shells start to fill. There are just not that many electrons in any actual atom.

SHELLS, SUBSHELLS, AND ORBITALS

4.15. What is the difference between (a) the $2s$ subshell and a $2s$ orbital and (b) the $2p$ subshell and a $2p$ orbital?

   Ans.   (a) Since the s subshell contains only one orbital, the $2s$ orbital is the $2s$ subshell.   (b) The $2p$ subshell contains three $2p$ orbitals—known as $2p_x$, $2p_y$, and $2p_z$.

4.16. (*a*) How many orbitals are there in the fourth shell of an atom? (*b*) How many electrons can be held in the fourth shell of an atom? (*c*) How many electrons can be held in the fourth shell of an atom before the fifth shell starts to fill?

Ans. (*a*) 16 (one *s* orbital, three *p* orbitals, five *d* orbitals, and seven *f* orbitals). (*b*) $2(4)^2 = 32$. (*c*) 8. (The 5*s* subshell starts to fill before the 4*d* subshell. Thus, only the 4*s* and the 4*p* subshells are filled before the fifth shell starts.)

4.17. How many electrons are permitted in each of the following subshells?

	n	l			n	l
(*a*)	3	1		(*c*)	3	2
(*b*)	3	0		(*d*)	4	0

Ans. (*a*) 6. This is a *p* subshell (with three orbitals). (*b*) 2. This is an *s* subshell. (*c*) 10. This is a *d* subshell, with five orbitals corresponding to m_l values of $-2, -1, 0, 1, 2$. Each orbital can hold a maximum of 2 electrons, and so the subshell can hold $5 \times 2 = 10$ electrons. (*d*) 2. This is an *s* subshell. [Compare to part (*b*).] The principal quantum number does not matter.

4.18. How many electrons are permitted in each of the following subshells? (*a*) 2*s* (*b*) 6*p*, and (*c*) 4*d*.

Ans. (*a*) 2 (*b*) 6 (*c*) 10.
Note that the principal quantum number does not affect the number of orbitals and thus the maximum number of electrons. The angular momentum quantum number is the only criterion of that.

4.19. In Chap. 3 why were electron dot diagrams drawn with four areas of electrons? Why are at least two of the electrons paired if at least two are shown?

Ans. The four areas represent the one *s* plus three *p* orbitals of the outermost shell. If there are at least two electrons in the outermost shell, the first two are paired because they are in the *s* subshell. The other electrons do not pair up until all have at least one electron in each area (orbital).

SHAPES OF ORBITALS

4.20. Draw an outline of the shape of the 1*s* orbital.

Ans. See Fig. 4-4.

4.21. How do the three 2*p* orbitals differ from one another?

Ans. They are oriented differently in space. The $2p_x$ orbital lies along the *x* axis; the $2p_y$ orbital lies along the *y* axis; the $2p_z$ orbital lies along the *z* axis. See Fig. 4-4.

4.22. Toward which direction, if any, is each of the following orbitals aligned: (*a*) 1*s*, (*b*) $2p_y$, and (*c*) $3d_{z^2}$?

Ans. (*a*) None; it is spherically symmetric (*b*) Along the *y* axis (*c*) Along the *z* axis

4.23. Which 2*p* orbital of a given atom would be expected to have the greatest interaction with another atom lying along the *x* axis of the first atom?

Ans. The $2p_x$ orbital. That one is oriented along the direction toward the second atom.

BUILDUP PRINCIPLE

4.24. Write detailed electronic configurations for the following atoms: (*a*) Li, (*b*) Ne, (*c*) C, and (*d*) S.

Ans. (*a*) $1s^2 2s^1$ (*c*) $1s^2 2s^2 2p^2$

(*b*) $1s^2 2s^2 2p^6$ (*d*) $1s^2 2s^2 2p^6 3s^2 3p^4$

ELECTRONIC STRUCTURE AND THE PERIODIC TABLE

4.25. Write detailed electronic configurations for (*a*) N, (*b*) P, (*c*) As, and (*d*) Sb. What makes their chemical properties similar?

   Ans.   (*a*) $1s^2 2s^2 2p^3$

   (*b*) $1s^2 2s^2 2p^6 3s^2 3p^3$

   (*c*) $1s^2 2s^2 2p^6 3s^2 3p^6 4s^2 3d^{10} 4p^3$

   (*d*) $1s^2 2s^2 2p^6 3s^2 3p^6 4s^2 3d^{10} 4p^6 5s^2 4d^{10} 5p^3$

   Their outermost electronic configurations are similar: $ns^2 np^3$.

4.26. What neutral atom is represented by each of the following configurations: (*a*) $1s^2 2s^2 2p^6 3s^2 3p^4$, (*b*) $1s^2 2s^2 2p^3$, and (*c*) $1s^2 2s^2 2p^6 3s^2 3p^6 4s^2$?

   Ans.   (*a*) S   (*b*) N   (*c*) Ca

4.27. Write the detailed electronic configuration for La (atomic number = 57) in shortened form.

   Ans.   La   [Xe] $6s^2 5d^1$

4.28. Write the electronic configuration of atoms of each of the following elements, using the periodic table as a memory aid: (*a*) Si, (*b*) V, (*c*) Br, (*d*) Rb.

   Ans.   (*a*) $1s^2 2s^2 2p^6 3s^2 3p^2$ (*c*) $1s^2 2s^2 2p^6 3s^2 3p^6 4s^2 3d^{10} 4p^5$

   (*b*) $1s^2 2s^2 2p^6 3s^2 3p^6 4s^2 3d^3$ (*d*) $1s^2 2s^2 2p^6 3s^2 3p^6 4s^2 3d^{10} 4p^6 5s^1$

4.29. Write the electronic configuration of atoms of each of the following elements, using the periodic table as a memory aid: (*a*) As, (*b*) Ni, (*c*) Pd, and (*d*) Ge.

   Ans.   (*a*) $1s^2 2s^2 2p^6 3s^2 3p^6 4s^2 3d^{10} 4p^3$ (*c*) $1s^2 2s^2 2p^6 3s^2 3p^6 4s^2 3d^{10} 4p^6 5s^2 4d^8$

   (*b*) $1s^2 2s^2 2p^6 3s^2 3p^6 4s^2 3d^8$ (*d*) $1s^2 2s^2 2p^6 3s^2 3p^6 4s^2 3d^{10} 4p^2$

Supplementary Problems

4.30. State the octet rule in terms described in this chapter.

   Ans.   A state of great stability is a state in which the outermost *s* and *p* subshells are filled and no other subshell of the outermost shell has any electrons.

4.31. List all the ways given in this chapter to determine the order of increasing energy of subshells.

   Ans.   The $n + l$ rule, the energy-level diagram (Fig. 4-5) and the periodic table. (There are also some mnemonics given by other texts.)

4.32. How many different drawings, like those in Fig. 4-4, are there for the $4f$ orbitals?

   Ans.   There are seven, corresponding to the seven *f* orbitals in a subshell (Table 4-5). General chemistry students are never asked to draw them, however.

4.33. What are the advantages of using the periodic table over the other ways of determining the order of increasing energy in subshells given in Problem 4.31?

   Ans.   The periodic table is generally available on examinations, it is easy to use after a little practice, it allows one to start at any noble gas, it reminds one of the $5d$ electron added before the $4f$ electrons (at La), and it tells by the number of elements in each block how many electrons are in the subshell.

4.34. Write the electronic configurations of K and Cu. What differences are there that could explain why K is so active and Cu so relatively inactive?

    Ans.   K $1s^2 2s^2 2p^6 3s^2 3p^6 4s^1$

                                   Cu $1s^2 2s^2 2p^6 3s^2 3p^6 4s^1 3d^{10}$

        The differences are the additional 10 protons in the nucleus and the additional ten $3d$ electrons that go along with them. Adding the protons and adding the electrons in an inner subshell make the outermost electron more tightly bound to the nucleus.

4.35. Starting at the first electron added to an atom, (*a*) what is the number of the first electron in the second shell of the atom? (*b*) What is the atomic number of the first element of the second period? (*c*) What is the number of the first electron in the third shell of the atom? (*d*) What is the atomic number of the first element of the third period? (*e*) What is the number of the first electron in the fourth shell of that atom? (*f*) What is the atomic number of the first element of the fourth period?

    Ans.   (*a*) and (*b*) 3   (*c*) and (*d*) 11   (*e*) and (*f*) 19
        There obviously is some relationship between the electronic configuration and the periodic table.

4.36. Check Table 4-3 to ensure that no two electrons have the same set of four quantum numbers.

4.37. Write the electronic configuration of atoms of each of the following elements, using the periodic table as a memory aid: (*a*) P, (*b*) Cl, (*c*) Mn, and (*d*) Zn.

    Ans.   (*a*) $1s^2 2s^2 2p^6 3s^2 3p^3$ (*c*) $1s^2 2s^2 2p^6 3s^2 3p^6 4s^2 3d^5$

                    (*b*) $1s^2 2s^2 2p^6 3s^2 3p^5$ (*d*) $1s^2 2s^2 2p^6 3s^2 3p^6 4s^2 3d^{10}$

4.38. Write the outer electronic configuration of atoms of each of the following elements, using the periodic table as a memory aid: (The atomic numbers are indicated for your convenience.)   (*a*) $_{57}$La, (*b*) $_{92}$U, (*c*) $_{88}$Ra, and (*d*) $_{82}$Pb.

    Ans.   (*a*)   La   [Xe] $6s^2 5d^1$ (*c*)   Ra   [Rn] $7s^2$

                    (*b*)   U   [Rn] $7s^2 6d^1 5f^3$ (*d*)   Pb   [Xe] $6s^2 5d^{10} 4f^{14} 6p^2$

4.39. Write the expected outer electronic configuration of atoms of each of the following elements, using the periodic table as a memory aid: (The atomic numbers are indicated for your convenience.)   (*a*) $_{64}$Gd, (*b*) $_{90}$Th, (*c*) $_{55}$Cs, and (*d*) $_{71}$Lu.

    Ans.   (*a*)   Gd   [Xe] $6s^2 5d^1 4f^7$ (*c*)   Cs   [Xe] $6s^1$

                    (*b*)   Th   [Rn] $7s^2 6d^1 5f^1$ (*d*)   Lu   [Xe] $6s^2 5d^1 4f^{14}$

CHAPTER 5

Chemical Bonding

5.1. INTRODUCTION

The vast bulk of materials found in nature are compounds or mixtures of compounds rather than free elements. On or near the earth's surface, the nonmetallic elements oxygen, nitrogen, sulfur, and carbon are sometimes found in the uncombined state, and the noble gases are always found in nature uncombined. Also, the metals copper, silver, mercury, and gold sometimes occur in the free state. It is thought that, except for the noble gases, these elements have been liberated from their compounds somewhat recently (compared with the age of the earth) by geological or biological processes. It is a rule of nature that the state which is most probably encountered corresponds to the state of lowest energy. For example, water flows downhill under the influence of gravity, and iron rusts when exposed to air. Since compounds are encountered more often than free elements, it can be inferred that the combined state must be the state of low energy compared with the state of the corresponding free elements. Indeed, those elements that do occur naturally as free elements must possess some characteristics that correspond to a relatively low energy state.

In this chapter, some aspects of chemical bonding will be discussed. It will be shown that chemical combination corresponds to the tendency of atoms to assume the most stable electronic configuration possible. Before we start to study the forces holding the particles together in a compound, however, we must first understand the meaning of a chemical formula (Sec. 5.2). Next we learn why ionic compounds have the formulas they have, for example, why sodium chloride is $NaCl$ and not $NaCl_2$ or $NaCl_3$.

5.2. CHEMICAL FORMULAS

Writing a formula implies that the atoms in the formula are bonded together in some way. The relative numbers of atoms of the elements in a compound are shown in a chemical formula by writing the symbols of the elements followed by appropriate *subscripts* to denote how many atoms of each element there are in the formula unit. A subscript *following* the symbol gives the number of atoms of that element per formula unit. If there is no subscript, *one* atom per formula unit is implied. For example, the formula H_3PO_4 describes a molecule containing three atoms of hydrogen and four atoms of oxygen, along with one atom of phosphorus. Sometimes groups of atoms which are bonded together within a molecule or within an ionic compound are grouped in the formula within parentheses. The number of such groups is indicated by a subscript following the closing parenthesis. For example, the 2 in $(NH_4)_2SO_4$ states that there are two NH_4 groups present per formula unit. There is only one SO_4 group; therefore parentheses are not necessary around it.

EXAMPLE 5.1. How many H atoms and how many P atoms are there per formula unit of $(NH_4)_3PO_3$?

Ans.   There are three NH_4 groups, each containing four H atoms, for a total of 12 H atoms per formula unit. There is only one P atom; the final 3 defines the number of O atoms.

In summary, chemical formulas yield the following information:

1. Which elements are present
2. The ratio of the number of atoms of each element to the number of atoms of each other element
3. The number of atoms of each element per formula unit of compound
4. The fact that all the atoms represented are bonded together in some way

You cannot tell from a formula how many atoms of each element are present in a given sample of substance, because there might be a little or a lot of the substance present. The formula tells the *ratio* of atoms of each element to all the others, and the ratio of atoms of each element to formula units as a whole.

EXAMPLE 5.2. (*a*) Can you tell how many ears and how many noses were present at the last Super Bowl football game? Can you guess how many ears there were per nose? How many ears per person? (*b*) Can you tell how many hydrogen and oxygen atoms there are in a sample of pure water? Can you tell how many hydrogen atoms there are per oxygen atom? per water molecule?

Ans.   (*a*)   Since the problem does not give the number of people at the game, it is impossible to tell the number of ears or noses from the information given. The ratios of ears to noses and ears to people are both likely to be 2 : 1.

(*b*)   Since the problem does not give the quantity of water, it is impossible to tell the number of hydrogen atoms or oxygen atoms from the information given. The ratios of hydrogen atoms to oxygen atoms and hydrogen atoms to water molecules are both 2 : 1.

The atoms of many nonmetals bond together into molecules when the elements are uncombined. For example, a pair of hydrogen atoms bonded together is a hydrogen molecule. Seven elements, when uncombined with other elements, form *diatomic* molecules. These elements are hydrogen, nitrogen, oxygen, fluorine, chlorine, bromine, and iodine. They are easy to remember because the last six form a large "7" in the periodic table, starting at element 7, nitrogen:

$$N \quad O \quad F$$
$$Cl$$
$$Br$$
$$I$$

5.3.   THE OCTET RULE

The elements helium, neon, argon, krypton, xenon, and radon—known as the *noble gases*—occur in nature as monatomic gases. Their atoms are not combined with atoms of other elements or with other atoms like themselves. Prior to 1962, no compounds of these elements were known. (Since 1962, some compounds of krypton, xenon, and radon have been prepared.) Why are these elements so stable, while the elements with atomic numbers 1 less or 1 more are so reactive? The answer lies in the electronic structures of their atoms. The electrons in atoms are arranged in shells, as described in Sec. 3.6 and in greater detail in Chap. 4.

EXAMPLE 5.3. (*a*) Arrange the 19 electrons of potassium into shells. (*b*) Arrange the 18 electrons of argon into shells.

Ans.   The first two electrons fill the first shell, the next eight fill the second shell, and the following eight occupy the third. That leaves one electron left in potassium for the fourth shell.

| | | Shell Number | | | | Detailed Configuration |
		1	2	3	4	(if you studied Chap. 4)
(*a*)	K	2	8	8	1	$1s^2 2s^2 2p^6 3s^2 3p^6 4s^1$
(*b*)	Ar	2	8	8		$1s^2 2s^2 2p^6 3s^2 3p^6$

The charge on the nucleus and the number of electrons in the valence shell determine the chemical properties of the atom. The electronic configurations of the noble gases (except for that of helium) correspond to a valence shell containing eight electrons—a very stable configuration called an *octet*. Atoms of other main group elements tend to react with other atoms in various ways to achieve the octet, as discussed in the next sections. The tendency to achieve an octet of electrons in the outermost shell is called the *octet rule*. If the outermost shell is the first shell, that is, if there is only one shell occupied, then the maximum number of electrons is two. A configuration of two electrons in the first shell, with no other shells occupied by electrons, is stable.

5.4.   IONS

The electronic configuration of a potassium atom is

$$\text{K} \quad 2 \quad 8 \quad 8 \quad 1 \quad (1s^2 2s^2 2p^6 3s^2 3p^6 4s^1)$$

It is readily seen that if a potassium atom were to lose one electron, the resulting species would have the configuration

$$\text{K}^+ \quad 2 \quad 8 \quad 8 \quad 0 \quad (1s^2 2s^2 2p^6 3s^2 3p^6 4s^0)$$

or more simply

$$\text{K}^+ \quad 2 \quad 8 \quad 8 \quad (1s^2 2s^2 2p^6 3s^2 3p^6)$$

The nucleus of a potassium atom contains 19 protons, and if there are only 18 electrons surrounding the nucleus after the atom has lost one electron, the atom will have a net charge of 1+. An atom (or group of atoms) that contains a net charge is called an *ion*. In chemical notation, an ion is represented by the symbol of the atom with the charge indicated as a superscript to the right. Thus, the potassium ion is written K^+. The potassium ion has the same configuration of electrons that an argon atom has (Example 5.3b). Ions that have the electronic configurations of noble gases are rather stable. Note the very important differences between a potassium ion and an argon atom—the different nuclear charges and the net 1+ charge on K^+. The K^+ ion is not as stable as the Ar atom.

EXAMPLE 5.4.   What is the electronic configuration of a chloride ion, obtained by adding an electron to a chlorine atom?

Ans.   The electronic configuration of a chlorine atom is

$$\text{Cl} \quad 2 \quad 8 \quad 7 \quad (1s^2 2s^2 2p^6 3s^2 3p^5)$$

Upon gaining an electron, the chlorine atom achieves the electronic configuration of argon:

$$\text{Cl}^- \quad 2 \quad 8 \quad 8 \quad (1s^2 2s^2 2p^6 3s^2 3p^6)$$

However, since the chlorine atom contains 17 protons in its nucleus and now contains 18 electrons outside the nucleus, it has a net negative charge; it is an ion. The ion is designated Cl^- and named the *chloride ion*.

Compounds—even ionic compounds—have no *net* charge. In the compound potassium chloride, there are potassium ions and chloride ions; the oppositely charged ions attract one another and form a regular geometric arrangement, as shown in Fig. 5-1. This attraction is called an *ionic bond*. There are equal numbers of K^+ ions and Cl^- ions, and the compound is electrically neutral. It would be inaccurate to speak of a molecule of solid potassium chloride or of a bond between a specific potassium ion and a specific chloride ion. The substance KCl is extremely stable because of (1) the stable electronic configurations of the ions and (2) the attractions between the oppositely charged ions.

The electronic configurations of ions of many main group elements and even a few transition elements can be predicted by assuming that the gain or loss of electrons by an atom results in a configuration analogous to that of a noble gas—a noble gas configuration, which contains an octet of electrons in the outermost shell. Not all the ions that could be predicted with this rule actually form. For example, few monatomic ions have charges of 4+, and no monatomic ions have charges of 4−. Moreover, nonmetal atoms bond to each other in another way (Sec. 5.6).

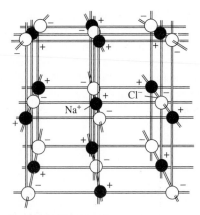

Fig. 5-1. Ball-and-stick model of the sodium chloride structure

The lines are not covalent bonds (Sec. 5.6), only indications of the positions of the ions. Sodium chloride is the most common compound with this structure, which is therefore called the *sodium chloride structure*. Potassium chloride also has this structure.

EXAMPLE 5.5. Predict the charge on a magnesium ion and that on a fluoride ion, and deduce the formula of magnesium fluoride.

Ans.   The electron configuration of a magnesium atom is

$$\text{Mg} \qquad 2 \qquad 8 \qquad 2 \qquad (1s^2 2s^2 2p^6 3s^2)$$

By losing two electrons, a magnesium atom attains the electronic configuration of a neon atom and thereby acquires a charge of 2+.

$$\text{Mg}^{2+} \qquad 2 \qquad 8 \qquad (1s^2 2s^2 2p^6)$$

A fluorine atom has the configuration

$$\text{F} \qquad 2 \qquad 7 \qquad (1s^2 2s^2 2p^5)$$

By gaining one electron, the fluorine atom attains the electronic configuration of neon and also attains a charge of 1−. The two ions expected are therefore Mg^{2+} and F^-. Since magnesium fluoride as a whole cannot have any net charge, there must be two fluoride ions for each magnesium ion; hence, the formula is MgF_2.

The ionic nature of these compounds can be shown by experiments in which the charged ions are made to carry an electric current. If a compound that consists of ions is dissolved in water and the solution is placed between electrodes in an apparatus like that shown in Fig. 5-2, the solution will conduct electricity when the electrodes are connected to the terminals of a battery. Each type of ion moves toward the electrode having a charge opposite to that of the ion. Positively charged ions are called *cations* (pronounced "cat'-ions"), and negatively charged ions are called *anions* (pronounced "an'-ions"). Thus, cations migrate to the negative electrode, and anions migrate to the positive electrode. In order to have conduction of electricity, the ions must be free to move.

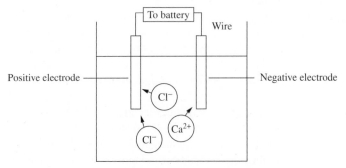

Fig. 5-2. Conducting by ions in solution

In the solid state, an ionic compound will not conduct. However, if the compound is heated until it melts or if it is dissolved in water, the resulting liquid will conduct electricity because in the liquid state the ions are free to move.

Periodic group IA and IIA metals always form ions with 1+ and 2+ charges, respectively. Most other metals form more than one type of ion, for example, iron forms Fe^{2+} and Fe^{3+} ions. Without further information (such as is given in Sec. 8.3), it is impossible to tell which of these ions will be formed in a given situation. The charges on these types of ions will be more fully presented in Sec. 6.3 and in Chap. 14.

5.5. ELECTRON DOT NOTATION

To represent the formation of bonds between atoms, it is convenient to use a system known as *electron dot notation*. In this notation, the symbol for an element is used to represent the nucleus of an atom of the element plus all the electrons except those in the outermost (valence) shell. The outermost electrons are represented by dots (or tiny circles or crosses). For example, the dot notations for the first 10 elements in the periodic table are as follows:

$$H\cdot \qquad He\colon \qquad Li\cdot \qquad Be\colon \qquad \colon\!B\cdot \qquad \colon\!\dot{C}\cdot \qquad \colon\!\dot{N}\cdot \qquad \colon\!\dot{O}\colon \qquad \colon\!\ddot{F}\colon \qquad \colon\!\ddot{Ne}\colon$$

Using electron dot notation, the production of sodium fluoride, calcium fluoride, and calcium oxide may be pictured as follows: A sodium atom and a fluorine atom react in a 1 : 1 ratio, since sodium has one electron to lose from its outermost shell and fluorine requires one more electron to complete its outermost shell.

$$Na\cdot \; + \; \cdot\ddot{F}\colon \; \longrightarrow \; Na^+ \; + \; \colon\!\ddot{F}\colon^-$$

To lose its entire outermost shell, a calcium atom must lose two electrons. Since each fluorine atom needs only one electron to complete its octet, it takes two fluorine atoms to react with one calcium atom:

$$Ca\colon \; + \; \begin{matrix} \cdot\ddot{F}\colon \\[4pt] \cdot\ddot{F}\colon \end{matrix} \; \longrightarrow \; Ca^{2+} \; + \; \begin{matrix} \colon\!\ddot{F}\colon^- \\[4pt] \colon\!\ddot{F}\colon^- \end{matrix}$$

The calcium atom has two electrons in its outermost shell. Each oxygen atom has six electrons in its outermost shell and requires two more electrons to attain its octet. Each oxygen atom therefore requires one calcium atom from which to obtain the two electrons, and calcium and oxygen react in a 1 : 1 ratio.

$$Ca\colon \; + \; \colon\!\dot{O}\colon \; \longrightarrow \; Ca^{2+} \; + \; \colon\!\ddot{O}\colon^{2-}$$

EXAMPLE 5.6. With the aid of the periodic table, use electron dot notation to determine the formula of the ionic compound formed between lithium and oxygen.

Ans.   Lithium is in group IA and oxygen is in group VIA, and so each lithium atom has one outermost electron and each oxygen atom has six. Therefore, it takes two lithium atoms to supply the two electrons needed by one oxygen atom.

$$\begin{matrix} Li\cdot \\[4pt] Li\cdot \end{matrix} \; + \; \colon\!\dot{O}\colon \; \longrightarrow \; \begin{matrix} Li^+ \\[4pt] Li^+ \end{matrix} \; + \; \colon\!\ddot{O}\colon^{2-}$$

The formula of lithium oxide is Li_2O.

Electron dot structures are not usually written for transition metal or inner transition metal atoms even though they do lose electrons, forming ions. If you are asked to draw an electron dot diagram for a compound containing a monatomic transition metal ion, show the ion with no outermost electrons.

5.6. COVALENT BONDING

The element hydrogen exists in the form of diatomic molecules, H_2. Since both hydrogen atoms are identical, they are not likely to have opposite charges. (Neither has more electron-attracting power than the other.) Each free hydrogen atom contains a single electron, and if the atoms are to achieve the same electronic configuration as atoms of helium, they must each acquire a second electron. If two hydrogen atoms are allowed to come sufficiently close to each other, their two electrons will effectively belong to both atoms. The positively charged hydrogen nuclei are attracted to the pair of electrons shared between them, and a bond is formed. The bond formed from the sharing of a pair of electrons (or more than one pair) between two atoms is called a *covalent bond*. The hydrogen molecule is more stable than two separate hydrogen atoms. By sharing a pair of electrons, each hydrogen atom acquires a configuration analogous to that of a helium atom. Other pairs of nonmetallic atoms share electrons in the same way.

The formation of covalent bonds between atoms can be conventionally depicted by means of the electron dot notation. The formation of some covalent bonds is shown in this manner below:

$$H\cdot + \cdot H \longrightarrow H\!:\!H$$

$$:\!\ddot{F}\!\cdot + \cdot\ddot{F}\!: \longrightarrow :\!\ddot{F}\!:\!\ddot{F}\!:$$

$$H\cdot + \cdot\ddot{F}\!: \longrightarrow H\!:\!\ddot{F}\!:$$

$$\cdot\dot{\underset{\cdot}{C}}\cdot + 4\cdot\ddot{F}\!: \longrightarrow \begin{array}{c} :\!\ddot{F}\!: \\ :\!\ddot{F}\!:\!\overset{\cdot\cdot}{C}\!:\!\ddot{F}\!: \\ :\!\ddot{F}\!: \end{array}$$

In these examples, it can be seen that the carbon and fluorine atoms can achieve octets of electrons by sharing pairs of electrons with other atoms. Hydrogen atoms attain "duets" of electrons because the first shell is complete when it contains two electrons. We note from Sec. 5.4 that many main group atoms lose all their valence electrons and their cations have none left in their valence shell.

Sometimes it is necessary for two atoms to share more than one pair of electrons to attain octets. For example, the nitrogen molecule N_2 can be represented as follows:

$$:\!N\!:::\!N\!:$$

Three pairs of electrons must be shared in order that each nitrogen atom, originally with five valence electrons, attains an octet of electrons. The formation of strong covalent bonds between nitrogen atoms in N_2 is responsible for the relative inertness of nitrogen gas.

Every group of electrons shared between two atoms constitutes a covalent bond. When one pair of electrons is involved, the bond is called a *single bond*. When two pairs of electrons unite two atoms, the bond is called a *double bond*. Three pairs of electrons shared between two atoms constitute a *triple bond*. Examples of these types of bonds are given below:

H:H	:Ö::C::Ö:	:N:::N:	:C̈l:S̈:C̈l:
Single bond	Two double bonds	Triple bond	Two single bonds

Two small points might be noted now that double bonds have been introduced. Hydrogen atoms rarely bond to oxygen atoms that are double-bonded to some other atom. Halogen atoms do not form (simple) double bonds.

There are exceptions to the so-called octet rule, but only a few will be encountered within the scope of this book. Such cases will be discussed in detail as they arise.

The constituent atoms in polyatomic ions are also linked by covalent bonds. In these cases, the net charge on the ion is determined by the total number of electrons and the total number of protons. For example, the ammonium ion, NH_4^+, formed from five atoms, contains one fewer electron than the number of protons. A nitrogen atom plus 4 hydrogen atoms contains a total of 11 protons and 11 electrons, but the ion has only 10 electrons, 8 of which are valence electrons:

$$\left[\begin{array}{c} \text{H} \\ \text{H:}\overset{\cdot\cdot}{\underset{\cdot\cdot}{\text{N}}}\text{:H} \\ \text{H} \end{array} \right]^{+}$$

Similarly, the charge on a hydroxide can be determined by counting the total numbers of protons and electrons. The hydroxide ion contains one valence electron more than the total in the two individual atoms—oxygen and hydrogen:

$$:\overset{\cdot\cdot}{\underset{\cdot\cdot}{\text{O}}}\text{:H}^{-}$$

Covalent Bonding of More Than Two Atoms

Writing electron dot diagrams for molecules or ions containing only two atoms is relatively straightforward. When *several* atoms are to be represented as being linked by means of covalent bonds, the following procedure may be used to determine precisely the total number of electrons which must be shared among the atoms. The procedure, useful only for compounds in which the atoms obey the octet rule, will be illustrated using sulfur dioxide as an example.

Steps		Example	
1.	Determine the number of valence electrons available.	S 2 O Total	6 $\underline{12}$ 18
2.	Determine the number of electrons necessary to satisfy the octet (or duet) rule with no electron sharing.	S 2 O Total	8 $\underline{16}$ 24
3.	The difference between the numbers obtained in steps 2 and 1 is the number of bonding electrons.	Required Available To be shared	24 $\underline{-18}$ 6
4.	Place the atoms as symmetrically as possible. (Note that a hydrogen atom cannot be bonded to more than one atom, since it is capable of sharing only two electrons.)	O S O	
5.	Place the number of electrons to be shared between the atoms, a pair at a time, at first one pair between each pair of atoms. Use as many pairs as remain to make double or triple bonds.	O::S:O	
6.	Add the remainder of the available electrons to complete the octets (or duets) of all the atoms. There should be just enough if the molecule or ion follows the octet rule.	$\overset{\cdot\cdot}{\text{O}}::\overset{\cdot\cdot}{\text{S}}:\overset{\cdot\cdot}{\underset{\cdot\cdot}{\text{O}}}:$	

EXAMPLE 5.7. Draw electron dot diagrams showing the bonding in the following compounds: (*a*) CO, (*b*) CO_2, (*c*) CH_4, and (*d*) $MgCl_2$.

Ans.

	(a)	C	O	Total		(b)	C	O	Total
Electrons needed		8	8	16			8	2×8	24
Available		4	6	10			4	2×6	16
To be shared				6					8

$$C:::O \qquad\qquad\qquad\qquad O::C::O$$

Add the rest of the electrons: $:C:::O: \qquad\qquad\qquad :\ddot{O}::C::\ddot{O}:$

	(c)	C	H	Total		(d)	Mg	Cl	Total
Electrons needed		8	4×2	16			0	2×8	16
Available		4	4×1	8			2	2×7	16
To be shared				8					0

$$\begin{array}{c} H \\ H:\ddot{C}:H \\ \ddot{H} \end{array} \qquad\qquad Mg^{2+} \quad 2 :\ddot{\underset{..}{C}l}:^{-}$$

Note that the main group cation needs no valence electrons. No covalent bonds exist in $MgCl_2$; there is no electron sharing.

In polyatomic *ions*, more or fewer electrons are available than the number that comes from the valence shells of the atoms in the ion. This gain or loss of electrons provides the charge.

EXAMPLE 5.8.   Draw an electron dot diagram for CO_3^{2-}.

Ans.

1. Valence electrons available

	C	4
	3 O	18
charge		2	(2− charge means 2 extra electrons)
Total		24

2. Valence electrons required

	C	8
	3 O	24
Total		32

3. To be shared $32 - 24 = 8$

5. With shared electrons $\left[\begin{array}{c} O::\underset{..}{C}:O \\ \ddot{O} \end{array} \right]^{2-}$

6. With all electrons $\left[\begin{array}{c} :\ddot{O}::\underset{..}{C}:\ddot{O}: \\ :\ddot{O}: \end{array} \right]^{2-}$

Scope of the Octet Rule

It must be emphasized that the octet rule does not describe the electronic configuration of all compounds. The very existence of any compounds of the noble gases is evidence that the octet rule does not apply in all cases. Other examples of compounds that do not obey the octet rule are BF_3, PF_5, and SF_6. But the octet rule does summarize, systematize, and explain the bonding in so many compounds that it is well worth learning and understanding. Compounds in which atoms attain the configuration of helium (the duets) are considered to obey the octet rule, despite the fact that they achieve only the duet characteristic of the complete first shell of electrons.

5.7.   DISTINCTION BETWEEN IONIC AND COVALENT BONDING

The word *bonding* applies to any situation in which two or more atoms are held together in such close proximity that they form a characteristic species which has distinct properties and which can be represented by a chemical formula. In compounds consisting of ions, bonding results from the attractions between the oppositely charged ions. In such compounds in the solid state, each ion is surrounded on all sides by ions of the opposite charge. (For example, see Fig. 5-1.) In a solid ionic compound, it is incorrect to speak of a bond between specific pairs of ions, and ionic compounds do not form molecules.

In contrast, covalent bonding involves the sharing of electron pairs between two specific atoms, and it is possible to speak of a definite bond. For example, in molecules of HCl and CH_4 there are one and four covalent bonds per molecule, respectively.

Polyatomic ions such as OH^-, ClO_3^-, and NH_4^+ possess covalent bonds as well as an overall charge.

$$:\ddot{O}:H^- \qquad \left[:\ddot{O}:\ddot{Cl}:\ddot{O}: \atop \quad :\ddot{O}: \right]^- \qquad \left[{H \atop H:\overset{\cdot\cdot}{N}:H} \atop {\;\;\ddot{H}} \right]^+$$

The charges on polyatomic ions cause ionic bonding between these groups of atoms and oppositely charged ions. In writing electron dot structures, the distinction between ionic and covalent bonds must be clearly indicated. For example, the electron dot diagram for the compound NH_4ClO_3 is

$$\left[{H \atop H:\overset{\cdot\cdot}{N}:H} \atop {\;\;\ddot{H}} \right]^+ \qquad \left[:\ddot{O}:\ddot{Cl}:\ddot{O}: \atop \quad :\ddot{O}: \right]^-$$

5.8. PREDICTING THE NATURE OF BONDING IN COMPOUNDS

Electronegativity

Electronegativity is a semiquantitative measure of the ability of an atom to attract electrons involved in covalent bonds. Atoms with higher electronegativities have greater electron-attracting ability. Selected values of electronegativity are given in Table 5-1. The greater the electronegativity difference between a pair of elements, the more likely they are to form an ionic compound; the lower the difference in electronegativity, the more likely that if they form a compound, the compound will be covalent.

Table 5-1   Selected Electronegativities

H 2.1							He
Li 1.0	Be 1.5		B 2.0	C 2.5	N 3.0	O 3.5	F 4.0	Ne
Na 0.9	Mg 1.2		Al 1.5	Si 1.8	P 2.1	S 2.5	Cl 3.0	Ar
K 0.8	Ca 1.0		Ga 1.6	Ge 1.8	As 2.0	Se 2.4	Br 2.8	Kr
Rb 0.8	Sr 1.0		In 1.7	Sn 1.8	Sb 1.9	Te 2.1	I 2.5	Xe
Cs 0.7	Ba 0.9		Tl 1.8	Pb 1.9	Bi 1.9	Po 2.0	At 2.2	Rn

You need not memorize values of electronegativity (although those of the second-period elements are very easy to learn). You may generalize and state that the greater the separation in the periodic table, the greater the electronegativity difference. Also, in general, electronegativity increases to the right and upward in the periodic table. Compounds are generally named and formulas are written for them with the less electronegative element first. (Hydrogen compounds are exceptions. The symbol for hydrogen is written first only for acids. Hydrogen combined with a halogen atom or with a polyatomic anion forms an acid. In NH_3, the H is written last despite its lower electronegativity because NH_3 is not an acid.) You may not even have to consider electronegativity in making deductions about compounds, for example, in naming compounds (Chap. 6). In fact, the following generalizations about bonding can be made without reference to electronegativity.

Most binary compounds (compounds of two elements) of metals with nonmetals are essentially ionic. All compounds involving only nonmetals are essentially covalent except for compounds containing the NH_4^+ ion. We do not consider compounds of metals with metals in this course.

Practically all tertiary compounds (compounds of three elements) contain covalent bonds. If one or more of the elements is a metal, there is likely to be ionic as well as covalent bonding in the compound.

Formation of Ions in Solution

When some molecules containing only covalent bonds are dissolved in water, the molecules react with the water to produce ions in solution. For example, pure hydrogen chloride, HCl, and pure ammonia, NH_3, consist of molecules containing only covalent bonds. When cooled to sufficiently low temperatures ($-33°C$ for NH_3, $-85°C$ for HCl), these substances condense to liquids. However, the liquids do not conduct electricity, since they are still covalent and contain no ions. In contrast, when HCl is dissolved in water, the resulting solution conducts electricity well. Aqueous solutions of ammonia also conduct, but poorly. In these cases, the following reactions occur to the indicated extent to yield ions:

$$H:\overset{\cdot\cdot}{\underset{\cdot\cdot}{Cl}}: \ + \ H:\overset{\cdot\cdot}{\underset{\cdot\cdot}{O}}:H \longrightarrow \left[\begin{array}{c} H \\ H:\overset{\cdot\cdot}{O}:H \end{array}\right]^+ \ + \ :\overset{\cdot\cdot}{\underset{\cdot\cdot}{Cl}}:^- \quad (100\%)$$

$$H:\overset{\cdot\cdot}{\underset{H}{N}}:H \ + \ H:\overset{\cdot\cdot}{\underset{\cdot\cdot}{O}}:H \longrightarrow \left[\begin{array}{c} H \\ H:\overset{}{\underset{H}{N}}:H \end{array}\right]^+ \ + \ :\overset{\cdot\cdot}{\underset{\cdot\cdot}{O}}:H^- \quad \text{(approximately 1\%)}$$

H_3O^+ is often abbreviated H^+.

5.9.   DETAILED ELECTRONIC CONFIGURATIONS OF IONS (OPTIONAL)

To get the electronic configuration of ions, a new rule is followed. First we write the electron configuration of the neutral atom (Chap. 4). Then, for positive ions, we remove the electrons in the subshell with highest principal quantum number first. Note that these electrons might not have been added last, because of the $n + l$ rule. Nevertheless, the electrons from the shell with highest principal quantum number are removed first. For negative ions, we add electrons to the shell of highest principal quantum number. (That shell has the electrons added last by the $n + l$ rule.)

EXAMPLE 5.9.   What is the electronic configuration of each of the following: (a) Mg^{2+}, (b) Cl^-, (c) Co^{2+}, and (d) Co^{3+}?

Ans.   (a)   The configuration of Mg is $1s^2 2s^2 2p^6 3s^2$. For the ion Mg^{2+}, the outermost ($3s$) electrons are removed, yielding

$$Mg^{2+} \qquad 1s^2 2s^2 2p^6 3s^0 \qquad \text{or} \qquad 1s^2 2s^2 2p^6$$

(b)   The configuration of Cl is $1s^2 2s^2 2p^6 3s^2 3p^5$. For the anion, we add an electron for the extra charge:

$$Cl^- \qquad 1s^2 2s^2 2p^6 3s^2 3p^6$$

(c)   The configuration of Co is $1s^2 2s^2 2p^6 3s^2 3p^6 4s^2 3d^7$. For Co^{2+}, we remove the two $4s$ electrons! They are the electrons in the outermost shell, despite the fact that they were not the last electrons added. The configuration is

$$Co^{2+} \qquad 1s^2 2s^2 2p^6 3s^2 3p^6 4s^0 3d^7 \qquad \text{or} \qquad 1s^2 2s^2 2p^6 3s^2 3p^6 3d^7$$

(d) The configuration of Co^{3+} is $1s^2 2s^2 2p^6 3s^2 3p^6 3d^6$. An inner electron is removed after both $4s$ electrons are removed.

Being able to write correct electronic configurations for transition metal ions becomes very important in discussions of coordination compounds (in the second semester of general chemistry).

Solved Problems

CHEMICAL FORMULAS

5.1.   Which seven elements form diatomic molecules when they are uncombined with other elements?

    Ans. H_2, N_2, O_2, F_2, Cl_2, Br_2, and I_2. Remember that the last six of these form a pattern resembling "7" in the periodic table, starting at element 7.

5.2.   How many atoms of each element are there in a formula unit of each of the following substances?

    (a) P_4 (c) $(NH_4)_2SO_4$ (e) I_2 (g) $Hg_2(ClO_3)_2$

    (b) Na_3PO_4 (d) H_2O_2 (f) K_3PO_4 (h) $NH_4H_2PO_4$

    Ans.   (a) 4 P atoms   (e) 2 I atoms

           (b) 3 Na, 1 P, 4 O atoms   (f) 3 K, 1 P, 4 O atoms

           (c) 2 N, 8 H, 1 S, 4 O atoms   (g) 2 Hg, 2 Cl, 6 O atoms

           (d) 2 H, 2 O atoms   (h) 1 N, 6 H, 1 P, 4 O atoms

THE OCTET RULE

5.3.   Arrange the electrons in each of the following atoms in shells: (a) F, (b) Cl, (c) Br, and (d) I.

    Ans.

		Shell Number				
		1	**2**	**3**	**4**	**5**
(a)	F	2	7			
(b)	Cl	2	8	7		
(c)	Br	2	8	18	7	
(d)	I	2	8	18	18	7

5.4.   Arrange the electrons in each of the following atoms in shells: (a) Li, (b) Na, (c) K, and (d) Rb.

    Ans.

		Shell Number				
		1	**2**	**3**	**4**	**5**
(a)	Li	2	1			
(b)	Na	2	8	1		
(c)	K	2	8	8	1	
(d)	Rb	2	8	18	8	1

5.5.   Which electron shell is most important to the bonding for each of the following atoms: (a) H, (b) Be, (c) Al, (d) Sr, (e) Bi, and (f) Ra?

    Ans.   In each case the outermost shell is the most important. (a) 1 (b) 2 (c) 3 (d) 5 (e) 6 (f) 7

5.6. How many electrons are there in the outermost shell of each of the following: (*a*) Ba, (*b*) P, (*c*) Pb, (*d*) F, and (*e*) Xe?

Ans.   (*a*) 2   (*b*) 5   (*c*) 4   (*d*) 7   (*e*) 8

5.7. Which elements acquire the electronic configuration of helium by covalent bonding?

Ans.   Only hydrogen. Lithium and beryllium are metals, which tend to lose electrons (and form ionic bonds) rather than share. The resulting configuration of two electrons in the first shell, with no other shells occupied, is stable. Second-period elements of higher atomic number tend to acquire the electronic configuration of neon. If the outermost shell of an atom is the first shell, the maximum number of electrons in the atom is 2.

5.8. How does beryllium achieve the octet configuration?

Ans.   Beryllium loses two electrons, leaving it with the electronic configuration of He. A configuration of two electrons in its outermost shell corresponds to the "octet" because the new outermost shell is the first shell, which can hold only two electrons.

5.9. How many electrons can fit into an atom in which the outermost shell is the (*a*) first shell, (*b*) second shell, (*c*) third shell, and (*d*) fourth shell?

Ans.   (*a*) 2. The first shell is filled.   (*b*) 10. The first and second shells are both filled.   (*c*) 18. The first and second shells are filled, and there are eight electrons in the third shell (the maximum number before the fourth shell starts filling).   (*d*) 36 (2 + 8 + 18 + 8)

5.10. Compare the number of electrons in Problem 5.9 with the atomic numbers of the first four noble gases.

Ans.   They are the same—2, 10, 18, and 36.

5.11. Explain why *uncombined* atoms of all the elements in a given main group of the periodic table will be represented by a similar electron dot notation.

Ans.   They all have the same number of valence electrons.

IONS

5.12. What is the difference between ClO_2 and $ClO_2{}^-$?

Ans.   The first is a compound and the second is an ion—a part of a compound.

5.13. (*a*) What is the charge on the aluminum atom? (*b*) What is the charge on the aluminum ion? (*c*) What is the charge on the aluminum nucleus?

Ans.   (*a*) 0   (*b*) 3+   (*c*) 13+
        Note how important it is to read the questions carefully.

5.14. When an aluminum atom loses three electrons to form Al^{3+}, how many electrons are there in what is now the outermost shell? in the valence shell?

Ans.   There are eight electrons in the second shell, which is now the outermost shell, since the three electrons in the third shell have been lost. There are now zero electrons in the valence shell.

5.15. (*a*) You have been given a stack of $2.00 gift certificates for a store which gives no change for these certificates. What is the minimum number of $3.00 items that you can buy without wasting money? How many certificates will you use? (*b*) Suppose the items were $4.00 each?

Ans.   (*a*) You can buy two items (for $6) with three certificates (worth $6).   (*b*) You can buy one item with two certificates.

5.16. What is the formula of the compound of (*a*) magnesium and nitrogen? (*b*) lead and oxygen?

    Ans.   (*a*) Mg_3N_2. Magnesium has two electrons in the valence shell of each atom, and nitrogen requires three. (*b*) PbO_2. This problem involves the same reasoning as the prior problem.

5.17. Arrange the electrons in each of the following ions in shells: (*a*) K^+, (*b*) Mg^{2+}, and (*c*) N^{3-}.

    Ans.

	Shell Number			
Ion	**1**	**2**	**3**	**4**
(*a*) K^+	2	8	8	0
(*b*) Mg^{2+}	2	8	0	
(*c*) N^{3-}	2	8		

5.18. Write formulas for the compounds formed by the reaction of (*a*) magnesium and nitrogen, (*b*) barium and bromine, (*c*) lithium and nitrogen, (*d*) sodium and sulfur, (*e*) magnesium and sulfur, (*f*) aluminum and fluorine, and (*g*) aluminum and oxygen.

    Ans.   (*a*) Mg_3N_2 (*b*) $BaBr_2$ (*c*) Li_3N (*d*) Na_2S (*e*) MgS (*f*) AlF_3 (*g*) Al_2O_3

5.19. What ions are present in each of the following compounds: (*a*) $MnCl_2$, (*b*) Cu_2S, (*c*) CuO, (*d*) $Ba(ClO_4)_2$, (*e*) $(NH_4)_2SO_4$, and (*f*) $MgCO_3$?

    Ans.   (*a*) Mn^{2+} and Cl^- (*b*) Cu^+ and S^{2-} (*c*) Cu^{2+} and O^{2-} (*d*) Ba^{2+} and ClO_4^- (*e*) NH_4^+ and SO_4^{2-} (*f*) Mg^{2+} and CO_3^{2-}
             With binary compounds of transition metals, determine the charge on the anion first and make the cation balance that charge.

5.20. Write formulas for the compounds formed by the following pairs of ions: (*a*) Li^+ and Cl^-, (*b*) Pb^{4+} and O^{2-}, (*c*) K^+ and S^{2-}, (*d*) Al^{3+} and S^{2-}, (*e*) Mg^{2+} and N^{3-}, and (*f*) Co^{2+} and ClO^-.

    Ans.   (*a*) $LiCl$ (*b*) PbO_2 (*c*) K_2S (*d*) Al_2S_3 (*e*) Mg_3N_2 (*f*) $Co(ClO)_2$

5.21. Write the formulas for the compounds of (*a*) Ca^{2+} and ClO^- and (*b*) K^+ and ClO_2^-. Explain why one of the formulas requires parentheses.

    Ans. $Ca(ClO)_2$ and $KClO_2$. The parentheses mean two ClO^- ions; no parentheses mean two O atoms in one ClO_2^- ion.

5.22. Complete the following table by writing the formula of the compound formed by the cation at the left and the anion at the top. NH_4Br is given as an example.

	Br^-	NO_2^-	NO_3^-	SO_4^{2-}	PO_4^{3-}
NH_4^+	NH_4Br				
Li^+					
Mn^{2+}					
Fe^{3+}					

    Ans.

	Br^-	NO_2^-	NO_3^-	SO_4^{2-}	PO_4^{3-}
NH_4^+	NH_4Br	NH_4NO_2	NH_4NO_3	$(NH_4)_2SO_4$	$(NH_4)_3PO_4$
Li^+	$LiBr$	$LiNO_2$	$LiNO_3$	Li_2SO_4	Li_3PO_4
Mn^{2+}	$MnBr_2$	$Mn(NO_2)_2$	$Mn(NO_3)_2$	$MnSO_4$	$Mn_3(PO_4)_2$
Fe^{3+}	$FeBr_3$	$Fe(NO_2)_3$	$Fe(NO_3)_3$	$Fe_2(SO_4)_3$	$FePO_4$

ELECTRON DOT NOTATION

5.23. How many electrons are "required" in the valence shell of atoms of each of the following elements in their compounds: (a) H, (b) O, (c) N, and (d) Ba?

Ans.   (a) 2   (b) 8   (c) 8   (d) 0

5.24. Draw electron dot diagrams for lithium atoms and sulfur atoms (a) before they react with each other and (b) after they react with each other.

Ans.

             Li· $\cdot$ Li$^+$ $\cdot\cdot$ $^{2-}$

(a)   :S: (b)   :S:

             Li·   Li$^+$ $\cdot\cdot$

5.25. Draw an electron dot diagram for each of the following: (a) Ca, (b) Cl, (c) P, (d) S^{2-}, (e) S, (f) P^{3-}, (g) Ne, and (h) Mg^{2+}.

Ans.   (a) Ca: (b) :Cl: (c) :P·   (d) :S:$^{2-}$ (e) :S: (f) :P:$^{3-}$ (g) :Ne: (h) Mg^{2+} (Both valence electrons have been lost.)

5.26. Draw an electron dot diagram for Mg_3N_2.

Ans.   3 Mg^{2+} 2 :N:$^{3-}$

COVALENT BONDING

5.27. Identify the octet of carbon by encircling the electrons that satisfy the octet rule for carbon in CO_2. Put rectangles around the electrons that satisfy the octet for the oxygen atoms.

Ans.

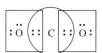

5.28. Draw electron dot diagrams for CH_2O and C_2H_4.

Ans.
          H   H H

      H:C::O   C::C

                       H H

5.29. Explain why hydrogen atoms cannot form double bonds.

Ans.   They cannot hold more than two electrons in their valence shell, because it is the first shell. A double bond includes four electrons.

5.30. Draw electron dot diagrams for oxygen atoms and sulfur atoms (a) before they react with each other and (b) after they react to form SO_3.

Ans.   (a)   ·O·   ·S·   ·O·   (b) :O::S:O:

                                                                            :O:

                                        ·O·

5.31. The elements hydrogen, nitrogen, and fluorine exist as diatomic molecules when they are not combined with other elements. Draw an electron dot structure for each molecule.

Ans.   H:H   :N:::N: :F:F:

5.32. Identify the bonding electrons and the nonbonding electrons on the sulfur atom in the electron dot diagram of SO_2.

Ans.

Nonbonding

$:\ddot{O}::\ddot{S}:\ddot{O}:$

Bonding

5.33. Draw electron dot diagrams for (*a*) KCl, (*b*) OF_2, (*c*) AsH_3, and (*d*) NH_4Cl.

Ans. (*a*) K^+ $:\ddot{\ddot{Cl}}:^-$ (*c*) $H:\ddot{As}:H$
                                                                          $\ddot{H}$

(*b*) $:\ddot{\ddot{F}}:\ddot{O}:\ddot{\ddot{F}}:$ (*d*) $\left[\begin{array}{c} H \\ H:\ddot{N}:H \\ H \end{array}\right]^+$ $:\ddot{\ddot{Cl}}:^-$

In part (*a*), and in other ionic compounds, do not draw the positive ion too near the negative ion; they are bonded by ionic bonding and do not share electrons.

5.34. Draw an electron dot diagram for $SOCl_2$ and $COCl_2$.

Ans. $:\ddot{\ddot{Cl}}:\ddot{S}:\ddot{\ddot{Cl}}:$ $:\ddot{\ddot{Cl}}:C:\ddot{\ddot{Cl}}:$
                                                    $:\ddot{O}:$ $:\ddot{O}:$

5.35. Draw electron dot diagrams for the following: (*a*) NH_3, (*b*) H_2O, (*c*) H_2O_2, (*d*) ClO_3^-, and (*e*) PCl_3.

Ans. (*a*) $H:\ddot{N}:H$ (*b*) $H:\ddot{O}:H$ (*c*) $H:\ddot{O}:\ddot{O}:H$ (*d*) $:\ddot{O}:\ddot{\ddot{Cl}}:\ddot{O}:^-$ (*e*) $:\ddot{\ddot{Cl}}:\ddot{P}:\ddot{\ddot{Cl}}:$
         $\ddot{H}$ $:\ddot{O}:$ $:\ddot{\ddot{Cl}}:$

5.36. Draw an electron dot diagram for each of the following: (*a*) SO_3, (*b*) SO_3^{2-}, (*c*) Na_2SO_3, and (*d*) H_2SO_3.

Ans. $:\ddot{O}:$
         (*a*) $:\ddot{O}::\ddot{S}:\ddot{O}:$ (*c*) $2\,Na^+$ $\left[\begin{array}{c} :\ddot{O}: \\ :\ddot{O}:\ddot{S}:\ddot{O}: \end{array}\right]^{2-}$

         (*b*) $\left[\begin{array}{c} :\ddot{O}: \\ :\ddot{O}:\ddot{S}:\ddot{O}: \end{array}\right]^{2-}$ (*d*) $H:\ddot{O}:\ddot{S}:\ddot{O}:H$
                                                                                    $:\ddot{O}:$

In (*a*) a double bond is needed to make the octet of sulfur. In (*b*), the extra pair of electrons corresponding to the charge makes the set of atoms an ion and it eliminates the need for a double bond. In (*c*), that same ion is present, along with the two sodium ions to balance the charge. In (*d*), because hydrogen is a nonmetal, the two hydrogen atoms are covalently bonded to oxygen atoms to complete the compound.

5.37. Draw electron dot structures for each of the following molecules: (*a*) CO, (*b*) CO_2, (*c*) HCN, and (*d*) N_2O (an unsymmetric molecule, with the two nitrogen atoms adjacent to each other).

Ans. (*a*) $:C:::O:$ (*b*) $:\ddot{O}::C::\ddot{O}:$ (*c*) $H:C:::N:$ (*d*) $:\ddot{N}::N::\ddot{O}:$

5.38. Draw the electron dot diagram for ammonium sulfide, $(NH_4)_2SO_3$.

Ans.
$$2\left[\begin{array}{c} H \\ H:\ddot{N}:H \\ H \end{array}\right]^+   \left[\begin{array}{c} :\ddot{O}: \\ :\ddot{O}:\ddot{S}:\ddot{O}: \end{array}\right]^{2-}$$

DISTINCTION BETWEEN IONIC AND COVALENT BONDING

5.39. Describe the bonding of the chlorine atoms in each of the following substances: (*a*) Cl_2, (*b*) SCl_2, and (*c*) $CaCl_2$.

   Ans.   (*a*) The chlorine atoms are bonded to each other with a covalent bond.   (*b*) The chlorine atoms are both bonded to the sulfur atom with covalent bonds.   (*c*) The chlorine atoms are changed to Cl^- ions, and are bonded to the calcium ion by ionic bonds.

5.40. Draw an electron dot diagram for (*a*) K_2SO_4 and (*b*) H_2SO_4. What is the major difference between them?

   Ans.

   In the potassium salt there is ionic bonding as well as covalent bonding; in the hydrogen compound, there is only covalent bonding.

5.41. What type of bonding is present in each of the following compounds: (*a*) $AlCl_3$, (*b*) NCl_3, and (*c*) $Al(ClO)_3$?

   Ans.   (*a*) Ionic   (*b*) Covalent   (*c*) Both ionic and covalent

PREDICTING THE NATURE OF BONDING IN COMPOUNDS

5.42. Which element of each of the following pairs has the higher electronegativity? Consult Table 5-1 only after writing down your answer. (*a*) K and Cl, (*b*) S and O, and (*c*) Cl and O.

   Ans.   (*a*) Cl. It lies farther to the right in the periodic table.   (*b*) O. It lies farther up in the periodic table.   (*c*) O. It lies farther up in the periodic table, and is more electronegative even though it lies one group to the left. (See Table 5-1.) O is an exception in this regard.

5.43. Which element is named first in the compound of each of the following pairs: (*a*) As and O, (*b*) As and Br, and (*c*) S and I?

   Ans.   (*a*) As, since it lies to the left of O and below it.   (*b*) As, since it lies left of Br.   (*c*) S, since it lies left of I, despite the fact that it is above I.

DETAILED ELECTRONIC CONFIGURATIONS OF IONS

5.44. What is the electronic configuration of H^+?

   Ans.   There are no electrons left in H^+, and so the configuration is $1s^0$. This is not really a stable chemical species. H^+ is a convenient abbreviation for a more complicated ion, H_3O^+.

5.45. Write detailed electronic configurations for the following ions: (*a*) K^+, (*b*) Ti^{3+}, (*c*) S^{2-}, and (*d*) Se^{2-}.

   Ans.   (*a*) K^+ $1s^22s^22p^63s^23p^6$ (*c*) S^{2-} $1s^22s^22p^63s^23p^6$

          (*b*) Ti^{3+} $1s^22s^22p^63s^23p^63d^1$ (*d*) Se^{2-} $1s^22s^22p^63s^23p^64s^23d^{10}4p^6$

5.46. What positive ion with a double charge is represented by each of the following configurations: (*a*) $1s^22s^22p^6$, (*b*) $1s^2$, and (*c*) $1s^22s^22p^63s^23p^6$?

   Ans.   (*a*) Mg^{2+} (*b*) Be^{2+} (*c*) Ca^{2+}

5.47. What is the electronic configuration of Pb^{2+}?

   Ans.   The configuration for Pb is

   $$Pb \quad 1s^22s^22p^63s^23p^64s^23d^{10}4p^65s^24d^{10}5p^66s^24f^{14}5d^{10}6p^2$$

The configuration for Pb^{2+} is the same except for the loss of the outermost two electrons. There are four electrons in the sixth shell; the two in the last subshell of that shell are lost first:

$$Pb^{2+} \quad 1s^2 2s^2 2p^6 3s^2 3p^6 4s^2 3d^{10} 4p^6 5s^2 4d^{10} 5p^6 6s^2 4f^{14} 5d^{10} 6p^0$$

In shortened terminology:

$$Pb \qquad [Xe] \quad 6s^2 4f^{14} 5d^{10} 6p^2$$
$$Pb^{2+} \qquad [Xe] \quad 6s^2 4f^{14} 5d^{10} 6p^0$$

5.48. What ion with a double negative charge is represented by each of the following configurations: (*a*) $1s^2 2s^2 2p^6$ and (*b*) $1s^2 2s^2 2p^6 3s^2 3p^6$?

Ans.   (*a*) O^{2-} (*b*) S^{2-}

5.49. What positive ion with a double charge is represented by each of the following configurations: (*a*) $1s^2 2s^2 2p^6 3s^2 3p^6$ and (*b*) $1s^2 2s^2 2p^6 3s^2 3p^6 3d^5$?

Ans.   (*a*) Ca^{2+} (*b*) Mn^{2+}

5.50. What is the difference in the electronic configurations of Fe and Ni^{2+}? (They both have 26 electrons.)

Ans.

Fe	$1s^2 2s^2 2p^6 3s^2 3p^6 4s^2 3d^6$
Ni	$1s^2 2s^2 2p^6 3s^2 3p^6 4s^2 3d^8$
Ni^{2+}	$1s^2 2s^2 2p^6 3s^2 3p^6 4s^0 3d^8$

The Ni atom has two more $3d$ electrons; the Ni^{2+} ion has lost its $4s$ electrons. Thus, the Ni^{2+} ion has two more $3d$ electrons and two fewer $4s$ electrons than the Fe atom has.

Supplementary Problems

5.51. (*a*) How many electrons are there in the outermost shell of an atom of phosphorus? (*b*) How many *additional* electrons is it necessary for an atom of phosphorus to share in order to attain an octet configuration? (*c*) How many additional electrons is it necessary for an atom of chlorine to share in order to attain an octet configuration? (*d*) Write the formula for a compound of phosphorus and chlorine. (*e*) Draw an electron dot structure showing the arrangement of electrons in a molecule of the compound.

Ans.   (*a*) 5 (It is in periodic group VA.)   (*b*) 3 $(8 - 5 = 3)$ (*c*) 1 $(8 - 7 = 1)$ (*d*) PCl_3 (*e*)

$$\overset{\displaystyle ..}{\underset{\displaystyle ..}{Cl}}:\overset{\displaystyle ..}{P}:\overset{\displaystyle ..}{\underset{\displaystyle ..}{Cl}}:$$

5.52. Write a formula for a binary compound formed between each of the following pairs of elements. (*a*) Na, Cl; (*b*) Mg, I; (*c*) K, P; (*d*) K, S; (*e*) Li, N; (*f*) Al, O; (*g*) Al, F; (*h*) Mg, N; (*i*) P, Cl; (*j*) Cl, Mg; (*k*) O, Mg; (*l*) Si, Cl; and (*m*) S, F.

Ans.   (*a*) NaCl   (*b*) MgI_2 (*c*) K_3P (*d*) K_2S (*e*) Li_3N (*f*) Al_2O_3 (*g*) AlF_3 (*h*) Mg_3N_2 (*i*) PCl_3 (*j*) $MgCl_2$ (The metal ion is written first.)   (*k*) MgO   (*l*) $SiCl_4$ (*m*) SF_2 (or SF_4 or SF_6)

5.53. Which of the following compounds involve covalent bonding? Which involve electron sharing? (*a*) $MgCl_2$, (*b*) SCl_2, and (*c*) $(NH_4)_2S$.

Ans. SCl_2 and $(NH_4)_2S$ involve covalent bonding and therefore, by definition, electron sharing. $(NH_4)_2S$ also exhibits ionic bonding. ($MgCl_2$ is entirely ionic.)

5.54. You can predict the formula of the compound between Na and Cl, but not between Fe and Cl. Explain why.

Ans.   Na is a main group element and forms Na^+ only in all of its compounds. Fe is a transition element and forms two different ions: Fe^{2+} and Fe^{3+}.

5.55. Draw an electron dot diagram for NH_4HS.

Ans.

$$\left[\begin{array}{c} H \\ H\!:\!\overset{\cdot\cdot}{N}\!:\!H \\ H \end{array} \right]^{+} \qquad H\!:\!\overset{\cdot\cdot}{\underset{\cdot\cdot}{S}}\!:^{-}$$

5.56. Phosphorus forms two covalent compounds with chlorine, PCl_3 and PCl_5. Discuss these compounds in terms of the octet rule.

Ans. PCl_3 obeys the octet rule; PCl_5 does not. PCl_5 has to bond five chlorine atoms around the phosphorus atom, each with a pair of electrons, for a total of 10 electrons around phosphorus.

5.57. Draw an electron dot diagram for NO. Explain why it cannot follow the octet rule.

Ans. $:\!\overset{\cdot}{N}\!:\!:\!\overset{\cdot\cdot}{O}\!:$

There are an odd number of electrons in NO; there is no way that there can be eight around each atom.

5.58. Distinguish between each of the following pairs: (*a*) an ion and a free atom; (*b*) an ion and an ionic bond; (*c*) a covalent bond and an ionic bond; (*d*) a triple bond and three single bonds on the same atom; and (*e*) a polyatomic molecule and a polyatomic ion.

Ans.   (*a*)   An ion is charged, and a free atom is uncharged.

(*b*)   An ion is a charged atom or group of atoms; an ionic bond is the attraction between ions.

(*c*)   A covalent bond involves sharing of electrons; an ionic bond involves electron transfer and as a result the formation of ions.

(*d*)   Although both involve three pairs of electrons, the triple bond has all three pairs of electrons between two atoms, and three single bonds have each pair of electrons between different pairs of atoms.

$$:\!N\!:\!:\!:\!N\!: \qquad\qquad H\!:\!\overset{\cdot\cdot}{\underset{\cdot\cdot}{N}}\!:\!H$$
$$ \qquad\qquad H$$

Triple bond   Three single bonds

(*e*)   Both have more than one atom. The polyatomic ion is charged and is only part of a compound; the polyatomic molecule is uncharged and represents a complete compound.

5.59. Write the formula for the compound formed by the combination of each of the following pairs of elements. State whether the compound is ionic or covalent. (*a*) Mg and Br, (*b*) Si and F, (*c*) Ca and O, and (*d*) Br and Cl.

Ans.   (*a*) $MgBr_2$ ionic   (*c*)   CaO   ionic

(*b*) SiF_4 covalent   (*d*)   BrCl   covalent

5.60. BF_3 and PF_5 are non-octet-rule compounds. Draw an electron dot diagram for each.

Ans.

$$:\!\overset{\cdot\cdot}{F}\!:\!B\!:\!\overset{\cdot\cdot}{F}\!:$$
$$:\!\overset{\cdot\cdot}{F}\!:$$

(Fluorine does not
form double bonds.)

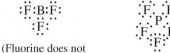

5.61. How many electrons remain in the valence shell after the loss of electrons by the neutral atom to form each of the following ions: (*a*) Pb^{2+} and (*b*) Pb^{4+}?

Ans.   (*a*) The Pb^{2+} still has two electrons in its sixth shell.   (*b*) The Pb^{4+} has no electrons left in that shell.

5.62. Complete the following table by writing the formula of the compound formed by the cation at the left and the anion at the top.

	ClO_2^{-}	ClO_3^{-}	$C_2O_4^{2-}$
Fe^{2+}			
Cr^{3+}			

Ans.

	ClO_2^-	ClO_3^-	$C_2O_4^{2-}$
Fe^{2+}	$Fe(ClO_2)_2$	$Fe(ClO_3)_2$	FeC_2O_4
Cr^{3+}	$Cr(ClO_2)_3$	$Cr(ClO_3)_3$	$Cr_2(C_2O_4)_3$

5.63. Draw an electron dot diagram for HNO_3. *Note*: As a rule, hydrogen atoms do not bond to double-bonded oxygen atoms.

Ans.

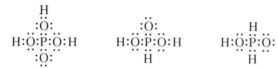

The hydrogen is bonded to one of the single-bonded oxygen atoms.

5.64. Draw electron dot diagrams for H_3PO_4, H_3PO_3, and H_3PO_2. *Hint*: Four atoms are bonded directly to the phosphorus atom in each case.

Ans.

5.65. Which of the following compounds do not obey the octet rule: (*a*) ICl_3, (*b*) SF_4, (*c*) H_2O_2, and (*d*) BrF_3?

Ans.   (*a*), (*b*), and (*d*).

Inorganic Nomenclature

6.1. INTRODUCTION

Naming and writing formulas for inorganic compounds are extremely important skills. For example, a physician might prescribe barium sulfate for a patient in preparation for a stomach X-ray. If barium sulfite or barium sulfide is given instead, the patient might die from barium poisoning. Such a seemingly small difference in the name makes a very big difference in the properties! (Barium sulfate is too insoluble to be toxic.)

There is a vast variety of inorganic compounds, and the compounds are named according to varying systems of nomenclature. The first job to do when you wish to name a compound is to determine which class it is in. Rules for the major classes will be given in this chapter. Compounds that are rarely encountered in general chemistry courses will not be covered.

Rules for writing formulas from names will also be presented. An outline of the classes that will be presented is given in Table 6-1, and rules for naming compounds in the different classes are illustrated in Fig. 6-1. These summaries are available if you want them, but they are not the only way to remember the various systems. Use either one or the other if you wish, but not both.

Table 6-1   Nomenclature Divisions for Inorganic Compounds

Binary nonmetal-nonmetal compounds	(Sec. 6.2)
Ionic compounds	(Sec. 6.3)
Cations	
Monatomic cations with constant charges	
Monatomic cations with variable charges	
Polyatomic cations	
Anions	
Monatomic anions	
Oxyanions	
Varying numbers of oxygen atoms	
Special anions	
Inorganic acids	(Sec. 6.4)
Acid salts	(Sec. 6.5)
Hydrates	(Sec. 6.6)

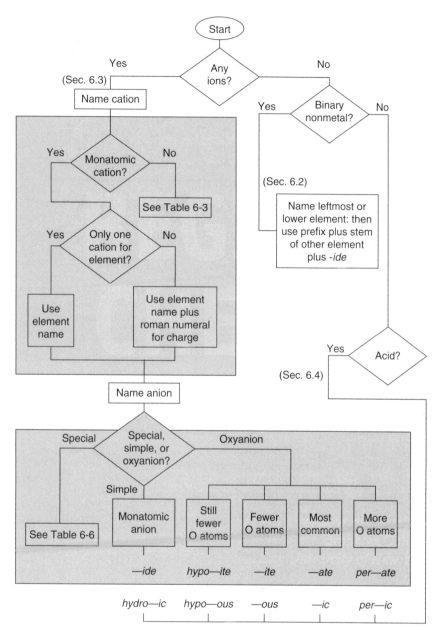

Fig. 6-1. Outline of nomenclature rules

6.2. BINARY COMPOUNDS OF NONMETALS

The first compounds to be discussed will be compounds of two nonmetals. These binary compounds are named with the element to the left or below in the periodic table named first. The other element is then named, with its ending changed to *-ide* and a prefix added to denote the number of atoms of that element present. If there is more than one atom of the first element present, a prefix is used with the first element also. If one of the elements is to the left and the other below, the one to the left is named first unless that element is oxygen or fluorine, in which case it is named last. The same order of elements is used in writing formulas for these compounds. (The element with the lower electronegativity is usually named first; refer to Table 5-1.) The prefixes are presented in Table 6-2. The first six prefixes are the most important to memorize.

Table 6-2   Prefixes for Nonmetal-Nonmetal Compounds

Number of Atoms	Prefix	Number of Atoms	Prefix
1	mono (or mon before names starting with a or o)	6	hexa
2	di	7	hepta
3	tri	8	octa
4	tetra (or tetr before names starting with a or o)	9	nona
5	penta (or pent before names starting with a or o)	10	deca

The systematic names presented for binary nonmetal-nonmetal compounds are not used for the hydrogen compounds of group III, IV, and V elements or for water. These compounds have common names which are used instead. Water and ammonia (NH_3) are the most important compounds in this class. See Sec. 6.4 for acid names.

EXAMPLE 6.1.   Name and write the formula for a compound containing two atoms of oxygen and one atom of sulfur in each molecule.

Ans.   This compound is a compound of two nonmetals. The sulfur is named first, since it lies below oxygen in the periodic table. Then the oxygen is named, with its ending changed to *-ide* and a prefix denoting the number of oxygen atoms present. The name is sulfur dioxide and the formula is SO_2.

EXAMPLE 6.2.   Name (*a*) CO and (*b*) CO_2.

Ans.   Both of these compounds are composed of two nonmetals. The carbon is named first, since it lies to the left of oxygen in the periodic table. Then the oxygen is named, with its ending changed to *-ide* and a prefix denoting the number of oxygen atoms present. (*a*) Carbon monoxide and (*b*) carbon dioxide.

EXAMPLE 6.3.   Name and write formulas for (*a*) a compound with one sulfur atom and two fluorine atoms per molecule and (*b*) a compound with six fluorine atoms and one sulfur atom per molecule.

Ans.   (*a*) Sulfur difluoride SF_2 (*b*) Sulfur hexafluoride SF_6
Sulfur is named first, since it lies below and to the left of fluorine in the periodic table. (The element that is first in the name is also first in the formula.) The prefixes tell how many atoms of the second element there are in each molecule.

EXAMPLE 6.4.   Name P_2O_3.

Ans.   Phosphorus lies to the left and below oxygen in the periodic table, so it is named first. The number of atoms of the first element is specified because it is greater than one: diphosphorus trioxide.

EXAMPLE 6.5.   Name a compound containing in each molecule (*a*) two oxygen atoms and a bromine atom and (*b*) one nitrogen atom and three chlorine atoms.

Ans.   (*a*) Bromine dioxide and (*b*) nitrogen trichloride. In BrO_2, O is above Br but to its left in the periodic table. In NCl_3, N is above Cl and to its left. When this situation arises with oxygen or fluorine, that element is written last; any other such pair has the element to the left written first. That is the reason for the order of naming in BrO_2 and NCl_3.

EXAMPLE 6.6.   Name NH_3.

Ans.   Ammonia. The common name is used.

6.3. NAMING IONIC COMPOUNDS

Ionic compounds are composed of *cations* (positive ions) and *anions* (negative ions). The cation is always named first, and then the anion is named. The name of the cation does not depend on the nature of the anion, and the name of the anion does not depend on the nature of the cation.

Naming Monatomic Cations

Naming of the positive ion depends on whether the cation is *monatomic* (has only one atom). If not, the special names given in the next subsection are used. If the cation is monatomic, the name depends on whether the element forms more than one positive ion in its compounds. For example, sodium forms only one positive ion in all its compounds—Na^+. Cobalt forms two positive ions—Co^{2+} and Co^{3+}. Cations of elements that form only one type of ion in all their compounds need not be further identified in the name. Thus, Na^+ may simply be called the sodium ion. Cations of metals that occur with two or more different charges must be further identified. $Co(NO_3)_2$ and $Co(NO_3)_3$ occur with Co^{2+} and Co^{3+} ions, respectively. If we just call the ion the "cobalt ion," we will not know which one it is. Therefore, for such cations, we use a Roman numeral in parentheses attached to the name to tell the charge on such ions. (Actually, oxidation numbers are used for this purpose, but if you have not yet studied oxidation numbers—Chap. 14—follow the rules given next.) Thus, Co^{2+} is called the cobalt(II) ion, and Co^{3+} is called the cobalt(III) ion.

The elements that form only one cation are the alkali metals (group IA), the alkaline earth metals (group IIA), zinc, cadmium, aluminum, and most often silver. The charge on the ions that these elements form in their compounds is always equal to their classical periodic table group number (or group number minus 10 for Ag, Cd, Zn, and Al in the modern labeling system in the periodic table).

EXAMPLE 6.7.   Name $NiCl_2$ and $MgCl_2$.

Ans.   Since Ni is not among the elements that always form ions of the same charge in all their compounds, the charge must be stated. The name is nickel(II) chloride. Since Mg is an alkaline earth element, the charge in its compounds is always 2+, so there is no need to mention the charge in the name. The compound is magnesium chloride.

EXAMPLE 6.8.   Write formulas for (*a*) copper(I) sulfide and (*b*) copper(II) sulfide.

Ans.   (*a*) Cu_2S and (*b*) CuS. Note carefully that the Roman numerals in the names mean one thing—the charge on the ion—and the Arabic numeral subscripts in the formulas mean another—number of atoms. Here the copper(I) has a charge of 1+, and therefore two copper(I) ions are required to balance the 2− charge on one sulfide ion. The copper(II) ion has a charge of 2+, and therefore one such ion is sufficient to balance the 2− charge on the sulfide ion.

EXAMPLE 6.9.   Name NF_3.

Ans.   Nitrogen trifluoride. (Although this is a binary compound of two nonmetals, it can be named with Roman numeral designations. It is indeed possible to call this nitrogen(III) fluoride in the most modern usage, but most chemists do not do that yet.)

Polyatomic Cations

Several cations that consist of more than one atom are important in general chemistry. There are few enough of these important ions to learn them individually. They are presented in Table 6-3. There are several ions like uranyl ion, also of limited importance in general chemistry. More will be said about these ions in the sections on oxidation states (Chap. 14).

Table 6-3   Several Polyatomic Cations

NH_4^+	Ammonium ion	Very important
H_3O^+	Hydronium ion	Very important in Chap. 17
Hg_2^{2+}	Mercury(I) ion or mercurous ion	Somewhat important
UO_2^{2+}	Uranyl ion	Not too important

Classical Nomenclature Systems

An older system for naming cations of elements having more than one possible cation uses the ending -ic for the ion with the higher charge and the ending -ous for the ion with the lower charge. In this system, the Latin names for some of the elements are used instead of the English names. Thus, this system is harder in two ways than the *Stock system*, described above. You must know whether a particular ion has a related ion of higher or lower charge, and you must know the Latin root. You should study this subsection if your text or your instructor uses this system; otherwise you may omit it. The names for common ions in this system are given in Table 6-4. Note that transition metals except the coinage metals do not have ions with charges of 1+. (This system was also used in the past for nonmetal-nonmetal compounds also, and still exists in the designation of nitrogen oxides. N_2O is called nitrous oxide, and NO is called nitric oxide.)

Table 6-4   Names of Cations in Classical System

Transition Metals			
Vanadous	V^{2+}	Vanadic	V^{3+}
Chromous	Cr^{2+}	Chromic	Cr^{3+}
Manganous	Mn^{2+}	Manganic	Mn^{3+}
Ferrous	Fe^{2+}	Ferric	Fe^{3+}
Cobaltous	Co^{2+}	Cobaltic	Co^{3+}
Nickelous	Ni^{2+}	Nickelic	Ni^{3+}
Cuprous	Cu^{+}	Cupric	Cu^{2+}
Argentous	Ag^{+}	Argentic	Ag^{2+} (rare)
Aurous	Au^{+}	Auric	Au^{3+}
Mercurous	Hg_2^{2+}	Mercuric	Hg^{2+}
Palladous	Pd^{2+}	Palladic	Pd^{4+}
Platinous	Pt^{2+}	Platinic	Pt^{4+}
Main Group Metals			
Stannous	Sn^{2+}	Stannic	Sn^{4+}
Plumbous	Pb^{2+}	Plumbic	Pb^{4+}
Inner Transition Element			
Cerous	Ce^{3+}	Ceric	Ce^{4+}

Naming Anions

Common anions may be grouped as follows: monatomic anions, oxyanions, and special anions. There are special endings for the first two groups; the third group is small enough to be memorized.

Monatomic Anions

If the anion is monatomic (has only one atom), the name of the element is amended by changing the ending to -ide. Note that this ending is also used for binary nonmetal-nonmetal compounds. All monatomic anions have names ending in -ide, but there are a few anions that consist of more than one atom which also end in -ide—the most important of these are OH^- and CN^-. OH^- is called the hydroxide ion, and CN^- is called the cyanide ion.

The charge on every monatomic anion is equal to the group number minus 8 (or 18 if the modern periodic table group numbering system is used).

EXAMPLE 6.10. What is the charge on (*a*) the sulfide ion, (*b*) the nitride ion, and (*c*) the fluoride ion?

Ans.   (*a*) Sulfur is in group VI (16), and so the charge is $6 - 8 = -2$ (or $16 - 18 = -2$).

     (*b*) Nitrogen is in group V (15), and so the charge is $5 - 8 = -3$.

     (*c*) Fluorine is in group VII (17), and so the charge is $7 - 8 = -1$.

Oxyanions

Oxyanions consist of an atom of an element plus some number of atoms of oxygen covalently bonded to it. The name of the anion is given by the name of the element with its ending changed to either *-ate* or *-ite*. In some cases, it is also necessary to add the prefix *per-* or *hypo-* to distinguish all the possible oxyanions from one another. For example, there are four oxyanions of chlorine, which are named as follows:

$$ClO_4^- \quad \text{perchlorate ion}$$
$$ClO_3^- \quad \text{chlorate ion}$$
$$ClO_2^- \quad \text{chlorite ion}$$
$$ClO^- \quad \text{hypochlorite ion}$$

One may think of the *-ite* ending as meaning "one fewer oxygen atom." The *per-* and *hypo-* prefixes then mean "one more oxygen atom" and "still one fewer oxygen atom," respectively. Thus, perchlorate means one more oxygen atom than chlorate has. Hypochlorite means still one fewer oxygen atom than chlorite has. Note that all four oxyanions have the same central atom (Cl) and the same charge ($1-$). The only difference in their constitutions is the number of oxygen atoms.

Other elements have similar sets of oxyanions, but not all have four different oxyanions. You should learn the names of the seven ions ending in *-ate* for the most common elements. These are the most important oxyanions. Use the rules given above for remembering the others. These ions are all presented in Table 6-5. Note that the ones with central atoms in odd periodic groups have odd charges and that those in even periodic groups have even charges.

Table 6-5   Common Oxyanions

ClO^-	Hypochlorite	ClO_2^-	Chlorite	ClO_3^-	Chlorate	ClO_4^-	Perchlorate
BrO^-	Hypobromite	BrO_2^-	Bromite	BrO_3^-	Bromate	BrO_4^-	Perbromate
IO^-	Hypoiodite	IO_2^-	Iodite	IO_3^-	Iodate	IO_4^-	Periodate
		NO_2^-	Nitrite	NO_3^-	Nitrate		
PO_2^{3-}	Hypophosphite	PO_3^{3-}	Phosphite	PO_4^{3-}	Phosphate		
		SO_3^{2-}	Sulfite	SO_4^{2-}	Sulfate		
				CO_3^{2-}	Carbonate		

Note that not all the possible oxyanions of these elements exist. If the name and formula are not given in Table 6-5, the ion is not known. If you learn the seven ions that end in *-ate* plus the meaning of the ending *-ite* and the prefixes, you will be able to write formulas for 20 oxyanions. You may double this number of names by learning an additional rule in Sec. 6.4. Note from Table 6-5 that for each central element, all the ions present have the same charge.

EXAMPLE 6.11. Name the following ions without consulting Table 6-5: (*a*) PO_3^{3-}, (*b*) ClO_4^-, and (*c*) IO^-.

Ans.   (*a*) Remembering that phosphate is PO_4^{3-}, we note that this ion has one fewer oxygen atom. It is the phosphite ion.   (*b*) Remembering that chlorate is ClO_3^-, we note that this ion has one more oxygen atom. This ion is the perchlorate ion.   (*c*) Remembering that iodate is IO_3^-, we note that this ion has two fewer oxygen atoms. It is the hypoiodite ion.

Special Anions

There are a few anions that seem rather unusual but are often used in general chemistry. Perhaps the best way to remember them is to memorize them, but some hints to help do that will be given below. The special anions are given in Table 6-6.

Table 6-6   Special Anions

ate endings		*ide* endings	
$CrO_4{}^{2-}$	Chromate	CN^-	Cyanide
$Cr_2O_7{}^{2-}$	Dichromate	OH^-	Hydroxide
$MnO_4{}^-$	Permanganate	$O_2{}^{2-}$	Peroxide
$C_2H_3O_2{}^-$	Acetate		

Note that chromate and permanganate have central atoms in periodic groups with the same numbers as those of sulfate and perchlorate, respectively, and they have analogous formulas to these ions. Dichromate has two Cr atoms and seven O atoms. The cyanide and hydroxide ions have already been discussed. The acetate ion is really the ion of an organic acid, which is why it looks so unusual. The peroxide ion has two oxygen atoms and a total charge of $2-$. (As usual, *per-* means one more O.)

Putting the Names of the Ions Together

Now that we know how to name the cations and anions, we merely have to put the two names together to get the names of ionic compounds. The cation is named first, and the anion is named next. The *number* of cations and anions per formula unit are not included in the name of the compound because anions have characteristic charges, and the charge on the cation has already been established by its name. Use as many cations and anions as needed to get a neutral compound with the lowest possible integral subscripts.

EXAMPLE 6.12.   Name (*a*) $Ca(NO_3)_2$ and $Cr(ClO_3)_3$.

Ans.   (*a*) The cation is the calcium ion. The anion is the nitrate ion. The compound is calcium nitrate. Note that we do not state anything to indicate the presence of two nitrate anions; we can deduce that from the fact that the calcium ion has a $2+$ charge and nitrate has a $1-$ charge.   (*b*) The cation is chromium(III). We know that it is chromium(III) because its charge must balance three chlorate ion charges, each $1-$. The compound is chromium(III) chlorate.

Writing Formulas for Ionic Compounds

Formulas of many ionic compounds are easily written by consideration of the charges on their ions. To do this, the charges on the ions must be memorized. The charges on most common ions are given here in Sec. 6.3. There are both positive and negative ions. The formula of a compound is written so that the ions are combined in ratios such that there is a net charge of 0 on the compound as a whole. For example, to write the formula for sodium chloride, one sodium ion (having a charge of $1+$) is combined with one chloride ion (having a charge of $1-$); hence, the formula is NaCl. The algebraic total of the charges is 0. To write the formula of aluminum chloride, an aluminum ion (having a charge of $3+$) is combined with three chloride ions; the resulting formula is $AlCl_3$. Again, the total algebraic sum of the charges is 0. The formula for ammonium sulfate includes two ammonium ions (each with a $1+$ charge) and one sulfate ion (with a charge of $2-$), written as $(NH_4)_2SO_4$. As a further example, the formula for aluminum oxide requires the combination of aluminum ions (with a charge $3+$) with oxygen ions (with a charge of $2-$) in such a manner that the net charge is 0. The formula is Al_2O_3, corresponding to $2 \times (3+)$ for aluminum plus $3 \times (2-)$ for oxygen, a total of 0.

EXAMPLE 6.13. Write the formulas for (a) lead(II) chlorite and (b) sodium sulfite.

Ans.   (a) The lead(II) ion is Pb^{2+}; chlorite is ClO_2^-, having one fewer oxygen atom than chlorate and a single minus charge. The formula is therefore $Pb(ClO_2)_2$. It takes two chlorite ions, each with a single negative charge, to balance one lead(II) ion.   (b) The sodium ion is Na^+. (Sodium forms no other positive ion.) Sulfite is SO_3^{2-}, having a $2-$ charge. It takes two sodium ions to balance the charge on one sulfite ion, and so the formula is Na_2SO_3.

EXAMPLE 6.14. Write the formulas for (a) lithium sulfite, (b) lithium iodite, (c) aluminum sulfate, and (d) iron(II) perchlorate.

Ans.   (a) Li_2SO_3 (b) $LiIO_2$ (c) $Al_2(SO_4)_3$ (d) $Fe(ClO_4)_2$

6.4. NAMING INORGANIC ACIDS

The anions described in the preceding sections may be formed by reaction of the corresponding acids with hydroxides:

$$HCl + NaOH \longrightarrow NaCl + H_2O$$

$$H_2SO_4 + 2\,NaOH \longrightarrow Na_2SO_4 + 2\,H_2O$$

$$H_3PO_4 + 3\,NaOH \longrightarrow Na_3PO_4 + 3\,H_2O$$

The salts formed by these reactions consist of cations and anions. The cation in each of these cases is Na^+, and the anions are Cl^-, SO_4^{2-}, and PO_4^{3-}, respectively. In these examples, the chloride ion, sulfate ion, and phosphate ion are formed from their parent acids. Thus the acids and anions are related, and so are their names.

When they are pure, acids are not ionic. When we put them into water solution, seven become fully ionized: HCl, $HClO_3$, $HClO_4$, HBr, HI, HNO_3, and H_2SO_4. They are called *strong acids*. All other acids ionize at least to some extent and are called *weak acids*. Strong acids react completely with water to form ions, and weak acids react to some extent to form ions; but both types react completely with hydroxides to form ions. Formulas for acids conventionally are written with the hydrogen atoms which can ionize placed first. Different names for some acids are given when the compound is pure and when it is dissolved in water. For example, HCl is called hydrogen chloride when it is in the gas phase, but in water it ionizes to give hydrogen ions and chloride ions and is called hydrochloric acid. The names for all the acids corresponding to the anions in Table 6-5 can be deduced by the following simple rule. Note that the number of hydrogen atoms in the acid is the same as the number of negative charges on the anion.

    Replace the -ate *ending of an anion with* -ic acid   or   *replace the* -ite *ending with* -ous acid.

This rule does not change if the anion has a prefix *per-* or *hypo-*; if the anion has such a prefix, so does the acid. If not, the acid does not either.

    If the anion ends in -ide, *add the prefix* hydro- *and change the ending to* -ic acid.

Ion Ending	Acid Name Components
-ate	-ic acid
-ite	-ous acid
-ide	hydro—ic acid

EXAMPLE 6.15. Name the following as acids: (a) HBr, (b) HNO_2, (c) H_3PO_4, (d) $HClO$, and (e) $HClO_4$.

Ans.   (a) HBr is related to Br^-, the bromide ion. The *-ide* ending is changed to *-ic acid* and the prefix *hydro-* is added. The name is hydrobromic acid.

    (b) HNO_2 is related to NO_2^-, the nitrite ion. The *-ite* ending is changed to *-ous acid*. The name is nitrous acid.

    (c) H_3PO_4 is related to PO_4^{3-}, the phosphate ion. The *-ate* ending is changed to *-ic acid*, so the name is phosphoric acid.

(d)   HClO is related to ClO^-, the hypochlorite ion. The prefix *hypo-* is not changed, but the *-ite* ending is changed to *-ous acid*. The name is hypochlorous acid.

(e) $HClO_4$ is related to ClO_4^-, the perchlorate ion. The prefix *per-* is not changed, but the ending is changed to *-ic acid*. The name is perchloric acid.

EXAMPLE 6.16.   Write formulas for the following acids: (a) nitric acid, (b) chloric acid, (c) hypophosphorous acid, and (d) perbromic acid.

Ans.   (a)   Nitric acid is related to the nitrate ion, NO_3^-. The acid has one hydrogen ion, corresponding to the one negative charge on the nitrate ion. The formula is HNO_3.

(b)   The acid is related to the chlorate ion, ClO_3^-. The acid has one hydrogen atom, because the ion has one negative charge. The formula is $HClO_3$.

(c)   The acid is related to hypophosphite, PO_2^{3-}. The acid has three hydrogen atoms, corresponding to the three negative charges on the ion. Its formula is H_3PO_2.

(d) $HBrO_4$.

6.5.   ACID SALTS

In Sec. 6.4 the reactions of hydroxides and acids were presented. It is possible for an acid with more than one ionizable hydrogen atom (with more than one hydrogen written first in the formula) to react with fewer hydroxide ions, and to form a product with some ionizable hydrogen atoms left:

$$H_2SO_4 + NaOH \longrightarrow NaHSO_4 + H_2O \qquad (Na^+ \text{ and } HSO_4^-)$$

$$H_3PO_4 + NaOH \longrightarrow NaH_2PO_4 + H_2O \qquad (Na^+ \text{ and } H_2PO_4^-)$$

$$H_3PO_4 + 2\,NaOH \longrightarrow Na_2HPO_4 + 2\,H_2O \qquad (Na^+ \text{ and } HPO_4^{2-})$$

The products formed are called *acid salts*, and each anion contains at least one ionizable hydrogen atom and at least one negative charge. The sum of the negative charges plus hydrogen atoms equals the original number of hydrogen atoms in the parent acid and also the number of negative charges in the normal anion. For example, HSO_4^- contains one hydrogen atom plus one negative charge, for a total of two. That is the number of hydrogen atoms in H_2SO_4 and also the number of negative charges in SO_4^{2-}.

The anions of acid salt are named with the word *hydrogen* placed before the name of the normal anion. Thus, HSO_4^- is the hydrogen sulfate ion. To denote two atoms, the prefix *di-* is used. HPO_4^{2-} is the hydrogen phosphate ion, while $H_2PO_4^-$ is the dihydrogen phosphate ion. In an older naming system, the prefix *bi-* was used instead of the word *hydrogen* when one of two hydrogen atoms was replaced. Thus, HCO_3^- was called the bicarbonate ion instead of the more modern name, hydrogen carbonate ion.

EXAMPLE 6.17.   Name HSO_4^-.

Ans.   Hydrogen sulfate ion.

EXAMPLE 6.18.   What are the formulas for the dihydrogen phosphate ion and ammonium dihydrogen phosphate?

Ans. $H_2PO_4^-$ and $NH_4H_2PO_4$. Note that the two hydrogen atoms plus the one charge total three, equal to the number of hydrogen atoms in phosphoric acid. The one charge is balanced by the ammonium ion in the complete compound.

6.6.   HYDRATES

Some stable ionic compounds are capable of bonding to a certain number of molecules of water per formula unit. Thus, copper(II) sulfate forms the stable $CuSO_4 \cdot 5H_2O$, with five molecules of water per $CuSO_4$ unit. This type of compound is called a *hydrate*. The name of the compound is the name of the *anhydrous* (without water) compound with a designation for the number of water molecules appended. Thus, $CuSO_4 \cdot 5H_2O$ is called copper(II) sulfate pentahydrate. The 5, written on line, multiplies everything after it until the next centered dot or the end of the formula. Thus, included in $CuSO_4 \cdot 5H_2O$ are ten H atoms and nine O atoms (five from the water and four in the sulfate ion).

Solved Problems

INTRODUCTION

6.1. Explain why the following three compounds, with such similar-looking formulas, have such different names:

KCl	Potassium chloride
CuCl	Copper(I) chloride
BrCl	Bromine monochloride

Ans.   The three compounds belong to different nomenclature classes. Potassium in its compounds always forms 1+ ions, and thus there is no need to state 1+ in the name. Copper forms 1+ and 2+ ions, and we need to designate which of these exists in this compound. BrCl is a binary nonmetal-nonmetal compound, using a prefix to denote the number of chlorine atoms.

BINARY COMPOUNDS OF NONMETALS

6.2. Name the following compounds: (*a*) ClF, (*b*) CO_2, (*c*) BrF_3, (*d*) CCl_4, (*e*) PF_5, and (*f*) SF_6.

Ans.   (*a*) Chlorine monofluoride   (*b*) Carbon dioxide   (*c*) Bromine trifluoride   (*d*) Carbon tetrachloride   (*e*) Phosphorus pentafluoride   (*f*) Sulfur hexafluoride

6.3. Name the following compounds: (*a*) P_4O_{10}, (*b*) Cl_2O, and (*c*) ClO_2.

Ans.   (*a*) Tetraphosphorus decoxide   (*b*) Dichlorine monoxide   (*c*) Chlorine dioxide

6.4. Name the following: (*a*) a compound with two oxygen atoms and one chlorine atom per molecule; (*b*) a compound with one iodine atom and five fluorine atoms per molecule; and (*c*) a compound with one sulfur atom and six fluorine atoms per molecule.

Ans.   (*a*) Chlorine dioxide   (*b*) Iodine pentafluoride   (*c*) Sulfur hexafluoride

6.5. Write formulas for each of the following compounds: (*a*) carbon tetrafluoride, (*b*) xenon hexafluoride, and (*c*) sulfur difluoride.

Ans.   (*a*) CF_4 (*b*) XeF_6 (*c*) SF_2

6.6. Name (*a*) H_2O and (*b*) NH_3.

Ans.   (*a*) Water   (*b*) Ammonia

6.7. Write formulas for each of the following compounds: (*a*) dinitrogen trioxide, (*b*) carbon disulfide, and (*c*) carbon tetrabromide.

Ans.   (*a*) N_2O_3 (*b*) CS_2 (*c*) CBr_4

6.8. Write formulas for each of the following compounds: (*a*) dichlorine monoxide, (*b*) dichlorine trioxide, and (*c*) chlorine dioxide.

Ans.   (*a*) Cl_2O (*b*) Cl_2O_3 (*c*) ClO_2

NAMING IONIC COMPOUNDS

6.9.   Name NH_3 and NH_4^+.

   Ans.   Ammonia and ammonium ion. Note carefully the differences between these names and these formulas.

6.10.   How can a beginning student recognize an ionic compound?

   Ans.   If the compound contains the NH_4^+ ion or a metal it is most likely ionic. (It might have *internal* covalent bonds.) If it is a strong acid in water solution, it is ionic. Otherwise, it is covalent.

6.11.   Name (*a*) CuO and (*b*) Cu_2O.

   Ans.   (*a*) Copper(II) oxide   (*b*) copper(I) oxide. This example again emphasizes the difference between the Arabic numerals in a formula and the Roman numerals in a name.

6.12.   In naming NO_3^-, professional chemists might say *nitrate* or *nitrate ion*, but in naming Na^+, they always say *sodium ion*. Explain the difference.

   Ans.   *Nitrate* is always an ion; *sodium* might refer to the element, the atom, or the ion, and so a distinction must be made.

6.13.   What is the difference in the rules for remembering charges on monatomic anions and oxyanions?

   Ans.   Monatomic anions have charges equal to the group number minus 8. The oxyanions have even charges for even-group central atoms and odd charges for odd-group central atoms, but the rule does not give a simple way to determine the charge definitely for oxyanions as it does for monatomic anions.

6.14.   Name the following ions: (*a*) Co^{3+}, (*b*) Fe^{2+}, (*c*) Pt^{2+}, and (*d*) Hg_2^{2+}.

   Ans.   (*a*) Cobalt(III) ion   (*b*) Iron(II) ion   (*c*) Platinum(II) ion   (*d*) Mercury(I) ion

6.15.   Which metals in each of the following periodic groups form ions of only one charge: (*a*) IA, (*b*) IIA, (*c*) IIIA, (*d*) IB, and (*e*) IIB?

   Ans.   (*a*) All the alkali metals (but not hydrogen)   (*b*) All the alkaline earth metals   (*c*) Aluminum ion   (*d*) Silver ion   (*e*) Zinc and cadmium ions

6.16.   What is the charge on each of the following ions? (*a*) cyanide, (*b*) hydroxide, (*c*) peroxide, (*d*) nitride, and (*e*) chloride.

   Ans.   (*a*) 1−   (*b*) 1−   (*c*) 2−   (*d*) 3−   (*e*) 1−

6.17.   Using the periodic table if necessary, write formulas for the following compounds: (*a*) hydrogen sulfide, (*b*) barium chloride, (*c*) ammonium phosphate, (*d*) aluminum sulfate, (*e*) calcium bromide, (*f*) lithium sulfide, and (*g*) sodium fluoride.

   Ans.   (*a*) H_2S (*b*) $BaCl_2$ (*c*) $(NH_4)_3PO_4$ (*d*) $Al_2(SO_4)_3$ (*e*) $CaBr_2$ (*f*) Li_2S (*g*) NaF

6.18.   Write formulas for copper(I) sulfide and (*b*) copper(II) sulfide.

   Ans.   (*a*) Cu_2S (*b*) CuS

6.19.   Name the following compounds: (*a*) $AgClO_3$ and (*b*) $Al(ClO)_3$.

   Ans.   (*a*) Silver chlorate and (*b*) aluminum hypochlorite. Note that parentheses enclose the ClO^- ions, because there is a subscript to show that there are three of them. In (*a*), there is only one anion that contains three oxygen atoms.

6.20. Name the following ions: (*a*) Pb^{2+}, (*b*) IO_3^-, (*c*) IO_4^-, and (*d*) N^{3-}.

Ans.   (*a*) Lead(II) ion   (*b*) Iodate ion   (*c*) Periodate ion   (*d*) Nitride ion

6.21. Name the following ions: (*a*) PO_3^{3-}, (*b*) S^{2-}, and (*c*) SO_4^{2-}.

Ans.   (*a*) Phosphite ion   (*b*) Sulfide ion   (*c*) Sulfate ion

6.22. Write formulas for each of the following compounds: (*a*) sodium chloride, (*b*) sodium chlorate, (*c*) sodium chlorite, (*d*) sodium hypochlorite, and (*e*) sodium perchlorate.

Ans.   (*a*) NaCl   (*b*) $NaClO_3$ (*c*) $NaClO_2$ (*d*) NaClO   (*e*) $NaClO_4$

6.23. What is the difference between SO_3 and SO_3^{2-}?

Ans.   The first is a compound, sulfur trioxide; the second is an ion, the sulfite ion. The sulfite ion has two extra electrons, as shown by the 2− charge.

6.24. (*a*) Write the formula for calcium sulfide, calcium sulfite, and calcium sulfate. (*b*) How many elements are implied in the compound by the *-ide* ending? by the *-ite* ending? by the *-ate* ending? (*c*) Name a particular element that is implied by the *-ate* ending.

Ans.   (*a*) CaS, $CaSO_3$, and $CaSO_4$.

   (*b*)   *-ide*: at least two (a monatomic anion); *-ite* and *-ate*: at least three.

   (*c*)   There is oxygen (plus another element) in the anion.

6.25. Name the following compounds: (*a*) $CrCl_2$, (*b*) $CrCl_3$, (*c*) CrO, (*d*) Cr_2O_3, and (*e*) $CrPO_4$.

Ans.   (*a*) Chromium(II) chloride   (*b*) Chromium(III) chloride   (*c*) Chromium(II) oxide   (*d*) Chromium(III) oxide   (*e*) Chromium(III) phosphate

6.26. Write formulas for the following compounds: (*a*) sodium chloride, (*b*) sodium sulfate, and (*c*) sodium phosphate.

Ans.   (*a*) NaCl   (*b*) Na_2SO_4 (*c*) Na_3PO_4
The only indication that there are one, two, and three sodium ions in the compound lies in the knowledge of the charges on the ions.

6.27. Write formulas for the following compounds: (*a*) ammonium chloride, (*b*) ammonium sulfide, and (*c*) ammonium phosphate.

Ans.   (*a*) NH_4Cl (*b*) $(NH_4)_2S$ (*c*) $(NH_4)_3PO_4$
Note that the last two formulas require parentheses around the ammonium ion, but that the first one does not since there is only one ammonium ion per formula unit.

6.28. Name the following compounds: (*a*) $Ca_3(PO_4)_2$ and (*b*) $Cr_2(SO_4)_3$.

Ans.   (*a*) Calcium phosphate   (*b*) Chromium(III) sulfate. We recognize that chromium has a 3+ charge in this compound because two chromium ions are needed to balance three sulfate ions, each of which has a 2− charge.

6.29. Fill in the table with the formula of the compound whose cation is named at the left and whose anion is named at the column head.

	Chloride	**Sulfate**	**Phosphate**	**Hydroxide**
Ammonium				
Lithium				
Lead(II)				
Vanadium(III)				

Ans.

	Chloride	Sulfate	Phosphate	Hydroxide
Ammonium	NH_4Cl	$(NH_4)_2SO_4$	$(NH_4)_3PO_4$	NH_4OH (unstable)
Lithium	$LiCl$	Li_2SO_4	Li_3PO_4	$LiOH$
Lead(II)	$PbCl_2$	$PbSO_4$	$Pb_3(PO_4)_2$	$Pb(OH)_2$
Vanadium(III)	VCl_3	$V_2(SO_4)_3$	VPO_4	$V(OH)_3$

6.30. Name the following ions: (*a*) ClO^-, (*b*) NO_2^-, and (*c*) SO_4^{2-}.

   Ans.   (*a*) Hypochlorite ion   (*b*) Nitrite ion   (*c*) Sulfate ion

6.31. Name the following ions: (*a*) MnO_4^-, (*b*) O_2^{2-}, (*c*) CrO_4^{2-}, (*d*) $Cr_2O_7^{2-}$, and (*e*) $C_2H_3O_2^-$.

   Ans.   (*a*) Permanganate ion   (*b*) Peroxide ion   (*c*) Chromate ion   (*d*) Dichromate ion   (*e*) Acetate ion

6.32. Name the following ions: (*a*) H^+, (*b*) OH^-, (*c*) H^-, (*d*) NH_4^+, and (*e*) CN^-.

   Ans.   (*a*) Hydrogen ion   (*b*) Hydroxide ion   (*c*) Hydride ion   (*d*) Ammonium ion   (*e*) Cyanide ion

6.33. Write formulas for each of the following compounds: (*a*) calcium cyanide, (*b*) calcium hydroxide, (*c*) calcium peroxide, and (*d*) calcium hydride.

   Ans.   (*a*) $Ca(CN)_2$ (*b*) $Ca(OH)_2$ (*c*) CaO_2 (from Table 6-6)   (*d*) CaH_2

6.34. What is the difference between the names *phosphorus* and *phosphorous*?

   Ans.   The first is the name of the element; the second is the name of the acid with fewer oxygen atoms than phosphoric acid—H_3PO_3, phosphorous acid.

6.35. Write formulas for each of the following compounds: (*a*) barium bromide, (*b*) copper(II) bromate, (*c*) titanium(III) fluoride, and (*d*) aluminum hydride.

   Ans.   (*a*) $BaBr_2$ (*b*) $Cu(BrO_3)_2$ (*c*) TiF_3 (*d*) AlH_3

6.36. Name and write formulas for each of the following: (*a*) the compound of potassium and bromine; (*b*) the compound of calcium and bromine; and (*c*) the compound of aluminum and bromine.

   Ans.   (*a*) Potassium bromide, KBr (*b*) Calcium bromide, $CaBr_2$ (*c*) Aluminum bromide, $AlBr_3$
            The bromide ion has a 1– charge. The potassium, calcium, and aluminum ions have charges of 1+, 2+, and 3+, respectively, and the subscripts given are the smallest possible to just balance the charges.

6.37. Name and write formulas for each of the following: (*a*) the compound of lithium and sulfur; (*b*) the compound of barium and sulfur; and (*c*) the compound of aluminum and sulfur.

   Ans.   (*a*) Lithium sulfide, Li_2S (*b*) Barium sulfide, BaS (*c*) Aluminum sulfide, Al_2S_3
            The sulfide ion has a 2– charge. The lithium, barium, and aluminum ions have charges of 1+, 2+, and 3+, respectively, and the subscripts given are the smallest possible to just balance the charges.

6.38. Name and write formulas for each of the following: (*a*) two compounds of copper and chlorine, (*b*) two compounds of platinum and fluorine, and (*c*) two compounds of cobalt and bromine. (If necessary, see Table 6-4 for data.)

   Ans.   (*a*) Copper(I) chloride and copper(II) chloride, $CuCl$ and $CuCl_2$ (*b*) Platinum(II) fluoride and platinum(IV) fluoride, PtF_2 and PtF_4 (*c*) Cobalt(II) bromide and cobalt(III) bromide, $CoBr_2$ and $CoBr_3$

6.39. Name and write formulas for each of the following: (*a*) two compounds of iron and bromine, (*b*) two compounds of palladium and bromine, and (*c*) two compounds of mercury and bromine.

Ans.   (*a*) Iron(II) bromide and iron(III) bromide, $FeBr_2$ and $FeBr_3$ (*b*) Palladium(II) bromide and palladium(IV) bromide, $PdBr_2$ and $PdBr_4$ (*c*) Mercury(I) bromide and mercury(II) bromide, Hg_2Br_2 and $HgBr_2$. (The charges on the cations can be obtained from Table 6-4.)

6.40. Name and write formulas for each of the following: (*a*) the compound of bromine and potassium (*b*) the compound of bromine and calcium, and (*c*) the compound of bromine and aluminum.

Ans.   This answer is exactly the same as that of Problem 6.36. The metal is named first even if it is given last in the statement of the problem.

6.41. Write the formula for each of the following compounds: (*a*) aluminum hydride, (*b*) calcium chloride, (*c*) lithium oxide, (*d*) silver nitrate, (*e*) iron(II) sulfite, (*f*) aluminum chloride, (*g*) ammonium carbonate, (*h*) zinc sulfate, (*i*) iron(III) oxide, (*j*) sodium phosphate, (*k*) iron(III) acetate, (*l*) ammonium chloride, and (*m*) copper(I) cyanide.

Ans.   (*a*) AlH_3 (*b*) $CaCl_2$ (*c*) Li_2O (*d*) $AgNO_3$ (*e*) $FeSO_3$ (*f*) $AlCl_3$ (*g*) $(NH_4)_2CO_3$ (*h*) $ZnSO_4$
(*i*) Fe_2O_3 (*j*) Na_3PO_4 (*k*) $Fe(C_2H_3O_2)_3$ (*l*) NH_4Cl (*m*) $CuCN$

NAMING INORGANIC ACIDS

6.42. Name the following acids: (*a*) HCl, (*b*) $HClO$, (*c*) $HClO_2$, (*d*) $HClO_3$, and (*e*) $HClO_4$.

Ans.   (*a*) Hydrochloric acid   (*b*) Hypochlorous acid   (*c*) Chlorous acid   (*d*) Chloric acid   (*e*) Perchloric acid

6.43. Which acids of the prior problem are strong acids?

Ans.   (*a*), (*d*), and (*e*)

6.44. Name the following compounds: (*a*) HCl (pure) and (*b*) HCl (in water solution).

Ans.   (*a*) Hydrogen chloride   (*b*) Hydrochloric acid

6.45. Write formulas for each of the following acids: (*a*) hydrobromic acid, (*b*) phosphorous acid, and (*c*) hypobromous acid.

Ans.   (*a*) HBr (*b*) H_3PO_3 (*c*) $HBrO$

6.46. Name H_2S in two ways.

Ans.   Hydrogen sulfide and hydrosulfuric acid.

6.47. How many ionizable hydrogen atoms are there in $HC_4H_7O_2$?

Ans.   One. Ionizable hydrogen atoms are written first. (The other seven hydrogen atoms will not react with metal hydroxides.)

ACID SALTS

6.48. Write formulas for each of the following compounds: (*a*) ammonium hydrogen sulfate, (*b*) sodium hydrogen sulfide, and (*c*) nickel(II) hydrogen carbonate.

Ans.   (*a*) NH_4HSO_4 (*b*) $NaHS$ (*c*) $Ni(HCO_3)_2$

6.49. Write formulas for each of the following compounds: (*a*) sodium hydrogen phosphate, (*b*) sodium dihydrogen phosphate, and (*c*) magnesium hydrogen phosphate.

Ans.   (*a*) Na_2HPO_4 (*b*) NaH_2PO_4 (*c*) $MgHPO_4$
(The charges must balance in each case.)

6.50. Write formulas for each of the following compounds: (*a*) calcium hydrogen carbonate, (*b*) disodium hydrogen phosphate, (*c*) sodium dihydrogen phosphate, and (*d*) calcium dihydrogen phosphate.

   Ans. (*a*) $Ca(HCO_3)_2$ (the single charge on one HCO_3^- ion is not sufficient to balance the 2+ charge on Ca^{2+}; two anions are necessary)   (*b*) Na_2HPO_4 (*c*) NaH_2PO_4 (*d*) $Ca(H_2PO_4)_2$
   [The same prefix is used to denote two atoms in (*b*), (*c*), and (*d*) as in nonmetal-nonmetal compounds.]

HYDRATES

6.51. Name the following compound, and state how many hydrogen atoms it contains per formula unit:

$$Ni(C_2H_3O_2)_2 \cdot 4H_2O$$

   Ans. Nickel(II) acetate tetrahydrate. It contains 14 H atoms per formula unit.

6.52. Write formulas for barium dihydrogen hypophosphite and barium dihydrogen hypophosphite monohydrate.

   Ans. $Ba(H_2PO_2)_2$ and $Ba(H_2PO_2)_2 \cdot H_2O$

Supplementary Problems

6.53. Name (*a*) H^+, (*b*) H^-, (*c*) NaH, and (*d*) BeH_2.

   Ans. (*a*) Hydrogen ion   (*b*) Hydride ion   (*c*) Sodium hydride   (*d*) Beryllium hydride

6.54. Explain why the formula for mercury(I) ion is Hg_2^{2+} rather than Hg^+.

   Ans. The actual formula shows that the two mercury atoms are covalently bonded.

6.55. Explain why mercury(I) ion, Hg_2^{2+}, has the Roman numeral I in it.

   Ans. The average charge on the two Hg atoms is 1+. (See Sec. 14.2.)

6.56. Write formulas for each of the following compounds: (*a*) ammonium nitrate, (*b*) mercury(I) cyanide, and (*c*) uranyl carbonate.

   Ans. (*a*) NH_4NO_3 (from Table 6-3)   (*b*) $Hg_2(CN)_2$ (from Tables 6-3 and 6-6)   (*c*) UO_2CO_3 (from Table 6-3)

6.57. Write a formula for each of the following compounds: (*a*) diphosphorus pentasulfide, (*b*) iodine heptafluoride, and (*c*) dinitrogen monoxide.

   Ans. (*a*) P_2S_5 (*b*) IF_7 (one of the few times the prefix *hepta-* is used)   (*c*) N_2O

6.58. State the meaning of each of the following terms: (*a*) *per-*, (*b*) *hypo-*, (*c*) *hydro-*, (*d*) hydrogen (as part of an ion).

   Ans. (*a*) *Per-* means "more oxygen atoms." For example, the perchlorate ion (ClO_4^-) has more oxygen atoms than does the chlorate ion (ClO_3^-).   (*b*) *Hypo-* means "fewer oxygen atoms." For example, the hypochlorite ion (ClO^-) has fewer oxygen atoms than does the chlorite ion (ClO_2^-).   (*c*) *Hydro-* means "no oxygen atoms." For example, hydrochloric acid (HCl) has no oxygen atoms, in contrast to chloric acid ($HClO_3$).   (*d*) Hydrogen signifies an acid salt, such as NaHS—sodium hydrogen sulfide.

6.59. What relationship is there in the meaning of the prefix *per-* when used with an oxyanion and when used in peroxide?

   Ans. In both cases, it means "one more oxygen atom."

6.60. What are the differences among the following questions? (*a*) What is the formula of the compound of sulfur and calcium? (*b*) What is the formula of the compound of Ca^{2+} and S^{2-}? (*c*) What is the formula for calcium sulfide?

Ans. In each case, the answer is CaS. Part (*b*) gives the ions and their changes and so is perhaps easiest to answer. Part (*a*) gives the elements, so it is necessary to know that periodic group IIA elements always form 2+ ions in all their compounds and that sulfur forms a 2– ion in its compounds with metals. It is also necessary to remember that the metal is named first. In part (*c*), the fact that there is only one compound of these two elements is reinforced by the fact that the calcium is stated with no Roman numeral, and that sulfide is a specific ion with a specific (2–) charge.

6.61. Name (*a*) Cu_2O and (*b*) CuO.

Ans. (*a*) Copper(I) oxide   (*b*) Copper(II) oxide.

6.62. Distinguish between ClO_2 and ClO_2^-.

Ans. ClO_2 is a compound, and ClO_2^- is an ion—part of a compound.

6.63. Use the periodic table relationships to write formulas for (*a*) the selenate ion and (*b*) the arsenate ion.

Ans. Selenium is below sulfur in the periodic table, and arsenic is below phosphorus. We write formulas analogous to those for sulfate and phosphate: (*a*) SeO_4^{2-} and (*b*) AsO_4^{3-}.

6.64. Complete the following table by writing the formula of the compound formed by the cation at the left and the anion at the top. NH_4Cl is given as an example.

	Cl^-	ClO^-	ClO_2^-	SO_3^{2-}	PO_4^{3-}
NH_4^+	NH_4Cl				
K^+					
Ba^{2+}					
Al^{3+}					

Ans.

	Cl^-	ClO^-	ClO_2^-	SO_3^{2-}	PO_4^{3-}
NH_4^+	NH_4Cl	NH_4ClO	NH_4ClO_2	$(NH_4)_2SO_3$	$(NH_4)_3PO_4$
K^+	KCl	$KClO$	$KClO_2$	K_2SO_3	K_3PO_4
Ba^{2+}	$BaCl_2$	$Ba(ClO)_2$	$Ba(ClO_2)_2$	$BaSO_3$	$Ba_3(PO_4)_2$
Al^{3+}	$AlCl_3$	$Al(ClO)_3$	$Al(ClO_2)_3$	$Al_2(SO_3)_3$	$AlPO_4$

6.65. From the data of Table 6-4, determine which charge is most common for first transition series elements (those elements in period 4 among the transition groups).

Ans. The 2+ charge is seen to be most common, occurring with every element in the first transition series that is listed. (Only Sc has no 2+ ion.)

6.66. Name the compounds of Problem 6.64.

Ans.

	Cl^-	ClO^-	ClO_2^-	SO_3^{2-}	PO_4^{3-}
NH_4^+	Ammonium chloride	Ammonium hypochlorite	Ammonium chlorite	Ammonium sulfite	Ammonium phosphate
K^+	Potassium chloride	Potassium hypochlorite	Potassium chlorite	Potassium sulfite	Potassium phosphate
Ba^{2+}	Barium chloride	Barium hypochlorite	Barium chlorite	Barium sulfite	Barium phosphate
Al^{3+}	Aluminum chloride	Aluminum hypochlorite	Aluminum chlorite	Aluminum sulfite	Aluminum phosphate

CHAPTER 7

Formula Calculations

7.1. INTRODUCTION

Atoms and their symbols were introduced in Chaps. 1 and 3. In Sec 5.2, the representation of compounds by their formulas was developed. The procedure for writing formulas from knowledge of the elements involved or from names was presented in Chaps. 5 and 6. The formula for a compound contains much information of use to the chemist, including the number of atoms of each element in a formula unit of a compound. Since atoms are so tiny, we will learn to use large groups of atoms—moles of atoms—to ease our calculations (Sec. 7.4). We will learn to calculate the percent by mass of each element in the compound (Sec. 7.5). We will learn how to calculate the simplest formula from percent composition data (Sec. 7.6) and molecular formulas from simplest formulas and molecular masses (Sec. 7.7).

7.2. MOLECULES AND FORMULA UNITS

Some elements combine by covalent bonding (Chap. 5) into units called *molecules*. Other elements combine by gaining or losing electrons to form *ions*, which are charged atoms (or groups of atoms). The ions are attracted to one another, a type of bonding called *ionic bonding* (Chap. 5), and can form combinations containing millions or more of ions of each element. To identify these ionic compounds, the simplest formula is generally used. We can tell that a compound is ionic if it contains at least one metal atom or the NH_4 group. If not, it is most likely bonded in molecules (Chap. 5).

Formulas for ionic compounds represent one *formula unit*. However, molecules, too, have formulas and thus formula units. Even uncombined atoms of an element have formulas. Thus, formula units may refer to uncombined atoms, molecules, or atoms combined in ionic compounds; one formula unit may be

One atom of uncombined element, for example, Pt

One molecule of a covalently bonded compound, for example, H_2O

One simple unit of an ionic compound, for example, NaCl or $(NH_4)_2SO_4$

EXAMPLE 7.1. What are the formula units of the element copper, the compound dinitrogen trioxide, and the compound sodium bromide?

Ans. The formula unit of copper is Cu, one atom of copper. The formula unit of dinitrogen trioxide is N_2O_3, one molecule of dinitrogen trioxide. The formula unit of sodium bromide is NaBr.

The word *molecule* or the general term *formula unit* may be applied to one unit of H_2O. The word *atom* or the term *formula unit* may be applied to one unit of uncombined Pt. However, there is no special name for one unit of NaCl. *Formula unit* is the best designation. (Some instructors and some texts refer to "molecules" of NaCl, and especially to "molecular mass" of NaCl, because the calculations done on formula units do not depend on the type of bonding involved. However, strictly speaking, the terms *molecule* and *molecular mass* should be reserved for substances bonded into molecules.)

7.3. FORMULA MASSES

The *formula mass* of a compound is the sum of the atomic masses (Chap. 3) of all the atoms (not merely each kind of atom) in the formula. Thus, in the same way that a symbol is used to represent an element, a formula is used to represent a compound or a molecule of an element, such as H_2, and also one unit of either. The formula mass of the substance or the mass of 1 mol of the substance is easily determined on the basis of the formula (Sec. 7.4). Note that just as *formula unit* may refer to uncombined atoms, molecules, or atoms combined in an ionic compound, the term *formula mass* may refer to the atomic mass of an atom, the molecular mass of a molecule, or the formula mass of a formula unit of an ionic compound.

Use at least three significant digits in formula mass calculations.

EXAMPLE 7.2. What is the formula mass of $CaSO_4$?

Ans.

$$\text{Atomic mass of calcium} = 40.08 \text{ amu}$$
$$\text{Atomic mass of sulfur} = 32.06 \text{ amu}$$
$$4 \times \text{Atomic mass of oxygen} = 4 \times 16.00 \text{ amu} = 64.00 \text{ amu}$$
$$\text{Formula mass} = \text{total} = 136.14 \text{ amu}$$

7.4. THE MOLE

Atoms and molecules are incredibly small. For hundreds of years after Dalton postulated their existence, no one was able to work with just one atom or molecule. (In recent times, with special apparatus, it has been possible to see the effects of individual atoms and molecules, but this subject will be developed later.) Just as the *dozen* is used as a convenient number of items in everyday life, the *mole* may be best thought of as as number of items. The mole is 6.02×10^{23} items, a number called *Avogadro's number*. This is a very large number: six hundred two thousand billion billion! The entire earth has a mass of 6×10^{24} kg. Thus, the earth has only 10 times as many kilograms as 1 mol of carbon has atoms. One can have a mole of any item, but it makes little sense to speak of moles of anything but the tiniest of particles, such as atoms and molecules. It might seem unusual to give a name to a number, but remember we do the same thing in everyday life; *dozen* is the name for 12 items. Just as a grocer finds selling eggs by the dozen more convenient than selling them individually, the chemist finds calculations more convenient with moles. The number of formula units (i.e., the number of uncombined atoms, of molecules of molecular elements or compounds, or of formula units of ionic compounds) can be converted to moles of the same substance, and vice versa, using Avogadro's number (Fig. 7-1).

Fig. 7-1. Avogadro's number conversions

Mole is abbreviated mol. Do not use *m* or *M* for mole; these symbols are used for other quantities related to moles, and so you will be confused if you use either of them. *Note*: A mole is referred to by some authors as a "gram molecular mass" because 1 mol of molecules has a mass in grams equal to its molecular mass. In this terminology, a "gram atomic mass" is 1 mol of atoms, and a "gram formula mass" is 1 mol of formula units.

The formula mass of a substance is equal to its number of grams per mole. Avogadro's number is the number of atomic mass units in 1 g. It is defined in that manner so that the atomic mass of an element (in amu) is numerically equal to the number of grams of the element per mole. Consider sodium, with atomic mass 23.0:

$$\frac{23.0 \text{ amu}}{\text{Na atom}} = \frac{23.0 \text{ amu}}{\text{Na atom}} \left(\frac{6.02 \times 10^{23} \text{ Na atoms}}{1 \text{ mol Na}} \right) \left(\frac{1 \text{ g}}{6.02 \times 10^{23} \text{ amu}} \right) = \frac{23.0 \text{ g}}{\text{mol Na}}$$

Avogadro's number appears in both the numerator and the denominator of this expression; the values reduce to a factor of 1 (they cancel), and the numeric value in grams per mole is equal to the numeric value of the atomic mass in amu per atom.

$$\frac{23.0 \text{ amu}}{\text{Na atom}} = \frac{23.0 \text{ g}}{\text{mol Na atoms}}$$

A similar argument leads to the conclusion that the formula mass of any element or compound is equal to the number of grams per mole of the element or compound.

EXAMPLE 7.3.   How many feet tall is a stack of a dozen shoe boxes, each 4 in. tall?

Ans.
$$\frac{4 \text{ in.}}{\text{box}} = \frac{4 \text{ in.}}{\text{box}} \left(\frac{12 \text{ box}}{1 \text{ dozen}} \right) \left(\frac{1 \text{ ft}}{12 \text{ in.}} \right) = \frac{4 \text{ ft}}{\text{dozen}}$$

The same number, but in different units, is obtained because the number of inches in 1 foot is the same as the number in 1 dozen. Compare this process to the one above for grams per mole.

Changing grams to moles and moles to grams is perhaps the most important calculation you will have to make all year (Fig. 7-2). We use the term *molar mass* for the mass of 1 mol of any substance. The units are typically grams per mole.

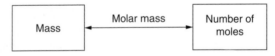

Fig. 7-2.   Molar mass conversions
The mass of a substance can be converted to moles, and vice versa, with the molar mass.

EXAMPLE 7.4.   Calculate the mass of 1.000 mol of SCl_2.

Ans.   The formula mass of SCl_2 is given by

$$S = \quad 32.06 \text{ amu}$$
$$2 \text{ Cl} = 2 \times 35.45 \text{ amu} = \quad 70.90 \text{ amu}$$
$$SCl_2 = \text{total} = 102.96 \text{ amu}$$

Thus, the mass of 1.000 mol of SCl_2 is 102.96 g.

EXAMPLE 7.5.   Calculate the mass of 2.50 mol $NaClO_4$.

Ans.   The formula mass of $NaClO_4$ is given by

Na	23.0 amu
Cl	35.5 amu
$4 \times O$	64.0 amu
$NaClO_4$	122.5 amu

$NaClO_4$ has a formula mass of 122.5 amu.

$$2.50 \text{ mol NaClO}_4 \left(\frac{122.5 \text{ g NaClO}_4}{1 \text{ mol NaClO}_4} \right) = 306 \text{ g NaClO}_4$$

EXAMPLE 7.6.   Calculate the mass in grams of 1 uranium atom.

Ans.   First we can calculate the number of moles of uranium, using Avogadro's number:

$$1 \text{ U atom}\left(\frac{1 \text{ mol U}}{6.02 \times 10^{23} \text{ U atoms}}\right) = 1.66 \times 10^{-24} \text{ mol U}$$

Then we calculate the mass from the number of moles and the molar mass:

$$1.66 \times 10^{-24} \text{ mol U}\left(\frac{238 \text{ g}}{1 \text{ mol U}}\right) = 3.95 \times 10^{-22} \text{ g}$$

Alternately, we can combine these expressions into one:

$$1 \text{ U atom}\left(\frac{1 \text{ mol U}}{6.02 \times 10^{23} \text{ U atoms}}\right)\left(\frac{238 \text{ g}}{1 \text{ mol U}}\right) = 3.95 \times 10^{-22} \text{ g}$$

Still another solution method:

$$1 \text{ U atom}\left(\frac{238 \text{ amu}}{1 \text{ U atom}}\right)\left(\frac{1.00 \text{ g}}{6.02 \times 10^{23} \text{ amu}}\right) = 3.95 \times 10^{-22} \text{ g}$$

How can we count such a large number of items as Avogadro's number? One way, which we also can use in everyday life, is to weigh a small number and the entire quantity. We can count the small number, and the ratio of the number of the small portion to the number of the entire quantity is equal to the ratio of their masses.

EXAMPLE 7.7.   A TV show requires contestants to guess the number of grains of rice in a gallon container. The closest contestant after 4 weeks will win a big prize. How could you prepare for such a contest, without actually counting the grains in 1 gal of rice?

Ans.   One way to get a good estimate is to count 100 grains of rice and weigh that sample. Weigh 1 gal of rice. Then the ratio of number of grains of rice in the small sample (100) to number of grains of rice in the large sample (which is the unknown) is equal to the ratio of masses. Suppose 100 grains of rice weighed 0.012 lb and 1 gal of rice weighed 8.80 lb. The unknown number of grains in 1 gal x can be calculated using a proportion:

$$\frac{x \text{ grains in 1 gallon}}{100 \text{ grains}} = \frac{8.80 \text{ lb}}{0.012 \text{ lb}}$$
$$x = 73\,000$$

Why not weigh just one grain? The calculation would be simpler, but weighing just one grain might be impossible with the balances available, and the grain that we choose might not have the average mass.

The situation is similar in counting atoms, but much more difficult. Individual atoms cannot be seen to be counted, nor can they be weighed in the ordinary manner. Still, if the mass of 1 atom can be determined (in amu, for example) the number of atoms in a mole can be calculated. Historically, what chemists have done in effect is to weigh very large numbers of atoms of different elements where the ratio of atoms of the elements is known; they have gotten the ratio of the masses of individual atoms from the ratio of the mass of the different elements and the relative numbers of atoms of the elements.

EXAMPLE 7.8.   If equal numbers of carbon and oxygen atoms in a certain sample have a ratio of masses 12.0 g to 16.0 g, what is the ratio of atomic masses?

Ans.   The individual atoms have the same ratio, $12.0/16.0 = 0.750$. (The atomic masses of carbon and oxygen are 12.0 amu and 16.0 amu, respectively.)

The number of moles of each element in a mole of compound is stated with subscripts in the chemical formula. Hence, the formula can be used to convert the number of moles of the compound to the number of moles of its component elements, and vice versa (Fig. 7-3).

EXAMPLE 7.9.   How many moles of hydrogen atoms are present in 1.25 mol of CH_4?

Ans.   The formula states that there is 4 mol H for every 1 mol CH_4. Therefore,

$$1.25 \text{ mol CH}_4\left(\frac{4 \text{ mol H}}{1 \text{ mol CH}_4}\right) = 5.00 \text{ mol H}$$

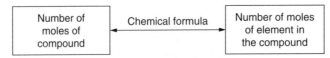

Fig. 7-3. Chemical formulas mole ratio

The number of moles of each element in a compound and the number of moles of the compound as a whole are related by the subscript of that element in the chemical formula.

7.5. PERCENT COMPOSITION OF COMPOUNDS

The term *percent* means the quantity or number of units out of 100 units total. Percentage is computed by finding the fraction of the total quantity represented by the quantity under discussion and multiplying by 100%. For example, if a group of 30 persons includes 6 females, the percent of females in the group is

$$\frac{6}{30} \times 100\% = 20\%$$

In other words, if there were 100 persons in the group and the ratio of females to males were the same, 20 of the group would be female. Percentage is a familiar concept to anyone who pays sales taxes.

The concept of percentage is often used to describe the composition of compounds. If the formula of a compound is known, the percent by mass of an element in the compound is determined by computing the fraction of the formula mass which is made up of that element, and multiplying that fraction by 100%. Thus, an element X with atomic mass 12.0 amu in a compound XY of formula mass 28.0 amu will be present in

$$\frac{12.0 \text{ amu}}{28.0 \text{ amu}} \times 100\% = 42.9\%$$

The chemical formula gives the number of moles of atoms of each element in each mole of the compound; then the number of grams of each element in the number of grams of compound in 1 mol of the compound can be directly computed. Knowing the mass of 1 mol of the compound and the mass of each element in that quantity of compound allows calculations of the percent by mass of each element.

EXAMPLE 7.10. Calculate the percent composition of $CaSO_4$, that is, the percent by mass of each element in the compound.

Ans. One mole of the compound contains 1 mol of calcium, 1 mol of sulfur, and 4 mol of oxygen atoms. The formula mass is 136.14 amu (Example 7.2); hence, there is 136.14 g/mol $CaSO_4$.

One mole of calcium has a mass of 40.08 g. The percent calcium is therefore

$$\% \text{ Ca} = \frac{40.08 \text{ g Ca}}{136.14 \text{ g CaSO}_4} \times 100\% = 29.44\% \text{ Ca}$$

One mole of sulfur has a mass of 32.06 g. The percent sulfur is given by

$$\% \text{ S} = \frac{32.06 \text{ g S}}{136.14 \text{ g CaSO}_4} \times 100\% = 23.55\% \text{ S}$$

Four moles of oxygen has a mass of 4×16.00 g. The percent oxygen is given by

$$\% \text{ O} = \frac{4 \times 16.00 \text{ g O}}{136.14 \text{ g CaSO}_4} \times 100\% = 47.01\% \text{ O}$$

The total of all the percentages in the compound is 100.00%. Within the accuracy of the calculation, the total of all the percentages must be 100%. This result may be interpreted to mean that if there were 100.00 g of $CaSO_4$, then 29.44 g would be calcium, 23.55 g would be sulfur, and 47.01 g would be oxygen.

In laboratory work, the identity of a compound may be established by determining its percent composition experimentally and then comparing the results with the percent composition calculated from its formula.

EXAMPLE 7.11. A compound was analyzed in the laboratory and found to contain 69.94% iron and 30.06% oxygen. Is the compound iron(II) oxide, FeO, or iron(III) oxide, Fe_2O_3?

Ans.   The percent composition of each compound may be determined:

FeO

$$\% \ Fe = \frac{55.85 \ g \ Fe}{71.85 \ g \ FeO} \times 100.0\% = 77.73\% \ Fe$$

$$\% \ O = \frac{16.00 \ g \ O}{71.85 \ g \ FeO} \times 100.0\% = 22.27\% \ O$$

Fe_2O_3

$$\% \ Fe = \frac{2 \times 55.85 \ g \ Fe}{159.70 \ g \ Fe_2O_3} \times 100.0\% = 69.94\% \ Fe$$

$$\% \ O = \frac{3 \times 16.00 \ g \ O}{159.70 \ g \ Fe_2O_3} \times 100.0\% = 30.06\% \ O$$

The compound is Fe_2O_3.

7.6.   EMPIRICAL FORMULAS

The formula of a compound gives the relative number of atoms of the different elements present. It also gives the relative number of moles of the different elements present. As was shown in Sec. 7.5, the percent by mass of each element in a compound may be computed from its formula. Conversely, if the formula is not known, it may be deduced from the experimentally determined composition. This procedure is possible because once the relative masses of the elements are found, the relative numbers of moles of each may be determined. Formulas derived in this manner are called *empirical formulas* or simplest formulas. In solving a problem in which percent composition is given, any size sample may be considered, since the percentage of each element does not depend on the size of the sample. The most convenient size to consider is 100 g, for with that size sample, the percentage of each element is equal to the same number of grams.

EXAMPLE 7.12.   What is the empirical formula of a compound which contains 52.96% carbon and 47.04% oxygen by mass?

Ans.   In each 100.0 g of the substance, there will be 52.96 g of carbon and 47.04 g of oxygen. The number of moles of each element in this sample is given by

$$52.96 \ g \ C\left(\frac{1 \ mol \ C}{12.01 \ g \ C}\right) = 4.410 \ mol \ C$$

$$47.04 \ g \ O\left(\frac{1 \ mol \ O}{16.00 \ g \ O}\right) = 2.940 \ mol \ O$$

Therefore, the ratio of moles of C to moles of O

$$\frac{4.410 \ mol \ C}{2.940 \ mol \ O} = \frac{1.500 \ mol \ C}{1.000 \ mol \ O} = \frac{3 \ mol \ C}{2 \ mol \ O}$$

is 3 to 2. The ratio of 3 mol of carbon to 2 mol of oxygen corresponds to the formula C_3O_2, a compound called carbon suboxide.

If more than two elements are present, divide all the numbers of moles by the smallest to attempt to get an integral ratio. Even after this step, it might be necessary to multiply every result by a small integer to get integral ratios, corresponding to the empirical formula.

EXAMPLE 7.13.   Determine the empirical formula of a compound containing 29.09% Na, 40.55% S, and 30.36% O.

Ans.

$$29.09 \ g \ Na\left(\frac{1 \ mol \ Na}{22.99 \ g \ Na}\right) = 1.265 \ mol \ Na$$

$$40.55 \ g \ S\left(\frac{1 \ mol \ S}{32.06 \ g \ S}\right) = 1.265 \ mol \ S$$

$$30.36 \ g \ O\left(\frac{1 \ mol \ O}{16.00 \ g \ O}\right) = 1.898 \ mol \ O$$

$$\frac{1.265 \ mol \ Na}{1.265} = 1.000 \ mol \ Na$$

$$\frac{1.265 \ mol \ S}{1.265} = 1.000 \ mol \ S$$

$$\frac{1.898 \ mol \ O}{1.265} = 1.500 \ mol \ O$$

Multiplying each value by 2 yields 2 mol Na, 2 mol S, and 3 mol O, corresponding to $Na_2S_2O_3$.

EXAMPLE 7.14.   Determine the empirical formula of a sample of a compound which contains 44.21 g Fe and 16.89 g O.

Ans.   Since the numbers of grams are given, the step changing percent to grams is not necessary.

$$44.21 \text{ g Fe}\left(\frac{1 \text{ mol Fe}}{55.85 \text{ g Fe}}\right) = 0.7916 \text{ mol Fe} \qquad 16.89 \text{ g O}\left(\frac{1 \text{ mol O}}{16.00 \text{ g O}}\right) = 1.056 \text{ mol O}$$

Dividing each by 0.7916 yields 1.000 mol Fe for every 1.333 mol O. This ratio is still not integral. Multiplying these values by 3 yields integers. (We can round when a value is within 1% or 2% of an integer, but not more.)

$$3.000 \text{ mol Fe} \qquad \text{and} \qquad 3.999 \text{ or 4 mol O}$$

The formula is Fe_3O_4.

7.7.   MOLECULAR FORMULAS

Formulas describe the composition of compounds. Empirical formulas give the mole ratio of the various elements. However, sometimes different compounds have the same ratio of moles of atoms of the same elements. For example, acetylene, C_2H_2, and benzene, C_6H_6, each have 1 : 1 ratios of moles of carbon atoms to moles of hydrogen atoms. That is, each has an empirical formula CH. Such compounds have the same percent compositions. However, they do not have the same number of atoms in each molecule. The *molecular formula* is a formula that gives all the information that the empirical formula gives (the mole ratios of the various elements) plus the information of how many atoms are in each molecule. In order to deduce molecular formulas from experimental data, the percent composition and the molar mass are usually determined. The molar mass may be determined experimentally in several ways, one of which will be described in Chap. 12.

It is apparent that the compounds C_2H_2 and C_6H_6 have different molecular masses. That of C_2H_2 is 26 amu; that of C_6H_6 is 78 amu. It straightforward to determine the molecular mass from the molecular formula, but how can the molecular formula be determined from the empirical formula and the molecular mass? The following steps are used, with benzene having a molecular mass of 78 amu and an empirical formula of CH serving as an example.

Step	Example
1. Determine the formula mass corresponding to 1 empirical formula unit.	CH has a formula mass of $12 + 1 = 13$ amu
2. Divide the molecular mass by the empirical formula mass. The result must be an integer.	Molecular mass $= 78$ amu Empirical mass $= \dfrac{78 \text{ amu}}{13 \text{ amu}} = 6$
3. Multiply the number of atoms of each element in the empirical formula by the whole number found in step 2.	$(CH)_6 = C_6H_6$

EXAMPLE 7.15.   A compound contains 85.7% carbon and 14.3% hydrogen and has a molar mass of 98.0 g/mol. What is its molecular formula?

Ans.   The first step is to determine the empirical formula from the percent composition data.

$$85.7 \text{ g C}\left(\frac{1 \text{ mol C}}{12.0 \text{ g C}}\right) = 7.14 \text{ mol C}$$

$$14.3 \text{ g H}\left(\frac{1 \text{ mol H}}{1.008 \text{ g H}}\right) = 14.2 \text{ mol H}$$

The empirical formula is CH_2.

Now the molecular formula is determined by using the steps outlined above:

1.   Mass of $CH_2 = 14.0$ amu

2.   Number of units $= (98.0 \text{ amu})/(14.0 \text{ amu}) = 7$

3.   Molecular formula $= (CH_2)_7 = C_7H_{14}$

Solved Problems

MOLECULES AND FORMULA UNITS

7.1. Why is the term *molecular mass* inappropriate for NaCl?

   Ans.   NaCl is an ionic compound; it does not form molecules and so does not have a *molecular mass*. The *formula mass* of a NaCl is 58.5 amu, calculated and used in exactly the same manner as the molecular mass for a molecular compound would be calculated and used.

7.2. Which of the following compounds occur in molecules? (a) C_6H_6, (b) CH_4O, (c) $C_6H_{12}O_6$, (d) $CoCl_2$, (e) $COCl_2$, (f) NH_4Cl, (g) CO, and (h) $FeCl_2$.

   Ans.   All but (d), (f), and (h) form molecules; (d), (f), and (h) are ionic.

7.3. The simplest type of base contains OH^- ions. Which of the following compounds is more apt to be a base, CH_3OH or KOH?

   Ans.   KOH is ionic and is a base. CH_3OH is covalently bonded and is not a base.

FORMULA MASSES

7.4. The standard for atomic masses is ^{12}C, at exactly 12 amu. What is the standard for formula masses of compounds?

   Ans.   The same standard, ^{12}C, is used for formula masses.

7.5. What is the difference between the atomic mass of bromine and the molecular mass of bromine?

   Ans.   The atomic mass of bromine is 79.90 amu, as seen in the periodic table or a table of atomic masses. The molecular mass of bromine, corresponding to Br_2, is twice that value, 159.8 amu.

7.6. What is the formula mass of each of the following? (a) P_4, (b) Na_2SO_4, (c) $(NH_4)_2SO_3$, (d) H_2O_2, (e) Br_2, (f) H_3PO_3, (g) $Hg_2(ClO_3)_2$, and (h) $(NH_4)_2HPO_3$.

   Ans.   (a) 4×30.97 amu $= 123.9$ amu   (d)   34.0 amu

   (b)   2 Na: 2×23.0 amu $= 46.0$ amu   (e)   159.8 amu

   S: 1×32.0 amu $= 32.0$ amu   (f)   82.0 amu

   4 O: 4×16.0 amu $= 64.0$ amu   (g)   568.1 amu

   total $= 142.0$ amu   (h)   116.0 amu

   (c)   116.1 amu

7.7. What is the mass, in amu, of 1 formula unit of $Al_2(SO_3)_3$?

   Ans.   There are 2 aluminum atoms, 3 sulfur atoms, and 9 oxygen atoms in the formula unit.

$$2 \times \text{atomic mass of aluminum} = 54.0 \text{ amu}$$

$$3 \times \text{atomic mass of sulfur} = 96.2 \text{ amu}$$

$$9 \times \text{atomic mass of oxygen} = 144.0 \text{ amu}$$

$$\text{Formula mass} = \text{total} = 294.2 \text{ amu}$$

The formula represents 294.2 amu of aluminum sulfite.

7.8. Calculate the formula mass of each of the following compounds: (a) NaCl, (b) NH_4NO_3, (c) SiF_4, (d) $Fe(CN)_2$, and (e) KCN.

   Ans.   (a) 58.44 amu   (b) 80.05 amu   (c) 104.1 amu   (d) 107.9 amu   (e) 65.12 amu

THE MOLE

7.9. How many H atoms are there in 1 molecule of H_2O_2? How many moles of H atoms are there in 1 mol of H_2O_2?

Ans. There are two H atoms per molecule of H_2O_2, and 2 mol H atoms per 1 mol H_2O_2. The chemical formula provides both these ratios.

7.10. How many atoms of K are there in 1.00 mol K? What is the mass of 1.00 mol K?

Ans. There are 6.02×10^{23} atoms in 1.00 mol K (Avogadro's number). There are 39.1 g of K in 1.00 mol K (equal to the atomic mass in grams). This problem requires use of two of the most important conversion factors involving moles. Note which one is used with masses and which one is used with numbers of atoms (or molecules or formula units). With *numbers* of atoms, molecules, or formula units, use Avogadro's *number*; with *mass*, use the formula *mass*.

7.11. What mass in grams is there in 1.000 mol of $Al_2(SO_3)_3$?

Ans. The molar mass has the same value in grams that the formula mass has in amu (Problem 7.7). Thus 1.000 mol represents 294.2 g of aluminum sulfite.

7.12. (*a*) Which contains more pieces of fruit, a dozen grapes or a dozen watermelons? Which weighs more? (*b*) Which contains more atoms, 1 mol of lithium or 1 mol of lead? Which has a greater mass?

Ans. (*a*) Both have the same number of fruits (12), but since each watermelon weighs more than a grape, the dozen watermelons weigh more than the dozen grapes. (*b*) Both have the same number of atoms (6.02×10^{23}), but since lead has a greater atomic mass (see the periodic table), 1 mol of lead has a greater mass.

Problems 7.13 through 7.16 are easier to do when both parts are worked together. Note the differences among the parts labeled (*a*) and also among the parts labeled (*b*). On examinations, you are likely to be asked only one such problem at a time, so you must read the problems carefully and recognize the difference between similar-sounding problems.

7.13. (*a*) How many socks are there in 10 dozen socks? (*b*) How many hydrogen atoms are there in 10.0 mol of hydrogen atoms?

Ans. (*a*) $10 \text{ dozen socks} \left(\dfrac{12 \text{ socks}}{1 \text{ dozen socks}} \right) = 120 \text{ socks}$

(*b*) $10.0 \text{ mol H} \left(\dfrac{6.02 \times 10^{23} \text{ H}}{1 \text{ mol H}} \right) = 6.02 \times 10^{24} \text{ H atoms}$

7.14. (*a*) How many pairs of socks are there in 10 dozen pairs of socks? (*b*) How many hydrogen molecules are there in 10.0 mol of hydrogen molecules?

Ans. (*a*) $10 \text{ dozen pairs socks} \left(\dfrac{12 \text{ pair socks}}{1 \text{ dozen pair socks}} \right) = 120 \text{ pairs socks}$

(*b*) $10.0 \text{ mol H}_2 \left(\dfrac{6.02 \times 10^{23} \text{ H}_2}{1 \text{ mol H}_2} \right) = 6.02 \times 10^{24} \text{ H}_2 \text{ molecules}$

7.15. (*a*) How many pairs of socks can be made with 10 dozen (identical) socks? (*b*) How many hydrogen molecules can be made with 10.0 mol of hydrogen atoms?

Ans. (*a*) $10 \text{ dozen socks} \underbrace{\left(\dfrac{1 \text{ dozen pairs socks}}{2 \text{ dozen socks}} \right)}_{\substack{\text{(from definition} \\ \text{of a pair)}}} \underbrace{\left(\dfrac{12 \text{ pairs socks}}{1 \text{ dozen pairs socks}} \right)}_{\substack{\text{(from definition} \\ \text{of a dozen)}}} = 60 \text{ pairs socks}$

(*b*) $10.0 \text{ mol H} \underbrace{\left(\dfrac{1 \text{ mol H}_2}{2 \text{ mol H}} \right)}_{\substack{\text{(from definition} \\ \text{of a molecule)}}} \underbrace{\left(\dfrac{6.02 \times 10^{23} \text{ H}_2}{1 \text{ mol H}_2} \right)}_{\substack{\text{(from definition} \\ \text{of a mole)}}} = 3.01 \times 10^{24} \text{ H}_2 \text{ molecules}$

7.16. (*a*) How many socks are there in 10 dozen pairs of socks? (*b*) How many hydrogen atoms are there in 10.0 mol of hydrogen molecules?

Ans.   (*a*)   10 dozen pairs socks $\left(\dfrac{12 \text{ pairs socks}}{1 \text{ dozen pairs socks}} \right)\left(\dfrac{2 \text{ socks}}{1 \text{ pair socks}} \right) = 240$ socks

(*b*)   10.0 mol $H_2 \left(\dfrac{6.02 \times 10^{23} \text{ } H_2}{1 \text{ mol } H_2} \right)\left(\dfrac{2 \text{ H}}{1 \text{ } H_2} \right) = 1.20 \times 10^{25}$ H atoms

7.17. What is the difference between 1 mol of nitrogen atoms and 1 mol of nitrogen molecules?

Ans.   The nitrogen molecules contain two nitrogen atoms each; hence 1 mol of nitrogen molecules contains twice as many atoms as 1 mol of nitrogen atoms. A mole of nitrogen molecules contains 2 mol of bonded N atoms.

7.18. Show that the ratio of the number of moles of two elements in a compound is equal to the ratio of the number of atoms of the two elements.

Ans.
$$\frac{x \text{ mol A}}{1 \text{ mol B}} = \frac{x \text{ mol A}\left(\dfrac{6.02 \times 10^{23} \text{ atoms A}}{1 \text{ mol A}} \right)}{1 \text{ mol B}\left(\dfrac{6.02 \times 10^{23} \text{ atoms B}}{1 \text{ mol B}} \right)} = \frac{x \text{ atoms A}}{1 \text{ atom B}}$$

7.19. How many moles of each substance are there in each of the following masses of substances? (*a*) 111 g F_2, (*b*) 6.50 g Na, (*c*) 44.1 g NaOH, (*d*) 0.0330 g NaCl, (*e*) 1.96 g $(NH_4)_2SO_3$, (*f*) 7.22 g NaH_2PO_4

Ans.   (*a*)   111 g $F_2 \left(\dfrac{1 \text{ mol } F_2}{38.0 \text{ g } F_2} \right) = 2.92$ mol F_2

(*b*)   6.50 g Na $\left(\dfrac{1 \text{ mol Na}}{23.0 \text{ g Na}} \right) = 0.283$ mol Na

(*c*)   44.1 g NaOH $\left(\dfrac{1 \text{ mol NaOH}}{40.0 \text{ g NaOH}} \right) = 1.10$ mol NaOH

(*d*)   0.0330 g NaCl $\left(\dfrac{1 \text{ mol NaCl}}{58.5 \text{ g NaCl}} \right) = 5.64 \times 10^{-4}$ mol NaCl

(*e*)   1.96 g $(NH_4)_2SO_3 \left(\dfrac{1 \text{ mol } (NH_4)_2SO_3}{116 \text{ g } (NH_4)_2SO_3} \right) = 0.0169$ mol $(NH_4)_2SO_3$

(*f*)   7.22 g $NaH_2PO_4 \left(\dfrac{1 \text{ mol } NaH_2PO_4}{120 \text{ g } NaH_2PO_4} \right) = 0.0602$ mol NaH_2PO_4

7.20. In 2.50 mol $Ba(NO_3)_2$, (*a*) how many moles of barium ions are present? (*b*) How many moles of nitrate ions are present? (*c*) How many moles of oxygen atoms are present?

Ans.   (*a*)   2.50 mol Ba^{2+} ions

(*b*)   5.00 mol NO_3^- ions

(*c*)   15.0 mol O atoms

7.21. Without doing actual calculations, determine which of the following have masses greater than 1 mg: (*a*) 1 CO_2 molecule, (*b*) 1 amu, (*c*) 6.02×10^{23} H atoms, (*d*) 1 mol CO_2, (*e*) 6.02×10^{23} amu, (*f*) $\frac{1}{6.02 \times 10^{23}}$ mol H atoms

Ans.   (*c*), (*d*), and (*e*) have masses greater than 1 mg.

7.22. What is the mass of each of the following? (*a*) 6.78 mol NaI, (*b*) 0.447 mol $NaNO_3$, (*c*) 0.500 mol BaO_2, (*d*) 5.44 mol C_8H_{18}

Ans.   (a) $6.78 \text{ mol NaI}\left(\dfrac{150 \text{ g NaI}}{1 \text{ mol NaI}}\right) = 1020 \text{ g NaI}$

(b) $0.447 \text{ mol NaNO}_3\left(\dfrac{85.0 \text{ g NaNO}_3}{1 \text{ mol NaNO}_3}\right) = 38.0 \text{ g NaNO}_3$

(c) $0.500 \text{ mol BaO}_2\left(\dfrac{169 \text{ g BaO}_2}{1 \text{ mol BaO}_2}\right) = 84.5 \text{ g BaO}_2$

(d) $5.44 \text{ mol C}_8\text{H}_{18}\left(\dfrac{114 \text{ g C}_8\text{H}_{18}}{1 \text{ mol C}_8\text{H}_{18}}\right) = 620 \text{ g C}_8\text{H}_{18}$

7.23.   In 0.125 mol $NaClO_4$, (*a*) how many moles of sodium ions are present? (*b*) how many moles of perchlorate ions are present?

Ans.   (*a*) 0.125 mol Na^+ ions   (*b*) 0.125 mol ClO_4^- ions. The subscript 4 refers to the number of O atoms per anion, not to the number of anions.

7.24.   (*a*) Calculate the formula mass of H_3PO_4. (*b*) Calculate the number of grams in 1.00 mol of H_3PO_4. (*c*) Calculate the number of grams in 2.00 mol of H_3PO_4. (*d*) Calculate the number of grams in 0.222 mol of H_3PO_4.

Ans.   (*a*) 3 H   3.02 amu   (*b*)   97.99 g
              P   30.97 amu
            4 O   64.00 amu   (*c*) $2.00 \text{ mol}\left(\dfrac{97.99 \text{ g}}{1 \text{ mol}}\right) = 196 \text{ g}$
          total   97.99 amu
                                             (*d*) $0.222 \text{ mol}\left(\dfrac{97.99 \text{ g}}{1 \text{ mol}}\right) = 21.8 \text{ g}$

7.25.   (*a*) Determine the number of moles of ammonia in 27.5 g of ammonia, NH_3. (*b*) Determine the number of moles of nitrogen atoms in 27.5 g of ammonia. (*c*) Determine the number of moles of hydrogen atoms in 27.5 g of ammonia.

Ans.   (*a*) $27.5 \text{ g NH}_3\left(\dfrac{1 \text{ mol NH}_3}{17.0 \text{ g NH}_3}\right) = 1.62 \text{ mol NH}_3$

(*b*) $1.62 \text{ mol NH}_3\left(\dfrac{1 \text{ mol N}}{1 \text{ mol NH}_3}\right) = 1.62 \text{ mol N}$

(*c*) $1.62 \text{ mol NH}_3\left(\dfrac{3 \text{ mol H}}{1 \text{ mol NH}_3}\right) = 4.86 \text{ mol H}$

7.26.   How many hydrogen atoms are there in 1.00 mol of methane, CH_4?

Ans. $1 \text{ mol CH}_4\left(\dfrac{6.02 \times 10^{23} \text{ CH}_4 \text{ molecules}}{1 \text{ mol CH}_4}\right)\left(\dfrac{4 \text{ H atoms}}{1 \text{ CH}_4 \text{ molecule}}\right) = 2.41 \times 10^{24} \text{ H atoms}$

7.27.   Determine the number of mercury atoms in 7.11 g of Hg_2Cl_2.

Ans. $7.11 \text{ g Hg}_2\text{Cl}_2\left(\dfrac{1 \text{ mol Hg}_2\text{Cl}_2}{472 \text{ g Hg}_2\text{Cl}_2}\right)\left(\dfrac{2 \text{ mol Hg}}{1 \text{ mol Hg}_2\text{Cl}_2}\right)\left(\dfrac{6.02 \times 10^{23} \text{ Hg atoms}}{1 \text{ mol Hg}}\right) = 1.81 \times 10^{22} \text{ atoms}$

7.28.   (*a*) If 1 dozen pairs of socks weighs 11 oz, how much would the same socks weigh if they were unpaired? (*b*) What is the mass of 1 mol O_2? What is the mass of the same number of oxygen atoms unbonded to one another?

Ans.   (*a*)   The socks would still weigh 11 oz; unpairing them makes no difference in their mass.

(*b*)   The mass of 1 mol O_2 is 32 g; the mass of 2 mol O is also 32 g. The mass does not depend on whether they are bonded.

7.29.   How many molecules are there in 1.00 mol F_2? How many atoms are there? What is the mass of 1.00 mol F_2?

Ans. There are 6.02×10^{23} molecules in 1.00 mol F_2 (Avogadro's number). Since there are two atoms per molecule, there are

$$6.02 \times 10^{23} \text{ molecules}\left(\frac{2 \text{ atoms}}{1 \text{ molecule}}\right) = 1.20 \times 10^{24} \text{ atoms F in 1.00 mol } F_2$$

The mass is that of 1.00 mol F_2 or 2.00 mol F:

$$1.00 \text{ mol } F_2\left(\frac{2 \times 19.0 \text{ g } F_2}{1 \text{ mol } F_2}\right) = 38.0 \text{ g } F_2$$

or

$$2.00 \text{ mol F}\left(\frac{19.0 \text{ g F}}{1 \text{ mol F}}\right) = 38.0 \text{ g F}$$

The *mass* is the same, no matter whether we focus on the atoms or molecules. (Compare the mass of 1 dozen pairs of socks rolled together to that of the same socks unpaired. Would the two masses differ? If so, which would be greater? See the prior problem.)

7.30. (*a*) Create one factor that will change 6.17 g of calcium carbonate to a number of formula units of calcium carbonate. (*b*) Is it advisable to learn and use such a factor?

Ans. (*a*)

$$\frac{1 \text{ mol } CaCO_3}{100 \text{ g } CaCO_3}\left(\frac{6.02 \times 10^{23} \text{ units } CaCO_3}{1 \text{ mol } CaCO_3}\right) = \frac{6.02 \times 10^{23} \text{ units } CaCO_3}{100 \text{ g } CaCO_3}$$

(*b*) It is possible to use such a conversion factor, but it is advisable while you are learning to use the factors involved with moles to use as few different ones as possible. That way, you have to remember fewer. Also, in each conversion you will change either the unit (mass → moles) or the chemical ($CaCO_3$ → O atoms) in a factor, but not both. Many texts do use such combined factors, however.

7.31. How many moles of Na are there in 7.20 mol Na_2SO_4?

Ans.

$$7.20 \text{ mol } Na_2SO_4\left(\frac{2 \text{ mol Na}}{1 \text{ mol } Na_2SO_4}\right) = 14.4 \text{ mol Na}$$

7.32. How many moles of water can be made with 1.76 mol H atoms (plus enough O atoms)?

Ans.

$$1.76 \text{ mol H}\left(\frac{1 \text{ mol } H_2O}{2 \text{ mol H}}\right) = 0.880 \text{ mol } H_2O$$

7.33. Make a table showing the number of molecules and the mass of each of the following: (*a*) 1.00 mol Cl_2, (*b*) 2.00 mol Cl_2, and (*c*) 0.135 mol Cl_2

Ans.

	Number of Molecules	Mass
(*a*) 1.00 mol Cl_2	6.02×10^{23}	71.0 g
(*b*) 2.00 mol Cl_2	1.20×10^{24}	142 g
(*c*) 0.135 mol Cl_2	8.13×10^{22}	9.59 g

Once you know the mass or number of molecules in 1 mol, you merely have to multiply to get the mass or number of molecules in any other given number of moles. Sometimes the calculation is easy enough to do in your head.

7.34. How can you measure the thickness of a sheet of notebook paper with a 10-cm ruler?

Ans. One way is to measure the combined thickness of many sheets and divide that distance by the number of sheets. For example, if 500 sheets is 5.05 cm thick, then each sheet is (5.05 cm)/500 = 0.0101 cm thick. Note that it is impossible to measure 0.0101 cm with a centimeter ruler, but we did accomplish the same purpose indirectly. (See also the following problem.)

7.35. How can you measure the mass of a carbon atom? Compare this problem with the prior problem.

Ans. We want to measure the mass of a large number of carbon atoms and divide the total mass by the number of atoms. However, we have the additional problem here, compared with counting sheets of paper (prior problem), that atoms are too small to count. We can "count" them by combining them with a known number of atoms of another element. For example, to count a number of carbon atoms, combine them with a known number of oxygen atoms to form CO, in which the ratio of atoms of carbon to oxygen is 1 : 1.

7.36. What mass of oxygen is combined with 4.13×10^{24} atoms of sulfur in Na_2SO_4?

Ans. $$4.13 \times 10^{24} \text{ atoms S} \left(\frac{1 \text{ mol S}}{6.02 \times 10^{23} \text{ atoms S}} \right) \left(\frac{4 \text{ mol O}}{1 \text{ mol S}} \right) \left(\frac{16.0 \text{ g O}}{1 \text{ mol O}} \right) = 439 \text{ g O}$$

7.37. A 106 g sample of an "unknown" element Q reacts with 32.0 g of O_2. Assuming the atoms of Q react in a 1 : 1 ratio with oxygen *molecules*, calculate the atomic mass of Q.

Ans. Since the 106 g of Q reacts with 1.00 mol of O_2 in a 1 : 1 mole ratio, there must be 1.00 mol of Q in 106 g. Therefore, 106 g is 1.00 mol, and the atomic mass of Q is 106 amu.

7.38. To form a certain compound, 29.57 g of oxygen reacts with 109.7 g of tin. What is the formula of the compound?

Ans. The formula is the mole ratio:

$$29.57 \text{ g O atoms} \left(\frac{1 \text{ mol O atoms}}{16.00 \text{ g O}} \right) = 1.848 \text{ mol O atoms}$$

$$109.7 \text{ g Sn atoms} \left(\frac{1 \text{ mol Sn atoms}}{118.7 \text{ g Sn}} \right) = 0.9242 \text{ mol Sn atoms}$$

The mole ratio is 1.848 mol O/0.9242 mol Sn = 2 mol O/1 mol Sn. The formula is SnO_2.

7.39. If 29.57 g O_2 were used in Problem 7.38, how would the problem change?

Ans. The answer would be the same; 29.57 g of O_2 is 29.57 g of O atoms (bonded in pairs). See Problem 7.28.

7.40. How many P atoms are there in 1.13 mol P_4?

Ans. $$1.13 \text{ mol } P_4 \left(\frac{6.02 \times 10^{23} \text{ molecules } P_4}{1 \text{ mol } P_4} \right) \left(\frac{4 \text{ atoms P}}{1 \text{ molecule } P_4} \right) = 2.72 \times 10^{24} \text{ P atoms}$$

7.41. How many hydrogen atoms are there in 1.36 mol NH_3?

Ans. $$1.36 \text{ mol } NH_3 \left(\frac{6.02 \times 10^{23} \text{ molecules } NH_3}{1 \text{ mol } NH_3} \right) \left(\frac{3 \text{ H atoms}}{1 \text{ molecule } NH_3} \right) = 2.46 \times 10^{24} \text{ H atoms}$$

PERCENT COMPOSITION OF COMPOUNDS

7.42. A 10.0-g sample of water has a percent composition of 88.8% oxygen and 11.2% hydrogen. (*a*) What is the percent composition of a 6.67-g sample of water? (*b*) Calculate the number of grams of oxygen in a 6.67-g sample of water.

Ans. (*a*) 88.8% O and 11.2% H. The percent composition does not depend on the sample size.

(*b*) $$6.67 \text{ g } H_2O \left(\frac{88.8 \text{ g O}}{100 \text{ g } H_2O} \right) = 5.92 \text{ g O}$$

7.43. Calculate the percent composition of each of the following: (*a*) C_3H_6 and (*b*) C_5H_{10}.

Ans. (*a*)   C $3 \times 12.0 = 36.0$ amu

H $6 \times 1.0 = 6.0$ amu

Total $= 42.0$ amu

The percent carbon is found by dividing the mass of carbon in one molecule by the mass of the molecule and multiplying the quotient by 100%:

$$\% \text{ C} = \left(\frac{36.0 \text{ amu C}}{42.0 \text{ amu total}} \right) \times 100\% = 85.7\% \text{C} \qquad \% \text{ H} = \left(\frac{6.0 \text{ amu H}}{42.0 \text{ amu total}} \right) \times 100\% = 14.3\% \text{ H}$$

The two percentages add up to 100.0%.

(b)
$$C \quad 5 \times 12.0 = 60.0 \text{ amu}$$
$$H \quad 10 \times 1.0 = 10.0 \text{ amu}$$
$$\text{Total} = 70.0 \text{ amu}$$

$$\% \text{ C} = \frac{60.0 \text{ amu C}}{70.0 \text{ amu total}} \times 100\% = 85.7\% \text{ C} \qquad \% \text{ H} = \frac{10.0 \text{ amu H}}{70.0 \text{ amu total}} \times 100\% = 14.3\% \text{ H}$$

The percentages are the same as those in part (a). That result might have been expected. Since the ratio of atoms of carbon to atoms of hydrogen is the same (1 : 2) in both compounds, the ratio of masses also ought to be the same, and their percent by mass ought to be the same. From another viewpoint, this result means that the two compounds cannot be distinguished from each other by their percent compositions alone.

7.44. Calculate the percent composition of DDT ($C_{14}H_9Cl_5$).

Ans.
$$C \quad 14 \times 12.01 = 168.1 \text{ amu}$$
$$H \quad 9 \times 1.008 = 9.07 \text{ amu}$$
$$Cl \quad 5 \times 35.45 = 177.2 \text{ amu}$$
$$\text{Total} = 354.4 \text{ amu}$$

The percent carbon is found by dividing the mass of carbon in one molecule by the mass of the molecule and multiplying the quotient by 100%:

$$\% \text{ C} = \left(\frac{168.1 \text{ amu C}}{354.4 \text{ amu total}} \right) \times 100\% = 47.43\% \text{ C}$$

$$\% \text{ H} = \left(\frac{9.07 \text{ amu H}}{354.4 \text{ amu total}} \right) \times 100\% = 2.56\% \text{ H}$$

$$\% \text{ Cl} = \left(\frac{177.2 \text{ amu Cl}}{354.4 \text{ amu total}} \right) \times 100\% = 50.00\% \text{ Cl}$$

The percentages add up to 99.99%. (The answer is correct within the accuracy of the number of significant figures used.)

7.45. A forensic scientist analyzes a drug and finds that it contains 80.22% carbon and 9.62% hydrogen. Could the drug be pure tetrahydrocannabinol ($C_{21}H_{30}O_2$)?

Ans.
$$21 \text{ C} \qquad 21 \times 12.01 = 252.2 \text{ amu}$$
$$30 \text{ H} \qquad 30 \times 1.008 = 30.24 \text{ amu}$$
$$2 \text{ O} \qquad 2 \times 16.00 = 32.00 \text{ amu}$$
$$\text{Formula mass} \qquad \qquad = 314.4 \text{ amu}$$

$$\% \text{ C} = \frac{252.2 \text{ amu}}{314.4 \text{ amu}} \times 100\% = 80.22\% \text{ C} \qquad \% \text{ H} = \frac{30.24 \text{ amu}}{314.4 \text{ amu}} \times 100\% = 9.618\% \text{ H}$$

Since the percentages are the same, the drug *could* be tetrahydrocannabinol. (It is not proved to be, however. If the percent composition were different, it would be proved *not* to be pure tetrahydrocannabinol.)

7.46. A certain mixture of salt (NaCl) and sugar ($C_{12}H_{22}O_{11}$) contains 40.0% chlorine by mass. Calculate the percentage of salt in the mixture.

Ans. In 100.0 g of sample (the size does not make any difference), there is 40.0 g of chlorine and therefore

$$40.0 \text{ g Cl} \left(\frac{58.5 \text{ g NaCl}}{35.5 \text{ g Cl}} \right) = 65.9 \text{ g NaCl}$$

The percentage of NaCl (in the 100.0 g sample) is therefore 65.9%. When using percentages, be careful to distinguish percentage of what in what!

EMPIRICAL FORMULAS

7.47. If each of the following mole ratios is obtained in an empirical formula problem, what should it be multiplied by to get an integer ratio? (a) 1.50 : 1, (b) 1.25 : 1, (c) 1.33 : 1, (d) 1.67 : 1, and (e) 1.75 : 1.

Ans. (*a*) 2, to get 3 : 2 (*b*) 4, to get 5 : 4 (*c*) 3, to get 4 : 3 (*d*) 3, to get 5 : 3 (*e*) 4, to get 7 : 4

7.48. Which do we use to calculate the empirical formula of an oxide, the atomic mass of oxygen (16 amu) or the molecular mass of oxygen (32 amu)?

Ans. The atomic mass. We are solving for a formula, which is a ratio of atoms. This type of problem has nothing to do with oxygen gas, O_2.

7.49. (*a*) Write a formula for a molecule with 4 phosphorus atoms and 6 oxygen atoms per molecule. (*b*) What is the empirical formula of this compound?

Ans. (*a*) P_4O_6 (*b*) P_2O_3

7.50. Which of the formulas in Problem 5.2 obviously are not empirical formulas?

Ans. (*a*), (*d*), (*e*), and (*g*). The subscripts in (*a*), (*d*), (*e*), and (*g*) can be divided by a small integer to give a simpler formula, so these cannot be empirical formulas. (They must have at least some covalent bonds.)

7.51. Calculate the empirical formula of a compound consisting of 92.26% C and 7.74% H.

Ans. Assume that 100.0 g of the compound is analyzed. Since the same percentages are present no matter what the sample size, we can consider any size sample we wish, and considering 100 g makes the calculations easier. The numbers of grams of the elements are then 92.26 g C and 7.74 g H.

$$92.26 \text{ g C}\left(\frac{1 \text{ mol C}}{12.01 \text{ g C}}\right) = 7.682 \text{ mol C} \qquad 7.74 \text{ g H}\left(\frac{1 \text{ mol H}}{1.008 \text{ g H}}\right) = 7.68 \text{ mol H}$$

The empirical formula (or any formula) must be in the ratio of small integers. Thus, we attempt to get the ratio of moles of carbon to moles of hydrogen into an integer ratio; we divide all the numbers of moles by the smallest number of moles:

$$\frac{7.682 \text{ mol C}}{7.68} = 1.00 \text{ mol C} \qquad \frac{7.68 \text{ mol H}}{7.68} = 1.00 \text{ mol H}$$

The ratio of moles of C to moles of H is 1 : 1, so the empirical formula is CH.

7.52. Calculate the empirical formula of a compound containing 69.94% Fe and the rest oxygen.

Ans. The oxygen must be 30.06%, to total 100.00%.

$$69.94 \text{ g Fe}\left(\frac{1 \text{ mol Fe}}{55.85 \text{ g Fe}}\right) = 1.252 \text{ mol Fe} \qquad 30.06 \text{ g O}\left(\frac{1 \text{ mol O}}{16.00 \text{ g O}}\right) = 1.879 \text{ mol O}$$

Dividing both of these numbers by the smaller yields

$$\frac{1.252 \text{ mol Fe}}{1.252} = 1.000 \text{ mol Fe} \qquad \frac{1.879 \text{ mol O}}{1.252} = 1.501 \text{ mol O}$$

This is still not a whole number ratio, since 1.501 is much too far from an integer to round. Since 1.501 is about $1\frac{1}{2}$, multiply *both* numbers of moles by 2:

$$2.000 \text{ mol Fe} \qquad \text{and} \qquad 3.002 \text{ mol O}$$

We can round when a value is within 1% or 2% of an integer, but not more. This is close enough to an integer ratio, so the empirical formula is Fe_2O_3.

7.53. Determine the empirical formula of a compound which has a percent composition 23.3% Mg, 30.7% S, and 46.0%, O.

Ans. In a 100-g sample, there are

$$23.3 \text{ g Mg}\left(\frac{1 \text{ mol Mg}}{24.3 \text{ g Mg}}\right) = 0.959 \text{ mol Mg} \qquad 46.0 \text{ g O}\left(\frac{1 \text{ mol O}}{16.0 \text{ g O}}\right) = 2.88 \text{ mol O}$$

$$30.7 \text{ g S}\left(\frac{1 \text{ mol S}}{32.0 \text{ g S}}\right) = 0.959 \text{ mol S}$$

To get integer mole ratios, divide by the smallest, 0.956:

$$\frac{0.959 \text{ mol Mg}}{0.959} = 1.00 \text{ mol Mg} \qquad \frac{2.88 \text{ mol O}}{0.959} = 3.00 \text{ mol O}$$

$$\frac{0.959 \text{ mol S}}{0.959} = 1.00 \text{ mol S}$$

The mole ratio is 1 mol Mg to 1 mol S to 3 mol O; the empirical formula is $MgSO_3$.

7.54. Calculate the empirical formula for each of the following compounds: (a) 42.1% Na, 18.9% P, and 39.0% O. (b) 55.0% K and 45.0% O.

Ans. (a) $\qquad 42.1 \text{ g Na}\left(\dfrac{1 \text{ mol Na}}{23.0 \text{ g Na}}\right) = 1.83 \text{ mol Na} \qquad 39.0 \text{ g O}\left(\dfrac{1 \text{ mol O}}{16.00 \text{ g O}}\right) = 2.44 \text{ mol O}$

$\qquad\qquad 18.9 \text{ g P}\left(\dfrac{1 \text{ mol P}}{31.0 \text{ g P}}\right) = 0.610 \text{ mol P}$

Dividing by 0.610 yields 3.00 mol Na, 1.00 mol P, 4.00 mol O. The empirical formula is Na_3PO_4.

(b) $\qquad 55.0 \text{ g K}\left(\dfrac{1 \text{ mol K}}{39.1 \text{ g K}}\right) = 1.41 \text{ mol K} \qquad 45.0 \text{ g O}\left(\dfrac{1 \text{ mol O}}{16.0 \text{ g O}}\right) = 2.81 \text{ mol O}$

Dividing by 1.41 yields 1.00 mol K and 2.00 mol O. The empirical formula is KO_2 (potassium super-oxide).

MOLECULAR FORMULAS

7.55. List five possible molecular formulas for a compound with empirical formula CH_2.

Ans. C_2H_4, C_3H_6, C_4H_8, C_5H_{10}, C_6H_{12} (and any other formula with a C-to-H ratio of 1 : 2).

7.56. Explain why we cannot calculate a molecular formula for a compound of phosphorus, potassium, and oxygen.

Ans. The compound is ionic; it does not form molecules.

7.57. Which one of the following could possibly be defined as "the ratio of moles of each of the given elements to moles of each of the others"? (a) percent composition by mass, (b) empirical formula, or (c) molecular formula.

Ans. Choice (b). This is a useful definition of empirical formula. The molecular formula gives the ratio of moles of each element to moles of the compound, plus the information given by the empirical formula. The percent composition does not deal with moles, but is a ratio of masses.

7.58. A compound consists of 92.26% C and 7.74% H. Its molecular mass is 65.0 amu. (a) Calculate its empirical formula. (b) Calculate its empirical formula mass. (c) Calculate the number of empirical formula units in one molecule. (d) Calculate its molecular formula.

Ans. (a) The empirical formula is calculated to be CH, as presented in problem 7.51.

(b) The empirical formula mass is 13.0 amu, corresponding to 1 C and 1 H atom.

(c) There are

$$\frac{65.0 \text{ amu/molecule}}{13.0 \text{ amu/empirical formula unit}} = \frac{5 \text{ empirical formula units}}{1 \text{ molecule}}$$

(d) The molecular formula is $(CH)_5$, or C_5H_5.

7.59. The percent composition of a certain compound is 85.7% C and 14.3% H. Its molecular mass is 70.0 amu. (a) Determine its empirical formula. (b) Determine its molecular formula.

Ans. (a) $\qquad 85.7 \text{ g C}\left(\dfrac{1 \text{ mol C}}{12.0 \text{ g C}}\right) = 7.14 \text{ mol C} \qquad 14.3 \text{ g H}\left(\dfrac{1 \text{ mol H}}{1.008 \text{ g H}}\right) = 14.2 \text{ mol H}$

The ratio is 1 : 2, and the empirical formula is CH_2.

(b)   The empirical formula mass is 14.0 amu. There are

$$\frac{70.0 \text{ g/mol}}{14.0 \text{ g/mol empirical formula unit}} = \frac{5 \text{ mol empirical formula units}}{1 \text{ mol}}$$

The molecular formula is C_5H_{10}.

Supplementary Problems

7.60.   Define or identify each of the following: molecule, ion, formula unit, formula mass, mole, molecular mass, Avogadro's number, percent, empirical formula, molecular formula, molar mass, empirical formula mass, molecular weight.

Ans.   See the text.

7.61.   How is molecular mass related to formula mass?

Ans.   They are the same for compounds which form molecules.

7.62.   Does the term *atomic mass* refer to uncombined atoms, atoms bonded in compounds, or both?

Ans.   Both

7.63.   A certain fertilizer is advertised to contain 10.5% K_2O. What percentage of the fertilizer is potassium?

Ans.
$$\frac{10.5 \text{ g } K_2O\left(\dfrac{78.2 \text{ g K}}{94.2 \text{ g } K_2O}\right)}{100 \text{ g sample}} = 8.72\% \text{ K}$$

7.64.   A compound consists of 92.26% C and 7.74% H. Its molecular mass is 65.0 amu. Calculate its molecular formula.

Ans. C_5H_5. This problem is exactly the same as Problem 7.58. The steps are the same even though they are not specified in the statement of this problem.

7.65.   (a) Calculate the percent composition of C_3H_6. (b) Calculate the percent composition of C_4H_8. (c) Compare the results and explain the reason for these results.

Ans.   (a, b) There is 85.7% C and 14.3% H in each.   (c) They are the same because they have the same ratio of moles of elements.

7.66.   Name and calculate the empirical formula of each of the following compounds.

(a)   36.77% Fe   21.10% S   42.13% O   (d)   71.05% Co   28.95% O
(b)   27.93% Fe   24.05% S   48.01% O   (e)   72.7% O   27.3% C
(c)   63.20% Mn   36.8% O   (f)   36.0% Al   64.0% S

Ans.   (a) $FeSO_4$ (b) $Fe_2S_3O_{12}$ (c) MnO_2 (d) Co_2O_3 (e) CO_2 (f) Al_2S_3
The names are (a) iron(II) sulfate, (b) iron(III) sulfate [$Fe_2(SO_4)_3$], (c) manganese(IV) oxide, (d) cobalt(III) oxide, (e) carbon dioxide, and (f) aluminum sulfide.

7.67.   What mass of oxygen is contained in 42.8 g $CaCO_3$?

Ans.
$$42.8 \text{ g } CaCO_3\left(\frac{1 \text{ mol } CaCO_3}{100 \text{ g } CaCO_3}\right)\left(\frac{3 \text{ mol O}}{1 \text{ mol } CaCO_3}\right)\left(\frac{16.0 \text{ g O}}{1 \text{ mol O}}\right) = 20.5 \text{ g O}$$

You can also do this problem by using percent composition (Sec. 7.5).

7.68.   Determine the molecular formula of a compound with molar mass between 105 and 115 g/mol which contains 88.8% C and 11.2% H.

Ans.

$$88.8 \text{ g C}\left(\frac{1 \text{ mol C}}{12.0 \text{ g C}}\right) = 7.40 \text{ mol C} \qquad 11.2 \text{ g H}\left(\frac{1 \text{ mol H}}{1.008 \text{ g H}}\right) = 11.1 \text{ mol H}$$

Dividing both numbers of moles by 7.40 yields 1.00 mol C and 1.50 mol H. Multiplying both of these by 2 yields the empirical formula C_2H_3. The empirical formula mass is thus 27.0 amu. The number of empirical formula units in 1 mol can be calculated by using 110 amu for the molecular mass. The number must be an integer.

$$\frac{110 \text{ amu/molecule}}{27.0 \text{ amu/empirical formula unit}} = \frac{4.07 \text{ empirical formula units}}{1 \text{ molecule}}$$

The answer must be integral—in this case 4. If we had used 105 amu or 115 amu, the answer would still have been closer to the integer 4 than to any other integer. The molecular formula is thus C_8H_{12}.

7.69. A compound has a molar mass of 90.0 g/mol and its percent composition is 2.22% H, 26.7% C, and 71.1% O. What is its molecular formula?

Ans. $H_2C_2O_4$

7.70. Combine Figures 7-1, 7-2, and 7-3 into one figure. List all the conversions possible using the combined figure.

Ans.   The figure is presented as Fig. 7-4. One can convert from mass to moles, moles of component elements, or number of formula units. Additionally, one can convert from number of formula units to moles, to moles of component elements, or to mass; also from moles of component elements to moles of compound, number of formula units of compound, or mass of compound; finally, from moles of compound to number of formula units, mass, or number of moles of component elements.

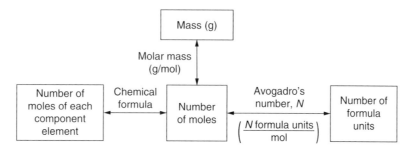

Fig. 7-4.   Conversions involving moles

7.71. To Fig. 7-4 (Problem 7.70), add a box for the mass of the substance in atomic mass units, and also add the factor labels with which that box can be connected to two others. Add another box to include the number of individual atoms of each element.

Ans.   The answer is shown on the factor-label conversion figure, page 348.

7.72. (*a*) Define the unit *millimole* (mmol). (*b*) How many gold atoms are there in 1 mmol of Au

Ans.   (*a*) 1 mmol = 0.001 mol   (*b*) 6.02×10^{20} Au atoms

Chemical Equations

8.1. INTRODUCTION

A chemical reaction is described by means of a shorthand notation called a *chemical equation*. One or more substances, called *reactants* or *reagents*, are allowed to react to form one or more other substances, called *products*. Instead of using words, equations are written using the formulas for the substances involved. For example, a reaction used to prepare oxygen may be described in words as follows:

Mercury(II) oxide, when heated, yields oxygen gas plus mercury.

Using the formulas for the substances involved, the process could be written

$$HgO \xrightarrow{\text{heat}} O_2 + Hg \qquad \text{(unfinished)}$$

A chemical equation describes a chemical reaction in many ways as an empirical formula describes a chemical compound. The equation describes not only which substances react, but also the relative number of moles of each reactant and product. Note especially that it is the mole ratios in which the substances *react*, not how much is *present*, that the equation describes. To show the quantitative relationships, the equation must be *balanced*. That is, it must have the same number of atoms of each element used up and produced, except for special equations that describe nuclear reactions (Chap. 19). The law of conservation of mass is thus obeyed as well as the "law of conservation of atoms." *Coefficients* are written in front of the formulas for elements and compounds to tell how many formula units of that substance are involved in the reaction. A coefficient does not imply any chemical bonding between units of the substance it is placed before. The number of atoms involved in each formula unit is multiplied by the coefficient to get the total number of atoms of each element involved. Later, when equations with individual ions are written (Chap. 9), the net charge on each side of the equation, as well as the numbers of atoms of each element, must be the same to have a balanced equation. The absence of a coefficient in a *balanced* equation implies a coefficient of 1.

The balanced equation for the decomposition of HgO is

$$2\,HgO \xrightarrow{\text{heat}} O_2 + 2\,Hg$$
$$\underline{\qquad\text{coefficients}\qquad}$$

EXAMPLE 8.1. Draw a ball-and-stick diagram depicting the reaction represented by the following equation:

$$2\,H_2 + O_2 \longrightarrow 2\,H_2O$$

Ans. The diagram is shown in Fig. 8-1.

120

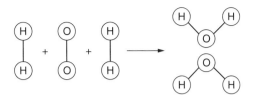

Fig. 8-1. The reaction of hydrogen and oxygen

EXAMPLE 8.2. Interpret the following equation:

$$2\,Na + Cl_2 \longrightarrow 2\,NaCl$$

Ans.   The equation states that elemental sodium reacts with elemental chlorine to produce sodium chloride, table salt. (The fact that chlorine is one of the seven elements that occur in diatomic molecules when not combined with other elements is indicated.) The numbers before Na and NaCl are coefficients, stating how many formula units of these substances are involved. If there is no coefficient in a *balanced equation*, a coefficient of 1 is implied, and so the absence of a coefficient before the Cl_2 implies one Cl_2 molecule. The equation thus states that when the two reagents react, they do so in a ratio of two atoms of sodium to one molecule of chlorine, to form two formula units of sodium chloride. In addition, it states that when these two reagents react, they do so in a ratio of 2 mol of sodium atoms to 1 mol of chlorine molecules, to form 2 mol of sodium chloride. The ratios of moles of each substance involved to every other substance are implied:

$$\frac{2\ \text{mol Na}}{1\ \text{mol Cl}_2} \qquad \frac{2\ \text{mol Na}}{2\ \text{mol NaCl}} \qquad \frac{1\ \text{mol Cl}_2}{2\ \text{mol NaCl}}$$

$$\frac{1\ \text{mol Cl}_2}{2\ \text{mol Na}} \qquad \frac{2\ \text{mol NaCl}}{2\ \text{mol Na}} \qquad \frac{2\ \text{mol NaCl}}{1\ \text{mol Cl}_2}$$

EXAMPLE 8.3. How many ratios are implied for a reaction in which four substances are involved?

Ans.   12. (Each of the four coefficients has a ratio with the other three, and three 4s equal 12.)

8.2. BALANCING SIMPLE EQUATIONS

If you know the reactants and products of a chemical reaction, you should be able to write an equation for the reaction and balance it. In writing the equation, *first write the correct formulas for all reactants and products.* After they are written, only then start to balance the equation. Do not balance the equation by changing the formulas of the substances involved. For simple equations, you should balance the equation "by inspection." (Balancing of oxidation-reduction equation will be presented in Chap. 14.) The following rules will help you to balance simple equations.

1.   Before the equation is balanced, there are no coefficients for any reactant or product; after the equation is balanced, the absence of a coefficient implies a coefficient of 1. Assume a coefficient of 1 for the most complicated substance in the equation. (Since you are getting ratios, you can assume any value for one of the substances.) Then work from this substance to figure out the coefficients of the others, one at a time. To prevent ambiguity while you are balancing the equation, place a question mark before every other substance.

2.   Replace each question mark as you figure out each real coefficient. After you get some practice, you will not need to use the question marks. If an element appears in more than one reactant or more than one product, balance that element last.

3.   Optionally, if a polyatomic ion that does not change during the reaction is involved, you may treat the whole thing as one unit, instead of considering the atoms that make it up.

4.   After you have provided coefficients for all the substances, if any fractions are present, multiply every coefficient by the same small integer to clear the fractions.

5. Eliminate any coefficients equal to 1.

6. Always check to see that you have the same number of atoms of each element on both sides of the equation after you finish.

EXAMPLE 8.4.   Sulfuric acid, H_2SO_4, reacts with excess sodium hydroxide to produce sodium sulfate and water. Write a balanced chemical equation for the reaction.

Ans.   Step 1:   Write down the formulas for reactants and products. Assume a coefficient of 1 for a complicated reactant or product. Add question marks:

$$1\ H_2SO_4 + ?\ NaOH \longrightarrow ?\ Na_2SO_4 + ?\ H_2O$$

Step 2:

Balance the S	$1\ H_2SO_4 + ?\ NaOH \longrightarrow 1\ Na_2SO_4 + ?\ H_2O$
Balance the Na	$1\ H_2SO_4 + 2\ NaOH \longrightarrow 1\ Na_2SO_4 + ?\ H_2O$
Balance the H	$1\ H_2SO_4 + 2\ NaOH \longrightarrow 1\ Na_2SO_4 + 2\ H_2O$

Since the coefficient of H_2SO_4 is 1, there is one sulfur atom on the left of the equation. Sulfur appears in only one product, and so that product must have a coefficient of 1. The one Na_2SO_4 has two Na atoms in it, and so there must be two Na atoms on the left; the NaOH gets a coefficient of 2. There are two H atoms in H_2SO_4 and two more in two NaOH, and so two water molecules are produced. The oxygen atoms are balanced, with six on each side.
Step 4 is not necessary.

Step 5:   We drop the coefficients of 1 to finish our equation.

Eliminate the 1s $H_2SO_4 + 2\ NaOH \longrightarrow Na_2SO_4 + 2\ H_2O$

Step 6:   *Check:* We find four H atoms, one S atom, two Na atoms, and six O atoms on each side. Alternatively (step 3), we count four H atoms, one SO_4 group, two Na atoms, and two other O atoms on each side of the equation.

EXAMPLE 8.5.   Magnesium metal reacts with HCl to produce $MgCl_2$ and hydrogen gas. Write a balanced equation for the process.

Ans.   Step 1: $?\ Mg + ?\ HCl \longrightarrow 1\ MgCl_2 + ?\ H_2$

We note that hydrogen is one of the seven elements that form diatomic molecules when in the elemental state.

Step 2: $1\ Mg + 2\ HCl \longrightarrow 1\ MgCl_2 + ?\ H_2$
                                  $1\ Mg + 2\ HCl \longrightarrow 1\ MgCl_2 + 1\ H_2$

Step 5: $Mg + 2\ HCl \longrightarrow MgCl_2 + H_2$

Step 6:   There are one Mg atom, two H atoms, and two Cl atoms on each side of the equation.

EXAMPLE 8.6.   Balance the following equation:

$$CoF_3 + NaI \longrightarrow NaF + CoI_2 + I_2$$

Ans.   Step 1: $1\ CoF_3 + ?\ NaI \longrightarrow ?\ NaF + ?\ CoI_2 + ?\ I_2$

Step 2: $1\ CoF_3 + 3\ NaI \longrightarrow 3\ NaF + 1\ CoI_2 + \frac{1}{2}\ I_2$
                          3rd   2nd   2nd   4th

Steps 4 and 5: $2\ CoF_3 + 6\ NaI \longrightarrow 6\ NaF + 2\ CoI_2 + I_2$

8.3.   PREDICTING THE PRODUCTS OF A REACTION

Before you can balance a chemical equation, you have to know the formulas for all the reactants and products. If the names are given for these substances, you have to know how to write formulas from the names (Chap. 6). If only reactants are given, you have to know how to predict the products from the reactants. This latter topic is the subject of this section.

To simplify the discussion, we will classify simple chemical reactions into five types:

Type 1: combination reactions   Type 4: double-substitution reactions
Type 2: decomposition reactions   Type 5: combustion reactions
Type 3: substitution reactions

More complex oxidation-reduction reactions will be discussed in Chap. 14.

Combination Reactions

A combination reaction is a reaction of two reactants to produce one product. The simplest combination reactions are the reactions of two elements to form a compound. After all, if two elements are treated with each other, they can either react or not. There generally is no other possibility, since neither can decompose. In most reactions like this, there will be a reaction. The main problem is to write the formula of the product correctly and then to balance the equation. In this process, first determine the formula of the product from the rules of chemical combination (Chap. 5). Only after the formulas of the reactants and products have all been written down, balance the equation by adjusting the coefficients.

EXAMPLE 8.7.   Complete and balance the following equations:

(a) $Na + F_2 \longrightarrow$ (b) $Mg + O_2 \longrightarrow$ (c) $K + S \longrightarrow$

Ans.   The products are determined first (from electron dot structures, if necessary) to be NaF, MgO, and K_2S, respectively. These are placed to the right of the respective arrows, and the equations are then balanced.

(a) $2\,Na + F_2 \longrightarrow 2\,NaF$ (b) $2\,Mg + O_2 \longrightarrow 2\,MgO$ (c) $2\,K + S \longrightarrow K_2S$

EXAMPLE 8.8.   Write a complete, balanced equation for the reaction of each of the following pairs of elements:

(a)   aluminum and sulfur   (b)   aluminum and iodine   (c)   aluminum and oxygen

Ans.   (a)   The reactants are Al and S. The Al can lose three electrons [it is in periodic group IIIA (3)], and each sulfur atom can gain two electrons [it is in periodic group VIA (16)]. The ratio of aluminum to sulfur atoms is thus $2:3$, and the compound which will be formed is Al_2S_3:

$$Al + S \longrightarrow Al_2S_3 \text{ (unbalanced)}$$
$$2\,Al + 3\,S \longrightarrow Al_2S_3$$

(b)   The reactants are Al and I_2. (In its elemental form, iodine is stable as diatomic molecules.) The combination of a group IIIA (13) metal and a group VIIA (17) nonmetal produces a salt with a $1:3$ ratio of atoms: AlI_3.

$$Al + I_2 \longrightarrow AlI_3 \quad \text{(unbalanced)}$$
$$2\,Al + 3\,I_2 \longrightarrow 2\,AlI_3$$

(c)

$$Al + O_2 \longrightarrow Al_2O_3 \quad \text{(unbalanced)}$$
$$4\,Al + 3\,O_2 \longrightarrow 2\,Al_2O_3$$

It is possible for an element and a compound of that element or for two compounds containing a common element to react by combination. For example,

$$MgO + SO_3 \longrightarrow MgSO_4$$
$$PtF_2 + F_2 \longrightarrow PtF_4$$

Decomposition Reactions

The second type of simple reaction is *decomposition*. This reaction is also easy to recognize. Typically, only one reactant is given. A type of energy, such as heat or electricity, may also be indicated. The reactant usually decomposes to its elements, to an element and a simpler compound, or to two simpler compounds.

Binary compounds may yield two elements or an element and a simpler compound. *Ternary* (three-element) *compounds* may yield an element and a compound or two simpler compounds. These possibilities are shown in Fig. 8-2.

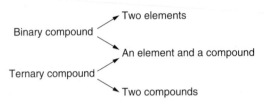

Fig. 8-2. Decomposition possibilities

A *catalyst* is a substance that speeds up a chemical reaction without undergoing a permanent change in its own composition. Catalysts are often but not always noted above or below the arrow in the chemical equation. Since a small quantity of catalyst is sufficient to cause a large quantity of reaction, the amount of catalyst need not be specified; it is not balanced as the reactants and products are. In this manner, the equation for a common laboratory preparation of oxygen is written as

$$2 \text{ KClO}_3 \xrightarrow{\text{MnO}_2} 2 \text{ KCl} + 3 \text{ O}_2$$

EXAMPLE 8.9. Write a complete, balanced equation for the reaction that occurs when (*a*) Ag_2O is heated, (*b*) H_2O is electrolyzed, and (*c*) $CaCO_3$ is heated.

Ans.   (*a*)   With only one reactant, what can happen? No simpler compound of Ag and O is evident, and the compound decomposes to its elements. Remember that oxygen occurs in diatomic molecules when it is uncombined:

$$Ag_2O \longrightarrow Ag + O_2 \qquad \text{(unbalanced)}$$
$$2\, Ag_2O \longrightarrow 4\, Ag + O_2$$

   (*b*)   Note that in most of these cases, energy of some type is added to make the compound decompose.

$$2 \text{ H}_2\text{O} \xrightarrow{\text{electricity}} 2 \text{ H}_2 + \text{ O}_2$$

   (*c*)   A ternary compound does not yield three elements; this one yields two simpler compounds.

$$CaCO_3 \longrightarrow CaO + CO_2$$

Substitution or Replacement Reactions

Elements have varying abilities to combine. Among the most reactive metals are the alkali metals and the alkaline earth metals. On the opposite end of the scale of reactivities, among the least active metals or the most stable metals are silver and gold, prized for their lack of reactivity. *Reactive* means the opposite of *stable*, but means the same as *active*.

When a free element reacts with a compound of different elements, the free element will replace one of the elements in the compound if the free element is more reactive than the element it replaces. In general, a free metal will replace the metal in the compound, or a free nonmetal will replace the nonmetal in the compound. A new compound and a new free element are produced. As usual, the formulas of the products are written according to the rules in Chap. 5. The formula of a product does not depend on the formula of the reacting element or compound. For example, consider the reactions of sodium with iron(II) chloride and of fluorine with aluminum oxide:

$$2 \text{ Na} + \text{FeCl}_2 \longrightarrow 2 \text{ NaCl} + \text{Fe}$$
$$6 \text{ F}_2 + 2 \text{ Al}_2\text{O}_3 \longrightarrow 4 \text{ AlF}_3 + 3 \text{ O}_2$$

Sodium, a metal, replaces iron, another metal. Fluorine, a nonmetal, replaces oxygen, another nonmetal. (In some high-temperature reactions, a nonmetal can displace a relatively inactive metal from its compounds.) The formulas are written on the basis of the rules of chemical bonding (Chap. 5).

You can easily recognize the possibility of a substitution reaction because you are given a free element and a compound of different elements.

EXAMPLE 8.10.   Look only at the reactants in the following equations. Tell which of the reactions represent substitution reactions.

(a) $3 Mg + N_2 \longrightarrow Mg_3N_2$ (c) $2 KClO_3 \longrightarrow 2 KCl + 3 O_2$

(b) $2 Li + MgO \longrightarrow Li_2O + Mg$ (d) $2 CrCl_2 + Cl_2 \longrightarrow 2 CrCl_3$

Ans.   Reaction (b) only. (a) is a combination, (c) is a decomposition, and (d) is a combination. Note that in (d) elemental chlorine is added to a compound of chlorine.

If the free element is less active than the corresponding element in the compound, no reaction will take place. A short list of metals and an even shorter list of nonmetals in order of their reactivities are presented in Table 8-1. The metals in the list range from very active to very stable; the nonmetals listed range from very active to fairly active. A more comprehensive list, a table of standard reduction potentials, is presented in general chemistry textbooks.

Table 8-1   Relative Reactivities of Some Metals and Nonmetals

	Metals	Nonmetals	
Most active metals	Alkali and alkaline earth metals	F	Most active nonmetal
		O	
	Al	Cl	
	Zn	Br	
	Fe	I	Less active nonmetals
	Pb		
	H		
	Cu		
Less active metals	Ag		
	Au		

EXAMPLE 8.11.   Complete and balance the following equations. If no reaction occurs, indicate that fact by writing "NR".

(a) $KCl + Fe \longrightarrow$ (b) $KF + Cl_2 \longrightarrow$

Ans.   (a) $KCl + Fe \longrightarrow NR$ (b) $KF + Cl_2 \longrightarrow NR$
In each of these cases, the free element is less active than the corresponding element in the compound, and cannot replace that element from its compound.

In substitution reactions, hydrogen in its compounds with nonmetals often acts as a metal; hence, it is listed among the metals in Table 8-1.

EXAMPLE 8.12.   Complete and balance the following equation. If no reaction occurs, indicate that fact by writing "NR".

$$Al + HCl \longrightarrow$$

Ans. $2 Al + 6 HCl \longrightarrow 2 AlCl_3 + 3 H_2$

Aluminum is more reactive than hydrogen (Table 8-1) and replaces it from its compounds. Note that free hydrogen is in the form H_2 (Sec. 5.2).

In substitution reactions with acids, metals that can form two different ions in their compounds generally form the one with the lower charge. For example, iron can form Fe^{2+} and Fe^{3+}. In its reaction with HCl, $FeCl_2$

is formed. In contrast, in combination with the free element, the higher-charged ion is often formed if sufficient nonmetal is available.

$$2\,Fe + 3\,Cl_2 \longrightarrow 2\,FeCl_3$$

See Table 6-4 for the charges on some common metal ions.

Double-Substitution or Double-Replacement Reactions

Double-substitution or *double-replacement reactions*, also called *double-decomposition reactions* or *metathesis reactions*, involve two ionic compounds, most often in aqueous solution. In this type of reaction, the cations simply swap anions. The reaction proceeds if a solid or a covalent compound is formed from ions in solution. All gases at room temperature are covalent. Some reactions of ionic solids plus ions in solution also occur. Otherwise, no reaction takes place. For example,

$$AgNO_3 + NaCl \longrightarrow AgCl + NaNO_3$$
$$HCl(aq) + NaOH \longrightarrow NaCl + H_2O$$
$$CaCO_3(s) + 2\,HCl \longrightarrow CaCl_2 + CO_2 + H_2O$$

In the first reaction, two ionic compounds in water are mixed. The AgCl formed by the swapping of anions is insoluble, causing the reaction to proceed. The solid AgCl formed from solution is an example of a *precipitate*. In the second reaction, a covalent compound, H_2O, is formed from its ions in solution, H^+ and OH^-, causing the reaction to proceed. In the third reaction, a solid reacts with the acid in solution to produce two covalent compounds.

Since it is useful to know what state each reagent is in, we often designate the state in the equation. The designation (s) for solid, (1) for liquid, (g) for gas, or (aq) for aqueous solution may be added to the formula. Thus, a reaction of silver nitrate with sodium chloride in aqueous solution, yielding solid silver chloride and aqueous sodium nitrate, may be written as

$$AgNO_3(aq) + NaCl(aq) \longrightarrow AgCl(s) + NaNO_3(aq)$$

Just as with replacement reactions, double-replacement reactions may or may not proceed. They need a driving force. In replacement reactions the driving force is reactivity; here it is insolubility or covalence. In order for you to be able to predict if a double-replacement reaction will proceed, you must know some solubilities of ionic compounds. A short list of solubilities is given in Table 8-2.

EXAMPLE 8.13.   Complete and balance the following equation. If no reaction occurs, indicate that fact by writing "NR."

$$NaCl + KNO_3 \longrightarrow$$

Ans.
$$NaCl + KNO_3 \longrightarrow NR$$

If a double-substitution reaction had taken place, $NaNO_3$ and KCl would have been produced. However, both of these are soluble and ionic; hence, there is no driving force and therefore no reaction.

Table 8-2   Some Solubility Classes

Soluble	Insoluble
Chlorates	$BaSO_4$
Acetates	Most sulfides
Nitrates	Most oxides
Alkali metal salts	Most carbonates
Ammonium salts	Most phosphates
Chlorides, except for	$AgCl$, $PbCl_2$, Hg_2Cl_2, $CuCl$

In double-replacement reactions, the charges on the metal ions (and indeed on nonmetal ions if they do not form covalent compounds) generally remain the same throughout the reaction.

EXAMPLE 8.14. Complete and balance the following equations. If no reaction occurs, indicate that fact by writing "NR".

(a) $FeCl_3 + AgNO_3 \longrightarrow$ (b) $FeCl_2 + AgNO_3 \longrightarrow$

Ans.   (a) $FeCl_3(aq) + 3\,AgNO_3(aq) \longrightarrow Fe(NO_3)_3(aq) + 3\,AgCl(s)$

          (b) $FeCl_2(aq) + 2\,AgNO_3(aq) \longrightarrow Fe(NO_3)_2(aq) + 2\,AgCl(s)$

          (a) If you start with Fe^{3+}, you wind up with Fe^{3+}.   (b) If you start with Fe^{2+}, you wind up with Fe^{2+}.

NH_4OH and H_2CO_3 are unstable. If one of these products were expected as a product of a reaction, either NH_3 plus H_2O or CO_2 plus H_2O would be obtained instead:

$$NH_4OH \longrightarrow NH_3 + H_2O \qquad H_2CO_3 \longrightarrow CO_2 + H_2O$$

Combustion Reactions

Reactions of elements and compounds with oxygen are so prevalent that they may be considered a separate type of reaction. Compounds of carbon, hydrogen, oxygen, sulfur, nitrogen, and other elements may be burned. Of greatest importance, if a reactant contains carbon, then carbon monoxide or carbon dioxide will be produced, depending upon how much oxygen is available. Reactants containing hydrogen always produce water on burning. SO_2 and NO are other products of burning in oxygen. (To produce SO_3 requires a catalyst in a combustion reaction with O_2.)

EXAMPLE 8.15. Complete and balance the following equations:

(a) $C_4H_8 + O_2(\text{limited amount}) \longrightarrow$ (b) $C_4H_8 + O_2(\text{excess amount}) \longrightarrow$

Ans.   (a) $C_4H_8 + 4\,O_2 \longrightarrow 4\,CO + 4\,H_2O$ (b) $C_4H_8 + 6\,O_2 \longrightarrow 4\,CO_2 + 4\,H_2O$

If sufficient O_2 is available (6 mol O_2 per mole C_4H_8), CO_2 is the product. In both cases, H_2O is produced.

Acids and Bases

Generally, acids react according to the rules for replacement and double-replacement reactions given above. They are so important, however, that a special nomenclature has developed for acids and their reactions. Acids were introduced in Sec. 6.4. They may be identified by their formulas that have the H representing hydrogen written first, and by their names that contain the word *acid*. An acid will react with a base to form a *salt* and water. The process is called *neutralization*. Neutralization reactions will be used as examples in Sec. 11.3, on titration.

$$HNO_3 + NaOH \longrightarrow NaNO_3 + H_2O$$
A salt

The driving force for such reactions is the formation of water, a covalent compound.

As pure compounds, acids are covalent. When placed in water, they react with the water to form ions; it is said that they *ionize*. If they react 100% with the water, they are said to be *strong acids*. The seven common strong acids are listed in Table 8-3. All the rest are *weak*; that is, the rest ionize only a few percent and largely stay in their covalent forms. Both strong and weak acids react 100% with metal hydroxides. All soluble metal hydroxides are ionic in water.

Table 8-3   The Seven Common Strong Acids

HCl, $HClO_3$, $HClO_4$, HBr, HI, HNO_3, H_2SO_4 (first proton only)

Solved Problems

INTRODUCTION

8.1. How many oxygen atoms are there in each of the following, perhaps part of a balanced chemical equation? (a) 7 H_2O, (b) 4 $Ba(NO_3)_2$, (c) 2 $CuSO_4 \cdot 5H_2O$, and (d) 4 $VO(ClO_3)_2$.

Ans.   (a) 7, (b) 24, (c) 18, and (d) 28.

BALANCING SIMPLE EQUATIONS

8.2. Balance the following equation: $C + Cu_2O \xrightarrow{\text{heat}} CO + Cu$

Ans. $C + Cu_2O \longrightarrow CO + 2\,Cu$

8.3. Balance the following equation:

$$Ca(HCO_3)_2 + H_3PO_4 \longrightarrow Ca_3(PO_4)_2 + H_2O + CO_2$$

Ans.
$$?\,Ca(HCO_3)_2 + ?\,H_3PO_4 \longrightarrow 1\,Ca_3(PO_4)_2 + ?\,H_2O + ?\,CO_2$$
$$3\,Ca(HCO_3)_2 + 2\,H_3PO_4 \longrightarrow 1\,Ca_3(PO_4)_2 + 6\,H_2O + 6\,CO_2$$

2nd   *2nd*   *3rd*   *3rd*

$$3\,Ca(HCO_3)_2 + 2\,H_3PO_4 \longrightarrow Ca_3(PO_4)_2 + 6\,H_2O + 6\,CO_2$$

8.4. Balance the following equation:

$$NH_4Cl + KOH \longrightarrow NH_3 + H_2O + KCl$$

Ans.
$$1\,NH_4Cl + ?\,KOH \longrightarrow ?\,NH_3 + ?\,H_2O + ?\,KCl$$

Balance N and Cl : $1\,NH_4Cl + ?\,KOH \longrightarrow 1\,NH_3 + ?\,H_2O + 1\,KCl$

Balance K : $1\,NH_4Cl + 1\,KOH \longrightarrow 1\,NH_3 + ?\,H_2O + 1\,KCl$

Balance H : $1\,NH_4Cl + 1\,KOH \longrightarrow 1\,NH_3 + 1\,H_2O + 1\,KCl$

Eliminate 1s : $NH_4Cl + KOH \longrightarrow NH_3 + H_2O + KCl$

8.5. Balance the following equation

$$NH_4Cl + NaOH + AgCl \longrightarrow Ag(NH_3)_2Cl + NaCl + H_2O$$

Ans.
$$?\,NH_4Cl + ?\,NaOH + ?\,AgCl \longrightarrow 1\,Ag(NH_3)_2Cl + ?\,NaCl + ?\,H_2O$$
$$2\,NH_4Cl + 2\,NaOH + 1\,AgCl \longrightarrow 1\,Ag(NH_3)_2Cl + 2\,NaCl + 2\,H_2O$$

2nd   *4th*   *2nd*   *3rd*   *5th*

$$2\,NH_4Cl + 2\,NaOH + AgCl \longrightarrow Ag(NH_3)_2Cl + 2\,NaCl + 2\,H_2O$$

8.6. Balance the following chemical equations:

(a) $NCl_3 + H_2O \longrightarrow HClO + NH_3$

(b) $NaOH + H_3PO_4 \longrightarrow Na_2HPO_4 + H_2O$

(c) $Al + HCl \longrightarrow AlCl_3 + H_2$

(d) $HCl + Mg \longrightarrow MgCl_2 + H_2$

(e) $SrCO_3 + HClO_4 \longrightarrow Sr(ClO_4)_2 + CO_2 + H_2O$

(f) $KC_2H_3O_2 + HBr \longrightarrow KBr + HC_2H_3O_2$

(g) $Ba(OH)_2 + H_3PO_4 \longrightarrow BaHPO_4 + H_2O$

(h) $HCl + Na_3PO_4 \longrightarrow NaCl + NaH_2PO_4$

Ans.   (a) $NCl_3 + 3\,H_2O \longrightarrow 3\,HClO + NH_3$

(b) $2\,NaOH + H_3PO_4 \longrightarrow Na_2HPO_4 + 2\,H_2O$

(c) $2\,Al + 6\,HCl \longrightarrow 2\,AlCl_3 + 3\,H_2$

(d) $2\,HCl + Mg \longrightarrow MgCl_2 + H_2$

(e) $SrCO_3 + 2\,HClO_4 \longrightarrow Sr(ClO_4)_2 + CO_2 + H_2O$

(f) $KC_2H_3O_2 + HBr \longrightarrow KBr + HC_2H_3O_2$

(g) $Ba(OH)_2 + H_3PO_4 \longrightarrow BaHPO_4 + 2\,H_2O$

(h) $2\,HCl + Na_3PO_4 \longrightarrow 2\,NaCl + NaH_2PO_4$

8.7. Write balanced equations for each of the following reactions:

(a) Sodium plus oxygen yields sodium peroxide.

(b) Mercury(II) oxide, when heated, yields mercury and oxygen.

(c) Carbon plus oxygen yields carbon monoxide.

(d) Sulfur plus oxygen yields sulfur dioxide.

(e) Propane (C_3H_8) plus oxygen yields carbon dioxide plus water.

(f) Ethane (C_2H_6) plus oxygen yields carbon monoxide plus water.

(g) Ethylene (C_2H_4) plus oxygen yields carbon dioxide plus water.

Ans. (a) $2\,Na + O_2 \longrightarrow Na_2O_2$ (e) $C_3H_8 + 5\,O_2 \longrightarrow 3\,CO_2 + 4\,H_2O$

(b) $2\,HgO \xrightarrow{\text{heat}} 2\,Hg + O_2$ (f) $2\,C_2H_6 + 5\,O_2 \longrightarrow 4\,CO + 6\,H_2O$

(c) $2\,C + O_2 \longrightarrow 2\,CO$ (g) $C_2H_4 + 3\,O_2 \longrightarrow 2\,CO_2 + 2\,H_2O$

(d) $S + O_2 \longrightarrow SO_2$

8.8. Write balanced chemical equations for the following reactions:

(a) Sodium plus fluorine yields sodium fluoride.

(b) Potassium chlorate, when heated, yields potassium chloride plus oxygen.

(c) Zinc plus copper(II) nitrate yields zinc nitrate plus copper.

(d) Magnesium hydrogen carbonate plus heat yields magnesium carbonate plus carbon dioxide plus water.

(e) Magnesium hydrogen carbonate plus hydrobromic acid yields magnesium bromide plus carbon dioxide plus water.

(f) Lead(II) acetate plus sodium chromate yields lead(II) chromate plus sodium acetate.

Ans. (a) $2\,Na + F_2 \longrightarrow 2\,NaF$ (Remember that free fluorine is F_2.)

(b) $2\,KClO_3 \longrightarrow 2\,KCl + 3\,O_2$ (Remember that free oxygen is O_2.)

(c) $Zn + Cu(NO_3)_2 \longrightarrow Zn(NO_3)_2 + Cu$

(d) $Mg(HCO_3)_2 \longrightarrow MgCO_3 + H_2O + CO_2$

(e) $Mg(HCO_3)_2 + 2\,HBr \longrightarrow MgBr_2 + 2\,H_2O + 2\,CO_2$

(f) $Pb(C_2H_3O_2)_2 + Na_2CrO_4 \longrightarrow PbCrO_4 + 2\,NaC_2H_3O_2$

8.9. Balance the following equation: $Ba(ClO_4)_2 + Na_2SO_4 \longrightarrow BaSO_4 + NaClO_4$

Ans. $Ba(ClO_4)_2 + Na_2SO_4 \longrightarrow BaSO_4 + 2\,NaClO_4$
The oxygen atoms need not be considered individually if the ClO_4^- and SO_4^{2-} ions are considered as groups. (See step 3, Sec. 8.2.)

8.10. Write balanced chemical equations for the following reactions:
(a) Phosphorus pentachloride plus water yields phosphoric acid plus hydrogen chloride.

(b) Sodium hydroxide plus sulfuric acid yields sodium hydrogen sulfate plus water.

(c) Ethane, C_2H_6, plus oxygen yields carbon dioxide plus water.

(d) Octane, C_8H_{18}, plus oxygen yields carbon monoxide plus water.

(e)   Copper(II) chloride plus hydrosulfuric acid yields copper(II) sulfide plus hydrochloric acid.

(f)   Barium hydroxide plus chloric acid yields barium chlorate plus water.

(g)   Copper(II) sulfate plus water yields copper(II) sulfate pentahydrate.

(h)   Copper(II) chloride plus sodium iodide yields copper(I) iodide plus iodine plus sodium chloride.

Ans.   (a) $PCl_5 + 4\,H_2O \longrightarrow H_3PO_4 + 5\,HCl$

     (b) $NaOH + H_2SO_4 \longrightarrow NaHSO_4 + H_2O$

     (c) $2\,C_2H_6 + 7\,O_2 \longrightarrow 4\,CO_2 + 6\,H_2O$

     (d) $2\,C_8H_{18} + 17\,O_2 \longrightarrow 16\,CO + 18\,H_2O$

     (e) $CuCl_2 + H_2S \longrightarrow CuS + 2\,HCl$

     (f) $Ba(OH)_2 + 2\,HClO_3 \longrightarrow Ba(ClO_3)_2 + 2\,H_2O$

     (g) $CuSO_4 + 5\,H_2O \longrightarrow CuSO_4 \cdot 5\,H_2O$

     (h) $2\,CuCl_2 + 4\,NaI \longrightarrow 2\,CuI + I_2 + 4\,NaCl$

8.11.   Balance the following equation: $Cr + CrCl_3 \longrightarrow CrCl_2$

    Ans.   Balance the Cl first, since the Cr appears in two reactants. Here, the Cr happens to be balanced automatically.

$$Cr + 2\,CrCl_3 \longrightarrow 3\,CrCl_2$$

8.12.   Write balanced chemical equations for the following reactions: (a) Hydrogen chloride is produced by the reaction of hydrogen and chlorine. (b) Hydrogen combines with chlorine to yield hydrogen chloride. (c) Chlorine reacts with hydrogen to give hydrogen chloride.

    Ans.   (a), (b), and (c). $H_2 + Cl_2 \longrightarrow 2\,HCl$

8.13.   Write balanced chemical equations for the following reactions: (a) Hydrogen fluoride is produced by the reaction of hydrochloric acid and sodium fluoride. (b) Hydrochloric acid combines with sodium fluoride to yield hydrogen fluoride. (c) Sodium fluoride reacts with hydrochloric acid to give hydrofluoric acid.

    Ans.   (a), (b), and (c). $HCl + NaF \longrightarrow NaCl + HF$

8.14.   Balance the following chemical equations:

    (a) $Na_2CO_3 + HClO_3 \longrightarrow NaClO_3 + CO_2 + H_2O$

    (b) $Ca(HCO_3)_2 + HCl \longrightarrow CaCl_2 + CO_2 + H_2O$

    (c) $BiCl_3 + H_2O \longrightarrow BiOCl + HCl$

    (d) $H_2S + O_2 \longrightarrow H_2O + SO_2$

    (e) $Cu_2S + O_2 \longrightarrow Cu + SO_2$

    (f) $NH_3 + O_2 \longrightarrow NO + H_2O$

    (g) $H_2O_2 \longrightarrow H_2O + O_2$

    Ans.   (a) $Na_2CO_3 + 2\,HClO_3 \longrightarrow 2\,NaClO_3 + CO_2 + H_2O$

     (b) $Ca(HCO_3)_2 + 2\,HCl \longrightarrow CaCl_2 + 2\,CO_2 + 2\,H_2O$

     (c) $BiCl_3 + H_2O \longrightarrow BiOCl + 2\,HCl$

     (d) $2\,H_2S + 3\,O_2 \longrightarrow 2\,H_2O + 2\,SO_2$

     (e) $Cu_2S + O_2 \longrightarrow 2\,Cu + SO_2$

     (f) $4\,NH_3 + 5\,O_2 \longrightarrow 4\,NO + 6\,H_2O$

     (g) $2\,H_2O_2 \longrightarrow 2\,H_2O + O_2$

8.15. Balance the following chemical equations:

(a) $Pb(NO_3)_2 + KI \longrightarrow PbI_2 + KNO_3$ (d) $NH_3 + CuCl_2 \longrightarrow Cu(NH_3)_4Cl$

(b) $H_2S + CuCl_2 \longrightarrow HCl + CuS$ (e) $As_2S_3 + Na_2S \longrightarrow NaAsS_2$

(c) $Hg_2Cl_2 + NH_3 \longrightarrow HgNH_2Cl + Hg + NH_4Cl$

Ans.   (a) $Pb(NO_3)_2 + 2\,KI \longrightarrow PbI_2 + 2\,KNO_3$ (d) $4\,NH_3 + CuCl_2 \longrightarrow Cu(NH_3)_4Cl$

  (b) $H_2S + CuCl_2 \longrightarrow 2\,HCl + CuS$ (e) $As_2S_3 + Na_2S \longrightarrow 2\,NaAsS_2$

  (c) $Hg_2Cl_2 + 2\,NH_3 \longrightarrow HgNH_2Cl + Hg + NH_4Cl$

8.16. Why is the catalyst not merely placed on both sides of the arrow, since it comes out of the reaction with the same composition as it started with?

Ans.   That would imply a certain mole ratio to the other reactants and products, which is not correct.

PREDICTING THE PRODUCTS OF A REACTION

8.17. In the list of reactivities of metals, Table 8-1, are all alkali metals more reactive than all alkaline earth metals, or are all elements of both groups of metals more active than any other metals?

Ans.   Both groups of metals are more active than any other metals. Actually, some alkaline earth metals are more active than some alkali metals, and vice versa.

8.18. (a) What type of reaction requires knowledge of reactivities of elements? (b) What type requires knowledge of solubility properties of compounds?

Ans.   (a) Substitution reaction   (b) Double-substitution reaction

8.19. Complete and balance the following equations:

(a) $CH_4 + O_2(\text{limited amount}) \longrightarrow$ (b) $CH_4 + O_2(\text{excess amount}) \longrightarrow$

Ans.   (a) $2\,CH_4 + 3\,O_2(\text{limited amount}) \longrightarrow 2\,CO + 4\,H_2O$
         If suffient O_2 is available, CO_2 is the product.

  (b) $CH_4 + 2\,O_2(\text{excess amount}) \longrightarrow CO_2 + 2\,H_2O$

8.20. What type of chemical reaction is represented by each of the following? Complete and balance the equation for each.

(a) $Cl_2 + AlBr_3 \longrightarrow$ (d) $FeCl_2 + AgC_2H_3O_2 \longrightarrow$

(b) $Cl_2 + K \longrightarrow$ (e) $C_3H_8 + O_2(\text{limited}) \longrightarrow$

(c) $Na + AlCl_3 \longrightarrow$

Ans.   (a) Substitution $3\,Cl_2 + 2\,AlBr_3 \longrightarrow 2\,AlCl_3 + 3\,Br_2$
  (b) Combination $Cl_2 + 2\,K \longrightarrow 2\,KCl$
  (c) Substitution $3\,Na + AlCl_3 \longrightarrow Al + 3\,NaCl$
  (d) Double substitution $FeCl_2 + 2\,AgC_2H_3O_2 \longrightarrow Fe(C_2H_3O_2)_2 + 2\,AgCl$
  (e) Combustion $2\,C_3H_8 + 7\,O_2 \longrightarrow 6\,CO + 8\,H_2O$

8.21. What type of chemical reaction is represented by each of the following? Complete and balance the equation for each.

(a) $Cl_2 + CrCl_2 \longrightarrow$ (d) $AlCl_3 + Cl_2 \longrightarrow$

(b) $CO + O_2 \longrightarrow$ (e) $C_5H_{12} + O_2 \text{ (limited quantity)} \longrightarrow$

(c) $MgCO_3 \overset{\text{heat}}{\longrightarrow}$

Ans. (*a*) Combination $Cl_2 + 2\,CrCl_2 \longrightarrow 2\,CrCl_3$

       (*b*) Combination (or combustion) $2\,CO + O_2 \longrightarrow 2\,CO_2$

       (*c*) Decomposition $MgCO_3 \overset{heat}{\longrightarrow} CO_2 + MgO$

       (*d*) No reaction $AlCl_3 + Cl_2 \longrightarrow NR$

       (*e*) Combustion $2\,C_5H_{12} + 11\,O_2 \longrightarrow 10\,CO + 12\,H_2O$

8.22. Which of the following is soluble in water, CuCl or $CuCl_2$?

    Ans. $CuCl_2$ is soluble; CuCl is one of the four chlorides that are listed as insoluble (Table 8-2).

8.23. Complete and balance an equation for the reaction of excess HCl and Na_3PO_4.

    Ans. $3\,HCl + Na_3PO_4 \longrightarrow H_3PO_4 + 3\,NaCl$
        The phosphoric acid produced is weak, that is, mostly covalent, and the formation of the H_3PO_4 is the driving force for this reaction. (HCl is one of the seven strong acids listed in Table 8-3.)

8.24. Is F_2 soluble in water?

    Ans. No, it reacts with water, liberating oxygen:

$$2\,F_2 + 2\,H_2O \longrightarrow 4\,HF + O_2$$

Supplementary Problems

8.25. What is the difference between dissolving and reacting?

    Ans. Dissolving is a physical change, and no set ratio of substances is required.

8.26. How can you tell that the following are combination reactions rather than replacement or double-replacement reactions.

$$CoCl_2 + Cl_2 \longrightarrow \qquad HgCl_2 + Hg \longrightarrow$$
$$CO + O_2 \quad \longrightarrow \qquad BaO + CO_2 \longrightarrow$$

    Ans. The same element appears in both reactants.

8.27. State why the equation of Problem 8.2 is unusual.

    Ans. It is a substitution reaction in which a nonmetal replaces a metal. Carbon, at high temperatures, can replace relatively inactive metals from their oxides.

8.28. Complete and balance each of the following. If no reaction occurs, write "NR."

    (*a*) $SO_2 + H_2O \longrightarrow$ (*l*) $C + O_2$ (excess) $\longrightarrow$

    (*b*) $BaO + H_2O \longrightarrow$ (*m*) $KOH + HNO_3 \longrightarrow$

    (*c*) $Al + O_2 \longrightarrow$ (*n*) $HClO_3 + ZnO \longrightarrow$

    (*d*) $Na + S \longrightarrow$ (*o*) $BaCl_2 + Na_2SO_4 \longrightarrow$

    (*e*) $MgO + CO_2 \longrightarrow$ (*p*) $AgNO_3 + KCl \longrightarrow$

    (*f*) $Al + Br_2 \longrightarrow$ (*q*) $NaOH + H_2SO_4 \longrightarrow$

    (*g*) $Ag + ZnCl_2 \longrightarrow$ (*r*) $Ba(OH)_2 + HClO_2 \longrightarrow$

    (*h*) $Cl_2 + KBr \longrightarrow$ (*s*) $C + O_2$ (limited) $\longrightarrow$

    (*i*) $BaSO_4 + NaCl \longrightarrow$ (*t*) $C_5H_{10} + O_2$ (excess) $\longrightarrow$

    (*j*) $NaCl(l) \overset{electricity}{\longrightarrow}$ (*u*) $C_5H_{10} + O_2$ (limited) $\longrightarrow$

    (*k*) $KCl + HNO_3 \longrightarrow$ (*v*) $C_5H_{10}O + O_2$ (excess) $\longrightarrow$

                                           (*w*) $C_5H_{10}O + O_2$ (limited) $\longrightarrow$

Ans.   (*a*) $SO_2 + H_2O \longrightarrow H_2SO_3$

(*b*) $BaO + H_2O \longrightarrow Ba(OH)_2$

(*c*) $4\,Al + 3\,O_2 \longrightarrow 2\,Al_2O_3$

(*d*) $2\,Na + S \longrightarrow Na_2S$

(*e*) $MgO + CO_2 \longrightarrow MgCO_3$

(*f*) $2\,Al + 3\,Br_2 \longrightarrow 2\,AlBr_3$

(*g*) $Ag + ZnCl_2 \longrightarrow NR$

(*h*) $Cl_2 + 2\,KBr \longrightarrow 2\,KCl + Br_2$

(*i*) $BaSO_4 + NaCl \longrightarrow NR$

(*j*) $2\,NaCl(l) \xrightarrow{\text{electricity}} 2\,Na + Cl_2$

(*k*) $KCl + HNO_3 \longrightarrow NR$

(*l*) $C + O_2 \text{ (excess)} \longrightarrow CO_2$

(*m*) $KOH + HNO_3 \longrightarrow KNO_3 + H_2O$

(*n*) $2\,HClO_3 + ZnO \longrightarrow Zn(ClO_3)_2 + H_2O$

(*o*) $BaCl_2 + Na_2SO_4 \longrightarrow BaSO_4 + 2\,NaCl$

(*p*) $AgNO_3 + KCl \longrightarrow AgCl + KNO_3$

(*q*) $NaOH + H_2SO_4 \longrightarrow NaHSO_4 + H_2O$
        or $2\,NaOH + H_2SO_4 \longrightarrow Na_2SO_4 + 2\,H_2O$

(*r*) $Ba(OH)_2 + 2\,HClO_2 \longrightarrow Ba(ClO_2)_2 + 2\,H_2O$

(*s*) $2\,C + O_2 \text{ (limited)} \longrightarrow 2\,CO$

(*t*) $2\,C_5H_{10} + 15\,O_2 \text{ (excess)} \longrightarrow$
        $10\,CO_2 + 10\,H_2O$

(*u*) $C_5H_{10} + 5\,O_2 \text{ (limited)} \longrightarrow 5\,CO + 5\,H_2O$

(*v*) $C_5H_{10}O + 7\,O_2 \text{ (excess)} \longrightarrow 5\,CO_2 + 5\,H_2O$

(*w*) $2\,C_5H_{10}O + 9\,O_2 \text{ (limited)} \longrightarrow$
        $10\,CO + 10\,H_2O$

Net Ionic Equations

9.1. INTRODUCTION

The chemist must know many thousands of facts. One way to remember such a wide variety of information is to systematize it. The periodic table, for example, allows us to learn data about whole groups of elements instead of learning about each element individually. Net ionic equations give chemists a different way of learning a lot of information with relatively little effort, rather than one piece at a time. In this chapter we will learn:

1. What a net ionic equation means
2. What it does not mean
3. How to write net ionic equations
4. How to use net ionic equations
5. What they cannot tell us

9.2. WRITING NET IONIC EQUATIONS

When a substance made up of ions is dissolved in water, the dissolved ions behave independently. That is, they undergo their own characteristic reactions regardless of what other ions may be present. For example, silver ions in solution, Ag^+, always react with chloride ions in solution, Cl^-, to form an insoluble ionic compound, $AgCl(s)$, no matter what other ions are present in the solution. If a solution of sodium chloride, NaCl, and a solution of silver nitrate, $AgNO_3$, are mixed, a white solid, silver chloride, is produced. The solid can be separated from the solution by filtration, and the resulting solution contains sodium nitrate, just as it would if solid $NaNO_3$ were added to water. In other words, when the two solutions are mixed, the following reaction occurs:

$$AgNO_3 + NaCl \longrightarrow AgCl(s) + NaNO_3$$
or
$$Ag^+ + NO_3^- + Na^+ + Cl^- \longrightarrow AgCl(s) + Na^+ + NO_3^-$$

Written in the latter manner, the equation shows that, in effect, the sodium ions and the nitrate ions have not changed. They began as ions in solution and wound up as those same ions in solution. They are called *spectator ions.* Since they have not reacted, it is really not necessary to include them in the equation. If they are left out, a *net ionic equation* results:

$$Ag^+ + Cl^- \longrightarrow AgCl(s)$$

134

This equation may be interpreted to mean that *any* soluble silver salt will react with *any* soluble ionic chloride to produce (insoluble) silver chloride.

EXAMPLE 9.1.   Write three equations that the preceding net ionic equation can represent.

Ans.   The following equations represent three of many possible equations:

$$AgNO_3 + NaCl \longrightarrow NaNO_3 + AgCl(s)$$
$$AgClO_3 + KCl \longrightarrow KClO_3 + AgCl(s)$$
$$AgC_2H_3O_2 + NH_4Cl \longrightarrow NH_4C_2H_3O_2 + AgCl(s)$$

Obviously, it is easier to remember the net ionic equation than the many possible overall equations that it represents. (See Problem 9.12.)

Net ionic equations may be written whenever reactions occur in solution in which some of the ions originally present are removed from solution or when ions not originally present are formed. Usually, ions are removed from solution by one or more of the following processes:

1.   Formation of an insoluble ionic compound (Table 8-2)
2.   Formation of molecules containing only covalent bonds
3.   Formation of new ionic species
4.   Formation of a gas (a corollary of 2)

Examples of these processes include

1. $AgClO_3 + NaCl \longrightarrow AgCl(s) + NaClO_3$ $Ag^+ + Cl^- \longrightarrow AgCl(s)$
2. $HI + NaOH \longrightarrow H_2O + NaI$ $H^+ + OH^- \longrightarrow H_2O$
3. $Cu + 2\,AgNO_3 \longrightarrow 2\,Ag + Cu(NO_3)_2$ $Cu + 2\,Ag^+ \longrightarrow Cu^{2+} + 2\,Ag$
4. $NH_4CO_3 + 2\,HCl \longrightarrow CO_2 + H_2O + 2\,NH_4Cl$ $CO_3^{2-} + 2\,H^+ \longrightarrow CO_2 + H_2O$

The question arises how the student can tell whether a compound is ionic or covalent. The following generalizations will be of some help in deciding:

1.   Binary compounds of two nonmetals are covalently bonded. However, strong acids (Table 8-3) in water form ions completely.
2.   Binary compounds of a metal and nonmetal are usually ionic.
3.   Ternary compounds are usually ionic, at least in part, except if they contain no metal atoms or ammonium ion.

EXAMPLE 9.2.   Predict which of the following will contain ionic bonds: (*a*) $NiCl_2$, (*b*) SO_2, (*c*)Al_2O_3, (*d*) NH_4NO_3, (*e*) H_2SO_4, (*f*) HCl, and (*g*) NCl_3.

Ans.   (*a*) $NiCl_2$, (*c*) Al_2O_3, and (*d*) NH_4NO_3 contain ionic bonds. NH_4NO_3 also has covalent bonds within each ion. (*e*) H_2SO_4 and (*f*) HCl would form ions if allowed to react with water.

When do we write compounds as separate ions, and when do we write them as complete compounds? Ions can act independently in solution, and so we write ionic compounds as separate ions only when they are soluble. We write compounds together when they are not ionic or when they are not in solution.

EXAMPLE 9.3.   Write each of the following compounds as it should be written in an ionic equation. (*a*) KCl, (*b*) $BaSO_4$, (*c*) SO_2, and (*d*) $Ca(HCO_3)_2$.

Ans.   (*a*) $K^+ + Cl^-$ (*b*) $BaSO_4$ (insoluble)   (*c*) SO_2 (not ionic)   (*d*) $Ca^{2+} + 2\,HCO_3{}^-$
The insoluble compound is written as one compound even though it is ionic. The covalent compound is written together because it is not ionic.

EXAMPLE 9.4.   Each of the following reactions produces 56 kilojoules (kJ) per mole of water produced. Is this just a coincidence? If not, explain why the same value is obtained each time.

$$KOH + HCl \longrightarrow KCl + H_2O \qquad LiOH + HBr \longrightarrow LiBr + H_2O$$
$$NaOH + HNO_3 \longrightarrow NaNO_3 + H_2O \qquad RbOH + HI \longrightarrow RbI + H_2O$$

Ans.   The same quantity of heat is generated per mole of water formed in each reaction because it is really the same reaction in each case:

$$OH^- + H^+ \longrightarrow H_2O$$

It does not matter whether it is a K^+ ion in solution that undergoes no reaction or an Na^+ ion in solution that undergoes no reaction. As long as the spectator ions undergo no reaction, they do not contribute anything to the heat of the reaction.

EXAMPLE 9.5.   Write a net ionic equation for the reaction of aqueous $Ba(OH)_2$ with aqueous HNO_3.

Ans.   The overall equation is

$$Ba(OH)_2(aq) \;+\; 2\,HNO_3 \longrightarrow Ba(NO_3)_2 + 2\,H_2O$$

In ionic form:

$$Ba^{2+} + 2\,OH^- + 2\,H^+ + 2\,NO_3{}^- \longrightarrow Ba^{2+} + 2\,NO_3{}^- + 2\,H_2O$$

Leaving out the spectator ions yields

$$2\,OH^- + 2\,H^+ \longrightarrow 2\,H_2O$$

Dividing each side by 2 yields the net ionic equation

$$OH^- + H^+ \longrightarrow H_2O$$

The net ionic equation is the same as that in Example 9.4.

Writing net ionic equations does not imply that any solution can contain only positive ions or only negative ions. For example, the net ionic equation

$$Ba^{2+} + SO_4{}^{2-} \longrightarrow BaSO_4(s)$$

does not imply that there is any solution containing Ba^{2+} ions with no negative ions, or any solution containing $SO_4{}^{2-}$ ions with no positive ions. It merely implies that whatever negative ion is present with the barium ion and whatever positive ion is present with the sulfate ion, these unspecified ions do not make any difference to the reaction that will occur.

Net ionic equations must always have the same net charge on each side of the equation. (The same number of each type of spectator ion must be omitted from both sides of the equation.) For example, the equation

$$Cu + Ag^+ \longrightarrow Cu^{2+} + Ag \qquad \text{(unbalanced)}$$

has the same number of each type of atom on its two sides, but it is still not balanced. (One cannot add just one nitrate ion to the left side of an equation and two to the right.) The net charge must also be balanced:

$$Cu + 2\,Ag^+ \longrightarrow Cu^{2+} + 2\,Ag$$

There is a net 2+ charge on each side of the balanced equation.

Solved Problems

WRITING NET IONIC EQUATIONS

9.1. How many types of ions are generally found in any ionic compound?

Ans. Two—one type of cation and one type of anion. (Alums are an exception. They are composed of two different cations and sulfate ions.)

9.2. Which of the following compounds are ionic? Which are soluble? Which would be written as separate ions in an ionic equation as written for the first equation in this section? Write the species as they would appear in an ionic equation. (*a*) CuCl, (*b*) $NH_4C_2H_3O_2$, (*c*) Hg_2Cl_2, (*d*) $CoCl_2$, (*e*) Na_3PO_4, and (*f*) $CH_3OH(aq)$.

Ans.	Ionic	Soluble	Written Separately	Formulas in Ionic Equation
(*a*) CuCl	Yes	No	No	CuCl
(*b*) $NH_4C_2H_3O_2$	Yes	Yes	Yes	$NH_4^+ + C_2H_3O_2^-$
(*c*) Hg_2Cl_2	Yes	No	No	Hg_2Cl_2
(*d*) $CoCl_2$	Yes	Yes	Yes	$Co^{2+} + 2\,Cl^-$
(*e*) Na_3PO_4	Yes	Yes	Yes	$3\,Na^+ + PO_4^{3-}$
(*f*) $CH_3OH(aq)$	No	Yes	No	CH_3OH

9.3. Write the formulas for the ions present in each of the following compounds: (*a*) $NaClO_4$, (*b*) $BaCl_2$, (*c*) $KClO$, (*d*) $Ba(NO_3)_2$, (*e*) $LiClO_3$, (*f*) $(NH_4)_3PO_4$, (*g*) $Co_3(PO_3)_2$, (*h*) $AgCl(s)$, and (*i*) $Mg(HCO_3)_2$.

Ans.
(*a*) Na^+ and ClO_4^- (*e*) Li^+ and ClO_3^- (*h*) Ag^+ and Cl^- (even though it is a solid)

(*b*) Ba^{2+} and Cl^- (*f*) NH_4^+ and PO_4^{3-}

(*c*) K^+ and ClO^- (*g*) Co^{2+} and PO_3^{3-} (*i*) Mg^{2+} and HCO_3^-

(*d*) Ba^{2+} and NO_3^-

9.4. Write a net ionic equation for the equation in each of the following parts:

(*a*) $NH_4I + AgClO_3 \longrightarrow AgI(s) + NH_4ClO_3$

(*b*) $H_2SO_4 + BaCl_2 \longrightarrow BaSO_4 + 2\,HCl$

Ans. (*a*) $I^- + Ag^+ \longrightarrow AgI$ (*b*) $SO_4^{2-} + Ba^{2+} \longrightarrow BaSO_4$

9.5. Write a net ionic equation for the equation in each of the following parts:

(*a*) $HNO_3 + NaHCO_3 \longrightarrow NaNO_3 + CO_2 + H_2O$

(*b*) $NaOH + NH_4ClO_3 \longrightarrow NH_3 + H_2O + NaClO_3$

(*c*) $HC_2H_3O_2 + KOH \longrightarrow KC_2H_3O_2 + H_2O$

(*d*) $Ba(OH)_2(aq) + 2\,H \longrightarrow BaI_2 + 2\,H_2O$

Ans. (*a*) $H^+ + HCO_3^- \longrightarrow CO_2 + H_2O$ (HNO$_3$ is strong)

(*b*) $OH^- + NH_4^+ \longrightarrow NH_3 + H_2O$

(*c*) $HC_2H_3O_2 + OH^- \longrightarrow C_2H_3O_2^- + H_2O$ (HC$_2$H$_3$O$_2$ is weak)

(*d*) $OH^- + H^+ \longrightarrow H_2O$

9.6. Given that $BaCO_3$ is insoluble in water and that NH_3 and H_2O are covalent compounds, write net ionic equations for the following processes:

(*a*) $NH_4Cl + NaOH \longrightarrow NH_3 + H_2O + NaCl$

(*b*) $BaCl_2 + Li_2CO_3 \longrightarrow BaCO_3 + 2\,LiCl$

Ans. (*a*) $NH_4^+ + OH^- \longrightarrow NH_3 + H_2O$ (*b*) $Ba^{2+} + CO_3^{2-} \longrightarrow BaCO_3$

9.7.   Write a net ionic equation for the following overall equation:

$$Cr + 2\,Cr(NO_3)_3 \longrightarrow 3\,Cr(NO_3)_2$$

Ans. $$Cr + 3\,Cr^{3+} \longrightarrow 3\,Cr^{2+}$$

Chromium ions appear on both sides of this equation, but they are not spectator ions since they are not identical. One is a 3+ ion and the other is a 2+ ion. The neutral atom is different from both of these.

9.8.   Write a net ionic equation for each of the following overall equations:

(a) $HNO_3 + NaOH \longrightarrow NaNO_3 + H_2O$ (d) $HNO_3 + RbOH \longrightarrow RbNO_3 + H_2O$

(b) $HCl + KOH \longrightarrow KCl + H_2O$ (e) $HClO_3 + KOH \longrightarrow KClO_3 + H_2O$

(c) $HClO_4 + LiOH \longrightarrow LiClO_4 + H_2O$

Ans.   In each case, the net ionic equation is

$$H^+ + OH^- \longrightarrow H_2O$$

9.9.   Write a net ionic equation for each of the following overall equations:

(a) $HClO_3 + LiOH \longrightarrow LiClO_3 + H_2O$

(b) $HClO_3 + RbOH \longrightarrow RbClO_3 + H_2O$

Ans.   In each case, the net ionic equation is

$$H^+ + OH^- \longrightarrow H_2O$$

9.10.   Write a net ionic equation for each of the following overall equations:

(a) $H_3PO_4 + 3\,NaOH \longrightarrow Na_3PO_4 + 3\,H_2O$

(b) $MgCO_3(s) + CO_2 + H_2O \longrightarrow Mg(HCO_3)_2$

(c) $KHCO_3 + NaOH \longrightarrow K_2CO_3 + H_2O$

Ans.   (a) $H_3PO_4 + 3\,OH^- \longrightarrow PO_4^{3-} + 3\,H_2O$ (c) $HCO_3^- + OH^- \longrightarrow CO_3^{2-} + H_2O$

(b) $MgCO_3 + CO_2 + H_2O \longrightarrow Mg^{2+} + 2\,HCO_3^-$

9.11.   Write a net ionic equation for each of the following overall equations:

(a) $2\,AgNO_3 + H_2S \longrightarrow Ag_2S(s) + 2\,HNO_3$ (d) $2\,Al + 3\,HgCl_2 \longrightarrow 2\,AlCl_3 + 3\,Hg$

(b) $Zn + 2\,HCl \longrightarrow ZnCl_2 + H_2$ (e) $Zn + FeCl_2 \longrightarrow ZnCl_2 + Fe$

(c) $Cu + 2\,AgNO_3 \longrightarrow Cu(NO_3)_2 + 2\,Ag$

Ans.   (a) $2\,Ag^+ + H_2S \longrightarrow Ag_2S + 2\,H^+$ (d) $2\,Al + 3\,Hg^{2+} \longrightarrow 2\,Al^{3+} + 3\,Hg$

(b) $Zn + 2\,H^+ \longrightarrow Zn^{2+} + H_2$ (Note the overall charge balance.)

(c) $Cu + 2\,Ag^+ \longrightarrow Cu^{2+} + 2\,Ag$ (e) $Zn + Fe^{2+} \longrightarrow Fe + Zn^{2+}$

9.12.   Write six more equations that can be represented by the net ionic equation of Example 9.1. Use the same reactants that are used in the equations in Example 9.1

Ans. $$AgNO_3 + KCl \longrightarrow KNO_3 + AgCl(s)$$
$$AgClO_3 + NH_4Cl \longrightarrow NH_4ClO_3 + AgCl(s)$$
$$AgC_2H_3O_2 + NaCl \longrightarrow NaC_2H_3O_2 + AgCl(s)$$
$$AgNO_3 + NH_4Cl \longrightarrow NH_4NO_3 + AgCl(s)$$
$$AgClO_3 + NaCl \longrightarrow NaClO_3 + AgCl(s)$$
$$AgC_2H_3O_2 + KCl \longrightarrow KC_2H_3O_2 + AgCl(s)$$

9.13. A student used H_3PO_4 to prepare a test solution that was supposed to contain phosphate ions. Criticize this choice.

Ans.   Phosphoric acid is weak, and relatively few ions are present. The student should have used an ionic phosphate—sodium phosphate or ammonium phosphate, for example.

9.14. Write one or more complete equations for each of the following net ionic equations:

(*a*) $NH_3 + H^+ \longrightarrow NH_4^+$ (*c*) $CO_2 + 2\,OH^- \longrightarrow CO_3^{2-} + H_2O$

(*b*) $Cu^{2+} + S^{2-} \longrightarrow CuS$

Ans.   (*a*) $NH_3 + HNO_3 \longrightarrow NH_4NO_3$ or $NH_3 + HCl \longrightarrow NH_4Cl$
or ammonia plus any other strong acid to yield the ammonium salt.

(*b*) $CuSO_4 + K_2S \longrightarrow CuS + K_2SO_4$ or $Cu(NO_3)_2 + BaS \longrightarrow CuS + Ba(NO_3)_2$
or $CuSO_4 + (NH_4)_2S \longrightarrow CuS + (NH_4)_2SO_4$
or any soluble copper(II) salt with any soluble sulfide.

(*c*) $CO_2 + 2\,KOH \longrightarrow K_2CO_3 + H_2O$
or CO_2 plus any other soluble hydroxide to give a soluble carbonate. Note that the following equation cannot be used, because $BaCO_3$ is insoluble: $CO_2 + Ba(OH)_2 \longrightarrow BaCO_3 + H_2O$

9.15. What chemical would you use to prepare a solution to be used for a test requiring the presence of Ca^{2+} ions?

Ans. $Ca(NO_3)_2$, $CaCl_2$, or any other soluble calcium salt.

9.16. What chemical would you use to prepare a solution to be used for a test requiring the presence of Br^- ions?

Ans. $NaBr$, $BaBr_2$, $HBr(aq)$, or any other soluble ionic bromide.

9.17. What chemical would you use to prepare a solution to be used for a test requiring the presence of SO_4^{2-} ions?

Ans. Na_2SO_4, $FeSO_4$, or any other soluble sulfate.

Supplementary Problems

9.18. Write a net equation for the following overall equation:

$$4\,Au + 16\,KCN + 6\,H_2O + 3\,O_2 \longrightarrow 4\,KAu(CN)_4(aq) + 12\,KOH$$

Ans. $4\,Au + 16\,CN^- + 6\,H_2O + 3\,O_2 \longrightarrow 4\,Au(CN)_4^- + 12\,OH^-$

9.19. Write a net ionic equation for each of the following equations:

(*a*) $CdS(s) + I_2 \longrightarrow CdI_2(aq) + S$

(*b*) $4\,KOH + 4\,KMnO_4(aq) \longrightarrow 4\,K_2MnO_4(aq) + O_2 + 2\,H_2O$

(*c*) $2\,HI + 2\,HNO_2 \longrightarrow 2\,NO + 2\,H_2O + I_2$

(*d*) $AlCl_3 + 4\,NaOH \longrightarrow NaAl(OH)_4(aq) + 3\,NaCl$

Ans.   (*a*) $CdS(s) + I_2 \longrightarrow Cd^{2+} + 2\,I^- + S$

(*b*) $4\,OH^- + 4\,MnO_4^- \longrightarrow 4\,MnO_4^{2-} + O_2 + 2\,H_2O$

(*c*) $2\,H^+ + 2\,I^- + 2\,HNO_2 \longrightarrow 2\,NO + 2\,H_2O + I_2$ (HNO$_2$ is weak)

(*d*) $Al^{3+} + 4\,OH^- \longrightarrow Al(OH)_4^-$

9.20. Balance the following net ionic equations:

(a) $Ag^+ + Fe \longrightarrow Fe^{2+} + Ag$

(c) $Cu^{2+} + I^- \longrightarrow CuI + I_2$

(b) $Fe^{3+} + I^- \longrightarrow Fe^{2+} + I_2$

(d) $Cd + Cr^{3+} \longrightarrow Cr^{2+} + Cd^{2+}$

Ans. In each part, the net charge as well as the number of each type of atom must balance.

(a) $2\,Ag^+ + Fe \longrightarrow Fe^{2+} + 2\,Ag$

(c) $2\,Cu^{2+} + 4\,I^- \longrightarrow 2\,CuI + I_2$

(b) $2\,Fe^{3+} + 2\,I^- \longrightarrow 2\,Fe^{2+} + I_2$

(d) $Cd + 2\,Cr^{3+} \longrightarrow 2\,Cr^{2+} + Cd^{2+}$

9.21. Try to write a complete equation corresponding to the unbalanced and the balanced net ionic equations of the prior problem. What do you find?

Ans. You cannot write a complete equation for an unbalanced net ionic equation. [In part (a), for example, you might have one acetate ion on the left and two on the right.] One complete equation for the balanced net ionic equation might be

$$2\,AgC_2H_3O_2 + Fe \longrightarrow Fe(C_2H_3O_2)_2 + 2\,Ag$$

9.22. Would the following reaction yield 56 kJ of heat per mole of water formed, as the reactions in Example 9.4 do? Explain.

$$HC_2H_3O_2 + NaOH \longrightarrow NaC_2H_3O_2 + H_2O$$

Ans. No. Since $HC_2H_3O_2$ is a weak acid, there is a different net ionic equation and thus a different amount of heat:

$$HC_2H_3O_2 + OH^- \longrightarrow C_2H_3O_2{}^- + H_2O$$

9.23. Per mole of water formed, how much heat is generated by the reaction of Example 9.5?

Ans. 56 kJ. It is the same reaction as that of Example 9.4.

9.24. Would 56 kJ per mole of water formed by generated by the following reaction? Compare your answer with that of the prior problem.

$$Ba(OH)_2(s) + 2\,HCl \longrightarrow BaCl_2 + 2\,H_2O$$

Ans. No. It is not represented by the same net ionic equation. Some heat is involved in dissolving the solid $Ba(OH)_2$.

9.25. Write a net ionic equation for each of the following overall equations:

(a) $Ca(OH)_2(s) + 2\,HClO_3 \longrightarrow Ca(ClO_3)_2 + 2\,H_2O$ (c) $ZnS(s) + 2\,HBr \longrightarrow H_2S + ZnBr_2$

(b) $CuSO_4(aq) + H_2S \longrightarrow CuS(s) + H_2SO_4$

Ans. (a) $Ca(OH)_2(s) + 2\,H^+ \longrightarrow Ca^{2+} + 2\,H_2O$ (c) $ZnS(s) + 2\,H^+ \longrightarrow H_2S + Zn^{2+}$

(b) $Cu^{2+} + H_2S \longrightarrow CuS + 2\,H^+$

9.26. Write 12 more equations represented by the net ionic equation given in Example 9.4, using only the reactants used in that example.

Ans.

$NaOH + HCl \longrightarrow NaCl + H_2O$	$NaOH + HBr \longrightarrow NaBr + H_2O$
$RbOH + NHO_3 \longrightarrow RbNO_3 + H_2O$	$KOH + HI \longrightarrow KI + H_2O$
$RbOH + HBr \longrightarrow RbBr + H_2O$	$LiOH + HCl \longrightarrow LiCl + H_2O$
$NaOH + HI \longrightarrow NaI + H_2O$	$KOH + HNO_3 \longrightarrow KNO_3 + H_2O$
$RbOH + HCl \longrightarrow RbCl + H_2O$	$KOH + HBr \longrightarrow KBr + H_2O$
$LiOH + HNO_3 \longrightarrow LiNO_3 + H_2O$	$LiOH + HI \longrightarrow LiI + H_2O$

9.27. Indicate how you could choose different compounds to enable you to write 100 additional equations in response to Problem 9.12.

> *Ans.* Choose 34 or more soluble, ionic chlorides—the other alkali metal chlorides (4), the alkaline earth chlorides (6), the first transition metal chlorides, most of the metals with two different charges (14), many second and third transition metal chlorides (about 10), aluminum and tin(II) chloride (2). Combine each of these with each of the three silver salts that you know are soluble (given in the example), and you have over 100 overall equations.

Stoichiometry

10.1. MOLE-TO-MOLE CALCULATIONS

In chemical work, it is important to be able to calculate how much raw material is needed to prepare a certain quantity of products. It is also useful to know if a certain reaction method can prepare more product from a given quantity of material than another reaction method. Analyzing material means finding out how much of each element is present. To do the measurements, often we convert parts of the material to compounds that are easy to separate, and then we measure those compounds. All these measurements involve chemical *stoichiometry*, the science of measuring how much of one thing can be produced from certain amounts of others.

From the practical viewpoint of a student, this chapter is extremely important. The calculations introduced here are also used in the chapters on gas laws, solution chemistry, equilibrium, and other topics.

In Chap. 8, the balanced chemical equation was introduced. The equation expresses the ratios of numbers of formula units of each chemical involved in the reaction. Thus, for the reaction of aluminum with oxygen to produce aluminum oxide

$$4\,Al + 3\,O_2 \longrightarrow 2\,Al_2O_3$$

the equation states that the chemicals react in the ratio of four atoms of aluminum with three molecules of oxygen (O_2) to produce two formula units of aluminum oxide. Thus, if eight atoms of aluminum react, they will react with six molecules of oxygen, and four formula units of aluminum will be produced.

The balanced chemical equation may also be used to express the ratios of *moles* of reactants and products involved. Thus, for the reaction whose equation is given above, 4 mol of Al reacts with 3 mol of O_2 to produce 2 mol of Al_2O_3. It is also true that 8 mol of aluminum can react with 6 mol of oxygen to produce 4 mol of Al_2O_3, and so on.

EXAMPLE 10.1. How many moles of Al_2O_3 can be prepared from the reaction of 0.450 mol of O_2 plus sufficient Al?

Ans. The first step in any stoichiometry problem is to write the balanced chemical equation:

$$4\,Al + 3\,O_2 \longrightarrow 2\,Al_2O_3$$

Then the coefficients in the balanced chemical equation can be used as factors in the factor-label method to convert from moles of one chemical to moles of any other in the equation:

$$0.450 \text{ mol } O_2\left(\frac{2 \text{ mol } Al_2O_3}{3 \text{ mol } O_2}\right) = 0.300 \text{ mol } Al_2O_3$$

In the problem given before Example 10.1 with 4 mol of aluminum, it was not necessary to use the factor-label method; the numbers were easy enough to work with. However, when the numbers get even slightly complicated, it is useful to use the factor-label method. Note that any of the following factors could be used for this equation,

142

but we used the one above because it is the one that changes moles of oxygen to moles of aluminum oxide.

$$\frac{2\ mol\ Al_2O_3}{3\ mol\ O_2} \qquad \frac{2\ mol\ Al_2O_3}{4\ mol\ Al} \qquad \frac{3\ mol\ O_2}{4\ mol\ Al} \qquad \frac{3\ mol\ O_2}{2\ mol\ Al_2O_3} \qquad \frac{4\ mol\ Al}{3\ mol\ O_2} \qquad \frac{4\ mol\ Al}{2\ mol\ Al_2O_3}$$

EXAMPLE 10.2.   How many moles of oxygen does it take to react completely with 1.48 mol of aluminum?

Ans.   According to the equation in Example 10.1, it takes

$$1.48\ mol\ Al\left(\frac{3\ mol\ O_2}{4\ mol\ Al}\right) = 1.11\ mol\ O_2$$

EXAMPLE 10.3.   How many moles of NO_2 are produced by the reaction at high temperature of 1.50 mol of O_2 with sufficient N_2?

Ans.   The balanced equation is

$$N_2 + 2\ O_2 \longrightarrow 2\ NO_2$$

$$1.50\ mol\ O_2\left(\frac{2\ mol\ NO_2}{2\ mol\ O_2}\right) = 1.50\ mol\ NO_2$$

A simple figure linking the quantities, with the factor label as a bridge, is shown in Fig. 10-1.

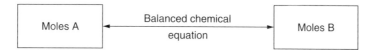

Fig. 10-1.   The conversion of moles of one reagent to moles of another, using a ratio of the coefficients of the balanced chemical equation as a factor label

In all the problems given above, a sufficient or excess quantity of a second reactant was stated in the problem. If nothing is stated about the quantity of a second (or third, etc.) reactant, it must be assumed to be present in sufficient quantity to allow the reaction to take place. Otherwise, no calculation can be done.

EXAMPLE 10.4.   How many moles of NaCl are produced by the reaction of 0.750 mol Cl_2 (with Na)?

Ans.   We must assume that there is enough sodium present. As long as we have enough sodium, we can base the calculation on the quantity of chlorine stated.

$$2\ Na + Cl_2 \longrightarrow 2\ NaCl$$

$$0.750\ mol\ Cl_2\left(\frac{2\ mol\ NaCl}{1\ mol\ Cl_2}\right) = 1.50\ mol\ NaCl$$

10.2.   CALCULATIONS INVOLVING OTHER QUANTITIES

The balanced equation expresses quantities in moles, but it is seldom possible to measure out quantities in moles directly. If the quantities given or required are expressed in other units, it is necessary to convert them to moles before using the factors of the balanced chemical equation. Conversion of mass to moles and vice versa was considered in Sec. 7.4. First we will use that knowledge to calculate the number of moles of reactant or product and use that value to calculate the numbers of moles of other reactants or products.

EXAMPLE 10.5.   How many moles of Na_2SO_4 are produced by reaction of 124 g of NaOH with sufficient H_2SO_4?

Ans.   Again, the first step is to write the balanced chemical equation:

$$2\ NaOH + H_2SO_4 \longrightarrow Na_2SO_4 + 2\ H_2O$$

Since the *mole ratio* is given by the equation, we must convert 124 g of NaOH to moles:

$$124 \text{ g NaOH}\left(\frac{1 \text{ mol NaOH}}{40.0 \text{ g NaOH}}\right) = 3.10 \text{ mol NaOH}$$

Now we can solve the problem just as we did those above:

$$3.10 \text{ mol NaOH}\left(\frac{1 \text{ mol Na}_2\text{SO}_4}{2 \text{ mol NaOH}}\right) = 1.55 \text{ mol Na}_2\text{SO}_4$$

The left two boxes of Fig. 10-2 illustrate the additional step required for this calculation.

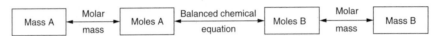

Fig. 10-2.   Conversion of mass of a reactant to mass of a product

It is also possible to calculate the mass of product from the number of moles of product.

EXAMPLE 10.6.   How many grams of Na_2SO_4 can be produced from 124 g NaOH with sufficient H_2SO_4?

Ans.   The first steps were presented in Example 10.5. It only remains to convert 1.55 mol Na_2SO_4 to grams (Fig. 10-2):

$$1.55 \text{ mol Na}_2\text{SO}_4\left(\frac{142 \text{ g Na}_2\text{SO}_4}{1 \text{ mol Na}_2\text{SO}_4}\right) = 220 \text{ g Na}_2\text{SO}_4$$

Not only mass, but any measurable quantity that can be converted to moles may be treated in this manner to determine the quantity of product or reactant involved in a reaction from the quantity of any other reactant or product. (In later chapters, the volumes of gases and the volumes of solutions of known concentrations will be used to determine the numbers of moles of a reactant or product.) We can illustrate the process with the following problem.

EXAMPLE 10.7.   How many grams of Na_3PO_4 can be produced by the reaction of 1.11×10^{23} formula units of NaOH with sufficient H_3PO_4?

Ans. $$3 \text{ NaOH} + \text{H}_3\text{PO}_4 \longrightarrow \text{Na}_3\text{PO}_4 + 3 \text{ H}_2\text{O}$$

Since the equation states the mole ratio, first we convert the number of formula units of NaOH to moles:

$$1.11 \times 10^{23} \text{ formula units NaOH}\left(\frac{1 \text{ mol NaOH}}{6.02 \times 10^{23} \text{ formula units NaOH}}\right) = 0.184 \text{ mol NaOH}$$

Next we convert that number of moles of NaOH to moles and then grams of Na_3PO_4:

$$0.184 \text{ mol NaOH}\left(\frac{1 \text{ mol Na}_3\text{PO}_4}{3 \text{ mol NaOH}}\right) = 0.0613 \text{ mol Na}_3\text{PO}_4$$

$$0.0613 \text{ mol Na}_3\text{PO}_4\left(\frac{164 \text{ g Na}_3\text{PO}_4}{1 \text{ mol Na}_3\text{PO}_4}\right) = 10.1 \text{ g Na}_3\text{PO}_4$$

Figure 10-3 illustrates the process.

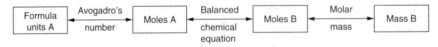

Fig. 10-3.   Conversion of number of formula units of a reactant to mass of a product

10.3.   LIMITING QUANTITIES

In Secs. 10.1 and 10.2, there was always sufficient (or excess) of all reactants except the one whose quantity was given. The quantity of only one reactant or product was stated in the problem. In this section, the quantities of more than one reactant will be stated. This type of problem is called a *limiting-quantities problem.*

How much Al_2O_3 can be prepared from 2.0 mol of O_2 and 0.0 mol of Al? The first step, as usual, is to write the balanced chemical equation:

$$4\ Al + 3\ O_2 \longrightarrow 2\ Al_2O_3$$

It should be obvious that with no Al, there can be no Al_2O_3 produced by this reaction. (This problem is not one which is likely to appear on examinations.)

How much sulfur dioxide is produced by the reaction of 1.00 g S and all the oxygen in the atmosphere of the earth? (If you strike a match outside, do you really have to worry about not having enough oxygen to burn all the sulfur in the match head?) This problem has the quantity of each of two reactants stated, but it is obvious that the sulfur will be used up before the oxygen. It is also obvious that not all the oxygen will react! (Otherwise, we are all in trouble.) The problem is solved just as the problems in Sec. 10.2.

To solve a limiting-quantities problem in which the reactant in excess is not obvious, do as follows: If the reagents appear in a 1 mol: 1 mol ratio in the balanced chemical equation, you can tell immediately that the reagent present in lower number of moles is the one in limiting quantity. If they are not in a 1 mol: 1 mol ratio, an easy way to determine which is in limiting quantity is to divide the number of moles of each by the corresponding coefficient in the balanced chemical equation. The reagent with the lower quotient is in limiting quantity. *Do not use these quotients for any further calculations. (In fact, it is useful to draw a line through them as soon as you see which reactant is in limiting quantity.)* We can use the reactant in limiting quantity to determine the number of moles of each product that will be produced and the number of moles of the other reactant(s) that will be used up in the reaction. These steps are illustrated in Example 10.8 below.

In a limiting-quantities problem, you might be asked the number of moles of every substance remaining after the reaction. A useful way to calculate all the quantities is to use a table to do the calculations.

Step 1: Use the balanced equation as the heads of the quantities in the table, and label the rows "Present initially," "Change due to reaction," and "Present finally." The initial quantities of reactants are entered in the first row. Zero is entered under the heading for any substances not present.

Step 2: Start the second row by entering the moles of limiting quantity—the same as the number of moles of that substance in the first row, since all of the limiting quantity is used up. The other entries in the second row are calculated using the technique of Sec. 10.1. *The entries in row 2 will always be in the same mole ratio as the coefficients in the balanced chemical equation.*

Step 3: Calculate the third row entries by adding the numbers of moles of products in the first two rows and subtracting the numbers of moles of each reactant in the second row from the corresponding number in the first row. (If you ever get a negative number of moles in row 3, you have made a mistake somewhere.)

EXAMPLE 10.8.   Calculate the number of moles of each substance present after 4.60 mol of aluminum is treated with 4.20 mol of oxygen gas and allowed to react.

Ans.   The balanced equation and the left table entries are written first, along with the initial quantities from the statement of the problem:

	4 Al	+ 3 O_2	$\longrightarrow$ 2 Al_2O_3
Present initially:	4.60 mol	4.20 mol	0.00 mol
Change due to reaction:			
Present finally:			

Dividing 4.60 mol of Al by its coefficient 4 yields 1.15 mol Al; dividing 4.20 mol O_2 by 3 yields 1.40 mol O_2. Since 1.15 is lower than 1.40, the Al is in limiting quantity. Therefore 4.60 mol is entered in the second row under Al. Note that the number of moles of Al present is greater than the number of moles of O_2, but the Al is still in limiting quantity. The quantity of O_2 that reacts and of Al_2O_3 that is formed is calculated as in Sec. 10.1:

$$4.60\ \text{mol Al}\left(\frac{3\ \text{mol } O_2}{4\ \text{mol Al}}\right) = 3.45\ \text{mol } O_2 \qquad 4.60\ \text{mol Al}\left(\frac{2\ \text{mol } Al_2O_3}{4\ \text{mol Al}}\right) = 2.30\ \text{mol } Al_2O_3$$

These quantities are entered in line 2 also.

	4 Al	+ 3 O$_2$	$\longrightarrow$ 2 Al$_2$O$_3$
Present initially:	4.60 mol	4.20 mol	0.00 mol
Change due to reaction:	4.60 mol	3.45 mol	2.30 mol
Present finally:			

The final line is calculated by subtraction for the reactants and addition for the product.

	4 Al	+ 3 O$_2$	$\longrightarrow$ 2 Al$_2$O$_3$
Present initially:	4.60 mol	4.20 mol	0.00 mol
Change due to reaction:	4.60 mol	3.45 mol	2.30 mol
Present finally:	0.00 mol	0.75 mol	2.30 mol

EXAMPLE 10.9.   How many moles of PbI$_2$ can be prepared by the reaction of 0.128 mol of Pb(NO$_3$)$_2$ and 0.206 mol NaI?

Ans.   The balanced equation is

$$Pb(NO_3)_2 + 2\,NaI \longrightarrow PbI_2 + 2\,NaNO_3$$

The limiting quantity is determined:

$$\frac{0.128 \text{ mol Pb(NO}_3)_2}{1} = 0.128 \text{ mol Pb(NO}_3)_2 \qquad \frac{0.206 \text{ mol NaI}}{2} = 0.103 \text{ mol NaI}$$

NaI is in limiting quantity.

$$0.206 \text{ mol NaI}\left(\frac{1 \text{ mol PbI}_2}{2 \text{ mol NaI}}\right) = 0.103 \text{ mol PbI}_2$$

Note especially that the number of moles of NaI exceeds the number of moles of Pb(NO$_3$)$_2$ present, but the NaI is still in limiting quantity.

EXAMPLE 10.10.   How many grams of Ca(ClO$_4$)$_2$ can be prepared by treatment of 12.0 g CaO with 102 g HClO$_4$? How many grams of excess reactant remains after the reaction?

Ans.   This problem gives the quantities of the two reactants in grams; we must first change them to moles:

$$12.0 \text{ g CaO}\left(\frac{1 \text{ mol CaO}}{56.0 \text{ g CaO}}\right) = 0.214 \text{ mol CaO}$$

$$102.0 \text{ g HClO}_4\left(\frac{1 \text{ mol HClO}_4}{100 \text{ g HClO}_4}\right) = 1.02 \text{ mol HClO}_4$$

Now the problem can be done as in Example 10.8. All quantities are in moles.
The balanced equation is

	CaO	+ 2 HClO$_4$	$\longrightarrow$ Ca(ClO$_4$)$_2$	+ H$_2$O
Present initially:	0.214	1.02	0.000	
Change due to reaction:	0.214	0.428	0.214	
Present finally:	0.000	0.59	0.214	

$$0.214 \text{ mol Ca(ClO}_4)_2\left(\frac{239 \text{ g Ca(ClO}_4)_2}{1 \text{ mol Ca(ClO}_4)_2}\right) = 51.1 \text{ g Ca(ClO}_4)_2 \text{ produced}$$

$$0.59 \text{ mol HClO}_4\left(\frac{100 \text{ g HClO}_4}{1 \text{ mol HClO}_4}\right) = 59 \text{ g HClO}_4$$

If the quantities of both reactants are in exactly the correct ratio for the balanced chemical equation, then *either* reactant may be used to calculate the quantity of product produced. (If on a quiz or examination it is obvious that they are in the correct ratio, you should state that they are, so that your instructor will understand that you recognize the problem to be a limiting-quantities type problem.)

EXAMPLE 10.11. How many moles of lithium nitride, Li_3N, can be prepared by the reaction of 0.600 mol Li and 0.100 mol N_2?

Ans. $6 Li + N_2 \longrightarrow 2 Li_3N$

Since the ratio of moles of lithium present to moles of nitrogen present is 6 : 1, just as is required for the balanced equation, either reactant may be used.

$$0.600 \text{ mol Li}\left(\frac{2 \text{ mol } Li_3N}{6 \text{ mol Li}}\right) = 0.200 \text{ mol } Li_3N \qquad or \qquad 0.100 \text{ mol } N_2\left(\frac{2 \text{ mol } Li_3N}{1 \text{ mol } N_2}\right) = 0.200 \text{ mol } Li_3N$$

(The first sentence of this answer should be stated on an examination.)

10.4. CALCULATIONS BASED ON NET IONIC EQUATIONS

The net ionic equation (Chap. 9), like all balanced chemical equations, gives the ratio of moles of each substance to moles of each of the others. It does not immediately yield information about the mass of the entire salt, however. (One cannot weight out only Ba^{2+} ions.) Therefore, when masses of reactants are required, the specific compound used must be included in the calculation.

EXAMPLE 10.12. (*a*) How many moles of silver ion are required to make 5.00 g AgCl? (*b*) What mass of silver nitrate is required to prepare 5.00 g AgCl?

Ans. $Ag^+ + Cl^- \longrightarrow AgCl$

(*a*)   The molar mass of AgCl, the sum of the atomic masses of Ag and Cl, is $108 + 35 = 143$ g/mol. In 5.00 g of AgCl, which is to be prepared, there is

$$5.00 \text{ g AgCl}\left(\frac{1 \text{ mol AgCl}}{143 \text{ g AgCl}}\right) = 0.0350 \text{ mol AgCl}$$

Hence, from the balanced chemical equation, 0.0350 mol of Ag^+ is required:

$$0.0350 \text{ mol AgCl}\left(\frac{1 \text{ mol } Ag^+}{1 \text{ mol AgCl}}\right) = 0.0350 \text{ mol } Ag^+$$

(*b*) $0.0350 \text{ mol } Ag^+\left(\dfrac{1 \text{ mol } AgNO_3}{1 \text{ mol } Ag^+}\right)\left(\dfrac{170 \text{ g } AgNO_3}{1 \text{ mol } AgNO_3}\right) = 5.95 \text{ g } AgNO_3$

Hence, when treated with enough chloride ion, 5.95 g of $AgNO_3$ will produce 5.00 g of AgCl.

EXAMPLE 10.13. What is the maximum mass of $BaSO_4$ that can be produced when a solution containing 4.35 g of Na_2SO_4 is added to another solution containing an excess of Ba^{2+}?

Ans. $Ba^{2+} + SO_4^{2-} \longrightarrow BaSO_4$

$$4.35 \text{ g } Na_2SO_4\left(\frac{1 \text{ mol } Na_2SO_4}{142 \text{ g } Na_2SO_4}\right)\left(\frac{1 \text{ mol } SO_4^{2-}}{1 \text{ mol } Na_2SO_4}\right)\left(\frac{1 \text{ mol } BaSO_4}{1 \text{ mol } SO_4^{2-}}\right)\left(\frac{233 \text{ g } BaSO_4}{1 \text{ mol } BaSO_4}\right) = 7.14 \text{ g } BaSO_4$$

10.5. HEAT CAPACITY AND HEAT OF REACTION

Heat is a reactant or product in most chemical reactions. Before we consider including heat in a balanced chemical equation, first we must learn how to measure heat. When heat is added to a system, in the absence of a chemical reaction the system may warm up, or a change of phase may occur. In this section change of phase will not be considered.

Temperature and heat are not the same. Temperature is a measure of the intensity of the heat in a system. Consider the following experiment: Hold a lit candle under a pot of water with 0.5 in. of water in the bottom. Hold an identical candle, also lit, under an identical pot full of water for the same length of time. To which sample of water is more heat added? Which sample of water gets hotter?

The same quantity of heat is added to each pot, since identical candles were used for the same lengths of time. However, the water in the pot with less water in it is heated to a higher temperature. The greater quantity of water would require more heat to get it to the same higher temperature.

The *specific heat capacity* of a substance is defined as the quantity of heat required to heat exactly 1 g of the substance 1°C. Specific heat capacity is often called *specific heat*. Lowercase c is used to represent specific heat. For example, the specific heat of water is 4.184 J/(g·°C). This means that 4.184 J will warm 1 g of water 1°C. To warm 2 g of water 1°C requires twice as much energy, or 8.368 J. To warm 1 g of water 2°C requires 8.368 J of energy also. In general, the heat required to effect a certain change in temperature in a certain sample of a given material is calculated with the following equation, where the Greek letter delta (Δ) means "change in."

$$\text{Heat required} = (\text{mass})(\text{specific heat})(\text{change in temperature}) = (m)(c)(\Delta t)$$

Heat capacities may be used as factors in factor-label method solutions to problems. Be aware that there are two units in the denominator, mass and temperature change. Thus, to get energy, one must multiply the heat capacity by both mass and temperature change.

EXAMPLE 10.14.   How much heat does it take to raise the temperature of 10.0 g of water 20.1°C?

Ans.
$$\text{Heat} = (m)(c)(\Delta t) = (10.0 \text{ g})\left(\frac{4.184 \text{ J}}{\text{g·°C}}\right)(20.1°C) = 841 \text{ J}$$

EXAMPLE 10.15.   How much heat does it take to raise the temperature of 10.0 g of water from 10.0°C to 30.1°C?

Ans.   This is the same problem as Example 10.14. In that problem the temperature *change* was specified. In this example, the initial and final temperatures are given, but the temperature change is the same 20.1°C. The answer is again 841 J.

EXAMPLE 10.16.   What is the final temperature after 945 J of heat is added to 60.0 g of water at 22.0°C?

Ans.
$$\Delta t = \frac{\text{heat}}{(m)(c)} = \frac{945 \text{ J}}{(60.0 \text{ g})[4.184 \text{ J/(g·°C)}]} = 3.76°C$$

Note that the problem is to find the final temperature; 3.76°C is the *temperature change*.

$$t_{\text{final}} = t_{\text{initial}} + \Delta t = 22.0°C + 3.76°C = 25.8°C$$

EXAMPLE 10.17.   What is the specific heat of a metal alloy if 412 J is required to heat 44.0 g of the metal from 19.5°C to 41.4°C?

Ans.
$$c = \frac{\text{heat}}{(m)(\Delta t)} = \frac{412 \text{ J}}{(44.0 \text{ g})(41.4°C - 19.5°C)} = \frac{0.428 \text{ J}}{\text{g·°C}}$$

EXAMPLE 10.18.   What is the final temperature of 229 g of water initially at 14.7°C from which 929 J of heat is removed?

Ans.
$$\Delta t = \frac{\text{heat}}{mC} = \frac{-929 \text{ J}}{(229 \text{ g})[4.184 \text{ J/(g·°C)}]} = -0.970°C$$

$$t_{\text{final}} = t_{\text{initial}} + \Delta t = 14.7°C + (-0.970°C) = 13.7°C$$

We note two things about this example. First, since the heat was removed, the value used in the equation was negative. Second, the final temperature is obviously lower than the initial temperature, since heat was removed.

As was mentioned earlier in this section, heat is a reactant or product in most chemical reactions. It is possible for us to indicate the quantity of heat in the balanced equation and to treat it with the rules of stoichiometry that we already know.

EXAMPLE 10.19.   How much heat will be produced by burning 50.0 g of carbon to carbon dioxide?

$$C + O_2 \longrightarrow CO_2 + 393 \text{ kJ}$$

Ans.
$$50.0 \text{ g C}\left(\frac{1 \text{ mol C}}{12.0 \text{ g C}}\right)\left(\frac{393 \text{ kJ}}{1 \text{ mol C}}\right) = 1640 \text{ kJ}$$

We can use specific heat calculations to measure heats of reaction.

EXAMPLE 10.20. What rise in temperature will occur if 24.5 kJ of heat is added to 175 g of a dilute aqueous solution of sodium chloride [$c = 4.10$ J/g·°C)] (*a*) by heating with a bunsen burner and (*b*) by means of a chemical reaction?

Ans.   (*a and b*) The source of the heat does not matter; the temperature rise will be the same in either case. Watch out for the units!

$$\Delta t = \frac{\text{heat}}{(m)(c)} = \frac{24\,500 \text{ J}}{(175 \text{ g})[4.10 \text{ J/(g·°C)}]} = 34.1°C$$

EXAMPLE 10.21. Calculate the heat of reaction per mole of water formed if 0.0500 mol of HCl and 0.0500 mol of NaOH are added to 15.0 g of water, all at 18.0°C. The solution formed is heated from 18.0°C to 54.3°C. The specific heat of the solution is 4.10 J/(g·°C).

Ans.   The law of conservation of mass allows us to calculate the mass of the solution:

$$15.0 \text{ g} + 1.82 \text{ g} + 2.00 \text{ g} = 18.8 \text{ g}$$
$$\text{Heat} = mc\Delta t = (18.8 \text{ g})[4.10 \text{ J/(g·°C)}](36.3°C) = 2800 \text{ J}$$
$$2.80 \text{ kJ/(0.0500 mol water formed)} = 56.0 \text{ kJ/mol}$$

Solved Problems

MOLE-TO-MOLE CALCULATIONS

10.1. Can the balanced chemical equation dictate to a chemist how much of each reactant to place in a reaction vessel?

> *Ans.*   The chemist can put in as little as is weighable or as much as the vessel will hold. For example, the fact that a reactant has a coefficient of 2 in the balanced chemical equation does not mean that the chemist must put 2 mol into the reaction vessel. The chemist might decide to add the reactants in the ratio of the balanced chemical equation, but that is not required. And even in that case, the numbers of moles of each reactant might be twice the respective coefficients or one-tenth those values, etc. The equation merely states the *reacting ratio*.

10.2. How many factor labels can be used corresponding to each of the following balanced equations?

(*a*) $2 \text{ K} + \text{Cl}_2 \longrightarrow 2 \text{ KCl}$

(*b*) $\text{NCl}_3 + 3 \text{ H}_2\text{O} \longrightarrow 3 \text{ HOCl} + \text{NH}_3$

Ans.   (*a*) 6:

$$\frac{2 \text{ mol K}}{1 \text{ mol Cl}_2} \qquad \frac{2 \text{ mol K}}{2 \text{ mol KCl}} \qquad \frac{1 \text{ mol Cl}_2}{2 \text{ mol K}} \qquad \frac{1 \text{ mol Cl}_2}{2 \text{ mol KCl}} \qquad \frac{2 \text{ mol KCl}}{2 \text{ mol K}} \qquad \frac{2 \text{ mol KCl}}{1 \text{ mol Cl}_2}$$

(*b*) 12: Each of the four compounds as numerators with the three others as denominators—$4 \times 3 = 12$.

10.3. Which of the factors of Problem 10.2*a* would be used to convert (*a*) the number of moles of Cl_2 to the number of moles of KCl, (*b*) K to Cl_2, and (*c*) Cl_2 to K?

Ans.   (*a*) $\dfrac{2 \text{ mol KCl}}{1 \text{ mol Cl}_2}$ (*b*) $\dfrac{1 \text{ mol Cl}_2}{2 \text{ mol K}}$ (*c*) $\dfrac{2 \text{ mol K}}{1 \text{ mol Cl}_2}$

10.4. How many moles of AlCl_3 can be prepared from 7.5 mol Cl_2 and sufficient Al?

Ans. $$2 \text{ Al} + 3 \text{ Cl}_2 \longrightarrow 2 \text{ AlCl}_3$$

$$7.5 \text{ mol Cl}_2 \left(\frac{2 \text{ mol AlCl}_3}{3 \text{ mol Cl}_2} \right) = 5.0 \text{ mol AlCl}_3$$

10.5. How many moles of H_2O will react with 2.25 mol PCl_5 to form HCl and H_3PO_4?

Ans. $$4 \text{ H}_2\text{O} + \text{PCl}_5 \longrightarrow 5 \text{ HCl} + \text{H}_3\text{PO}_4$$

$$2.25 \text{ mol PCl}_5 \left(\frac{4 \text{ mol H}_2\text{O}}{1 \text{ mol PCl}_5} \right) = 9.00 \text{ mol H}_2\text{O}$$

10.6. Balance the following equation. Calculate the number of moles of CO_2 that can be prepared by the reaction of 2.50 mol of $Mg(HCO_3)_2$.

$$Mg(HCO_3)_2 + HCl \longrightarrow MgCl_2 + CO_2 + H_2O$$

Ans. $$Mg(HCO_3)_2 + 2\,HCl \longrightarrow MgCl_2 + 2\,CO_2 + 2\,H_2O$$

$$2.50 \text{ mol } Mg(HCO_3)_2 \left[\frac{2 \text{ mol } CO_2}{1 \text{ mol } Mg(HCO_3)_2} \right] = 5.00 \text{ mol } CO_2$$

10.7. (*a*) How many moles of $CaCl_2$ can be prepared by the reaction of 2.50 mol HCl with excess $Ca(OH)_2$? (*b*) How many moles of NaCl can be prepared by the reaction of 2.50 mol HCl with excess NaOH?

Ans.   (*a*) $Ca(OH)_2 + 2\,HCl \longrightarrow CaCl_2 + 2\,H_2O$

$$2.50 \text{ mol } HCl \left(\frac{1 \text{ mol } CaCl_2}{2 \text{ mol } HCl} \right) = 1.25 \text{ mol } CaCl_2$$

   (*b*) $NaOH + HCl \longrightarrow NaCl + H_2O$

$$2.50 \text{ mol } HCl \left(\frac{1 \text{ mol } NaCl}{1 \text{ mol } HCl} \right) = 2.50 \text{ mol } NaCl$$

10.8. How many moles of H_2O are prepared along with 0.750 mol Na_3PO_4 in a reaction of NaOH and H_3PO_4?

Ans. $$H_3PO_4 + 3\,NaOH \longrightarrow Na_3PO_4 + 3\,H_2O$$

$$0.750 \text{ mol } Na_3PO_4 \left(\frac{3 \text{ mol } H_2O}{1 \text{ mol } Na_3PO_4} \right) = 2.25 \text{ mol } H_2O$$

10.9. Consider the following equation:

$$KMnO_4 + 5\,FeCl_2 + 8\,HCl \longrightarrow MnCl_2 + 5\,FeCl_3 + 4\,H_2O + KCl$$

How many moles of $FeCl_3$ will be produced by the reaction of 0.968 mol of HCl?

Ans.   No matter how complicated the equation, the reacting ratio is still given by the coefficients. The coefficients of interest are 8 for HCl and 5 for $FeCl_3$.

$$0.968 \text{ mol } HCl \left(\frac{5 \text{ mol } FeCl_3}{8 \text{ mol } HCl} \right) = 0.605 \text{ mol } FeCl_3$$

Note: The hard part of this problem is balancing the equation, which will be presented in Chap. 14. Since the balanced equation was given in the statement of the problem, the problem is as easy to solve as the previous ones.

CALCULATIONS INVOLVING OTHER QUANTITIES

10.10. Figure 10-2 is a combination of which two earlier figures?

   Ans.   Figures 10-1 and 7-2.

10.11. Which earlier sections must be understood before mass-to-mass conversions can be studied profitably?

   Ans.   Section 2.2, factor-label method; Sec. 7.3, calculation of formula masses; Sec. 7.4, changing moles to grams and vice versa; Sec. 7.4, Avogadro's number; and/or Sec. 8.2, balancing chemical equations.

10.12. In a stoichiometry problem, (*a*) if the mass of a reactant is given, what conversions (if any) should be made? (*b*) If a number of molecules is given, what conversions (if any) should be made? (*c*) If a number of moles is given, what conversions (if any) should be made?

   Ans.   (*a*) The mass should be converted to moles.   (*b*) The number of molecules should be converted to moles. (*c*) No conversion need be done; the quantity is given in moles.

10.13. Phosphoric acid reacts with sodium hydroxide to produce sodium phosphate and water. (*a*) Write a balanced chemical equation for the reaction. (*b*) Determine the number of moles of phosphoric acid in 50.0 g of the acid. (*c*) How many moles of sodium phosphate will be produced by the reaction of this number of moles of phosphoric acid? (*d*) How many grams of sodium phosphate will be produced? (*e*) How many moles of sodium hydroxide will it take to react with this quantity of phosphoric acid? (*f*) How many grams of sodium hydroxide will be used up?

Ans.   (*a*) $H_3PO_4 + 3\,NaOH \longrightarrow Na_3PO_4 + 3\,H_2O$

　　　(*b*) $50.0\text{ g }H_3PO_4\left(\dfrac{1\text{ mol }H_3PO_4}{98.0\text{ g }H_3PO_4}\right) = 0.510\text{ mol }H_3PO_4$

　　　(*c*) $0.510\text{ mol }H_3PO_4\left(\dfrac{1\text{ mol }Na_3PO_4}{1\text{ mol }H_3PO_4}\right) = 0.510\text{ mol }Na_3PO_4$

　　　(*d*) $0.510\text{ mol }Na_3PO_4\left(\dfrac{164\text{ g }Na_3PO_4}{1\text{ mol }Na_3PO_4}\right) = 83.6\text{ g }Na_3PO_4$

　　　(*e*) $0.510\text{ mol }H_3PO_4\left(\dfrac{3\text{ mol }NaOH}{1\text{ mol }Na_3PO_4}\right) = 1.53\text{ mol }NaOH$

　　　(*f*) $1.53\text{ mol }NaOH\left(\dfrac{40.0\text{ g }NaOH}{1\text{ mol }NaOH}\right) = 61.2\text{ g }NaOH$

10.14. (*a*) Write the balanced chemical equation for the reaction of sodium with chlorine. (*b*) How many moles of Cl_2 are there in 7.650 g chlorine? (*c*) How many moles of NaCl will that number of moles of chlorine produce? (*d*) What mass of NaCl is that number of moles of NaCl?

Ans.   (*a*) $2\,Na + Cl_2 \longrightarrow 2\,NaCl$

　　　(*b*) $7.650\text{ g }Cl_2\left(\dfrac{1\text{ mol }Cl_2}{70.90\text{ g }Cl_2}\right) = 0.1079\text{ mol }Cl_2$

　　　(*c*) $0.1079\text{ mol }Cl_2\left(\dfrac{2\text{ mol }NaCl}{1\text{ mol }Cl_2}\right) = 0.2158\text{ mol }NaCl$

　　　(*d*) $0.2158\text{ mol }NaCl\left(\dfrac{58.45\text{ g }NaCl}{1\text{ mol }NaCl}\right) = 12.61\text{ g }NaCl$

10.15. How many formula units of sodium hydroxide, along with H_2SO_4, does it take to make 7.50×10^{22} formula units of Na_2SO_4?

Ans. $2\,NaOH + H_2SO_4 \longrightarrow Na_2SO_4 + 2\,H_2O$

$$7.50 \times 10^{22}\text{ units }Na_2SO_4\left(\frac{1\text{ mol }Na_2SO_4}{6.02 \times 10^{23}\text{ units }Na_2SO_4}\right)\left(\frac{2\text{ mol }NaOH}{1\text{ mol }Na_2SO_4}\right)\left(\frac{6.02 \times 10^{23}\text{ units }NaOH}{1\text{ mol }NaOH}\right)$$

$$= 1.50 \times 10^{23}\text{ units }NaOH$$

Since the balanced chemical equation also relates the numbers of formula units of reactants and products, the problem can be solved by converting directly with the factor label from the balanced equation:

$$7.50 \times 10^{22}\text{ units }Na_2SO_4\left(\frac{2\text{ units }NaOH}{1\text{ unit }Na_2SO_4}\right) = 1.50 \times 10^{23}\text{ units }NaOH$$

10.16. How many grams of barium hydroxide will be used up in the reaction with hydrogen chloride (hydrochloric acid) to produce 16.70 g of barium chloride plus some water?

Ans. $Ba(OH)_2 + 2\,HCl \longrightarrow BaCl_2 + 2\,H_2O$

$$16.70\text{ g }BaCl_2\left(\frac{1\text{ mol }BaCl_2}{208.2\text{ g }BaCl_2}\right)\left(\frac{1\text{ mol }Ba(OH)_2}{1\text{ mol }BaCl_2}\right)\left(\frac{171.3\text{ g }Ba(OH)_2}{1\text{ mol }Ba(OH)_2}\right) = 13.74\text{ g }Ba(OH)_2$$

10.17. Draw a figure like Fig. 10-2 for Problem 10.15.

Ans.   See Fig. 10-4.

10.18. (*a*) What reactant may be treated with phosphoric acid to produce 6.00 mol of potassium hydrogen phosphate (plus some water)? (*b*) How many moles of phosphoric acid will it take? (*c*) How many moles of the other reactant are required? (*d*) How many grams?

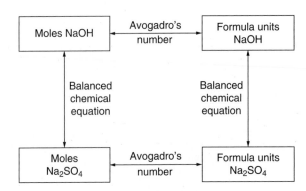

Fig. 10-4. Conversion of formula units of a reactant to formula units of a product

Ans. (a) KOH may be used: $H_3PO_4 + 2\,KOH \longrightarrow K_2HPO_4 + 2\,H_2O$

(b) $6.00\ \text{mol K}_2\text{HPO}_4\left(\dfrac{1\ \text{mol H}_3\text{PO}_4}{1\ \text{mol K}_2\text{HPO}_4}\right) = 6.00\ \text{mol H}_3\text{PO}_4$

(c) $6.00\ \text{mol K}_2\text{HPO}_4\left(\dfrac{2\ \text{mol KOH}}{1\ \text{mol K}_2\text{HPO}_4}\right) = 12.0\ \text{mol KOH}$

(d) $12.0\ \text{mol KOH}\left(\dfrac{56.1\ \text{g KOH}}{1\ \text{mol KOH}}\right) = 673\ \text{g KOH}$

10.19. Determine the number of grams of hydrochloric acid that will just react with 20.0 g of calcium carbonate to produce carbon dioxide, water, and calcium chloride.

Ans. $CaCO_3 + 2\,HCl \longrightarrow CaCl_2 + H_2O + CO_2$

$$20.0\ \text{g CaCO}_3\left(\frac{1\ \text{mol CaCO}_3}{100\ \text{g CaCO}_3}\right)\left(\frac{2\ \text{mol HCl}}{1\ \text{mol CaCO}_3}\right)\left(\frac{36.5\ \text{g HCl}}{1\ \text{mol HCl}}\right) = 14.6\ \text{g HCl}$$

10.20. How many grams of Hg_2Cl_2 can be prepared from 15.0 mL of mercury (density 13.6 g/mL)?

Ans. $2\,Hg + Cl_2 \longrightarrow Hg_2Cl_2$

$$15.0\ \text{mL Hg}\left(\frac{13.6\ \text{g Hg}}{1\ \text{mL Hg}}\right)\left(\frac{1\ \text{mol Hg}}{200.6\ \text{g Hg}}\right)\left(\frac{1\ \text{mol Hg}_2\text{Cl}_2}{2\ \text{mol Hg}}\right)\left(\frac{471\ \text{g Hg}_2\text{Cl}_2}{1\ \text{mol Hg}_2\text{Cl}_2}\right) = 239\ \text{g Hg}_2\text{Cl}_2$$

10.21. How many grams of methyl alcohol, CH_3OH, can be obtained in an industrial process from 5.00 metric tons (5.00×10^6 g) of CO plus hydrogen gas? To calculate the answer: (a) Write a balanced chemical equation for the process. (b) Calculate the number of moles of CO in 5.00×10^6 g CO. (c) Calculate the number of moles of CH_3OH obtainable from that number of moles of CO. (d) Calculate the number of grams of CH_3OH obtainable.

Ans. (a) $CO + 2\,H_2 \xrightarrow{\ \text{special conditions}\ } CH_3OH$

(b) $5.00 \times 10^6\ \text{g CO}\left(\dfrac{1\ \text{mol CO}}{28.0\ \text{g CO}}\right) = 1.79 \times 10^5\ \text{mol CO}$

(c) $1.79 \times 10^5\ \text{mol CO}\left(\dfrac{1\ \text{mol CH}_3\text{OH}}{1\ \text{mol CO}}\right) = 1.79 \times 10^5\ \text{mol CH}_3\text{OH}$

(d) $1.79 \times 10^5\ \text{mol CH}_3\text{OH}\left(\dfrac{32.0\ \text{g CH}_3\text{OH}}{1\ \text{mol CH}_3\text{OH}}\right) = 5.73 \times 10^6\ \text{g CH}_3\text{OH}$

10.22. Determine the number of grams of barium hydroxide it would take to neutralize (just react completely, with none left over) 14.7 g of phosphoric acid.

Ans.

$$3\,Ba(OH)_2 + 2\,H_3PO_4 \longrightarrow Ba_3(PO_4)_2 + 6\,H_2O$$

$$14.7\,g\,H_3PO_4\left(\frac{1\,mol\,H_3PO_4}{98.0\,g\,H_3PO_4}\right)\left(\frac{3\,mol\,Ba(OH)_2}{2\,mol\,H_3PO_4}\right)\left(\frac{171\,g\,Ba(OH)_2}{1\,mol\,Ba(OH)_2}\right) = 38.5\,g\,Ba(OH)_2$$

10.23. Calculate the number of moles of NaOH required to remove the SO_2 from 3.50 metric tons (3.50×10^6 g) of atmosphere if the SO_2 is 0.10% by mass. (Na_2SO_3 and water are the products.)

Ans. $$2\,NaOH + SO_2 \longrightarrow Na_2SO_3 + H_2O$$

$$3.50 \times 10^6\,g\,atm\left(\frac{0.10\,g\,SO_2}{100\,g\,atm}\right)\left(\frac{1\,mol\,SO_2}{64.0\,g\,SO_2}\right)\left(\frac{2\,mol\,NaOH}{1\,mol\,SO_2}\right) = 110\,mol\,NaOH$$

10.24. Calculate the number of moles of Al_2O_3 needed to prepare 4.00×10^6 g of Al metal in the Hall process:

$$Al_2O_3 + 3\,C \xrightarrow[\text{special solvent}]{\text{electricity}} 2\,Al + 3\,CO$$

Ans. $$4.00 \times 10^6\,g\,Al\left(\frac{1\,mol\,Al}{27.0\,g\,Al}\right)\left(\frac{1\,mol\,Al_2O_3}{2\,mol\,Al}\right) = 7.41 \times 10^4\,mol\,Al_2O_3$$

10.25. How much AgCl can be prepared with 50.0 g $CaCl_2$ and excess $AgNO_3$?

Ans. $$CaCl_2 + 2\,AgNO_3 \longrightarrow Ca(NO_3)_2 + 2\,AgCl$$

$$50.0\,g\,CaCl_2\left(\frac{1\,mol\,CaCl_2}{111\,g\,CaCl_2}\right)\left(\frac{2\,mol\,AgCl}{1\,mol\,CaCl_2}\right)\left(\frac{143\,g\,AgCl}{1\,mol\,AgCl}\right) = 129\,g\,AgCl$$

10.26. In chemistry recitation, a student hears (incorrectly) the instructor say, "Hydrogen chloride reacts with $Ba(OH)_2$" when the instructor has actually said, "Hydrogen fluoride reacts with $Ba(OH)_2$." The instructor then asked how much product is formed. The student answers the question correctly. Which section, 10.1 or 10.2, were they discussing? Explain.

Ans.   They were discussing Sec. 10.1. Since the student got the answer correct despite hearing the wrong name, they must have been discussing the number of moles of reactants and products. The numbers of moles of HF and HCl would be the same in the reaction, but since they have different formula masses, their masses would be different.

10.27. Consider the equation

$$KMnO_4 + 5\,FeCl_2 + 8\,HCl \longrightarrow MnCl_2 + 5\,FeCl_3 + 4\,H_2O + KCl$$

How many grams of $FeCl_3$ will be produced by the reaction of 2.72 g of $KMnO_4$?

Ans.   The reacting ratio is given by the coefficients. The coefficients of interest are 1 for $KMnO_4$ and 5 for $FeCl_3$.

$$2.72\,g\,KMnO_4\left(\frac{1\,mol\,KMnO_4}{158\,g\,KMnO_4}\right)\left(\frac{5\,mol\,FeCl_3}{1\,mol\,KMnO_4}\right)\left(\frac{162\,g\,FeCl_3}{1\,mol\,FeCl_3}\right) = 13.9\,g\,FeCl_3$$

10.28. How many grams of NaCl can be produced from 7.650 g chlorine?

Ans.   This problem is the same as Problem 10.14. Problem 10.14 was stated in steps, and this problem is not, but you must do the same steps whether or not they are explicitly stated.

10.29. How much $KClO_3$ must be decomposed thermally to produce 14.6 g O_2?

Ans. $$2\,KClO_3 \longrightarrow 2\,KCl + 3\,O_2$$

$$14.6\,g\,O_2\left(\frac{1\,mol\,O_2}{32.0\,g\,O_2}\right)\left(\frac{2\,mol\,KClO_3}{3\,mol\,O_2}\right)\left(\frac{122.6\,g\,KClO_3}{1\,mol\,KClO_3}\right) = 37.3\,g\,KClO_3\,decomposed$$

LIMITING QUANTITIES

10.30. How many sandwiches, each containing 1 slice of salami and 2 slices of bread, can you make with 42 slices of bread and 25 slices of salami?

Ans. With 42 slices of bread, the maximum number of sandwiches you can make is 21. The bread is the limiting quantity.

10.31. How can you recognize a limiting-quantities problem?

Ans. The quantities of two reactants (or products) are given in the problem. They might be stated in any terms—moles, mass, etc.—but they must be given for the problem to be a limiting-quantities problem.

10.32. How much sulfur dioxide is produced by the reaction of 5.00 g S and all the oxygen in the atmosphere of the earth?

Ans. In this problem, it is obvious that the oxygen in the entire earth's atmosphere is in excess, so that no preliminary calculation need be done.

$$S + O_2 \longrightarrow SO_2$$

$$5.00 \text{ g S} \left(\frac{1 \text{ mol S}}{32.06 \text{ g S}} \right) \left(\frac{1 \text{ mol SO}_2}{1 \text{ mol S}} \right) \left(\frac{64.1 \text{ g SO}_2}{1 \text{ mol SO}_2} \right) = 10.0 \text{ g SO}_2$$

10.33. (*a*) The price of pistachio nuts is $5.00 per pound. If a grocer has 17 lb for sale and a buyer has $45.00 to buy nuts with, what is the maximum number of pounds that can be sold? (*b*) Consider the reaction

$$\text{KMnO}_4 + 5 \text{ FeCl}_2 + 8 \text{ HCl} \longrightarrow \text{MnCl}_2 + 5 \text{ FeCl}_3 + 4 \text{ H}_2\text{O} + \text{KCl}$$

If 45.0 mol of FeCl_2 and 17.0 mol of KMnO_4 are mixed with excess HCl, how many moles of MnCl_2 can be formed?

Ans. (*a*) With $45, the buyer can buy

$$45 \text{ dollars} \left(\frac{1 \text{ lb}}{5 \text{ dollars}} \right) = 9.0 \text{ lb}$$

Since the seller has more nuts than that, the money is in limiting quantity and controls the amount of the sale.

(*b*) With 45.0 mol FeCl_2,

$$45.0 \text{ mol FeCl}_2 \left(\frac{1 \text{ mol KMnO}_4}{5 \text{ mol FeCl}_2} \right) = 9.00 \text{ mol KMnO}_4 \text{ required}$$

Since the number of moles of KMnO_4 present (17.0 mol) exceeds that number, the limiting quantity is the number of moles of FeCl_2.

$$45.0 \text{ mol FeCl}_2 \left(\frac{1 \text{ mol MnCl}_2}{5 \text{ mol FeCl}_2} \right) = 9.00 \text{ mol MnCl}_2$$

10.34. In each of the following cases, determine which reactant is present in excess, and tell how many moles in excess it is.

Equation	Moles Present
(*a*) $2 \text{ Na} + \text{Cl}_2 \longrightarrow 2 \text{ NaCl}$	1.20 mol Na, 0.400 mol Cl_2
(*b*) $\text{P}_4\text{O}_{10} + 6 \text{ H}_2\text{O} \longrightarrow 4 \text{ H}_3\text{PO}_4$	0.25 mol P_4O_{10}, 1.5 mol H_2O
(*c*) $\text{HNO}_3 + \text{NaOH} \longrightarrow \text{NaNO}_3 + \text{H}_2\text{O}$	0.90 mol acid, 0.85 mol base
(*d*) $\text{Ca(HCO}_3)_2 + 2 \text{ HCl} \longrightarrow \text{CaCl}_2 + 2 \text{ CO}_2 + 2 \text{ H}_2\text{O}$	2.5 mol HCl, 1.0 mol $\text{Ca(HCO}_3)_2$
(*e*) $\text{H}_3\text{PO}_4 + 3 \text{ NaOH} \longrightarrow \text{Na}_3\text{PO}_4 + 3 \text{ H}_2\text{O}$	0.70 mol acid, 2.2 mol NaOH

Ans.   (*a*) $0.400 \text{ mol Cl}_2\left(\dfrac{2 \text{ mol Na}}{1 \text{ mol Cl}_2}\right) = 0.800 \text{ mol Na required}$

There is 0.40 mol more Na present than is required.

(*b*) $0.25 \text{ mol P}_4\text{O}_{10}\left(\dfrac{6 \text{ mol H}_2\text{O}}{1 \text{ mol P}_4\text{O}_{10}}\right) = 1.5 \text{ mol H}_2\text{O required}$

Neither reagent is in excess; there is just enough H_2O to react with all the P_4O_{10}.

(*c*) $0.85 \text{ mol NaOH}\left(\dfrac{1 \text{ mol HNO}_3}{1 \text{ mol NaOH}}\right) = 0.85 \text{ mol HNO}_3 \text{ required}$

There is not enough NaOH present; HNO_3 is in excess.
There is 0.05 mol HNO_3 in excess.

(*d*) $1.0 \text{ mol Ca(HCO}_3)_2\left(\dfrac{2 \text{ mol HCl}}{1 \text{ mol Ca(HCO}_3)_2}\right) = 2.0 \text{ mol HCl required}$

There is 0.5 mol HCl in excess.

(*e*) $0.70 \text{ mol H}_3\text{PO}_4\left(\dfrac{3 \text{ mol NaOH}}{1 \text{ mol H}_3\text{PO}_4}\right) = 2.1 \text{ mol NaOH required}$

There is $2.2 - 2.1 = 0.1$ mol NaOH in excess.

10.35. For the following reaction,

$$2 \text{ NaOH} + \text{ H}_2\text{SO}_4 \longrightarrow \text{Na}_2\text{SO}_4 + 2 \text{ H}_2\text{O}$$

(*a*)   How many moles of NaOH would react with 0.250 mol H_2SO_4? How many moles of Na_2SO_4 would be produced?

(*b*)   If 0.250 mol of H_2SO_4 and 0.750 mol NaOH were mixed, how much NaOH would react?

(*c*)   If 24.5 g H_2SO_4 and 30.0 g NaOH were mixed, how many grams of Na_2SO_4 would be produced?

Ans.   (*a*)

$$0.250 \text{ mol H}_2\text{SO}_4\left(\dfrac{2 \text{ mol NaOH}}{1 \text{ mol H}_2\text{SO}_4}\right) = 0.500 \text{ mol NaOH}$$

$$0.250 \text{ mol H}_2\text{SO}_4\left(\dfrac{1 \text{ mol Na}_2\text{SO}_4}{1 \text{ mol H}_2\text{SO}_4}\right) = 0.250 \text{ mol Na}_2\text{SO}_4$$

(*b*)   0.500 mol NaOH, as calculated in part (*a*).

(*c*)   This is really the same problem as part (*b*), except that it is stated in grams, because 24.5 g H_2SO_4 is 0.250 mol and 30.0 g NaOH is 0.750 mol NaOH.

To finish: $$0.250 \text{ mol Na}_2\text{SO}_4\left(\dfrac{142 \text{ g Na}_2\text{SO}_4}{1 \text{ mol Na}_2\text{SO}_4}\right) = 35.5 \text{ g Na}_2\text{SO}_4$$

10.36. For the reaction

$$3 \text{ HCl} + \text{ Na}_3\text{PO}_4 \longrightarrow \text{H}_3\text{PO}_4 + 3 \text{ NaCl}$$

a chemist added 2.55 mol of HCl and a certain quantity of Na_3PO_4 to a reaction vessel, which produced 0.750 mol H_3PO_4. Which one of the reactants was in excess?

Ans. $$2.55 \text{ mol HCl}\left(\dfrac{1 \text{ mol H}_3\text{PO}_4}{3 \text{ mol HCl}}\right) = 0.850 \text{ mol H}_3\text{PO}_4$$

would be produced by reaction of all the HCl. Since the actual quantity of H_3PO_4 produced is 0.750 mol, not all the HCl was used up and the Na_3PO_4 must be the limiting quantity. The HCl was in excess.

10.37. (*a*) How many moles of Cl_2 will react with 1.22 mol Na to produce NaCl? (*b*) If 0.880 mol Cl_2 is treated with 1.22 mol Na, how much Cl_2 will react? (*c*) What is the limiting quantity in this problem?

Ans.   (*a*) $2 \text{ Na} + \text{Cl}_2 \longrightarrow 2 \text{ NaCl}$

$$1.22 \text{ mol Na}\left(\frac{1 \text{ mol Cl}_2}{2 \text{ mol Na}}\right) = 0.610 \text{ mol Cl}_2 \text{ reacts}$$

(*b*)   We calculated in part (*a*) that 0.610 mol of Cl_2 is needed to react. As long as we have at least 0.610 mol present, 0.610 mol Cl_2 will react.

(*c*)   Since we have more Cl_2 than that, the Na is in limiting quantity.

10.38. What mass of CO_2 can be produced by the complete combustion of 2.00 kg of octane, C_8H_{18}, a major component of gasoline, in 8.00 kg of oxygen?

Ans. $2 \text{ C}_8\text{H}_{18} + 25 \text{ O}_2 \longrightarrow 16 \text{ CO}_2 + 18 \text{ H}_2\text{O}$

$$2.00 \text{ kg C}_8\text{H}_{18}\left(\frac{1000 \text{ g}}{1 \text{ kg}}\right)\left(\frac{1 \text{ mol C}_8\text{H}_{18}}{114 \text{ g C}_8\text{H}_{18}}\right) = 17.5 \text{ mol C}_8\text{H}_{18} \text{ present}$$

$$8.00 \text{ kg O}_2\left(\frac{1000 \text{ g}}{1 \text{ kg}}\right)\left(\frac{1 \text{ mol O}_2}{32.0 \text{ g O}_2}\right) = 250 \text{ mol O}_2 \text{ present}$$

$$17.5 \text{ mol C}_8\text{H}_{18}\left(\frac{25 \text{ mol O}_2}{2 \text{ mol C}_8\text{H}_{18}}\right) = 219 \text{ mol O}_2 \text{ required}$$

Since 219 mol of O_2 is required and 250 mol O_2 is present, C_8H_{18} is in limiting quantity.

$$17.5 \text{ mol C}_8\text{H}_{18}\left(\frac{16 \text{ mol CO}_2}{2 \text{ mol C}_8\text{H}_{18}}\right)\left(\frac{44.0 \text{ g CO}_2}{1 \text{ mol CO}_2}\right) = 6160 \text{ g} = 6.16 \text{ kg CO}_2 \text{ produced}$$

10.39. If you were treating two chemicals, one very cheap and one expensive, to produce a product, which chemical would you use in excess, if one had to be in excess?

Ans.   Economically, it would be advisable to use the cheap one in excess, since more product could be obtained per dollar by using up all the expensive reactant.

CALCULATIONS BASED ON NET IONIC EQUATIONS

10.40. What mass of each of the following silver salts would be required to react completely with a solution containing 35.5 g of chloride ion to form (insoluble) silver chloride? (*a*) AgNO_3, (*b*) Ag_2SO_4, and (*c*) $\text{AgC}_2\text{H}_3\text{O}_2$.

Ans. $35.5 \text{ g Cl}^-\left(\dfrac{1 \text{ mol Cl}^-}{35.5 \text{ g Cl}^-}\right)\left(\dfrac{1 \text{ mol Ag}^+}{1 \text{ mol Cl}^-}\right) = 1.00 \text{ mol Ag}^+$

(*a*) $1.00 \text{ mol Ag}^+\left(\dfrac{1 \text{ mol AgNO}_3}{1 \text{ mol Ag}^+}\right)\left(\dfrac{169.9 \text{ g AgNO}_3}{1 \text{ mol AgNO}_3}\right) = 170 \text{ g AgNO}_3$

(*b*) $1.00 \text{ mol Ag}^+\left(\dfrac{1 \text{ mol Ag}_2\text{SO}_4}{2 \text{ mol Ag}^+}\right)\left(\dfrac{311.8 \text{ g Ag}_2\text{SO}_4}{1 \text{ mol Ag}_2\text{SO}_4}\right) = 156 \text{ Ag}_2\text{SO}_4$

(*c*) $1.00 \text{ mol Ag}^+\left(\dfrac{1 \text{ mol AgC}_2\text{H}_3\text{O}_2}{1 \text{ mol Ag}^+}\right)\left(\dfrac{166.9 \text{ g AgC}_2\text{H}_3\text{O}_2}{1 \text{ mol AgC}_2\text{H}_3\text{O}_2}\right) = 167 \text{ g AgC}_2\text{H}_3\text{O}_2$

HEAT CAPACITY AND HEAT OF REACTION

10.41. The student political science society decides to take a trip to Washington, D.C. The hostel rate is $20 per student each night. How much will it cost for 23 students to stay for three nights?

Ans. $23 \text{ students}\left(\dfrac{20 \text{ dollars}}{\text{student} \cdot \text{night}}\right)3 \text{ nights} = 1380 \text{ dollars}$

10.42. A system is initially at 15°C. What will be the final temperature if the system (*a*) is warmed 35°C and (*b*) is warmed to 35°C? (*c*) What is the difference between final temperature and temperature change?

Ans.   (*a*) 50°C   (*b*) 35°C   (*c*) The temperature difference is the final temperature minus the initial temperature. The final temperature is merely a single temperature. Be sure to read the problems carefully so that you do not mistake temperature change for initial or final temperature.

10.43. How much heat is required to raise 43.2 g of iron from $20.0°C$ to $35.1°C$? [$c = 0.447$ J/(g·°C)]

Ans.
$$\Delta t = 35.1°C - 20.0°C = 15.1°C$$
$$\text{Heat} = (m)(c)(\Delta t) = (43.2 \text{ g})\left(\frac{0.447 \text{ J}}{\text{g·°C}}\right)(15.1°C) = 292 \text{ J}$$

10.44. Calculate the heat produced by incomplete combustion of carbon, producing 45.7 g of CO according to the following equation:

$$2 \text{ C} + \text{O}_2 \longrightarrow 2 \text{ CO} + 220 \text{ kJ}$$

Ans.
$$45.7 \text{ g CO}\left(\frac{1 \text{ mol CO}}{28.0 \text{ g CO}}\right)\left(\frac{220 \text{ kJ}}{2 \text{ mol CO}}\right) = 180 \text{ kJ}$$

10.45. Calculate the temperature change produced by the addition of 225 J of heat to 15.3 g of water.

Ans.
$$t = \frac{225 \text{ J}}{[4.184 \text{ J/(g·°C)}](15.3 \text{ g})} = 3.51°C$$

10.46. Calculate the final temperature t_f if 225 J of energy is added to 15.3 g of water at $19.0°C$.

Ans. Δt was calculated in the prior problem.
$$t_f = 19.0°C + 3.51°C = 22.5°C$$

10.47. Calculate the specific heat of a 125 g metal bar which rose in temperature from $18.0°C$ to $33.0°C$ on addition of 2.37 kJ of heat.

Ans.
$$c = \frac{\text{heat}}{(m)(t)} = \frac{2370 \text{ J}}{(125 \text{ g})(15.0°C)} = 1.26 \text{ J/(g·°C)}$$

Supplementary Problems

10.48. Read each of the other supplementary problems and state which ones are limiting-quantities problems.

Ans.   10.57, 10.58, 10.59, 10.64, 10.65, 10.66, 10.67, 10.71.

10.49. Calculate the number of grams of methyl alcohol, CH_3OH, obtainable in an industrial process from 5.00 metric tons $(5.00 \times 10^6 \text{ g})$ of CO plus hydrogen gas.

Ans.   See Problem 10.21.

10.50. Make up your own stoichiometry problem, using the equation

$$CaCl_2 + 2 \text{ AgNO}_3 \longrightarrow \text{Ca(NO}_3)_2 + 2 \text{ AgCl}$$

10.51. What percentage of 50.0 g of $KClO_3$ must be decomposed thermally to produce 14.6 g O_2?

Ans.   The statement of the problem implies that not all the $KClO_3$ is decomposed; it must be in excess for the purpose of producing the 14.6 g O_2. Hence we can base the solution on the quantity of O_2. In Problem 10.29 we found that 37.3 g of $KClO_3$ decomposed. The percent $KClO_3$ decomposed is then

$$\frac{37.3 \text{ g}}{50.0 \text{ g}} \times 100\% = 74.6\%$$

10.52. Percent yield is defined as 100 times the amount of a product actually prepared during a reaction divided by the amount theoretically possible to be prepared according to the balanced chemical equation. (Some reactions are slow, and sometimes not enough time is allowed for their completion; some reactions are accompanied by side reactions which consume a portion of the reactants; some reactions never get to completion.) If 5.00 g $C_6H_{11}Br$ is prepared by treating 5.00 g C_6H_{12} with excess Br_2, what is the percent yield? The equation is

$$C_6H_{12} + Br_2 \longrightarrow C_6H_{11}Br + HBr$$

Ans.

$$5.00 \text{ g } C_6H_{12}\left(\frac{1 \text{ mol } C_6H_{12}}{84.2 \text{ g } C_6H_{12}}\right)\left(\frac{1 \text{ mol } C_6H_{11}Br}{1 \text{ mol } C_6H_{12}}\right)\left(\frac{163 \text{ g } C_6H_{11}Br}{1 \text{ mol } C_6H_{11}Br}\right) = 9.68 \text{ g } C_6H_{11}Br$$

$$\frac{5.00 \text{ g } C_6H_{11}Br \text{ obtained}}{9.68 \text{ g } C_6H_{11}Br \text{ possible}} \times 100\% = 51.7\% \text{ yield}$$

10.53. In a certain experiment, 225 g of ethane, C_2H_6, is burned. (*a*) Write a balanced chemical equation for the combustion of ethane to produce CO_2 and water. (*b*) Determine the number of moles of ethane in 225 g of ethane. (*c*) Determine the number of moles of CO_2 produced by the combustion of that number of moles of ethane. (*d*) Determine the mass of CO_2 that can be produced by the combustion of 225 g of ethane.

Ans.   (*a*) $2 C_2H_6 + 7 O_2 \longrightarrow 4 CO_2 + 6 H_2O$

(*b*) $225 \text{ g } C_2H_6\left(\dfrac{1 \text{ mol } C_2H_6}{30.1 \text{ g } C_2H_6}\right) = 7.48 \text{ mol } C_2H_6$

(*c*) $7.48 \text{ mol } C_2H_6\left(\dfrac{4 \text{ mol } CO_2}{2 \text{ mol } C_2H_6}\right) = 15.0 \text{ mol } CO_2$

(*d*) $15.0 \text{ mol } CO_2\left(\dfrac{44.0 \text{ g } CO_2}{1 \text{ mol } CO_2}\right) = 660 \text{ g } CO_2$

10.54. Determine the number of kilograms of SO_3 produced by treating excess SO_2 with 50.0 g of oxygen.

Ans. $2 SO_2 + O_2 \longrightarrow 2 SO_3$

$$50.0 \text{ g } O_2\left(\frac{1 \text{ mol } O_2}{32.0 \text{ g } O_2}\right)\left(\frac{2 \text{ mol } SO_3}{1 \text{ mol } O_2}\right)\left(\frac{80.0 \text{ g } SO_3}{1 \text{ mol } SO_3}\right) = 250 \text{ g } SO_3 = 0.250 \text{ kg } SO_3$$

10.55. How many grams of K_2CO_3 will be produced by thermal decomposition of 4.00 g $KHCO_3$?

Ans. $2 KHCO_3 \xrightarrow{\text{heat}} K_2CO_3 + CO_2 + H_2O$

$$4.00 \text{ g } KHCO_3\left(\frac{1 \text{ mol } KHCO_3}{100 \text{ g } KHCO_3}\right)\left(\frac{1 \text{ mol } K_2CO_3}{2 \text{ mol } KHCO_3}\right)\left(\frac{138 \text{ g } K_2CO_3}{1 \text{ mol } K_2CO_3}\right) = 2.76 \text{ g } K_2CO_3$$

10.56. What is the difference in mass between reactions involving 2.25 g Cl^- and 2.25 g NaCl?

Ans.   The former might be part of any ionic chloride, but more importantly, the former has a greater mass of chlorine, because 2.25 g NaCl has less than 2.25 g Cl^-.

10.57. How many moles of H_2 can be prepared by treating 5.00 g Na with 2.80 g H_2O? *Caution*: This reaction is very energetic and can even cause an explosion.

Ans. $2 Na + 2 H_2O \longrightarrow 2 NaOH + H_2$

$$5.00 \text{ g Na}\left(\frac{1 \text{ mol Na}}{23.0 \text{ g Na}}\right) = 0.217 \text{ mol Na}$$

$$2.80 \text{ g } H_2O\left(\frac{1 \text{ mol } H_2O}{18.0 \text{ g } H_2O}\right) = 0.156 \text{ mol } H_2O$$

Since they react in a 1 : 1 mole ratio, H_2O is obviously in limiting quantity.

$$0.156 \text{ mol } H_2O\left(\frac{1 \text{ mol } H_2}{2 \text{ mol } H_2O}\right) = 0.0780 \text{ mol } H_2$$

10.58. How many grams of NaCl can be prepared by treating 18.9 g $BaCl_2$ with 14.8 g Na_2SO_4?

Ans. $BaCl_2 + Na_2SO_4 \longrightarrow BaSO_4 + 2 NaCl$

$$18.9 \text{ g } BaCl_2\left(\frac{1 \text{ mol } BaCl_2}{208 \text{ g } BaCl_2}\right) = 0.0909 \text{ mol } BaCl_2$$

$$14.8 \text{ g } Na_2SO_4\left(\frac{1 \text{ mol } Na_2SO_4}{142 \text{ g } Na_2SO_4}\right) = 0.104 \text{ mol } Na_2SO_4$$

Since they react in a 1:1 mole ratio, $BaCl_2$ is obviously in limiting quantity.

$$0.0909 \text{ mol } BaCl_2 \left(\frac{2 \text{ mol NaCl}}{1 \text{ mol } BaCl_2} \right) \left(\frac{58.5 \text{ g NaCl}}{1 \text{ mol NaCl}} \right) = 10.6 \text{ g NaCl}$$

10.59. If 4.00 g Cu is treated with 15.0 g $AgNO_3$, how many grams of Ag metal can be prepared?

$$2 \, AgNO_3 + Cu \longrightarrow 2 \, Ag + Cu(NO_3)_2$$

Ans.
$$4.00 \text{ g Cu} \left(\frac{1 \text{ mol Cu}}{63.5 \text{ g Cu}} \right) = 0.0630 \text{ mol Cu present}$$

$$15.0 \text{ g } AgNO_3 \left(\frac{1 \text{ mol } AgNO_3}{170 \text{ g } AgNO_3} \right) = 0.0882 \text{ mol } AgNO_3 \text{ present}$$

$$\frac{0.0882 \text{ mol } AgNO_3}{2} = 0.0441 \text{ mol } AgNO_3$$

$$\frac{0.0630 \text{ mol Cu}}{1} = 0.0630 \text{ mol Cu}$$

$AgNO_3$ is present in limiting quantity, and we base the calculation on the $AgNO_3$ present.

$$0.0882 \text{ mol } AgNO_3 \left(\frac{1 \text{ mol Ag}}{1 \text{ mol } AgNO_3} \right) \left(\frac{108 \text{ g Ag}}{1 \text{ mol Ag}} \right) = 9.53 \text{ g Ag}$$

10.60. What, if any, is the difference between the following three exam questions?

How much NaCl is produced by treating 10.00 g NaOH with excess HCl?

How much NaCl is produced by treating 10.00 g NaOH with sufficient HCl?

How much NaCl is produced by treating 10.00 g NaOH with HCl?

Ans.   There is no difference.

10.61. What mass of water can be heated from 20.0°C to 45.0°C by the combustion of 4.00 g of carbon to carbon dioxide?

$$C + O_2 \longrightarrow CO_2 + 393 \text{ kJ}$$

Ans.
$$4.00 \text{ g C} \left(\frac{1 \text{ mol C}}{12.0 \text{ g C}} \right) \left(\frac{393 \text{ kJ}}{1 \text{ mol C}} \right) = 131 \text{ kJ produced}$$

Hence 131 kJ = 131 000 J is added to the water.

$$m = \frac{\text{heat}}{(c)(\Delta t)} = \frac{131\,000 \text{ J}}{[4.184 \text{ J}/(g\cdot°C)](25.0°C)} = 1250 \text{ g} = 1.25 \text{ kg}$$

Note that the heat *given off* by the reaction is *absorbed* by the water.

10.62. Some pairs of substances undergo different reactions if one or the other is in excess. For example,

$$\text{With excess } O_2: \qquad C + O_2 \longrightarrow CO_2$$
$$\text{With excess } C: \qquad 2\,C + O_2 \longrightarrow 2\,CO$$

Does this fact make the limiting-quantities calculations in the text untrue for such pairs?

Ans.   No. The principles are true for *each* possible reaction.

10.63. A manufacturer produces impure ammonium phosphate as a fertilizer. Treatment of 56.0 g of product with excess NaOH produces 18.3 g NH_3. What percentage of the fertilizer is pure $(NH_4)_3PO_4$? (Assume that any impurity contains no ammonium compound.)

Ans.
$$3 \, NaOH + (NH_4)_3PO_4 \longrightarrow 3 \, NH_3 + 3 \, H_2O + Na_3PO_4$$

$$18.3 \text{ g } NH_3 \left(\frac{1 \text{ mol } NH_3}{17.0 \text{ g } NH_3} \right) \left(\frac{1 \text{ mol } (NH_4)_3PO_4}{3 \text{ mol } NH_3} \right) \left(\frac{149 \text{ g } (NH_4)_3PO_4}{1 \text{ mol } (NH_4)_3PO_4} \right) = 53.5 \text{ g } (NH_4)_3PO_4$$

$$\left(\frac{53.5 \text{ g } (NH_4)_3PO_4}{56.0 \text{ g product}} \right) \times 100\% = 95.5\% \text{ pure}$$

10.64. (*a*) When 5.00 g NaOH reacts with 5.00 g HCl, how much NaCl is produced? (*b*) When 6.00 g NaOH reacts with 5.00 g HCl, how much NaCl is produced?

Ans.
$$HCl + NaOH \longrightarrow NaCl + H_2O$$

(*a*)
$$5.00 \text{ g HCl}\left(\frac{1 \text{ mol HCl}}{36.5 \text{ g HCl}}\right) = 0.137 \text{ mol HCl}$$

$$5.00 \text{ g NaOH}\left(\frac{1 \text{ mol NaOH}}{40.0 \text{ g NaOH}}\right) = 0.125 \text{ mol NaOH}$$

Since the reactants react in a 1:1 ratio, NaOH is the limiting quantity:

$$0.125 \text{ mol NaOH}\left(\frac{1 \text{ mol NaCl}}{1 \text{ mol NaOH}}\right)\left(\frac{58.5 \text{ g NaCl}}{1 \text{ mol NaCl}}\right) = 7.31 \text{ g NaCl}$$

(*b*)
$$6.00 \text{ g NaOH}\left(\frac{1 \text{ mol NaOH}}{40.0 \text{ g NaOH}}\right) = 0.150 \text{ mol NaOH}$$

$$5.00 \text{ g HCl}\left(\frac{1 \text{ mol HCl}}{36.5 \text{ g HCl}}\right) = 0.137 \text{ mol HCl}$$

Since the reactants react in a 1:1 ratio, HCl is the limiting quantity:

$$0.137 \text{ mol HCl}\left(\frac{1 \text{ mol NaCl}}{1 \text{ mol HCl}}\right)\left(\frac{58.5 \text{ g NaCl}}{1 \text{ mol NaCl}}\right) = 8.01 \text{ g NaCl}$$

10.65. Calculate the number of moles of each solute in the final solution after 1.75 mol of aqueous $BaCl_2$ and 2.70 mol of aqueous $AgNO_3$ are mixed.

Ans.

	$BaCl_2$	+ 2 $AgNO_3$	$\longrightarrow$	$Ba(NO_3)_2$	+ 2 AgCl(s)
Initial:	1.75 mol	2.70 mol		0.00 mol	
Change:	1.35 mol	2.70 mol		1.35 mol	
Final:	0.40 mol	0.00 mol		1.35 mol	

10.66. Calculate the number of moles of each ion in the final solution after 1.75 mol of aqueous $BaCl_2$ and 2.70 mol of aqueous $AgNO_3$ are mixed.

Ans. The Ba^{2+} ion and the NO_3^- ion do not react (they are spectator ions), so there are 1.75 mol of Ba^{2+} ion and 2.70 mol of NO_3^- ion in the final solution. The silver ion and chloride ion concentrations are calculated using the net ionic equation:

	Cl^-	+ Ag^+	$\longrightarrow$	AgCl(s)
Initial:	3.50 mol	2.70 mol		
Change:	2.70 mol	2.70 mol		
Final:	0.80 mol	0.00 mol		

These results are the same as those of the prior problem.

10.67. Consider the reaction

$$Ba(OH)_2 + 2 HCl \longrightarrow BaCl_2 + 2 H_2O$$

If exactly 20.0 g of $Ba(OH)_2$ reacts according to this equation, do you know how much $Ba(OH)_2$ was added to the HCl? Do you know how much HCl was added to the $Ba(OH)_2$? Explain.

Ans. You cannot tell. Either reactant might have been in excess. The information given allows calculations of how much reacted and how much of the products were produced, but not how much was added in the first place.

10.68. Redo Problem 10.18 without bothering to solve for intermediate answers.

10.69. Determine the number of grams of SO_3 produced by treating excess SO_2 with 50.0 g of oxygen.

Ans. 250 g.

10.70. A sample of a hydrocarbon (a compound of carbon and hydrogen only) is burned, and 1.10 g CO_2 and 0.450 g H_2O are produced. What is the empirical formula of the hydrocarbon?

   Ans.   The empirical formula can be determined from the ratio of moles of carbon atoms to moles of hydrogen atoms. From the masses of products, we can get the numbers of moles of products, from which we can get the numbers of moles of C and H. The mole ratio of these two elements is the same in the products as in the original compound.

$$1.10 \text{ g } CO_2\left(\frac{1 \text{ mol } CO_2}{44.0 \text{ g } CO_2}\right) = 0.0250 \text{ mol } CO_2$$

$$0.450 \text{ g } H_2O\left(\frac{1 \text{ mol } H_2O}{18.0 \text{ g } H_2O}\right) = 0.0250 \text{ mol } H_2O$$

0.0250 mol CO_2 contains 0.0250 mol C

0.0250 mol H_2O contains 0.500 mol H

   The mole ratio is 1:2, and the empirical formula is CH_2.

10.71. How many moles of each substance are present in solution after 0.100 mol of NaOH is added to 0.200 mol of $HC_2H_3O_2$?

   Ans.   The balanced equation is

$$NaOH + HC_2H_3O_2 \longrightarrow NaC_2H_3O_2 + H_2O$$

   The limiting quantity is NaOH, and so 0.100 mol NaOH reacts with 0.100 mol $HC_2H_3O_2$ to produce 0.100 mol $NaC_2H_3O_2$ + 0.100 mol H_2O. There is also 0.100 mol excess $HC_2H_3O_2$ left in the solution.

10.72. Combine Figs. 7-5 and 10-4 to show all the conversions learned so far.

   Ans.   The answer is shown on the factor-label conversion figure, page 348. It includes the boxes numbered 3 to 8, 11 to 14, 16 and 17, plus all the conversion factors on the arrows between them.

Molarity

CHAPTER 11

11.1. INTRODUCTION

Many quantitative chemical reactions are carried out in solution because volumes are easier to measure than masses. If you dissolve a certain mass of a substance in a measured volume of solution, you can take measured fractions of the total volume of the solution and know how much of the substance it contains.

Solute is defined as the material that is dissolved in a *solvent*. For example, if you dissolve sugar in water, the sugar is the solute and the water is the solvent. In the discussions of this chapter, the solvent will be water; that is, we will discuss *aqueous* solutions only.

Perhaps the most useful measure of concentration is *molarity*. Molarity is defined as the number of moles of solute per liter of solution:

$$\text{Molarity} = \frac{\text{number of moles of solute}}{\text{liter of solution}}$$

Note particularly that liters of *solution* are used, not liters of *solvent*. Chemists often abbreviate the definition to merely "moles per liter," but the shortening does not change the way the unit is actually defined. The unit of molarity is *molar*, symbolized *M*. Special care must be taken with abbreviations in this chapter: *M* stands for molar; mol for mole(s). (Some authors use *M* to stand for molarity as well as molar. They may use italics for one and not the other. Check your text and follow its conventions.)

EXAMPLE 11.1. What is the molarity of the solution produced by dissolving 6.00 mol of solute in enough water to make 1.50 L of solution?

Ans.
$$\text{Molarity} = \frac{6.00 \text{ mol}}{1.50 \text{ L}} = 4.00 \text{ } M$$

The answer is stated out loud as "The molarity is 4.00 molar," or more commonly, "The solution is 4.00 molar." Note the difference between the words *molarity* (the quantity) and *molar* (the unit of molarity).

11.2. MOLARITY CALCULATIONS

Molarity, being a ratio, can be used as a factor. Anywhere *M* (for molar) is used, it can be replaced by mol/L. The reciprocal of molarity can also be used.

EXAMPLE 11.2. How many moles of solute are contained in 3.0 L of 2.0 *M* solution?

Ans. As usual, the quantity given is put down first and multiplied by a ratio—in this case, the molarity. It is easy to visualize the solution, pictured in Fig. 11-1.

$$3.0 \text{ L} \left(\frac{2.0 \text{ mol}}{1 \text{ L}} \right) = 6.0 \text{ mol}$$

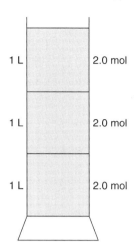

Fig. 11-1.   3.0 L of 2.0 *M* solution

EXAMPLE 11.3.   What volume of 1.50 *M* solution contains 6.00 mol of solute?

Ans.
$$6.00 \text{ mol}\left(\frac{1 \text{ L}}{1.50 \text{ mol}}\right) = 4.00 \text{ L}$$

EXAMPLE 11.4.   What is the molarity of a solution prepared by dissolving 63.0 g NaF in enough water to make 250.0 mL of solution?

Ans.   Molarity is defined in moles per liter. We convert each of the quantities given to those used in the definition:

$$63.0 \text{ g NaF}\left(\frac{1 \text{ mol NaF}}{42.0 \text{ g NaF}}\right) = 1.50 \text{ mol NaF} \qquad 250.0 \text{ mL}\left(\frac{1 \text{ L}}{1000 \text{ mL}}\right) = 0.2500 \text{ L}$$

$$\text{Molarity} = \frac{1.50 \text{ mol}}{0.2500 \text{ L}} = 6.00 \ M$$

EXAMPLE 11.5.   What is the concentration of a solution containing 1.23 mmol in 1.00 mL?

Ans.
$$\frac{1.23 \text{ mmol}\left(\dfrac{1 \text{ mol}}{1000 \text{ mmol}}\right)}{1.00 \text{ mL}\left(\dfrac{1 \text{ L}}{1000 \text{ mL}}\right)} = \frac{1.23 \text{ mol}}{1 \text{ L}} = 1.23 \ M$$

The numeric value is the same in mmol/mL as in mol/L. Thus, ***molarity can be defined as the number of millimoles of solute per milliliter of solution***, which is an advantage because chemists more often use volumes measured in milliliters.

EXAMPLE 11.6.   Do we know how much water was added to the solute in Example 11.1?

Ans.   We have no way of knowing how much water was added. We know the final volume of the solution, but not the volume of the solvent.

When water is added to a solution, the volume increases but the numbers of moles of the solutes do not change. The molarity of every solute in the solution therefore decreases.

EXAMPLE 11.7.   What is the final concentration of 2.00 L of 3.00 *M* solution if enough water is added to dilute the solution to 10.0 L?

Ans.   The concentration is the number of moles of solute per liter of solution, equal to the number of moles of solute divided by the total volume in liters. The original number of moles of solute does not change:

$$\text{Number of moles} = 2.00 \text{ L}\left(\frac{3.00 \text{ mol}}{1 \text{ L}}\right) = 6.00 \text{ mol}$$

The final volume is 10.0 L, as stated in the problem.

$$\text{Molarity} = \frac{6.00 \text{ mol}}{10.0 \text{ L}} = 0.600 \; M$$

EXAMPLE 11.8. What is the final concentration of 2.0 L of 0.30 M solution if 5.0 L of water is added to dilute the solution?

Ans. Note the difference between the wording of this example and the prior one. Here the final volume is 7.0 L. (When you mix solutions, unless they have identical compositions, the final volume might not be exactly equal to the sum of the individual volumes. When only dilute aqueous solutions and water are involved, the volumes are very nearly additive, however.)

$$\text{Molarity} \cong \frac{0.60 \text{ mol}}{7.0 \text{ L}} = 0.086 \; M$$

EXAMPLE 11.9. (*a*) A car was driven 20 mi/h for 1.0 h and then 40 mi/h for 1.0 h. What was the average speed over the whole trip? (*b*) If 1.0 L of 2.0 M NaCl solution is added to 1.0 L of 4.0 M NaCl solution, what is the final molarity?

Ans. (*a*) The average speed is equal to the total distance divided by the total time. The total time is 2.0 h. The total distance is

$$\text{Distance} = 1.0 \text{ h}\left(\frac{20 \text{ mi}}{1 \text{ h}}\right) + 1.0 \text{ h}\left(\frac{40 \text{ mi}}{1 \text{ h}}\right) = 60 \text{ mi}$$

$$\text{Average speed} = \frac{60 \text{ mi}}{2.0 \text{ h}} = \frac{30 \text{ mi}}{1 \text{ h}}$$

Note that we cannot merely add the speeds here to get the speed for the entire trip.

(*b*) The final concentration is the total number of moles of NaCl divided by the total volume. The total volume is about 2.0 L. The total number of moles is

$$1.0 \text{ L}\left(\frac{2.0 \text{ mol}}{1 \text{ L}}\right) + 1.0 \text{ L}\left(\frac{4.0 \text{ mol}}{1 \text{ L}}\right) = 6.0 \text{ mol}$$

The final concentration is (6.0 mol)/(2.0 L) = 3.0 M. Note that we cannot merely add the concentrations to get the final concentration. (If equal *volumes* are combined, the concentration is the *average* of those of the initial solutions.)

EXAMPLE 11.10. Calculate the final concentration if 3.00 L of 4.00 M NaCl and 4.00 L of 2.00 M NaCl are mixed.

Ans. $$\text{Final volume} = 7.00 \text{ L}$$

$$\text{Final number of moles} = 3.00 \text{ L}\left(\frac{4.00 \text{ mol}}{1 \text{ L}}\right) + 4.00 \text{ L}\left(\frac{2.00 \text{ mol}}{1 \text{ L}}\right) = 20.0 \text{ mol}$$

$$\text{Molarity} = \frac{20.0 \text{ mol}}{7.00 \text{ L}} = 2.86 \; M$$

Note that the answer is reasonable; the final concentration is between the concentrations of the two original solutions.

EXAMPLE 11.11. Calculate the final concentration if 3.00 L of 4.00 M NaCl, 4.00 L of 2.00 M NaCl, and 3.00 L of water are mixed.

Ans. The final volume is about 10.0 L. The final number of moles of NaCl is 20.0 mol, the same as in Example 11.10, since there was no NaCl in the water. Hence, the final concentration is

$$\text{Molarity} = \frac{20.0 \text{ mol}}{10.0 \text{ L}} = 2.00 \; M$$

Note the the concentration is lower than it was in Example 11.10 despite the presence of the same number of moles of NaCl, since there is a greater volume.

11.3. TITRATION

To determine the concentration of a solute in a solution (e.g., HCl) we treat the unknown solution with a solution of known concentration and volume (e.g., NaOH) until the mole ratio is exactly what is required by the balanced chemical equation. Then from the known volumes of both solutions, the unknown concentration of the

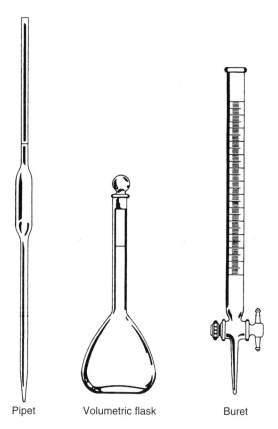

Pipet   Volumetric flask   Buret

Fig. 11-2.   Volumetric glassware.

The pipet is designed to deliver an exact volume of liquid; the volumetric flask is designed to hold an exact volume of liquid; and the buret is designed to deliver precisely measurable volumes of liquid. (Not drawn to scale.)

solute can be calculated. In this manner, we can determine the concentrations of solutions without weighing out the solute, which is often difficult or even impossible. (For example, some reactants absorb water from the air so strongly that you do not know how much reactant and how much water you are weighing.) The procedure used is called *titration*. In a typical titration experiment, 25.00 mL of 0.7500 *M* HCl is pipetted into an Erlenmeyer flask. A *pipet* (Fig. 11-2) is a piece of glassware that is calibrated to deliver an exact volume of liquid. A solution of NaOH of unknown concentration is placed in *buret* (Fig. 11-2), and some is allowed to drain out the bottom to ensure that the portion below the stopcock is filled. The buret volume is read (say, 5.27 mL) before any NaOH is added from it to the HCl and again after the NaOH is added (say, 46.92 mL). The volume of added NaOH is merely the difference in readings (46.92 mL − 5.27 mL = 41.65 mL). The concentration of NaOH may now be calculated because the exact number of moles of NaOH has been added to react entirely with the HCl.

$$25.00 \text{ mL HCl}\left(\frac{0.7500 \text{ mmol HCl}}{1 \text{ mL HCl}}\right) = 18.75 \text{ mmol HCl}$$

The number of millimoles of NaOH is exactly the same, since the addition was stopped when the reaction was just complete. Therefore, there is 18.75 mmol NaOH in the volume of NaOH added,

$$\frac{18.75 \text{ mmol NaOH}}{41.65 \text{ mL}} = 0.4502 \text{ } M \text{ NaOH}$$

Two questions should immediately arise. (1) How can you know exactly when to stop the titration so that the number of moles of NaOH is equal to the number of moles of HCl? (2) What is the use of determining the concentration of a solution of NaOH when the NaOH has now been used up by reaction with the HCl?

(1) An *indicator* is used to tell us when to stop the titration. Typically an indicator is a compound that changes color as the acidity or basicity level of a solution changes. Thus, we add NaOH slowly, drop by drop

toward the end, until a permanent color change takes place. At that point, the *endpoint* has been reached, and the titration is complete. The HCl has just been used up. (2) The purpose of doing a titration is to determine the concentration of a solution. If the concentration of 1 L of NaOH is to be determined, only a small portion of it is used in the titration. The rest has the same concentration, and although the part used in the titration is no longer useful, the concentration of most of the solution is now known.

EXAMPLE 11.12. What is the concentration of a $Ba(OH)_2$ solution if it takes 43.12 mL to neutralize 25.00 mL of 2.000 M HCl?

Ans.   The number of moles of HCl is easily calculated.

$$25.00 \text{ mL} \left(\frac{2.000 \text{ mmol}}{1 \text{ mL}} \right) = 50.00 \text{ mmol} \quad \text{or} \quad 0.025\ 00 \text{ L} \left(\frac{2.000 \text{ mol}}{1 \text{ L}} \right) = 0.050\ 00 \text{ mol HCl}$$

The balanced chemical equation shows that the ratio of moles of HCl to $Ba(OH)_2$ is 2 : 1.

$$2 \text{ HCl} + Ba(OH)_2 \longrightarrow BaCl_2 + 2 \text{ H}_2\text{O}$$

$$50.00 \text{ mmol HCl} \left[\frac{1 \text{ mmol } Ba(OH)_2}{2 \text{ mmol HCl}} \right] \quad \text{or} \quad 0.050\ 00 \text{ mol HCl} \left[\frac{1 \text{ mol } Ba(OH)_2}{2 \text{ mol HCl}} \right]$$
$$= 25.00 \text{ mmol } Ba(OH)_2 \qquad\qquad = 0.025\ 00 \text{ mol } Ba(OH)_2$$

The molarity is given by

$$\frac{25.00 \text{ mmol } Ba(OH)_2}{43.12 \text{ mL}} = 0.5798 \ M \quad \text{or} \quad \frac{0.025\ 00 \text{ mol } Ba(OH)_2}{0.043\ 12 \text{ L}} = 0.5798 \ M \ Ba(OH)_2$$

Note that you *cannot* calculate this concentration with the equation

$$M_1 V_1 = M_2 V_2 \qquad \text{(limited applications)}$$

where M is molarity and V is volume. This equation is given in some texts for simple dilution problems. The equation reduces to $\text{moles}_1 = \text{moles}_2$, which is not true in cases in which there is not a 1 : 1 ratio in the balanced chemical equation.

11.4.   STOICHIOMETRY IN SOLUTION

With molarity and volume of solution, numbers of moles can be calculated. The number of moles may be used in stoichiometry problems just as moles calculated in any other way are used. Also, the number of moles calculated as in Chap. 10 can be used to calculate molarities or volumes of solution.

EXAMPLE 11.13. Calculate the number of moles of AgCl that can be prepared by mixing 0.740 L of 1.25 M $AgNO_3$ with excess NaCl.

Ans. $NaCl + AgNO_3 \longrightarrow AgCl(s) + NaNO_3$
$$(0.740 \text{ L})(1.25 \text{ mol } AgNO_3/\text{L}) = 0.925 \text{ mol } AgNO_3$$
$$(0.925 \text{ mol } AgNO_3)(1 \text{ mol } AgCl/1 \text{ mol } AgNO_3) = 0.925 \text{ mol } AgCl$$

EXAMPLE 11.14. Calculate the number of moles of AgCl that can be prepared by mixing 0.740 L of 1.25 M $AgNO_3$ with 1.50 L of 0.900 M NaCl.

Ans. $NaCl + AgNO_3 \longrightarrow AgCl(s) + NaNO_3$
$$(0.740 \text{ L})(1.25 \text{ mol } AgNO_3/\text{L}) = 0.925 \text{ mol } AgNO_3 \text{ present}$$
$$(1.50 \text{ L})(0.900 \text{ mol NaCl/L}) = 1.35 \text{ mol NaCl present}$$

$AgNO_3$ is in limiting quantity, and the problem is completed just as the prior example was done.

EXAMPLE 11.15. (*a*) Calculate the concentration of Zn^{2+} produced when excess solid zinc is treated with 100.0 mL of 6.000 *M* HCl. Assume no change in volume. (*b*) Repeat the problem with solid aluminum.

Ans.   (*a*)

$$Zn + 2\,H^+ \longrightarrow Zn^{2+} + H_2$$

$$100.0\text{ mL}\left(\frac{6.000\text{ mmol H}^+}{1\text{ mL}}\right)\left(\frac{1\text{ mmol Zn}^{2+}}{2\text{ mmol H}^+}\right) = 300.0\text{ mmol Zn}^{2+}$$

$$\frac{300.0\text{ mmol Zn}^{2+}}{100.0\text{ mL}} = 3.000\ M\ Zn^{2+}$$

(*b*)

$$2\,Al + 6\,H^+ \longrightarrow 2\,Al^{3+} + 3\,H_2$$

$$100.0\text{ mL}\left(\frac{6.000\text{ mmol H}^+}{1\text{ mL}}\right)\left(\frac{2\text{ mmol Al}^{3+}}{6\text{ mmol H}^+}\right) = 200.0\text{ mmol Al}^{3+}$$

$$\frac{200.0\text{ mmol Al}^{3+}}{100.0\text{ mL}} = 2.000\ M\ Al^{3+}$$

Solved Problems

INTRODUCTION

11.1. Which, if either, has more sugar in it, (*a*) a half cup of tea with one lump of sugar or (*b*) a whole cup of tea with two lumps of sugar?

   Ans.   Two lumps are more than one; the whole cup has more sugar. Note the difference between *quantity* of sugar and *concentration* of sugar.

11.2. Which, if either, of the following tastes sweeter, (*a*) a half cup of tea with one lump of sugar or (*b*) a whole cup of tea with two lumps of sugar?

   Ans.   They both taste equally sweet, since their concentrations are equal.

MOLARITY CALCULATIONS

11.3. Calculate the molarity of each of the following solutions: (*a*) 4.00 mol solute in 2.50 L of solution, (*b*) 0.200 mol solute in 0.240 L of solution, (*c*) 0.0500 mol solute in 25.0 mL of solution, and (*d*) 0.240 mol solute in 750.0 mL of solution.

   Ans.   (*a*) $\dfrac{4.00\text{ mol}}{2.50\text{ L}} = 1.60\ M$ (*c*) $\dfrac{0.0500\text{ mol}}{0.0250\text{ L}} = 2.00\ M$ or $\dfrac{50.0\text{ mmol}}{25.0\text{ mL}} = 2.00\ M$

   (*b*) $\dfrac{0.200\text{ mol}}{0.240\text{ L}} = 0.833\ M$ (*d*) $\dfrac{0.240\text{ mol}}{0.750\text{ L}} = 0.320\ M$ or $\dfrac{240\text{ mmol}}{750.0\text{ mL}} = 0.320\ M$

   (Note: moles per liter or millimoles per milliliter, not moles per milliliter)

11.4. Calculate the number of moles of solute in each of the following solutions: (*a*) 1.50 L of 0.800 *M* solution, (*b*) 1.66 L of 0.150 *M* solution, (*c*) 45.0 mL of 0.600 *M* solution, and (*d*) 25.0 mL of 2.00 *M* solution.

   Ans.   (*a*) $1.50\text{ L}\left(\dfrac{0.800\text{ mol}}{1\text{ L}}\right) = 1.20\text{ mol}$ (*c*) $0.0450\text{ L}\left(\dfrac{0.600\text{ mol}}{1\text{ L}}\right) = 0.0270\text{ mol}$

   (*b*) $1.66\text{ L}\left(\dfrac{0.150\text{ mol}}{1\text{ L}}\right) = 0.249\text{ mol}$ (*d*) $0.0250\text{ L}\left(\dfrac{2.00\text{ mol}}{1\text{ L}}\right) = 0.0500\text{ mol}$

11.5. How can you make 2.25 L of 4.00 *M* sugar solution?

    Ans.   The solution will contain 9.00 mol sugar:

$$2.25 \text{ L}\left(\frac{4.00 \text{ mol}}{1 \text{ L}}\right) = 9.00 \text{ mol}$$

        Thus, place 9.00 mol sugar in a liter or two of water, mix until dissolved, dilute the resulting solution to 2.25 L, and mix thoroughly.

11.6. Calculate the volume of 1.75 *M* solution required to contain 4.20 mol of solute.

    Ans.
$$4.20 \text{ mol}\left(\frac{1 \text{ L}}{1.75 \text{ mol}}\right) = 2.40 \text{ L}$$

11.7. A 10.0-mL solution contains 2.40 mmol of solute. What is its molarity?

    Ans.
$$\frac{2.40 \text{ mmol}}{10.0 \text{ mL}} = 0.240 \ M$$

    Molarity can be calculated by dividing millimoles by milliliters.

11.8. How many moles of solute are present in 29.4 mL of 0.606 *M* solution?

    Ans.   (0.0294 L)(0.606 mol/L) = 0.0178 mol

11.9. Calculate the number of milliliters of 1.25 *M* solution required to contain 0.622 mol of solute.

    Ans.
$$0.622 \text{ mol}\left(\frac{1 \text{ L}}{1.25 \text{ mol}}\right) = 0.498 \text{ L} = 498 \text{ mL}$$

11.10. What volume of 1.45 *M* NaCl solution contains 71.3 g NaCl?

    Ans.
$$71.3 \text{ g NaCl}\left(\frac{1 \text{ mol NaCl}}{58.5 \text{ g NaCl}}\right) = 1.22 \text{ mol NaCl}$$

$$1.22 \text{ mol NaCl}\left(\frac{1 \text{ L solution}}{1.45 \text{ mol NaCl}}\right) = 0.841 \text{ L} = 841 \text{ mL}$$

11.11. What is the concentration of a solution prepared by dissolving 22.2 g NaCl in sufficient water to make 86.9 mL of solution?

    Ans.   Molarity is in moles per liter. We must change the grams of NaCl to moles and the milliliters of solution to liters.

$$22.2 \text{ g NaCl}\left(\frac{1 \text{ mol NaCl}}{58.5 \text{ g NaCl}}\right) = 0.379 \text{ mol}$$

$$0.379 \text{ mol}/0.0869 \text{ L} = 4.36 \ M$$

11.12. How many grams of NaCl are present in 72.1 mL of 1.03 *M* NaCl?

    Ans.   (0.0721 L)(1.03 mol/L) = 0.0743 mol

$$(0.0743 \text{ mol})(58.5 \text{ g NaCl/mol}) = 4.35 \text{ g NaCl}$$

11.13. Calculate the volume of 0.900 *M* solution required to contain 1.84 mol of solute.

    Ans.
$$1.84 \text{ mol}\left(\frac{1 \text{ L}}{0.900 \text{ mol}}\right) = 2.04 \text{ L}$$

11.14. How many milligrams of NaOH are present in 35.0 mL of 2.18 *M* NaOH?

    Ans.
$$35.0 \text{ mL}\left(\frac{2.18 \text{ mmol}}{1 \text{ mL}}\right)\left(\frac{40.0 \text{ mg}}{1 \text{ mmol}}\right) = 3050 \text{ mg}$$

11.15. What is the concentration of a solution prepared by diluting 3.0 L of 2.5 M solution to 8.0 L with water?

Ans. The number of moles of solute is not changed by addition of the water. The number of moles in the original solution is

$$3.0\,\text{L}\left(\frac{2.5\,\text{mol}}{1\,\text{L}}\right) = 7.5\,\text{mol}$$

That 7.5 mol is now dissolved in 8.0 L, and its concentration is

$$\frac{7.5\,\text{mol}}{8.0\,\text{L}} = 0.94\,M$$

11.16. What is the concentration of a solution prepared by diluting 0.500 L of 1.80 M solution to 1.50 L with solvent?

Ans. The number of moles of solute is not changed by addition of the solvent. The number of moles in the original solution is

$$0.500\,\text{L}\left(\frac{1.80\,\text{mol}}{1\,\text{L}}\right) = 0.900\,\text{mol}$$

That 0.900 mol is now dissolved in 1.50 L (compare Problem 11.17), and its concentration is

$$\frac{0.900\,\text{mol}}{1.50\,\text{L}} = 0.600\,M$$

11.17. What is the concentration of a solution prepared by diluting 0.500 L of 1.80 M solution with 1.50 L of solvent?

Ans. The number of moles of solute is not changed by addition of the solvent; it is 0.900 mol. What is the final volume of the solution? If we add 1.50 L of solvent, it will be about 2.00 L. Compare this wording with that of Problem 11.16. The 1.50 mol is now dissolved in 2.00 L, and its concentration is

$$\frac{0.900\,\text{mol}}{2.00\,\text{L}} = 0.450\,M$$

11.18. What is the concentration of a solution prepared by diluting 25.0 mL of 3.00 M solution to 60.0 mL?

Ans. The number of moles of solute is not changed by addition of the solvent. The number of moles in the original solution is

$$(0.0250\,\text{L})(3.00\,\text{mol/L}) = 0.0750\,\text{mol}$$

That 0.0750 mol is now dissolved in 0.0600 L, and its concentration is

$$\frac{0.0750\,\text{mol}}{0.0600\,\text{L}} = 1.25\,M$$

Alternatively, the number of millimoles is given by

$$(25.0\,\text{mL})(3.00\,\text{mmol/mL}) = 75.0\,\text{mmol}$$

The concentration is

$$\frac{75.0\,\text{mmol}}{60.0\,\text{mL}} = 1.25\,M$$

The use of millimoles and milliliters saves conversions of milliliters to liters.

11.19. What concentration of salt is obtained by mixing 20.0 mL of 3.0 M salt solution with 30.0 mL of 2.0 M salt solution?

Ans. The final concentration is the total number of millimoles divided by the total number of milliliters. The volume is 20.0 mL + 30.0 mL = 50.0 mL. The total number of millimoles is given by

$$(20.0\,\text{mL})(3.0\,\text{mmol/mL}) + (30.0\,\text{mL})(2.0\,\text{mmol/mL}) = 120\,\text{mmol}$$

The concentration is 120 mmol/50.0 mL = 2.4 M.

11.20. What concentration of salt is obtained by mixing 20.0 mL of 3.0 M salt solution with 30.0 mL of 2.0 M salt solution and diluting with water to 100.0 mL?

> *Ans.* The final concentration is the total number of moles divided by the total number of liters. The volume is 100.0 mL. Since there is no solute in the water, the total number of moles is the same as that in the prior problem. The concentration is 120 mmol/100.0 mL = 1.2 M. The concentration is lower than that in Problem 11.19 despite the same number of moles of solute, because of the greater volume.

11.21. Calculate the concentration of sugar in a solution prepared by mixing 3.0 L of 2.0 M sugar with 2.5 L of 1.0 M salt.

> *Ans.* The sugar concentration is given by dividing the number of moles of sugar by the total volume. The number of moles of salt makes no difference in this problem because the problem does not ask about salt concentration and the salt does not react. This is simply a dilution problem for the sugar.
>
> $$(3.0 \text{ L})(2.0 \text{ mol/L}) = 6.0 \text{ mol sugar}$$
>
> $$\frac{6.0 \text{ mol sugar}}{5.5 \text{ L}} = 1.1 \text{ } M \text{ sugar}$$

TITRATION

11.22. (*a*) A solid acid containing one hydrogen atom per molecule is titrated with 1.000 M NaOH. If 27.21 mL of base is used in the titration, how many moles of base is present? (*b*) How many moles of acid? (*c*) If the mass of the acid was 3.494 g, what is the molar mass of the acid?

> *Ans.* (*a*) (27.21 mL)(1.000 mmol/mL) = 27.21 mmol base
>
> (*b*) Since the acid has one hydrogen atom per molecule, it will react in a 1:1 ratio with the base. The equation might be written as
>
> $$\text{HX} + \text{NaOH} \longrightarrow \text{NaX} + \text{H}_2\text{O}$$
>
> where X stands for the anion of the acid, whatever it might be (just as x is often used for an unknown in algebra). The quantity of acid is therefore 27.21 mmol, or 0.027 21 mol.
>
> (*c*) $$\text{Molar mass} = \frac{3.494 \text{ g}}{0.027 \text{ 21 mol}} = 128.4 \text{ g/mol}$$

11.23. A 25.00-mL sample of 1.000 M HCl is titrated with 31.72 mL of NaOH. What is the concentration of the base?

> *Ans.* $$(25.00 \text{ mL HCl})(1.000 \text{ mmol/mL}) = 25.00 \text{ mmol HCl}$$
>
> Since the reagents react in a 1:1 ratio, there is 25.00 mmol NaOH in the 31.72 mL of base.
>
> $$\frac{25.00 \text{ mmol NaOH}}{31.72 \text{ mL}} = 0.7881 \text{ } M \text{ NaOH}$$

11.24. A 25.00-mL sample of 2.000 M H$_2$SO$_4$ is titrated with 16.54 mL of NaOH until both hydrogen atoms of each molecule of the acid are just neutralized. What is the concentration of the base?

> *Ans.* $$(25.00 \text{ mL H}_2\text{SO}_4)(2.000 \text{ mmol/mL}) = 50.00 \text{ mmol H}_2\text{SO}_4$$
>
> $$\text{H}_2\text{SO}_4 + 2 \text{ NaOH} \longrightarrow \text{Na}_2\text{SO}_4 + 2 \text{ H}_2\text{O}$$
>
> $$50.00 \text{ mmol H}_2\text{SO}_4 \left(\frac{2 \text{ mmol NaOH}}{1 \text{ mmol H}_2\text{SO}_4} \right) = 100.0 \text{ mmol NaOH}$$
>
> $$\frac{100.0 \text{ mmol NaOH}}{16.54 \text{ mL}} = 6.046 \text{ } M \text{ NaOH}$$

STOICHIOMETRY IN SOLUTION

11.25. What mass of $H_2C_2O_4$ can react with 35.0 mL of 1.50 M $KMnO_4$ according to the following equation?

$$5\,H_2C_2O_4 + 2\,KMnO_4 + 3\,H_2SO_4 \longrightarrow 2\,MnSO_4 + K_2SO_4 + 10\,CO_2 + 8\,H_2O$$

Ans. $(0.0350\text{ L})(1.50\text{ mol/L}) = 0.0525\text{ mol } KMnO_4$

$$0.0525\text{ mol } KMnO_4 \left(\frac{5\text{ mol } H_2C_2O_4}{2\text{ mol } KMnO_4}\right)\left(\frac{90.0\text{ g } H_2C_2O_4}{1\text{ mol } H_2C_2O_4}\right) = 11.8\text{ g } H_2C_2O_4$$

11.26. Calculate the number of grams of $BaSO_4$ that can be prepared by treating 35.0 mL of 0.479 M $BaCl_2$ with excess Na_2SO_4.

Ans. $BaCl_2 + Na_2SO_4 \longrightarrow BaSO_4 + 2\,NaCl$

$$(35.0\text{ mL})(0.479\text{ mmol } BaCl_2/\text{mL}) = 16.8\text{ mmol } BaCl_2$$

$$16.8\text{ mmol } BaCl_2 \left(\frac{1\text{ mmol } BaSO_4}{1\text{ mmol } BaCl_2}\right)\left(\frac{233\text{ mg } BaSO_4}{1\text{ mmol } BaSO_4}\right) = 3910\text{ mg} = 3.91\text{ g } BaSO_4$$

11.27. When 20.0 mL of 1.71 M $AgNO_3$ is added to 35.0 mL of 0.444 M $CuCl_2$, how many grams of AgCl will be produced?

Ans. $0.0200\text{ L}\left(\frac{1.71\text{ mol } AgNO_3}{1\text{ L}}\right)\left(\frac{1\text{ mol } Ag^+}{1\text{ mol } AgNO_3}\right) = 0.0342\text{ mol } Ag^+$ present

$$0.0350\text{ L}\left(\frac{0.444\text{ mol } CuCl_2}{1\text{ L}}\right)\left(\frac{2\text{ mol } Cl^-}{1\text{ mol } CuCl_2}\right) = 0.0311\text{ mol } Cl^-\text{ present}$$

$$Ag^+ + Cl^- \longrightarrow AgCl$$

The chloride is limiting. It will produce 0.0311 mol AgCl.

$$(0.0311\text{ mol } AgCl)(143\text{ g } AgCl/1\text{ mol } AgCl) = 4.45\text{ g } AgCl$$

11.28. What concentration of NaCl will be produced when 1.00 L of 1.11 M HCl and 250 mL of 4.05 M NaOH are mixed? Assume that the volume of the final solution is the sum of the two initial volumes.

Ans. $(1.00\text{ L})(1.11\text{ mol/L}) = 1.11\text{ mol } HCl$

$$(0.250\text{ L})(4.05\text{ mol/L}) = 1.01\text{ mol } NaOH$$

$$NaOH + HCl \longrightarrow NaCl + H_2O$$

So 1.01 mol NaCl will be produced, in 1.25 L:

$$\frac{1.01\text{ mol}}{1.25\text{ L}} = 0.808\ M$$

Supplementary Problems

11.29. Describe in detail how you would prepare 250.0 mL of 4.000 M NaCl solution.

Ans.   First, figure out how much NaCl you need:

$$(0.2500\text{ L})(4.000\text{ mol/L}) = 1.000\text{ mol}$$

Since the laboratory balance does not weigh out in moles, convert this quantity to grams:

$$(1.000\text{ mol } NaCl)(58.45\text{ g/mol}) = 58.45\text{ g } NaCl$$

Weight out 58.45 g of NaCl and dissolve it in a portion of water in a 250-mL volumetric flask (Fig. 11-2). After the salt has dissolved, dilute the solution with water until the volume reaches the calibration mark on the flask (250.0 mL). Mix the solution thoroughly by inverting and shaking the stoppered flask several times.

11.30. What is the percent by mass of NaCl in 1.60 M NaCl solution? Assume that the solution has a density of 1.06 g/mL.

Ans. Percent by mass is the number of grams of NaCl in 100.0 g solution.

$$\frac{1.60 \text{ mol NaCl}}{1 \text{ L}} \left(\frac{58.5 \text{ g NaCl}}{1 \text{ mol NaCl}} \right) \left(\frac{1 \text{ L}}{1060 \text{ g solution}} \right) = \frac{0.0883 \text{ g NaCl}}{1 \text{ g solution}}$$

For 100.0 g of solution, multiply the numerator and denominator by 100:

$$\frac{0.0883 \text{ g NaCl}}{1 \text{ g solution}} = \frac{8.83 \text{ g NaCl}}{100 \text{ g solution}} = 8.83\% \text{ NaCl}$$

11.31. Which boxes of the conversion diagram (page 348) indicate the method of converting volumes of solutions to moles of reactants and/or products and vice versa?

Ans. Boxes 2, 6, 12, and 15 with the intervening arrows.

CHAPTER 12

Gases

12.1. INTRODUCTION

Long before the science of chemistry was established, materials were described as existing in one of three physical states. There are rigid, *solid* objects, having a definite volume and a fixed shape; there are nonrigid *liquids*, having no fixed shape other than that of their containers but having definite volumes; and there are *gases*, that have neither fixed shape nor fixed volume.

The techniques used for handling various materials depend on their physical states as well as their chemical properties. While it is comparatively easy to handle liquids and solids, it is not as convenient to measure out a quantity of a gas. Fortunately, except under rather extreme conditions, all gases have similar physical properties, and the chemical identity of the substance does not influence those properties. For example, all gases expand when they are heated in a nonrigid container and contract when they are cooled or subjected to increased pressure. They readily diffuse through other gases. Any quantity of gas will occupy the entire volume of its container, regardless of the size of the container.

12.2. PRESSURE OF GASES

Pressure is defined as force per unit area. All *fluids* (liquids and gases) exert pressure at all points within them in all directions. For example, in an inflated balloon, the gas inside pushes against the interior walls of the balloon with such force that the walls stretch. The pressure of the gas is merely the force exerted on the interior surface of the balloon divided by the area. The pressure *of* a gas is equal to the pressure *on* the gas. For example, if the atmosphere presses on a piston against a gas with a pressure of 14.7 pounds per square inch (abbreviated lb/in.^{2}), then the pressure of the gas must also be 14.7 lb/in.^{2}. A way of measuring the pressure of the atmosphere is by means of a *barometer* (Fig. 12-1). The *standard atmosphere* (abbreviated atm) is defined is the pressure that will support a column of mercury to a vertical height of exactly 760 mm at a temperature of 0°C. It is convenient to express the measured gas pressure in terms of the vertical height of a mercury column that the gas is capable of supporting. Thus, if the gas supports a column of mercury to a height of only 152 mm, the gas is exerting a pressure of 0.200 atm:

$$152 \text{ mm}\left(\frac{1 \text{ atm}}{760 \text{ mm}}\right) = 0.200 \text{ atm}$$

EXAMPLE 12.1. What is the pressure in atmospheres of a gas that supports a column of mercury to a height of 882 mm?

Ans.
$$\text{Pressure} = 882 \text{ mm}\left(\frac{1 \text{ atm}}{760 \text{ mm}}\right) = 1.16 \text{ atm}$$

173

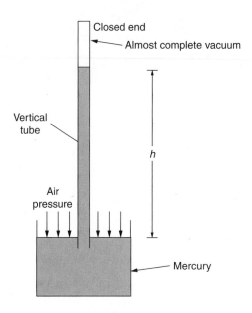

Fig. 12-1. Simple barometer

Air pressure on the surface of the open dish is balanced by the extra pressure caused by the weight of the mercury in the closed tube above the mercury level in the dish. The greater the air pressure, the higher the mercury stands in the vertical tube.

Note that the dimension 1 atmosphere (1 atm) is not the same as *atmospheric pressure*. The atmospheric pressure—the pressure of the atmosphere—varies widely from day to day and from place to place, whereas the dimension 1 atm has a fixed value by definition.

The unit *torr* is defined as the pressure necessary to support mercury to a vertical height of exactly 1 mm. Thus, 1 atm is by definition equal to 760 torr.

The SI unit of pressure is the pascal (Pa).

$$1.000 \text{ atm} = 1.013 \times 10^5 \text{ Pa} = 101.3 \text{ kPa}$$

12.3. BOYLE'S LAW

Robert Boyle (1627–1691) studied the effect of changing the pressure of a gas on its volume at constant temperature. He measured the volume of a given quantity of gas at a given pressure, changed its pressure, and measured the volume again. He obtained data similar to the data shown in Table 12-1. After repeating the process many times with several different gases, he concluded that

At constant temperature, the volume of a given sample of a gas is inversely proportional to its pressure.

This statement is known as *Boyle's Law*.

**Table 12-1   Typical Set of Data
Illustrating Boyle's Law**

Pressure P (atm)	Volume V (L)
4.0	2.0
2.0	4.0
1.0	8.0
0.50	16.0

The term *inversely proportional* in the statement of Boyle's law means that as the pressure increases, the volume becomes smaller by the same factor. That is, if the pressure is doubled, the volume is halved; if the pressure goes up to 5 times its previous value, the volume goes down to $\frac{1}{5}$ of its previous value. This relationship can be represented mathematically by any of the following:

$$P \alpha \frac{1}{V} \qquad P = \frac{k}{V} \qquad PV = k \quad \text{(at constant temperature)}$$

where P represents the pressure, V represents the volume, and k is a constant.

EXAMPLE 12.2.   What is the value of constant k for the sample of gas for which data are given in Table 12-1?

Ans.   For each case, multiplying the observed pressure by the volume gives the value 8.0 L·atm. Therefore, the constant k is 8.0 L·atm.

If for a given sample of a gas at a given temperature, the product PV is a constant, then changing the pressure from some initial value P_1 to a new value P_2 will cause a corresponding change in the volume from the original volume V_1 to a new volume V_2 such that

$$P_1 V_1 = k = P_2 V_2$$

or

$$P_1 V_1 = P_2 V_2$$

Using the last equation makes it unnecessary to solve numerically for k.

EXAMPLE 12.3.   A 2.00-L sample of gas at 675 torr pressure is changed at constant temperature until its pressure is 925 torr. What is its new volume?

Ans.   A useful first step in doing this type of problem is to tabulate clearly all the information given in terms of initial and final conditions:

$$P_1 = 675 \text{ torr} \qquad P_2 = 925 \text{ torr}$$
$$V_1 = 2.00 \text{ L} \qquad V_2 = ?$$

Let the new, unknown volume be represented by V_2. Since the temperature is constant, Boyle's law is used:

$$P_1 V_1 = P_2 V_2 = (675 \text{ torr})(2.00 \text{ L}) = (925 \text{ torr}) V_2$$

Solving for V_2 by dividing each side by 925 torr yields

$$V_2 = \frac{(675 \text{ torr})(2.00 \text{ L})}{925 \text{ torr}} = 1.46 \text{ L}$$

The final volume is 1.46 L. The answer is seen to be reasonable by noting that since the pressure increases, the volume must decrease.

Note that the units of pressure must be the same on both sides of the equation, as must be the units of volume.

EXAMPLE 12.4.   To what pressure must a sample of gas be subjected at constant temperature in order to compress it from 122 mL to 105 mL if its original pressure is 1.71 atm?

Ans.
$$V_1 = 122 \text{ mL} \qquad V_2 = 105 \text{ mL}$$
$$P_1 = 1.71 \text{ atm} \qquad P_2 = ?$$
$$P_1 V_1 = P_2 V_2 = (1.71 \text{ atm})(122 \text{ mL}) = P_2(105 \text{ mL})$$
$$P_2 = 1.99 \text{ atm}$$

The pressure necessarily is increased to make the volume smaller.

12.4.   GRAPHICAL REPRESENTATION OF DATA

Often in scientific work it is useful to report data in the form of a graph to enable immediate visualization of general trends and relationships. Another advantage of plotting data in the form of a graph is to be able to estimate values for points between and beyond the experimental points. For example, in Fig. 12-2 the data of

Table 12-1 are plotted on a graph using P as a vertical axis (y axis) and V as the horizontal axis (x axis). Note that in plotting a graph based on experimental data, the numerical scales of the axes should be chosen so that the scales can be read to the same number of significant figures as was used in reporting the measurements. It can be seen that as the magnitude of the pressure decreases, the magnitude of the volume increases. It is possible to obtain values of the volume at intermediate values of the pressure merely by reading from points on the curve (interpolating).

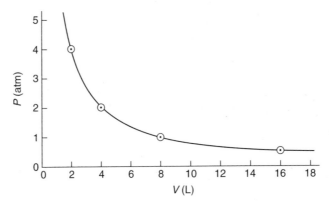

Fig. 12-2.　Plot of P-V data for the gas in Table 12-1

EXAMPLE 12.5.　For the sample of gas described in Fig. 12-2, what pressure is required to make the volume 6.0 L?

Ans.　It is apparent from the graph that a pressure of 1.3 atm is required.

If the graph of one variable against another should happen to be a straight line, the relationship between the variables can be expressed by a simple algebraic equation. If the data fall on a straight line that goes through the *origin* [the (0, 0) point], then the two variables are *directly proportional*. As one goes up, the other goes up by the same factor. For example, as one doubles, the other doubles. When one finds such a direct proportionality, for example, the distance traveled at constant velocity by an automobile and the time of travel, one can immediately write a mathematical equation relating the two variables:

$$\text{Distance} = \text{speed} \times \text{time}$$

In this case, the longer the time spent traveling at constant speed, the greater the distance traveled.

Often it is worthwhile to plot data in several ways until a straight-line graph is obtained. Consider the plot of volume versus pressure shown in Fig. 12-2, which is definitely not a straight line. What will the pressure of the sample be at a volume of 19 L? It is somewhat difficult to estimate past the data points (to extrapolate) in this case, because the curve is not a straight line, and it is difficult to know how much it will bend. However, these data can be replotted in the form of a straight line by using the reciprocal of one of the variables. The straight line can be extended past the experimental points (extrapolated) rather easily, and the desired number can be estimated. For example, if the data of Table 12-1 are retabulated as in Table 12-2 and an additional column is added with the reciprocal of pressure $1/P$, it is possible to plot $1/P$ as the vertical axis against V on the horizontal axis, and a straight line through the origin is obtained (Fig. 12-3). Thus, it is proper to say that the quantity $1/P$ is directly proportional to V. When the *reciprocal* of a quantity is *directly* proportional to a second quantity, the first quantity itself is *inversely* proportional to the second.

If $1/P$ is directly proportional to V, that is, $(1/P)k = V$, then P is inversely proportional to V, that is, $k = PV$. The straight line found by plotting $1/P$ versus V can easily be extended to the point where $V = 19$ L. The $1/P$ point is 2.4/atm; therefore,

$$P = \frac{1}{2.4/\text{atm}} = 0.42 \text{ atm}$$

**Table 12-2   Reciprocal of
Pressure Data**

P (atm)	$1/P$ (1/atm)	V (L)
4.0	0.25	2.0
2.0	0.50	4.0
1.0	1.0	8.0
0.50	2.0	16.0

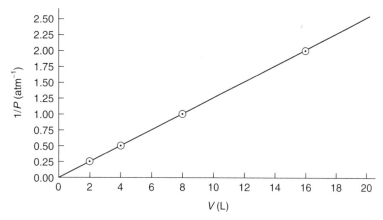

Fig. 12-3.   Plot of $1/P$ versus V for the gas in Table 12-2 to show proportionality

12.5.   CHARLES' LAW

If a given quantity of gas is heated at constant pressure in a container that has a movable wall, such as a piston (Fig. 12-4), the volume of the gas will increase. If a given quantity of gas is heated in a container that has a fixed volume (Fig. 12-5), its pressure will increase. Conversely, cooling a gas at constant pressure causes a decrease in its volume, while cooling it at constant volume causes a decrease in its pressure.

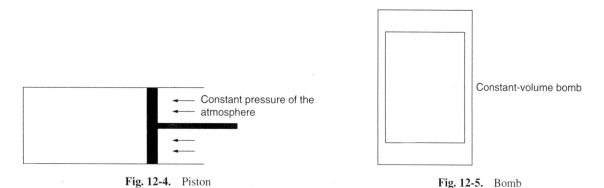

Fig. 12-4.   Piston

Fig. 12-5.   Bomb

J. A. C. Charles (1746–1823) observed, and J. L. Gay-Lussac (1778–1850) confirmed, that when a given sample of gas is cooled at constant pressure, it shrinks by $\frac{1}{273}$ times its volume at $0°C$ for every degree Celsius that it is cooled. Conversely, when the sample of gas is heated at constant pressure, it expands by $\frac{1}{273}$ times its volume at $0°C$ for every degree Celsius that it is heated. The changes in volume with temperature of two different-sized samples of a gas are shown in Fig. 12-6.

The chemical identity of the gas has no influence on the volume changes as long as the gas does not liquefy in the range of temperatures studied. It is seen in Fig. 12-6 that for each sample, the volume of the

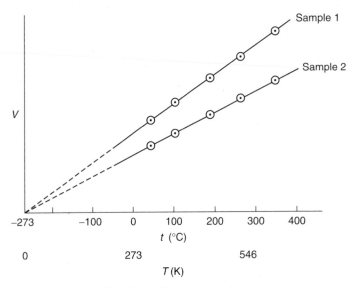

Fig. 12-6. Two samples of gas

gas changes linearly with temperature. A straight line can be drawn through all the points. If it were assumed that the gas does not liquefy at very low temperatures, each sample would have zero volume at $-273°C$. Of course, any real gas could never have zero volume. (Gases liquefy before this very cold temperature is reached.) Nevertheless, $-273°C$ is the temperature at which a sample of gas would theoretically have zero volume. This type of graph can be plotted for any sample of any gas, and in every case the observed volume will change with temperature at such a rate that at $-273°C$ the volume would be zero. Therefore, the temperature $-273°C$ can be regarded as the *absolute zero* of temperature. Since there cannot be less than zero volume, there can be no colder temperature than $-273°C$. The temperature scale that has been devised using this fact is called the *Kelvin*, or *absolute*, temperature scale (Sec. 2.7). A comparison of the Kelvin scale and the Celsius scale is shown in Fig. 12-7. It is seen that any temperature in degrees Celsius $(t, °C)$ may be converted to kelvins (T, K) by adding $273°$. It is customary to use capital T to represent Kelvin temperatures and small t to represent Celsius temperatures. No degree sign $(°)$ is used with Kelvin temperatures, and the units are called kelvins.

$$T = t + 273°$$

In Fig. 12-6, the volume can be seen to be directly proportional to the Kelvin temperature, since the lines go through the zero-volume point at 0 K.

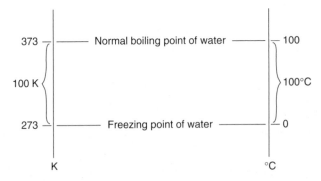

Fig. 12-7. Comparison of Kelvin and Celsius temperature scales

EXAMPLE 12.6. What are the freezing point of water and the normal boiling point of water on the Kelvin scale?

Ans.

Freezing point of water $= 0°C + 273° = 273$ K

Normal boiling point of water $= 100°C + 273° = 373$ K

The fact that the volume of a gas varies linearly with temperature is combined with the concept of absolute temperature to give a statement of *Charles' law*:

At constant pressure, the volume of a given sample of gas is directly proportional to its absolute temperature.

Expressed mathematically,

$$V = kT \qquad \text{(at constant pressure)}$$

This equation can be rearranged to give

$$\frac{V}{T} = k$$

Since V/T is a constant, this ratio for a given sample of gas at one volume and temperature is equal to the same ratio at any other volumes and temperatures. See Table 12-3. That is, for a given sample at constant pressure,

$$\frac{V_1}{T_1} = \frac{V_2}{T_2} \qquad \text{(at constant pressure)}$$

One can see from the data of Table 12-3 that *absolute* temperatures must be used.

Table 12-3   Volumes and Temperatures of a Given Sample of Gas

Volume V (L)	Temperature T (K)	Temperature t (°C)	V/T (L/K)	V/t (L/°C)
1.0	125	−148	8.0×10^{-3}	-6.8×10^{-3}
2.0	250	−23	8.0×10^{-3}	-8.7×10^{-2}
3.0	375	102	8.0×10^{-3}	$+2.9 \times 10^{-2}$
4.0	500	227	8.0×10^{-3}	$+1.8 \times 10^{-2}$

EXAMPLE 12.7. A 22.5-mL sample of gas is warmed at constant pressure from 291 K to 309 K. What is its final volume?

Ans.

	1	2
V	22.5 mL	V_2
T	291 K	309 K

$$\frac{V_1}{T_1} = \frac{V_2}{T_2} = \frac{22.5 \text{ mL}}{291 \text{ K}} = \frac{V_2}{309 \text{ K}}$$
$$V_2 = 23.9 \text{ mL}$$

EXAMPLE 12.8. A 22.5-mL sample of gas is warmed at constant pressure from 18°C to 36°C. What is its final volume?

Ans.   This example is a restatement of Example 12.7. The conditions are precisely the same; the only difference is that the temperatures are expressed in degrees Celsius and first must be converted to kelvins.

$$18°C = 18°C + 273° = 291 \text{ K}$$
$$36°C = 36°C + 273° = 309 \text{ K}$$

The example is solved as shown above. Note again that V is directly proportional to T, but not to t. In this example, t doubles but T does not double, and so V does not double.

12.6.   THE COMBINED GAS LAW

Suppose it is desired to calculate the final volume V_2 of a gas originally at volume V_1 when its temperature is changed from T_1 to T_2 at the same time its pressure is changed from P_1 to P_2. One might consider the two effects separately, for example, that first the pressure is changed at constant temperature T_1 and calculate a new volume V_{new} using Boyle's law. Then, using Charles' law, one can calculate how the new volume V_{new} changes to V_2 when the temperature is changed from T_1 to T_2 at the constant pressure P_2 (boxes 1, 3, and 4 in Fig. 12-8).

It would be equally correct to consider that first the temperature of the gas was changed from T_1 to T_2 at the constant pressure P_1, for which a new volume V_{new} could be calculated using Charles' law. Then, assuming that the temperature is held constant at T_2, calculate how the volume changes as the pressure is changed from P_1 to P_2 (boxes 1, 2, and 4 in Fig. 12-8).

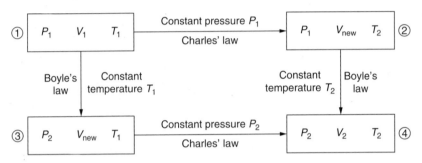

Fig. 12-8.   Change in gas volume with both pressure and temperature

However, the fact that the volume V of a given mass of gas is inversely proportional to its pressure P and directly proportional to its absolute temperature T can be combined mathematically to give the single equation

$$V = k\left(\frac{T}{P}\right)$$

where k is the proportionality constant. Rearranging the variables gives the following equation:

$$\frac{PV}{T} = k$$

That is, for a given sample of gas, the ratio PV/T remains constant, and therefore

$$\frac{P_1V_1}{T_1} = \frac{P_2V_2}{T_2} \qquad \text{(a given sample of gas)}$$

This expression is a mathematical statement of the *combined* (or general) gas law. In words, the volume of a given sample of gas is inversely proportional to its pressure and directly proportional to its absolute temperature.

Note that if the temperature is constant, $T_1 = T_2$, then the expression reduces to the equation for Boyle's law, $P_1V_1 = P_2V_2$. Alternatively, if the pressure is constant, $P_1 = P_2$, the expression is equivalent to Charles' law $V_1/T_1 = V_2/T_2$.

EXAMPLE 12.9.   A sample of gas is pumped from a 1.50-L vessel at 77°C and 760-torr pressure to a 0.950-L vessel at 12°C. What is its final pressure?

Ans.

	1	2
P	760 torr	P_2
V	1.50 L	0.950 L
T	77°C = 350 K	12°C = 285 K

$$\frac{P_1V_1}{T_1} = \frac{P_2V_2}{T_2} = \frac{(760 \text{ torr})(1.50 \text{ L})}{350 \text{ K}} = \frac{P_2(0.950 \text{ L})}{285 \text{ K}}$$

$$P_2 = 977 \text{ torr}$$

Standard Conditions

According to the combined gas law, the volume of a given sample of gas can have any value, depending on its temperature and pressure. To compare the quantities of gas present in two different samples, it is useful to adopt a set of standard conditions of temperature and pressure. By universal agreement, the standard temperature is chosen as 273 K (0°C), and the standard pressure is chosen as exactly 1 atm (760 torr). Together, these conditions are referred to as *standard conditions* or as *standard temperature and pressure* (*STP*). Many textbooks and instructors find it convenient to use this short notation for this particular temperature and pressure.

EXAMPLE 12.10.   A sample of gas occupies a volume of 1.88 L at 22°C and 0.979 atm pressure. What is the volume of this sample at STP?

Ans.

$$P_1 = 0.979 \text{ atm} \qquad\qquad P_2 = 1.00 \text{ atm}$$
$$V_1 = 1.88 \text{ L} \qquad\qquad\quad V_2 = ?$$
$$T_1 = 22°\text{C} + 273° = 295 \text{ K} \qquad T_2 = 273 \text{ K}$$

$$\frac{P_1 V_1}{T_1} = \frac{P_2 V_2}{T_2} = \frac{(0.979 \text{ atm})(1.88 \text{ L})}{295 \text{ K}} = \frac{(1.00 \text{ atm}) V_2}{273 \text{ K}}$$
$$V_2 = 1.70 \text{ L}$$

12.7.   THE IDEAL GAS LAW

All the gas laws described so far apply only to a given sample of gas. If a gas is produced during a chemical reaction or some of the gas under study escapes during processing, these gas laws are not used because the gas sample has only one temperature and one pressure. The *ideal gas law* works (at least approximately) for *any* sample of gas. Consider a given sample of gas, for which

$$\frac{PV}{T} = k \qquad \text{(a constant)}$$

If we increase the number of moles of gas at constant pressure and temperature, the volume must also increase. Thus, we can conclude that the constant k can be regarded as a product of two constants, one of which represents the number of moles of gas. We then get

$$\frac{PV}{T} = nR \qquad \text{or} \qquad PV = nRT$$

where P, V, and T have their usual meanings; n is the number of moles of gas *molecules*; and R is a new constant that is valid for any sample of any gas. This equation is known as the *ideal gas law*. You should remember the following value of R; values of R in other units will be introduced later.

$$R = 0.0821 \text{ L·atm/(mol·K)} \qquad \text{(Note that there is a zero after the decimal point.)}$$

In the simplest ideal gas law problems, values for three of the four variables are given and you are asked to calculate the value of the fourth. As usual with the gas laws, the temperature must be given as an absolute temperature, in kelvins. The units of P and V are most conveniently given in atmospheres and liters, respectively, because the units of R with the value given above are in terms of these units. If other units are given for pressure or volume, convert each to atmospheres or liters, respectively.

EXAMPLE 12.11.   How many moles of O_2 are present in a 0.500-L sample at 25°C and 1.09 atm?

Ans.

$$T = 25°\text{C} + 273 = 298 \text{ K}$$
$$n = \frac{PV}{RT} = \frac{(1.09 \text{ atm})(0.500 \text{ L})}{[0.0821 \text{ L·atm/(mol·K)}](298 \text{ K})} = 0.0223 \text{ mol } O_2$$

Note that T must be in kelvins.

EXAMPLE 12.12.   What is the volume of 1.00 mol of gas at STP?

Ans.
$$V = \frac{nRT}{P} = \frac{(1.00 \text{ mol})(0.0821 \text{ L·atm/mol·K})(273 \text{ K})}{1.00 \text{ atm}} = 22.4 \text{ L}$$

The volume of 1.00 mol of gas at STP is called the *molar volume* of a gas. This value should be memorized.

EXAMPLE 12.13.   How many moles of SO_2 are present in a 765-mL sample at 37°C and 775 torr?

Ans.   Since R is defined in terms of liters and atmospheres, the pressure and volume are first converted to those units. Temperature as usual is given in kelvins. First, change the pressure to atmospheres and the volume to liters:

$$775 \text{ torr} \left(\frac{1 \text{ atm}}{760 \text{ torr}} \right) = 1.02 \text{ atm} \qquad 765 \text{ mL} \left(\frac{1 \text{ L}}{1000 \text{ mL}} \right) = 0.765 \text{ L}$$

$$n = \frac{PV}{RT} = \frac{(1.02 \text{ atm})(0.765 \text{ L})}{[0.0821 \text{ L·atm/(mol·K)}](310 \text{ K})} = 0.0307 \text{ mol}$$

Alternately, we can do the entire calculation, including the conversions, with one equation:

$$n = \frac{PV}{RT} = \frac{(775 \text{ torr})(1.00 \text{ atm}/760 \text{ torr})(765 \text{ mL})(1 \text{ L}/1000 \text{ mL})}{[0.0821 \text{ L·atm/(mol·K)}](310 \text{ K})} = 0.0307 \text{ mol}$$

EXAMPLE 12.14.   At what temperature will 0.0750 mol of CO_2 occupy 2.75 L at 1.11 atm?

Ans.
$$T = \frac{PV}{nR} = \frac{(1.11 \text{ atm})(2.75 \text{ L})}{[0.0821 \text{ L·atm/(mol·K)}](0.0750 \text{ mol})} = 496 \text{ K}$$

EXAMPLE 12.15.   What use was made of the information about the chemical identity of the gas in Examples 12.13 and 12.14?

Ans.   None. The ideal gas law works no matter what gas is being used.

EXAMPLE 12.16.   What volume will 7.00 g of Cl_2 occupy at STP?

Ans.   The value of n is not given explicitly in the problem, but the mass is given, from which we can calculate the number of moles:

$$7.00 \text{ g Cl}_2 \left(\frac{1 \text{ mol Cl}_2}{71.0 \text{ g Cl}_2} \right) = 0.0986 \text{ mol Cl}_2$$

$$V = \frac{nRT}{P} = \frac{(0.0986 \text{ mol})[0.0821 \text{ L·atm/(mol·K)}](273 \text{ K})}{1.00 \text{ atm}} = 2.21 \text{ L}$$

The identity of the gas is important here to determine the number of moles.

EXAMPLE 12.17.   (*a*) How many moles of O atoms were present in Example 12.11? (*b*) How many moles of He atoms are present in 465 mL at 25°C and 1.00 atm?

Ans.   (*a*)
$$0.0223 \text{ mol O}_2 \left(\frac{2 \text{ mol O}}{1 \text{ mol O}_2} \right) = 0.0446 \text{ mol O atoms}$$

Note that n in the ideal gas equation in the example refers to moles of O_2 molecules.

(*b*)
$$n = \frac{PV}{RT} = \frac{(1.00 \text{ atm})(0.465 \text{ L})}{[0.0821 \text{ L·atm/(mol·K)}](298 \text{ K})} = 0.0190 \text{ mol He}$$

The He molecules are individual He atoms, and thus there are 0.0190 mol He atoms present.

As soon as you recognize P, V, and T data in a problem, you can calculate n. (In a complicated problem, if you are given P, V, and T, but you do not see how to do the whole problem, first calculate n and then see what you can do with it.)

EXAMPLE 12.18.   If 4.58 g of a gas occupies 3.33 L at 27°C and 808 torr, what is the molar mass of the gas?

Ans.   If you do not see at first how to solve this problem to completion, at least you can recognize that P, V, and T data are given. First calculate the number of moles of gas present:

$$808 \text{ torr} \left(\frac{1 \text{ atm}}{760 \text{ torr}} \right) = 1.06 \text{ atm}$$

$$27°C + 273° = 300 \text{ K}$$

$$n = \frac{PV}{RT} = \frac{(1.06 \text{ atm})(3.33 \text{ L})}{[0.0821 \text{ L·atm/(mol·K)}](300 \text{ K})} = 0.143 \text{ mol}$$

We now know the mass of the gas and the number of moles. That is enough to calculate the molar mass:

$$\frac{4.58 \text{ g}}{0.143 \text{ mol}} = 32.0 \text{ g/mol}$$

EXAMPLE 12.19.   What volume is occupied by the oxygen liberated by heating 0.250 g of $KClO_3$ until it completely decomposes to KCl and oxygen? The gas is collected at STP.

Ans.   Although the temperature and pressure of the gas are given, the number of moles of gas is not. Can we get it somewhere? The chemical reaction of $KClO_3$ yields the oxygen, and the rules of stoichiometry (Chap. 10) may be used to calculate the number of moles of gas. Note that the number of moles of $KClO_3$ is *not* used in the ideal gas law equation.

$$2 \text{ KClO}_3 \xrightarrow{\text{heat}} 2 \text{ KCl} + 3 \text{ O}_2$$

$$0.250 \text{ g KClO}_3 \left(\frac{1 \text{ mol KClO}_3}{122 \text{ g KClO}_3} \right) \left(\frac{3 \text{ mol O}_2}{2 \text{ mol KClO}_3} \right) = 0.00307 \text{ mol O}_2$$

Now that we know the number of moles, the pressure, and the temperature of O_2, we can calculate its volume:

$$V = \frac{nRT}{P} = \frac{(0.00307 \text{ mol})[0.0821 \text{ L·atm/(mol·K)}](273 \text{ K})}{1.00 \text{ atm}} = 0.0688 \text{ L}$$

12.8.   DALTON'S LAW OF PARTIAL PRESSURES

When two or more gases are mixed, they each occupy the entire volume of the container. They each have the same temperature as the other(s). However, each gas exerts its own pressure, independent of the other gases. Moreover, according to Dalton's law of partial pressures, their pressures must add up to the total pressure of the gas mixture.

EXAMPLE 12.20.   (*a*) If I try to put a 1.00-L sample of O_2 at 300 K and 1.00 atm plus a 1.00-L sample of N_2 at 300 K and 1.00 atm into a rigid 1.00-L container at 300 K, will they fit? (*b*) If so, what will be their total volume and total pressure?

Ans.   (*a*) The gases will fit; gases expand or contract to fill their containers. (*b*) The total volume is the volume of the container—1.00 L. The temperature is 300 K, given in the problem. The total pressure is the sum of the two partial pressures. *Partial pressure* is the pressure of each gas (as if the other were not present). The oxygen pressure is 1.00 atm. The oxygen has been moved from a 1.00-L container at 300 K to another 1.00-L container at 300 K, and so its pressure does not change. The nitrogen pressure is 1.00 atm for the same reason. The total pressure is 1.00 atm + 1.00 atm = 2.00 atm.

EXAMPLE 12.21.   A 1.00-L sample of O_2 at 300 K and 1.00 atm plus a 0.500-L sample of N_2 at 300 K and 1.00 atm are put into a rigid 1.00-L container at 300 K. What will be their total volume, temperature, and total pressure?

Ans.   The total volume is the volume of the container—1.00 L. The temperature is 300 K, given in the problem. The total pressure is the sum of the two partial pressures. The oxygen pressure is 1.00 atm. (See Example 12.20.) The nitrogen pressure is 0.500 atm, since it was moved from 0.500 L at 1.00 atm to 1.00 L at the same temperature (Boyle's law). The total pressure is

$$1.00 \text{ atm} + 0.500 \text{ atm} = 1.50 \text{ atm}$$

EXAMPLE 12.22. If the N_2 of the last example were added to the O_2 in the container originally containing the O_2, how would the problem be affected?

Ans.   It would not change; the final volume would still be 1.00 L.

EXAMPLE 12.23. If the O_2 of Example 12.21 were added to the N_2 in the container originally containing the N_2, how would the problem change?

Ans.   The pressure would be doubled, because the final volume would be 0.500 L.

The ideal gas law applies to each individual gas in a gas mixture as well as to the gas mixture as a whole. Thus, in a mixture of nitrogen and oxygen, one can apply the ideal gas law to the oxygen, to the nitrogen, and to the mixture as a whole.

$$PV = nRT$$

Variables V and T, as well as R, refer to each gas and to the total mixture. If we want the number of moles of O_2, we use the pressure of O_2. If we want the number of moles of N_2, we use the pressure of N_2. If we want the total number of moles, we use the total pressure.

EXAMPLE 12.24. Calculate the number of moles of O_2 in Example 12.21, both before and after mixing.

Ans.   Before mixing:

$$n = \frac{PV}{RT} = \frac{(1.00 \text{ atm})(1.00 \text{ L})}{[0.0821 \text{ L·atm/(mol·K)}](300 \text{ K})} = 0.0406 \text{ mol}$$

After mixing:

$$n_{O_2} = \frac{P_{O_2}V}{RT} = \frac{(1.00 \text{ atm})(1.00 \text{ L})}{[0.0821 \text{ L·atm/(mol·K)}](300 \text{ K})} = 0.0406 \text{ mol}$$

Water Vapor

At 25°C, water is ordinarily a liquid. However, even at 25°C, water evaporates. In a closed container at 25°C, water evaporates enough to get a 24-torr water vapor pressure in its container. The pressure of the gaseous water is called its *vapor pressure* at that temperature. At different temperatures, it evaporates to different extents to give different vapor pressures. As long as there is liquid water present, however, the vapor pressure above pure water depends on the temperature alone. Only the nature of the liquid and the temperature affect the vapor pressure; the volume of the container does not affect the final pressure.

The water vapor mixes with any other gas(es) present, and the mixture is governed by Dalton's law of partial pressures, just as any other gas mixture is.

EXAMPLE 12.25. O_2 is collected in a bottle over water at 25°C at 1.00-atm barometric pressure. (*a*) What gas(es) is (are) in the bottle? (*b*) What is (are) the pressure(s)?

Ans.   (*a*) Both O_2 and water vapor are in the bottle. (*b*) The total pressure is the barometric pressure, 760 torr. The water vapor pressure is 24 torr, given in the first paragraph of this subsection for the gas phase above liquid water at 25°C. The pressure of the O_2 is therefore

$$760 \text{ torr} - 24 \text{ torr} = 736 \text{ torr}$$

EXAMPLE 12.26. How many moles of oxygen are contained in a 1.00-L vessel over water at 25°C and a barometric pressure of 1.00 atm?

Ans.   The barometric pressure is the total pressure of the gas mixture. The pressure of O_2 is 760 torr − 24 torr = 736 torr. Since we want to know about moles of O_2, we need to use the pressure of O_2 in the ideal gas law:

$$n_{O_2} = \frac{P_{O_2}V}{RT} = \frac{[(736/760)\text{atm}](1.00 \text{ L})}{[0.0821 \text{ L·atm/(mol·K)}](298 \text{ K})} = 0.0396 \text{ mol}$$

Solved Problems

INTRODUCTION

12.1. What is the difference between gas and gasoline?

Ans. Gas is a state of matter. Gasoline is a liquid, used mainly for fuel, with a nickname *gas*. Do not confuse the two. This chapter is about the gas phase, not about liquid gasoline.

12.2. What is a fluid?

Ans. A gas or a liquid.

PRESSURE OF GASES

12.3. Change 806 mmHg to (*a*) torr, (*b*) atmospheres, and (*c*) kilopascals.

Ans. (*a*) 806 torr   (*b*) $806 \text{ torr}\left(\dfrac{1 \text{ atm}}{760 \text{ torr}}\right) = 1.06 \text{ atm}$ (*c*) $1.06 \text{ atm}\left(\dfrac{101.3 \text{ kPa}}{1 \text{ atm}}\right) = 107 \text{ kPa}$

12.4. Change (*a*) 703 torr to atmospheres, (*b*) 1.25 atm to torr, (*c*) 743 mmHg to torr, and (*d*) 1.01 atm to millimeters of mercury (mmHg).

Ans. (*a*) $703 \text{ torr}\left(\dfrac{1 \text{ atm}}{760 \text{ torr}}\right) = 0.925 \text{ atm}$ (*c*) $743 \text{ mmHg}\left(\dfrac{1 \text{ torr}}{1 \text{ mmHg}}\right) = 743 \text{ torr}$

(*b*) $1.25 \text{ atm}\left(\dfrac{760 \text{ torr}}{1 \text{ atm}}\right) = 950 \text{ torr}$ (*d*) $1.01 \text{ atm}\left(\dfrac{760 \text{ mmHg}}{1 \text{ atm}}\right) = 768 \text{ mmHg}$

12.5. How many pounds force does 1 atm pressure exert on the side of a metal can that measures 6.0 in. by 9.0 in.? (1 atm = 14.7 lb/in.²)

Ans. $$\text{Area} = 6.0 \text{ in.} \times 9.0 \text{ in.} = 54 \text{ in.}^2$$

$$54 \text{ in.}^2\left(\frac{14.7 \text{ lb}}{1 \text{ in.}^2}\right) = 790 \text{ lb} = 7.9 \times 10^2 \text{ lb}$$

BOYLE'S LAW

12.6. What law may be stated qualitatively as "When you squeeze a gas, it gets smaller"?

Ans. Boyle's law

12.7. Calculate the product of the pressure and volume for each point in Table 12-1. What can you conclude?

Ans. $PV = 8.0 \text{ L·atm}$ in each case. You can conclude that PV is a constant for this sample of gas and that Boyle's law is obeyed.

12.8. If 4.00 L of gas at 1.22 atm is changed to 876 torr at constant temperature, what is its final volume?

Ans. $$1.22 \text{ atm}\left(\frac{760 \text{ torr}}{1 \text{ atm}}\right) = 927 \text{ torr}$$

$$P_1 V_1 = P_2 V_2$$

$$V_2 = \frac{P_1 V_1}{P_2} = \frac{(927 \text{ torr})(4.00 \text{ L})}{876 \text{ torr}} = 4.23 \text{ L}$$

12.9. If 1.25 L of gas at 780 torr is changed to 965 mL at constant temperature, what is its final pressure?

Ans. $$P_2 = \frac{P_1 V_1}{V_2} = \frac{(780 \text{ torr})(1250 \text{ mL})}{965 \text{ mL}} = 1010 \text{ torr}$$

12.10. A 1.00-L sample of gas at 25°C and 1.00-atm pressure is changed to 3.00-atm pressure at 25°C. What law may be used to determine its final volume?

Ans.   Boyle's law may be used because the temperature is unchanged. Alternately, the combined gas law may be used, with 298 K, the Kelvin equivalent of 25°C, used for both T_1 and T_2.

12.11. A sample of gas occupies 2.48 L. What will be its new volume if its pressure is doubled at constant temperature?

Ans.   According to Boyle's law, doubling the pressure will cut the volume in half; the new volume will be 1.24 L. A second method allows us to use unknown variables for the pressure:

	1	2
P	P_1	$P_2 = 2P_1$
V	2.48 L	V_2

$$V_2 = \frac{P_1 V_1}{P_2} = \frac{P_1 V_1}{2P_1} = \frac{V_1}{2} = \frac{2.48\text{ L}}{2} = 1.24\text{ L}$$

GRAPHICAL REPRESENTATION OF DATA

12.12. On a graph, what criteria represent direct proportionality?

Ans.   The plot is a straight line and it passes through the origin.

12.13. What is the pressure of the gas described in Table 12-1 at 10.0 L? Answer first by calculating with Boyle's law, second by reading from Fig. 12-2, and third by reading from Fig. 12-3. Which determination is easiest (assuming that the graphs have already been drawn)?

Ans.   The pressure of the gas is 0.80 atm. The second method involves merely reading a point from a graph. To use Fig. 12-3, you have to calculate the reciprocal of the pressure. None of the methods is difficult, however.

12.14. Plot the following data:

V (L)	P (atm)
1.50	4.00
3.00	2.00
6.00	1.00
12.00	0.500

Replot the data, using the volume and the reciprocal of the pressure. Do these values fall on a straight line? Are volume and the reciprocal of pressure directly proportional? Are volume and pressure directly proportional?

Ans.   The points of the second plot fall on a straight line through the origin, and so the volume and reciprocal of pressure are directly proportional to each other, making the volume inversely proportional to the pressure.

CHARLES' LAW

12.15. Plot the following data:

V (L)	t (°C)
4.92	100
4.26	50
3.60	0
2.94	−50

Do the values fall on a straight line? Are volume and Celsius temperature directly proportional? Replot the volume versus the Kelvin temperature. Are volume and Kelvin temperature directly proportional?

Ans.   The points of the first plot fall on a straight line, but that line does not go through the origin (the point at 0 L and 0°C), and so these quantities are *not* directly proportional. When volume is replotted versus Kelvin temperature, the resulting straight line goes through the origin, and thus volume and *absolute temperature* are directly proportional.

12.16. If 42.3 mL of gas at 22°C is changed to 44°C at constant pressure, what is its final volume?

    Ans.   Absolute temperatures must be used:

$$22°C + 273° = 295 \text{ K}$$
$$44°C + 273° = 317 \text{ K}$$

According to Charles' law:

$$\frac{V_1}{T_1} = \frac{V_2}{T_2}$$

$$V_2 = \frac{V_1 T_2}{T_1} = \frac{(42.3 \text{ mL})(317 \text{ K})}{295 \text{ K}} = 45.5 \text{ mL}$$

Note that the volume is *not* doubled by doubling the Celsius temperature.

12.17. If 0.979 L of gas at 0°C is changed to 737 mL at constant pressure, what is its final temperature?

    Ans.   Using the reciprocal of the equation usually used for Charles' law:

$$\frac{T_2}{V_2} = \frac{T_1}{V_1}$$

and solving for T_2:

$$T_2 = \frac{T_1 V_2}{V_1} = \frac{(273 \text{ K})(737 \text{ mL})}{979 \text{ mL}} = 206 \text{ K}$$

The temperature was lowered to reduce the volume.

THE COMBINED GAS LAW

12.18. Calculate the missing value for each set of data in the following table:

	P_1	V_1	T_1	P_2	V_2	T_2
(a)	—	29.1 L	45°C	780 torr	2.22 L	77°C
(b)	12.0 atm	—	28°C	12.0 atm	750 mL	53°C
(c)	721 torr	200 mL	—	1.21 atm	0.850 L	100°C
(d)	1.00 atm	4.00 L	273 K	1.00 atm	2.00 L	—
(e)	7.00 atm	—	333 K	3.10 atm	6.00 L	444 K
(f)	1.00 atm	3.65 L	130°C	—	5.43 L	130°C

    Ans.   Each problem is solved by rearranging the equation

$$\frac{P_1 V_1}{T_1} = \frac{P_2 V_2}{T_2}$$

All temperatures must be in kelvins.

(a) $P_1 = \dfrac{P_2 V_2 T_1}{T_2 V_1} = \dfrac{(780 \text{ torr})(2.22 \text{ L})(318 \text{ K})}{(350 \text{ K})(29.1 \text{ L})} = 54.1 \text{ torr}$

(b) $V_1 = \dfrac{P_2 V_2 T_1}{P_1 T_2} = \dfrac{(12.0 \text{ atm})(750 \text{ mL})(301 \text{ K})}{(12.0 \text{ atm})(326 \text{ K})} = 692 \text{ mL}$

Since P did not change, Charles' law could have been used in the form

$$V_1 = \frac{V_2 T_1}{T_2}$$

(c) $T_1 = \dfrac{P_1 V_1 T_2}{P_2 V_2} = \dfrac{(721 \text{ torr})(0.200 \text{ L})(373 \text{ K})}{(1.21 \text{ atm})(760 \text{ torr/atm})(0.850 \text{ L})} = 68.8 \text{ K}$

The units of P and V each must be the same in state 1 and state 2. Since each of them is given in diffferent units, one of each must be changed.

(d) $T_2 = \dfrac{T_1 P_2 V_2}{P_1 V_1} = \dfrac{(273\ \text{K})(1.00\ \text{atm})(2.00\ \text{L})}{(1.00\ \text{atm})(4.00\ \text{L})} = 137\ \text{K}$

(e) $V_1 = \dfrac{P_2 V_2 T_1}{P_1 T_2} = \dfrac{(3.10\ \text{atm})(6.00\ \text{L})(333\ \text{K})}{(7.00\ \text{atm})(444\ \text{K})} = 1.99\ \text{L}$

(f) $P_2 = \dfrac{T_2 P_1 V_1}{T_1 V_2} = \dfrac{(403\ \text{K})(1.00\ \text{atm})(3.65\ \text{L})}{(403\ \text{K})(5.43\ \text{L})} = 0.672\ \text{atm}$

Since $T_1 = T_2$, we could have used $P_2 = P_1 V_1 / V_2$ and arrived at the same answer.

12.19. A 9.00-L sample of gas has its pressure tripled while its absolute temperature is increased by 50%. What is its new volume?

Ans.   Tripling P reduces the volume to 3.00 L. Then increasing T by 50% (multiplying it by 1.50) increases the volume by a factor of 1.50 to 4.50 L.

Alternatively:
$$V_2 = \left(\frac{P_1 V_1}{T_1}\right)\left(\frac{T_2}{P_2}\right) = \left(\frac{P_1}{P_2}\right)\left(\frac{T_2}{T_1}\right)(V_1)$$

$$= \left(\frac{1}{3}\right)\left(\frac{1.50}{1}\right)(9.00\ \text{L}) = 4.50\ \text{L}$$

THE IDEAL GAS LAW

12.20. Calculate R, the gas law constant, (a) in units of L·torr/(mol·K) and (b) in units of mL·atm/(mol·K).

Ans.   (a) $R = \dfrac{0.0821\ \text{L·atm}}{\text{mol·K}}\left(\dfrac{760\ \text{torr}}{1\ \text{atm}}\right) = \dfrac{62.4\ \text{L·torr}}{\text{mol·K}}$

   (b) $R = \dfrac{0.0821\ \text{L·atm}}{\text{mol·K}}\left(\dfrac{1000\ \text{mL}}{1\ \text{L}}\right) = \dfrac{82.1\ \text{mL·atm}}{\text{mol·K}}$

12.21. How can you recognize an ideal gas law problem?

Ans.   Ideal gas law problems involve moles. If the number of moles of gas is given or asked for, or if a quantity that involves moles is given or asked for, the problem is most likely an ideal gas law problem. Thus, any problem involving masses of gas (which can be converted to moles of gas) or molar masses (grams per mole) or numbers of individual molecules (which can be converted to moles) and so forth is an ideal gas law problem. Problems that involve more than one temperature and/or one pressure of gas are most likely not ideal gas law problems.

12.22. Calculate the absolute temperature of 0.118 mol of a gas that occupies 10.0 L at 0.933 atm.

Ans. $T = \dfrac{PV}{nR} = \dfrac{(0.933\ \text{atm})(10.0\ \text{L})}{(0.118\ \text{mol})[0.0821\ \text{L·atm/(mol·K)}]} = 963\ \text{K}$

12.23. Calculate the pressure of 0.0303 mol of a gas that occupies 1.24 L at 22°C.

Ans. $P = \dfrac{nRT}{V} = \dfrac{(0.0303\ \text{mol})[0.0821\ \text{L·atm/(mol·K)}](295\ \text{K})}{1.24\ \text{L}} = 0.592\ \text{atm}$

12.24. Calculate the value of R if 1.00 mol of gas occupies 22.4 L at STP.

Ans.   STP means 1.00 atm and 273 K (0°C). Thus,

$$R = \frac{PV}{nT} = \frac{(1.00\ \text{atm})(22.4\ \text{L})}{(1.00\ \text{mol})(273\ \text{K})} = 0.0821\ \text{L·atm/(mol·K)}$$

12.25. Calculate the volume of 0.193 mol of a gas at 27°C and 825 torr.

Ans.
$$V = \frac{nRT}{P} = \frac{(0.193 \text{ mol})[0.0821 \text{ L·atm/(mol·K)}](300 \text{ K})}{(825 \text{ torr})(1 \text{ atm/760 torr})} = 4.38 \text{ L}$$

12.26. Calculate the volume of 1.00 mol of H_2O at 1.00-atm pressure and a temperature of 25°C.

Ans.   Water (H_2O) is not a gas under these conditions, and so the equation $PV = nRT$ does not apply. (The ideal gas law can be used for water vapor, e.g., water over 100°C at 1 atm or water at lower temperatures mixed with air.) At 1-atm pressure and 25°C, water is a liquid with a density about 1.00 g/mL.

$$1.00 \text{ mol} \left(\frac{18.0 \text{ g}}{1 \text{ mol}} \right) \left(\frac{1 \text{ mL}}{1.00 \text{ g}} \right) = 18.0 \text{ mL}$$

12.27. Calculate the number of moles of a gas that occupies 4.00 L at 303 K and 1.12 atm.

Ans.
$$n = \frac{PV}{RT} = \frac{(1.12 \text{ atm})(4.00 \text{ L})}{[0.0821 \text{ L·atm/(mol·K)}](303 \text{ K})} = 0.180 \text{ mol}$$

12.28. Calculate the number of moles of gas present in Problem 12.18*d*.

Ans.   Since the number of moles of gas does not change during the process of changing from state 1 to state 2, either set of data can be used to calculate the number of moles of gas. Since all three values are given for state 1, it is easier to use that state:

$$n = \frac{PV}{RT} = \frac{(1.00 \text{ atm})(4.00 \text{ L})}{[0.0821 \text{ L·atm/(mol·K)}](273 \text{ K})} = 0.178 \text{ mol}$$

12.29. (*a*) Compare qualitatively the volumes at STP of 6.8 mol N_2 and 6.8 mol H_2. (*b*) Compare the volumes at STP of 6.8 g N_2 and 6.8 g H_2.

Ans.   (*a*) Since the pressures, temperature, and numbers of moles are the same, the volumes also must be the same. (*b*) The number of moles of nitrogen is less than the number of moles of hydrogen because its molar mass is greater. Since the number of moles of nitrogen is less, so is its volume.

12.30. Calculate the volume of 7.07 g of helium at 27°C and 1.00 atm.

Ans.
$$7.07 \text{ g He} \left(\frac{1 \text{ mol He}}{4.00 \text{ g He}} \right) = 1.77 \text{ mol He}$$

$$V = \frac{nRT}{P} = \frac{(1.77 \text{ mol})[0.0821 \text{ L·atm/(mol·K)}](300 \text{ K})}{1.00 \text{ atm}} = 43.6 \text{ L}$$

12.31. Repeat the prior problem, using hydrogen gas instead of helium. Explain why the occurrence of hydrogen gas in diatomic molecules is so important.

Ans. $n = 3.51 \text{ mol}$ $V = 86.5 \text{ L}$

The gas laws work with moles of *molecules*, not atoms. It is necessary to know that hydrogen gas occurs in diatomic molecules so that the proper number of moles of gas may be calculated from the mass of the gas.

DALTON'S LAW OF PARTIAL PRESSURES

12.32. What is the total pressure of a gas mixture containing He at 0.173 atm, Ne at 0.201 atm, and Ar at 0.550 atm?

Ans. $P_{\text{total}} = P_{\text{He}} + P_{\text{Ne}} + P_{\text{Ar}} = 0.173 \text{ atm} + 0.201 \text{ atm} + 0.550 \text{ atm} = 0.924 \text{ atm}$

12.33. What is the pressure of H_2 if 0.250 mol of H_2 and 0.120 mol of He are placed in a 10.0-L vessel at 27°C?

Ans.   The presence of He makes no difference. The pressure of H_2 is calculated with the ideal gas law, using the number of moles of H_2.

$$P = \frac{nRT}{V} = \frac{(0.250 \text{ mol})[0.0821 \text{ L·atm/(mol·K)}](300 \text{ K})}{10.0 \text{ L}} = 0.616 \text{ atm}$$

12.34. What is the total pressure of a gas mixture containing H_2 at 0.173 atm, N_2 at 0.201 atm, and NO at 0.550 atm?

Ans. $P_{total} = P_{H_2} + P_{N_2} + P_{NO} = 0.924$ atm

12.35. What is the difference between Problems 12.32 and 12.34?

   Ans.   In Problem 12.34, molecules are involved. In Problem 12.32, uncombined atoms are involved. In both cases, the same laws apply, and it is best to regard an uncombined atom of He, for example, as a monatomic molecule.

12.36. The total pressure of a 125-mL sample of oxygen collected over water at 25°C is 1.030 atm. (*a*) How many moles of gas are present? (*b*) How many moles of water vapor are present? (*c*) How many moles of oxygen are present?

   Ans.   (*a*) $n = \dfrac{PV}{RT} = \dfrac{(1.030 \text{ atm})(0.125 \text{ L})}{[0.0821 \text{ L·atm/(mol·K)}](298 \text{ K})} = 5.26 \times 10^{-3}$ mol total

   (*b*)   The gas is composed of O_2 and water vapor. The pressure of the water vapor at 25°C, given in the last part of Sec. 12.8 or in tables in your text, is 24 torr. In atmospheres:

$$24 \text{ torr}\left(\frac{1 \text{ atm}}{760 \text{ torr}}\right) = 0.032 \text{ atm}$$

$$n_{H_2O} = \frac{PV}{RT} = \frac{(0.032 \text{ atm})(0.125 \text{ L})}{[0.0821 \text{ L·atm/(mol·K)}](298 \text{ K})} = 1.6 \times 10^{-4} \text{ mol } H_2O$$

   (*c*)   The pressure of the O_2 is therefore

$$1.030 \text{ atm} - 0.032 \text{ atm} = 0.998 \text{ atm}$$

$$n_{O_2} = \frac{PV}{RT} = \frac{(0.998 \text{ atm})(0.125 \text{ L})}{[0.0821 \text{ L·atm/(mol·K)}](298 \text{ K})} = 5.10 \times 10^{-3} \text{ mol } O_2$$

   Check : 5.10×10^{-3} mol $+ 0.16 \times 10^{-3}$ mol $= 5.26 \times 10^{-3}$ mol

12.37. In Dalton's law problems, what is the difference in behavior of water vapor mixed with air compared to helium mixed with air?

   Ans.   The pressure of water vapor, if it is in contact with liquid water, is governed by the temperature only. More water can evaporate or some water vapor can condense if the pressure is not equal to the tabulated vapor pressure at the given temperature. Helium is a gas under most conditions and is not capable of adjusting its pressure in the same way.

Supplementary Problems

12.38. Explain why gas law problems are not given with data to four or five significant figures.

   Ans.   The laws are only approximate, and having better data would not necessarily yield more accurate answers.

12.39. A mixture of gases contains He, Ne, and Ar. The pressure of He is 0.300 atm. The volume of Ne is 4.00 L. The temperature of Ar is 27°C. What value can be calculated from these data? Explain.

   Ans.   The temperature of Ar is the temperature of all the gases, since they are all mixed together. The volume of Ne is the volume of each of the gases and the total volume, too. Therefore, the number of moles of He can be calculated because its pressure is also known:

$$n = \frac{PV}{RT} = \frac{(0.300 \text{ atm})(4.00 \text{ L})}{[0.0821 \text{ L·atm/(mol·K)}](300 \text{ K})} = 0.0487 \text{ mol He}$$

12.40. Under what conditions of temperature and pressure do the gas laws work best?

Ans.   Under low-pressure and high-temperature conditions (far from the possibility of change to the liquid state).

12.41. If 0.223 g of a gas occupies 2.87 L at 17°C and 700 torr, what is the identity of the gas?

Ans.   If you do not see at first how to solve this problem to completion, at least you can recognize that P, V, and T data are given. First calculate the number of moles of gas present:

$$700 \text{ torr}\left(\frac{1 \text{ atm}}{760 \text{ torr}}\right) = 0.921 \text{ atm} \qquad 17°C = (17 + 273) \text{ K} = 290 \text{ K}$$

$$n = \frac{PV}{RT} = \frac{(0.921 \text{ atm})(2.87 \text{ L})}{[0.0821 \text{ L·atm/(mol·K)}](290 \text{ K})} = 0.111 \text{ mol}$$

We now know the mass of the gas and the number of moles. That is enough to calculate the molar mass:

$$\frac{0.223 \text{ g}}{0.111 \text{ mol}} = 2.01 \text{ g/mol}$$

The gas has a molar mass of 2.01 g/mol. It could only be hydrogen, H_2, because no other gas has a molar mass that low.

12.42. Which temperature scale must be used in (*a*) Boyle's law problems, (*b*) ideal gas law problems, (*c*) combined gas law problems, and (*d*) Charles' law problems?

Ans.   (*a*) No temperature is used in the calculations for Boyle's law problems since the temperature must be constant. (*b*) through (*d*) Kelvin.

12.43. The total pressure of a mixture of gases is 1.50 atm. The mixture contains 0.10 mol of N_2 and 0.20 mol of O_2. What is the partial pressure of O_2?

Ans.   The oxygen in the mixture is two-thirds of the number of moles, so it exerts two-thirds of the total pressure—1.00 atm. Or

$$\frac{n_{N_2}}{n_{O_2}} = \frac{0.10 \text{ mol } N_2}{0.20 \text{ mol } O_2} = \frac{P_{N_2} V/(RT)}{P_{O_2} V/(RT)} = \frac{P_{N_2}}{P_{O_2}} = 0.50$$

$$P_{N_2} = 0.50 P_{O_2}$$

$$P_{N_2} + P_{O_2} = 1.50 \text{ atm}$$

$$0.50 P_{O_2} + P_{O_2} = 1.50 P_{O_2} = 1.50 \text{ atm}$$

$$P_{O_2} = \frac{1.50 \text{ atm}}{1.50} = 1.00 \text{ atm}$$

12.44. Calculate the mass of $KClO_3$ required to decompose to provide 0.728 L of O_2 at 20°C and 1.02 atm.

Ans.   The volume, temperature, and pressure given allow us to calculate the number of moles of *oxygen*:

$$n_{O_2} = \frac{PV}{RT} = \frac{(1.02 \text{ atm})(0.728 \text{ L})}{[0.0821 \text{ L·atm/(mol·K)}](293 \text{ K})} = 0.0309 \text{ mol}$$

The number of moles of $KClO_3$ may be calculated from the number of moles of O_2 by means of the balanced chemical equation, and that value is then converted to mass.

$$2 \text{ KClO}_3 \longrightarrow 2 \text{ KCl} + 3 \text{ O}_2$$

$$0.0309 \text{ mol O}_2\left(\frac{2 \text{ mol KClO}_3}{3 \text{ mol O}_2}\right)\left(\frac{122 \text{ g KClO}_3}{1 \text{ mol KClO}_3}\right) = 2.51 \text{ g KClO}_3$$

12.45. Two samples of gas at equal pressures and temperatures are held in containers of equal volume. What can be stated about the comparative number of molecules in each gas sample?

Ans.   Since the volumes, temperatures, and pressures are the same, the numbers of moles of the two gases are the same. Therefore, there are equal numbers of molecules of the two gases.

12.46. P is inversely proportional to V. Write three mathematical expressions that relate this fact.

Ans.
$$P \propto \frac{1}{V} \qquad P = \frac{k}{V} \qquad PV = k$$

12.47. Write an equation for the combined gas law, using temperature in degrees Celsius. Explain why the Kelvin scale is more convenient.

Ans.
$$\frac{P_1 V_1}{t_1 + 273} = \frac{P_2 V_2}{t_2 + 273}$$

The law is simpler with the Kelvin temperatures.

12.48. Calculate the molar mass of a gas if 12.6 g occupies 5.11 L at 1.12 atm and 22°C.

Ans. From the pressure, volume, and temperature data, we can calculate the number of moles of gas present. From the number of moles and the mass, we can calculate the molar mass.
$$n = \frac{PV}{RT} = \frac{(1.12 \text{ atm})(5.11 \text{ L})}{[0.0821 \text{ L·atm/(mol·K)}](295 \text{ K})} = 0.236 \text{ mol}$$
Molar mass $= 12.6 \text{ g}/0.236 \text{ mol} = 53.4 \text{ g/mol}$

12.49. What volume of a CO_2 and H_2O mixture at 1.00 atm and 450 K can be prepared by the thermal decomposition of 0.950 g $NaHCO_3$?
$$2 \text{ NaHCO}_3 \xrightarrow{\text{heat}} \text{Na}_2\text{CO}_3 + \text{CO}_2 + \text{H}_2\text{O(g)}$$

Ans.
$$0.950 \text{ g NaHCO}_3 \left(\frac{1 \text{ mol NaHCO}_3}{84.0 \text{ g NaHCO}_3} \right) \left(\frac{1 \text{ mol CO}_2}{2 \text{ mol NaHCO}_3} \right) = 5.65 \times 10^{-3} \text{ mol CO}_2$$

The same number of moles of gaseous water (at 450 K) is obtained. The total number of moles of gas is 0.0113 mol. The volume is given by
$$V = \frac{nRT}{P} = \frac{(0.0113 \text{ mol})[0.0821 \text{ L·atm/(mol·K)}](450 \text{ K})}{1.00 \text{ atm}} = 0.417 \text{ L}$$

Note: At 450 K and 1.00 atm, water is completely in the gas phase. (The vapor pressure of water at 450 K is greater than 1.00 atm.)

12.50. What volume of O_2 at STP can be prepared by the thermal decomposition of 0.500 g of Hg_2O?

Ans.
$$2 \text{ Hg}_2\text{O} \xrightarrow{\text{heat}} 4 \text{ Hg} + \text{O}_2$$
$$0.500 \text{ g Hg}_2\text{O} \left(\frac{1 \text{ mol Hg}_2\text{O}}{416 \text{ g Hg}_2\text{O}} \right) \left(\frac{1 \text{ mol O}_2}{2 \text{ mol Hg}_2\text{O}} \right) = 6.01 \times 10^{-4} \text{ mol O}_2$$
$$V = \frac{nRT}{P} = \frac{(6.01 \times 10^{-4} \text{ mol})[0.0821 \text{ L·atm/(mol·K)}](273 \text{ K})}{1.00 \text{ atm}} = 0.0135 \text{ L}$$

12.51. What volume will 14.3 g of CH_4 occupy at 38°C and 0.888 atm?

Ans.
$$14.3 \text{ g CH}_4 \left(\frac{1 \text{ mol CH}_4}{16.0 \text{ g CH}_4} \right) = 0.894 \text{ mol CH}_4$$
$$V = \frac{nRT}{P} = \frac{(0.894 \text{ mol})[0.0821 \text{ L·atm/(mol·K)}](311 \text{ K})}{0.950 \text{ atm}} = 24.0 \text{ L}$$

12.52. Show that the volumes of individual gases involved in a chemical reaction (before they are mixed or after they are separated), all measured at the same temperature and pressure, are in proportion to their numbers of moles. Use the following reaction as an example:
$$2 \text{ CO} + \text{O}_2 \longrightarrow 2 \text{ CO}_2$$

Ans.
$$\frac{n_{O_2}}{n_{CO}} = \frac{PV_{O_2}/(RT)}{PV_{CO}/(RT)} = \frac{V_{O_2}}{V_{CO}}$$

If gas *mixtures* were involved, their volumes would necessarily be the same and their *pressures* would be in the ratios of their numbers of moles.

12.53. What is the ratio of volume of CO_2 produced to O_2 used up, both at the same temperature and pressure, for the reaction in the prior problem?

Ans.   The ratio is 2 : 1, as in the balanced equation.

12.54. (*a*)   Calculate the initial volume of 3.00 L of gas whose pressure has been increased from 1.00 atm to 2.00 atm.

   (*b*)   Calculate the final volume of 3.00 L of gas whose pressure has been increased from 1.00 atm to 2.00 atm.

Ans.   Read the problem carefully. What must be assumed in part (*a*)? in part (*b*)? Since no mention was made of temperature, we must assume that the temperature is constant in both parts, or we cannot do the problem.

   (*a*)   The initial volume was asked for, so we must assume that 3.00 L is the final volume.

$$V_1 = \frac{V_2 P_2}{P_1} = \frac{(3.00 \text{ L})(2.00 \text{ atm})}{1.00 \text{ atm}} = 6.00 \text{ L}$$

   (*b*)   The final volume was asked for, so 3.00 L must be the initial volume.

$$V_2 = \frac{V_1 P_1}{P_2} = \frac{(3.00 \text{ L})(1.00 \text{ atm})}{2.00 \text{ atm}} = 1.50 \text{ L}$$

   (In each case, doubling the pressure reduced the volume to one-half its initial volume.)

12.55. Calculate the final volume of 1.20 L of gas whose pressure is halved at constant temperature.

Ans.   The final volume is 2.40 L. According to Boyle's law, halving the pressure causes V to be doubled.

12.56. In a certain experiment, when 2.500 g of $KClO_3$ was heated, some O_2 was driven off. After the experiment, 2.402 g of solid was left. (Not all the $KClO_3$ decomposed.) (*a*) Write a balanced chemical equation for the reaction. (*b*) What compounds make up the solid? (*c*) What causes the loss of mass of the sample? (*d*) Calculate the volume of oxygen produced at STP.

Ans.   (*a*) $2 \text{ KClO}_3 \longrightarrow 2 \text{ KCl} + 3 \text{ O}_2$

   (*b*)   The solid is KCl and unreacted $KClO_3$.

   (*c*)   The loss of mass is caused solely by the escape of O_2.

   (*d*) $0.098 \text{ g O}_2 \left(\dfrac{1 \text{ mol O}_2}{32.0 \text{ g O}_2} \right) = 0.0031 \text{ mol O}_2$

$$V = \frac{nRT}{P} = \frac{(0.0031 \text{ mol})[0.0821 \text{ L·atm/(mol·K)}](273 \text{ K})}{1.00 \text{ atm}} = 0.069 \text{ L} = 69 \text{ mL}$$

12.57. Calculate the ratios of volume to Celsius temperature for the data in Table 12-3. Are the ratios the same for all the temperatures?

Ans.   No

12.58. Replot the data of Table 12-1, using P and $1/V$. Is the result a straight line? Explain.

Ans.   The result is still a straight line. $PV = k$ so $P = k(1/V)$ and also $V = k(1/P)$.

12.59. What is the final temperature or pressure in each of the following parts: (*a*) The pressure of a sample of gas at STP is raised 2 atm. (*b*) The pressure of a sample of gas at STP is raised to 2 atm. (*c*) The temperature of a sample of gas at STP is raised 20°C. (*d*) The temperature of a sample of gas at STP is raised to 20°C?

Ans.   (*a*) 3 atm   (*b*) 2 atm   (*c*) 20°C   (*d*) 20°C

12.60. For a mixture of two gases, gas 1 and gas 2, put in equals signs or plus signs in the appropriate places in the equations below:

$$P_{\text{total}} = P_1 \quad P_2 \qquad T_{\text{total}} = T_1 \quad T_2$$
$$V_{\text{total}} = V_1 \quad V_2 \qquad n_{\text{total}} = n_1 \quad n_2$$

Ans.
$$P_{\text{total}} = P_1 + P_2 \qquad T_{\text{total}} = T_1 = T_2$$
$$V_{\text{total}} = V_1 = V_2 \qquad n_{\text{total}} = n_1 + n_2$$

12.61. Draw a figure to represent all the conversions learned so far, including volumes of gases. Start with the figure you did for Problem 11.31.

Ans.   The figure is presented on p. 348.

Kinetic Molecular Theory

13.1. INTRODUCTION

In Chap. 12 laws governing the behavior of gases were presented. The fact that gases exert pressure was stated, but no reasons why gases should exhibit such behavior were given. The *kinetic molecular theory* (KMT) explains all the gas laws that we have studied and some additional ones. It describes gases in terms of the behavior of the molecules that make them up. (The noble gases exist as individual atoms, but for purposes of explaining the KMT they will be included and treated as molecules. They may be thought of as monatomic molecules.)

13.2. POSTULATES OF THE KINETIC MOLECULAR THEORY

All gases, under ordinary conditions of temperature and pressure, are made of molecules (including one-atom molecules such as are present in samples of the noble gases). That is, ionic substances do not form gases under conditions prevalent on earth. The molecules of a gas act according to the following postulates:

1. Molecules are in constant random motion.

2. The molecules exhibit negligible intermolecular attractions or repulsions except when they collide.

3. Molecular collisions are elastic, which means that although the molecules transfer energy from one to another, as a whole they do not lose kinetic energy when they collide with one another or with the walls of their container.

4. The molecules occupy a negligible fraction of the volume occupied by the gas as a whole.

5. The *average kinetic energy* of the gas molecules is directly proportional to the *absolute* temperature of the gas.

$$\overline{\text{KE}} = \tfrac{3}{2}kT = \tfrac{1}{2}\overline{mv^2}$$

The overbars mean "average." The k in the proportionality constant is called the *Boltzmann constant*. It is equal to R, the ideal gas law constant (with unfamiliar units), divided by Avogadro's number. Note that this k is the same for all gases.

Postulate 1 means that molecules move in any direction whatsoever until they collide with another molecule or a wall, whereupon they bounce off and move in another direction until their next collision. Postulate 2 means that the molecules move in a straight line at constant speed between collisions. Postulate 3 means that there is no friction in molecular collisions. The molecules have the same total kinetic energy after each collision as before. Postulate 4 concerns the volume of the molecule themselves versus the volume of the container they occupy. The individual particles do not occupy the entire container. If the molecules of gas had zero volumes and zero intermolecular attractions and repulsions, the gas would obey the ideal gas law exactly. Postulate 5 means that if two gases are at the same temperature, their molecules will have the same average kinetic energies.

EXAMPLE 13.1.   Calculate the value for k, the Boltzmann constant, using the following value for R:

$$R = 8.31 \text{ J/(mol·K)}$$

Ans.

$$k = \frac{R}{N} = \frac{8.31 \text{ J/(mol·K)}}{6.02 \times 10^{23} \text{ molecules/mol}} = \frac{1.38 \times 10^{-23} \text{ J}}{\text{molecule·K}}$$

EXAMPLE 13.2.   Calculate the average kinetic energy of H_2 molecules at 1.00 atm and 300 K.

Ans.

$$\overline{\text{KE}} = \tfrac{3}{2}kT = 1.5[1.38 \times 10^{-23} \text{ J/(molecule·K)}](300 \text{ K}) = 6.21 \times 10^{-21} \text{ J}$$

13.3.   EXPLANATION OF GAS PRESSURE, BOYLE'S LAW, AND CHARLES' LAW

Kinetic molecular theory explains why gases exert pressure. The constant bombardment of the walls of the vessel by the gas molecules, like the hitting of a target by machine gun bullets, causes a constant force to be applied to the wall. The force applied, divided by the area of the wall, is the pressure of the gas.

Boyle's law may be explained using the kinetic molecular theory by considering the box illustrated in Fig. 13-1. If a sample of gas is placed in the left half of the box shown in the figure, it will exert a certain pressure. If the volume is doubled by extending the right wall to include the entire box shown in the figure, the pressure should fall to one-half its original value. Why should that happen? In an oversimplified picture, the molecules bouncing back and forth between the right and left walls now have twice as far to travel, and thus they hit each wall only one-half as often in a given time. Therefore, the pressure is only one-half what it was. How about the molecules that are traveling up and down or in and out? There are as many such molecules as there were before, and they hit the walls as often; but they are now striking an area twice as large, and so the pressure is one-half what it was originally. Thus, doubling the volume halves the pressure. This can be shown to be true no matter what the shape of the container.

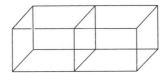

Fig. 13-1.   Explanation of Boyle's law

Charles' law governs the volume of a gas at constant pressure when its temperature is changed. When the absolute temperature of a gas is multiplied by 4, for example, the average kinetic energy of its molecules is also multiplied by 4 (postulate 5). The kinetic energy of any particle is given by $\text{KE} = \tfrac{1}{2}mv^2$, where m is the mass of the particle and v is its velocity. When the kinetic energy is multiplied by 4, what happens to the velocity? It is doubled. (Since the v term is being squared, to effect a fourfold increase, you need only to double the velocity: $2^2 = 4$.)

$$\text{KE}_1 = \tfrac{1}{2}mv_1{}^2$$

$$\text{KE}_2 = \tfrac{1}{2}mv_2{}^2 = 4\text{KE}_1 = \tfrac{1}{2}m(2v_1)^2$$

The velocity v_2 is equal to $2v_1$. On average, in a sample of gas the molecules are going twice as fast at the higher temperature. They therefore hit the walls (1) twice as often per unit time and (2) twice as hard each time they hit, for a combined effect of 4 times the pressure (in a given volume). If we want a constant pressure, we have to expand the volume to 4 times what it was before, and we see that multiplying the absolute temperature by 4 must be accompanied by a fourfold increase in volume if the pressure is to be kept constant.

13.4. GRAHAM'S LAW

An experimental law not yet discussed is *Graham's law*, which states that the rate of effusion or diffusion of a gas is inversely proportional to the square root of its molar mass. *Effusion* is the passage of a gas through small holes in its container, such as the pores in a porous cup. The deflation of a helium-filled party balloon over several days results from the helium atoms effusing through the tiny pores of the balloon wall. *Diffusion* is the passage of a gas through another gas. For example, if a bottle of ammonia water is spilled in one corner of a room, the odor of ammonia is soon apparent throughout the room. The ammonia molecules have diffused through the air molecules. Consider two gases with molar masses MM_1 and MM_2. The ratio of their rates of effusion or diffusion is given by

$$\frac{r_1}{r_2} = \sqrt{\frac{MM_2}{MM_1}}$$

That is, the heavier a molecule of the gas, the more slowly it effuses or diffuses.

The rate of effusion or diffusion of a gas is directly proportional to the "average" velocity of its molecules.

EXAMPLE 13.3.   A sample of nitrogen and a sample of neon are both at the same temperature. What is the ratio of the "average" velocities of their molecules?

Ans.   Since the temperatures are the same, so are the average kinetic energies of their molecules. From Graham's law,

$$\frac{v_{Ne}}{v_{nitrogen}} = \sqrt{\frac{MM_{nitrogen}}{MM_{Ne}}} = \sqrt{\frac{28.0}{20.2}} = 1.18$$

The neon atoms are moving 1.18 times as fast as the (heavier) nitrogen molecules.

Graham's law may be explained in terms of the kinetic molecular theory as follows: Since two gases are at the same temperature, their average kinetic energies are the same:

$$\overline{KE_1} = \overline{KE_2} = \tfrac{1}{2}m_1\overline{v_1^2} = \tfrac{1}{2}m_2\overline{v_2^2}$$

Multiplying the last of these equations by 2 yields

$$m_1\overline{v_1^2} = m_2\overline{v_2^2} \qquad \text{or} \qquad \frac{m_1}{m_2} = \frac{\overline{v_2^2}}{\overline{v_1^2}}$$

Since the masses of the molecules are proportional to their molar masses, and the average velocity of the molecules is a measure of the rate of effusion or diffusion, all we have to do to this equation to get Graham's law is to take its square root. (The square root of $\overline{v^2}$ is not quite equal to the average velocity, but is a quantity called the *root mean square velocity*. See Problem 13.17.)

Solved Problems

POSTULATES OF THE KINETIC MOLECULAR THEORY

13.1.   (*a*) Calculate the volume at 100°C of 18.0 g of liquid water, assuming the density to be 1.00 g/mL. (*b*) Calculate the volume of 18.0 g of water vapor at 100°C and 1.00-atm pressure, using the ideal gas law. (*c*) Assuming that the volume of the liquid is the total volume of the molecules themselves, calculate the percentage of the gas volume occupied by molecules.

Ans.   (*a*) $18.0 \text{ g}\left(\dfrac{1.00 \text{ mL}}{1.00 \text{ g}}\right) = 18.0 \text{ mL} = 0.0180 \text{ L}$

   (*b*) $V = \dfrac{nRT}{P} = \dfrac{(1.00 \text{ mol})[0.0821 \text{ L·atm/(mol·K)}](373 \text{ K})}{1.00 \text{ atm}} = 30.6 \text{ L}$

   (*c*)   The percentage occupied by the molecules is

$$\frac{0.0180 \text{ L}}{30.6 \text{ L}} \times 100\% = 0.0588\%$$

13.2. If two different gases are at the same temperature, which of the following must also be equal, (*a*) their pressures, (*b*) their average molecular velocities, or (*c*) the average kinetic energies of their molecules?

   Ans.   (*c*) Their average kinetic energies must be equal since the temperatures are equal.

13.3. Does the kinetic molecular theory state that all the molecules of a given sample of gas have the same velocity since they are all at one temperature?

   Ans.   No. The kinetic molecular theory states that the *average* kinetic energy is related to the temperature, not the velocity or kinetic energy of any one molecule. The velocity of each individual molecule changes as it strikes other molecules or the walls.

13.4. Calculate the temperature at which CO_2 molecules have the same "average" velocity as nitrogen molecules have at 273 K.

   Ans.   Let

$$\bar{v}^2 = \overline{v_{N_2}^2} = \overline{v_{CO_2}^2}$$

$$\frac{T_{CO_2}}{T_{N_2}} = \frac{\overline{KE_{CO_2}}}{\overline{KE_{N_2}}} = \frac{m_{CO_2}\overline{v^2}}{m_{N_2}\overline{v^2}} = \frac{m_{CO_2}}{m_{N_2}} = \frac{44.0}{28.0} = 1.57$$

$$T_{CO_2} = 1.57T_{N_2} = 1.57(273 \text{ K}) = 429 \text{ K}$$

13.5. If the molecules of a gas are compressed so that their average distance of separation gets smaller, what should happen to the forces between them? To their ideal behavior?

   Ans.   As the average distance gets smaller, the intermolecular forces increase. As the intermolecular forces increase and the volume of the molecules themselves becomes a more significant fraction of the total gas volume, their behavior becomes further from ideal.

EXPLANATION OF GAS PRESSURE, BOYLE'S LAW, AND CHARLES' LAW

13.6. Suppose that we double the length of each side of a rectangular box containing a gas. (*a*) What will happen to the volume? (*b*) What will happen to the pressure? (*c*) Explain the effect on the pressure on the basis of the kinetic molecular theory.

   Ans.   (*a*) The volume will increase by a factor of $(2)^3 = 8$.   (*b*) The pressure will fall to one-eighth its original value.   (*c*) In each direction, the molecules will hit the wall only one-half as often, and the force on each wall will drop to one-half of what it was originally because of this effect. Each wall has 4 times the area, and so the pressure will be reduced to one-fourth its original value because of this effect. The total reduction in pressure is $\frac{1}{2} \times \frac{1}{4} = \frac{1}{8}$, in agreement with Boyle's law.

GRAHAM'S LAW

13.7. (*a*) If the velocity of a single gas molecule doubles, what happens to its kinetic energy? (*b*) If the average velocity of the molecules of a gas doubles, what happens to the temperature of the gas?

   Ans.   (*a*)

$$v_2 = 2v_1$$

$$KE_2 = \tfrac{1}{2}mv_2^2 = \tfrac{1}{2}m(2v_1)^2 = 4\left(\tfrac{1}{2}mv_1^2\right) = 4\,KE_1$$

   The kinetic energy is increased by a factor of 4.

   (*b*)   The absolute temperature is increased by a factor of 4.

13.8. Would it be possible to separate isotopes by using the principle of Graham's law? Explain what factors would be important.

   Ans.   Since the molecules containing different isotopes have different masses, it is possible to separate them on the basis of their different "average" molecular velocities. One would have to have gaseous molecules in which the element being separated into its isotopes was the only element present in more than one isotope. For example, if uranium is being separated, a gaseous compound of uranium is needed. The compound could be made with chlorine perhaps, but chlorine exists naturally in two isotopes of its own, and many different masses of molecules could be prepared with the naturally occurring mixture. Naturally occurring fluorine exists 100% as ^{19}F, and molecules of its gaseous compound with uranium, UF_6, will have two different masses, corresponding to $^{235}UF_6$ and $^{238}UF_6$. Uranium has been separated into its isotopes by repeated effusion of UF_6 through towers of porous dividers. Each process enriches the individual isotopes a little, and very many repetitions are required to get relatively pure isotopes.

13.9. Calculate the ratio of rates of effusion of $^{238}UF_6$ and $^{235}UF_6$. Fluorine is 100% ^{19}F.

   Ans.   The masses of the molecules are 352 and 349 amu. The relative rates of effusion are

$$\sqrt{\frac{352}{349}} = 1.004$$

   The molecules of $^{235}UF_6$ will travel on average 1.004 times as fast as those of $^{238}UF_6$.

13.10. List the different molecular masses possible in UCl_3 with ^{238}U and ^{235}U as well as ^{35}Cl and ^{37}Cl.

   Ans.

$^{238}U(^{35}Cl)_3$	343 amu	$^{235}U(^{35}Cl)_3$	340 amu
$^{238}U(^{35}Cl)_2(^{37}Cl)$	345 amu	$^{235}U(^{35}Cl)_2(^{37}Cl)$	342 amu
$^{238}U(^{35}Cl)(^{37}Cl)_2$	347 amu	$^{235}U(^{35}Cl)(^{37}Cl)_2$	344 amu
$^{238}U(^{37}Cl)_3$	349 amu	$^{235}U(^{37}Cl)_3$	346 amu

13.11. What possible complications would there be in trying to separate hydrogen into ^1H and ^2H by gaseous diffusion?

   Ans.   Hydrogen occurs as diatomic molecules, and it would be easy to separate $^1H_2$, $^1H^2H$, and $^2H_2$, but not the individual atoms. There would be very little $^2H_2$ since the heavy isotope accounts for only 0.015% of naturally occurring hydrogen atoms.

Supplementary Problems

13.12. (*a*) Is the ratio

$$\frac{\text{Total volume of gas molecules}}{\text{Volume of gas sample}}$$

smaller for a given sample of gas at constant pressure at 300 K or at 400 K? (*b*) Will the gas exhibit more ideal behavior at 300 K or at 400 K?

   Ans.   (*a*)   The ratio is smaller at 400 K. The volume of the molecules themselves does not change appreciably between the two temperatures, but the gas volume changes according to Charles' law.

   (*b*)   Since the gas volume is larger at 400 K, the gas molecules are farther apart at that temperature and exhibit lower intermolecular forces. The gas is therefore more ideal at the higher temperature.

13.13. (*a*) Is the ratio

$$\frac{\text{Total volume of gas molecules}}{\text{Volume of gas sample}}$$

smaller for a given sample of gas at constant temperature at 1.00 atm or at 2.00 atm? (*b*) Will the gas exhibit more ideal behavior at 1.00 atm or at 2.00 atm?

   Ans.   (*a*)   The ratio is smaller at 1.00 atm. The volume of the molecules themselves does not change appreciably between the two pressures, but the gas volume changes according to Boyle's law.

   (*b*)   Since the gas volume is larger at 1.00 atm, the gas molecules are farther apart at that temperature, and exhibit lower intermolecular forces. The gas is therefore more ideal at the lower pressure.

13.14. (*a*) Calculate the average kinetic energy of O_2 molecules at 1.00 atm and 300 K. (*b*) Does the pressure matter? (*c*) Does the identity of the gas matter?

   Ans.   (*a*) $\overline{KE} = \frac{3}{2}kT = 1.5[1.38 \times 10^{-23}\ \text{J/(molecule·K)}](300\ \text{K}) = 6.21 \times 10^{-21}\ \text{J}$

   (*b*) and (*c*) The pressure and the identity of the gas do not matter.

13.15. (*a*) Calculate the "average" velocity of O_2 molecules at 1.00 atm and 300 K. (*b*) Does the pressure matter? (*c*) Does the identity of the gas matter?

Ans.   (*a*) $\frac{1}{2}m\overline{v^2} = 6.21 \times 10^{-21}$ J   (from the prior problem)

$$m_{O_2} = 32.0 \text{ amu}\left(\frac{1 \text{ g}}{6.02 \times 10^{23} \text{ amu}}\right)\left(\frac{1 \text{ kg}}{1000 \text{ g}}\right) = 5.32 \times 10^{-26} \text{ kg}$$

$$\overline{v^2} = 2.33 \times 10^5 \text{ (m/s)}^2$$

$$v_{rms} = 483 \text{ m/s (about 0.30 miles/s)}$$

The square root of the average of the square of the velocity, v_{rms}, is not the average velocity, but a quantity called the *root mean square velocity*.

(*b*)   The pressure of the gas does not matter.

(*c*)   The identity of the gas is important, because the mass of the molecule is included in the calculation. (Contrast this conclusion with that of the prior problem.)

13.16. Explain why neon atoms obey the gas laws the same as nitrogen molecules.

Ans.   The gas laws work for unbonded atoms as well as for multiatom molecules.

13.17. (*a*) Calculate the square of each of the following numbers: 1, 2, 3, 4, and 5. (*b*) Calculate the average of the numbers. (*c*) Calculate the average of the squares. (*d*) Is the square root of the average of the squares equal to the average of the numbers? (*e*) Explain why quotation marks are used with "average" velocity in the text for the velocity of molecules with average kinetic energies.

Ans.   (*a*)

Number	Square
1	1
2	4
3	9
4	16
5	25

(*b*) 3

(*c*) 11

(*d*) The square root of the average of the squares ($\sqrt{11}$) is not equal to the average of the numbers (3).

(*e*) The velocity equal to the square root of the quotient of twice the average kinetic energy divided by the molecular mass, is not really the average velocity. That is, the square root of $\overline{v^2}$ does not give $\overline{v}$, but the root mean square velocity.

13.18. Contrast the motions of the molecules of a sample of gas at rest to those in a hurricane wind.

Ans.   At rest, as many molecules are traveling on average in any given direction as in the opposite direction, and at the same average speeds. In a hurricane, the molecules on average are traveling somewhat faster in the direction of the wind than in the opposite direction.

13.19. Oxygen gas and sulfur dioxide gas are at the same temperature. What is the ratio of the "average" velocities of their molecules?

Ans.   Since the temperatures are the same, so are the average kinetic energies of their molecules. From Graham's law,

$$\frac{v_1}{v_2} = \sqrt{\frac{MM_2}{MM_1}} = \sqrt{\frac{64.06}{32.00}} = 1.415$$

The oxygen molecules are moving, on average, 1.415 times as fast as the SO_2 molecules.

CHAPTER 14

Oxidation and Reduction

14.1. INTRODUCTION

We learned to write formulas of ionic compounds in Chaps. 5 and 6. We balanced the charges to determine the number of each ion to use in the formula. We could not do the same thing for atoms of elements in covalent compounds, because in these compounds the atoms do not have charges. In order to overcome this difficulty, we define *oxidation numbers*, also called *oxidation states*.

In Sec. 14.2 we will learn to determine oxidation numbers from the formulas of compounds and ions. We will learn how to assign oxidation numbers from electron dot diagrams and more quickly from a short set of rules. In Sec. 14.3 we will learn to predict oxidation numbers for the elements from their positions in the periodic table in order to predict formulas for their compounds and ions. We use these oxidation numbers for naming the compounds or ions (Chap. 6 and Sec. 14.4) and to balance equations for oxidation-reduction reactions (Sec. 14.5). In addition, we will use oxidation-reduction reactions to discuss electrochemical reactions in Sec. 14.5.

14.2. ASSIGNING OXIDATION NUMBERS

In Sec. 5.5, electron dot diagrams were introduced. The electrons shared between atoms were counted as "belonging" to both atoms. We thus counted more valence electrons than we actually had. For oxidation numbers, however, we can count each electron only once. If electrons are shared, we arbitrarily assign "control" of them to the more electronegative atom. For atoms of the same element, each atom is assigned one-half of the shared electrons. The oxidation number is then defined as the number of valence electrons in the free atom minus the number "controlled" by the atom in the compound. If we actually transfer the electrons from one atom to another, the oxidation number equals the resulting charge. If we share the electrons, the oxidation number does not equal the charge; there may be no charge. In this case, *control* is not meant literally, but is just a term to describe the counting procedure. For example, the electron dot diagram of CO_2 may be written as

$$:\ddot{O}::C::\ddot{O}:$$

Since O is to the right of C in the second period of the periodic table, O is more electronegative, and we assign control of all eight shared electrons to the two O atoms. (It does not really have complete control of the electrons; if it did, the compound would be ionic.) Thus, the oxidation number of each atom is calculated as follows:

	C	Each O atom
Number of valence electrons in free atom	4	6
− Number of valence electrons "controlled"	−0	−8
Oxidation number	+4	−2

Like the charge on an ion, *each* atom is assigned an oxidation number. Do not say that oxygen in CO_2 has an oxidation number of −4 because the two oxygen atoms together control four more electrons in the compound than they would in the free atoms. *Each* oxygen atom has an oxidation number of −2.

Although the assignment of control of electrons is somewhat arbitrary, the total number of electrons is accurately counted, which leads to a main principle of oxidation numbers:

> *The total of the oxidation numbers of all the atoms* (not *just all the elements*) *is equal to the net charge on the molecule or ion.*

For example, the total of the three oxidation numbers in CO_2 is $4 + 2(−2) = 0$, and the charge on CO_2 is 0.

One principal source of student errors is due to confusion between the charge and oxidation number. Do not confuse them. In naming compounds or ions, use Roman numerals to represent positive oxidation numbers. (The Romans did not have negative numbers.) In this book, charges have the numeral first, followed by the sign; oxidation numbers have the sign first, followed by the numeral. In formulas, Arabic numeral superscripts represent charges. While working out answers, you might want to write oxidation numbers encircled and under the symbol for the element. Individual covalently bonded atoms do not have easily calculated charges, but they do have oxidation numbers. For example, the oxidation number of each element and the charge on the ion for $SO_3^{2−}$ and for $Cl^−$ are shown below.

$$\text{Charges}$$

Oxidation numbers for $SO_3^{2−}$: $\fbox{+4}$ $\fbox{−2}$ and for $Cl^−$: $\fbox{−1}$

It is too time consuming to calculate oxidation numbers by drawing electron dot diagrams each time. We can speed up the process by learning the following simple rules:

1. The sum of all the oxidation numbers in a species is equal to the charge on the species.

2. The oxidation number of uncombined elements is equal to 0.

3. The oxidation number of every monatomic ion is equal to its charge.

4. In its compounds, the oxidation number of every alkali metal and alkaline earth metal is equal to its group number.

5. The oxidation number of hydrogen in compounds is +1 except when the hydrogen is combined with active metals; then it is −1.

6. The oxidation number of oxygen in its compounds is −2, with some much less important exceptions. The oxidation number of oxygen in peroxides is −1, in superoxides it is $−\frac{1}{2}$, and in OF_2 and O_2F_2 it is positive. The peroxides and superoxides generally occur only with other elements in their maximum oxidation states. You will be able to recognize peroxides or superoxides by the presence of pairs of oxygen atoms and by the fact that if the compounds were normal oxides, the other element present would have too high an oxidation number (Sec. 14.3).

Na_2O_2	sodium peroxide	oxidation number of sodium = 1
BaO_2	barium peroxide	oxidation number of barium = 2
SnO_2	tin(IV) oxide	oxidation number of tin = 4 (permitted)
KO_2	potassium superoxide	oxidation number of potassium = 1
H_2O_2	hydrogen peroxide	oxidation number of hydrogen = 1

7. The oxidation number of every halogen atom in its compounds is -1 except for a chlorine, bromine, or iodine atom combined with oxygen or a halogen atom higher in the periodic table. For example, the chlorine atoms in each of the following compounds have oxidation numbers of -1:

$$SnCl_2 \qquad SiCl_4 \qquad KCl \qquad PCl_3 \qquad HCl$$

The chlorine atom in each of the following has an oxidation number different from -1:

$$Cl_2O_3 \qquad ClO_2{}^- \qquad ClF_3 \text{ (F has an oxidation number of } -1.)$$

With these rules, we can quickly and easily calculate the oxidation numbers of an element most of the time from the formulas of its compounds.

EXAMPLE 14.1.   What are the oxidation numbers of fluorine and iodine in IF_7?

Ans.   F has an oxidation state of -1 (rule 7). I has an oxidation number of $+7$ (rule 1). (I is not -1, because it is combined with a halogen higher in the periodic table.)

EXAMPLE 14.2.   Calculate the oxidation number of S in (a) SO_2, (b) $SO_3{}^{2-}$, and (c) SO_3.

Ans.   Let x = oxidation number of S in each case:

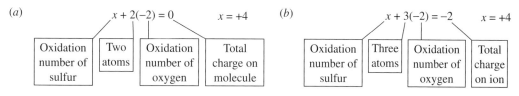

(c) $x + 3(-2) = 0 \qquad x = +6$

EXAMPLE 14.3.   Calculate the oxidation number of (a) Cr in $Cr_2O_7{}^{2-}$ and (b) S in $S_2O_3{}^{2-}$.

Ans.   (a)

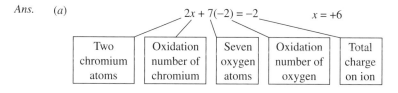

(b) $2x + 3(-2) = -2 \qquad x = +2$

Oxidation numbers are most often, but not always, integers. If there are nonintegral oxidation numbers, there must be multiple atoms so that the number of electrons is an integer.

EXAMPLE 14.4.   Calculate the oxidation number of N in NaN_3, sodium azide.

Ans.   Na has an oxidation number of $+1$. Therefore, the total charge on the three nitrogen atoms is $1-$, and the average charge, which is equal to the oxidation number, is $\frac{1}{3}-$. (Three nitrogen atoms, times $-\frac{1}{3}$ each, have a total of oxidation numbers equal to an integer, -1.)

14.3. PERIODIC RELATIONSHIPS OF OXIDATION NUMBERS

Oxidation numbers are very useful in correlating and systematizing a lot of inorganic chemistry. For example, the metals in very high oxidation states behave as nonmetals. They form oxyanions like $MnO_4{}^-$, but do not form highly charged monatomic ions, for example. A few simple rules allow the prediction of the formulas of covalent compounds using oxidation numbers, just as predictions were made for ionic compounds in Chap. 5 by using the charges on the ions. We can learn more than 200 possible oxidation numbers with relative ease by learning the following rules. We will learn other oxidation numbers as we progress.

1. All elements when uncombined have oxidation numbers equal to 0. (Some atoms also have oxidation numbers equal to 0 in some of their compounds, by the way.)

2. The maximum oxidation number of most atoms in its compounds is equal to its periodic group number. There are three groups that have atoms in excess of the group number and thus are exceptions to this rule. The coinage metals have the following maximum oxidation numbers: Cu, $+2$; Ag, $+2$ (rare); and Au, $+3$. Some of the noble gases (group 0) have positive oxidation numbers. Some lanthanide and actinide element oxidation numbers exceed $+3$, their nominal group number.

3. The minimum oxidation number of hydrogen is -1. That of any other nonmetallic atom is equal to its group number minus 8. That of any metallic atom is 0.

EXAMPLE 14.5. Give three possible oxidation numbers for chlorine.

Ans. Cl is in periodic group VIIA, and so its maximum oxidation number is $+7$ and its minimum oxidation number is $7 - 8 = -1$. It also has an oxidation number of 0 when it is a free element.

EXAMPLE 14.6. Give the possible oxidation numbers for potassium.

Ans. K can have an oxidation number of 0 when it is a free element and $+1$ in all its compounds. (See rule 4, Sec. 14.2.)

EXAMPLE 14.7. What is the maximum oxidation number of (*a*) Mn, (*b*) Os, (*c*) Ba, and (*d*) P?

Ans. (*a*) $+7$ (group VIIB)   (*b*) $+8$ (group VIII)   (*c*) $+2$ (group IIA)   (*d*) $+5$ (group VA)

EXAMPLE 14.8. Can titanium (Ti) exist in an oxidation state $+5$?

Ans. No. Its maximum oxidation state is $+4$ since it is in group IVB in the periodic table.

EXAMPLE 14.9. What is the minimum oxidation state of (*a*) P, (*b*) Br, and (*c*) K?

Ans. (*a*) -3 (group number $- 8 = -3$)   (*b*) -1 (group number $- 8 = -1$)   (*c*) 0 (Metallic atoms do not have negative oxidation states.)

EXAMPLE 14.10. Name one possible binary compound of (*a*) S and F and (*b*) P and O.

Ans. The more electronegative element will take the negative oxidation state.   (*a*) The maximum oxidation state of sulfur is $+6$; the only oxidation number of fluorine in its compounds is -1. Therefore, it takes six fluorine atoms to balance one sulfur atom, and the formula is SF_6.   (*b*) The maximum oxidation state of phosphorus is $+5$; the most common negative oxidation number of oxygen is -2. Therefore, it takes five oxygen atoms to balance two phosphorus atoms, and the formula is P_2O_5.

EXAMPLE 14.11. What is the formula for the phosphorus fluoride which has phosphorus in its maximum oxidation state?

Ans. The maximum oxidation number that phosphorus can have is $+5$ (from group VA), and so the formula is PF_5.

The rules above gave maximum and minimum oxidation numbers, but those might not be the only oxidation numbers or even the most important oxidation numbers for an element. Elements of the last six groups of the periodic table, for example, may have several oxidation numbers in their compounds, most of which vary from one another in steps of 2. For example, the major oxidation states of chlorine in its compounds are $-1, +1, +3, +5,$ and $+7$. The transition metals have oxidation numbers that may vary from one another in steps of 1. The inner transition elements mostly form oxidation states of $+3$, but the first part of the actinide series acts more as transition elements and the elements have maximum oxidation numbers that increase from $+4$ for Th to $+6$ for U. These generalizations are not absolute rules, but allow students to make educated guesses about possible compound formation without exhaustive memorization. These possibilities are illustrated in Fig. 14-1.

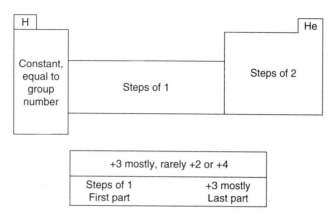

Fig. 14-1. Possible oxidation numbers

EXAMPLE 14.12.   Determine the formula of two oxides of sulfur.

Ans.   The oxygen must exist in a -2 oxidation state, because it is more electronegative than sulfur. Therefore, sulfur must exist in two different positive oxidation states in the two compounds. Its maximum oxidation state is $+6$, corresponding to its position in periodic group VIA. It also has an oxidation state of $+4$, which is 2 less than its maximum (see Fig. 14-1). The formulas therefore are SO_3 and SO_2.

14.4. OXIDATION NUMBERS IN INORGANIC NOMENCLATURE

In Chap. 6 we placed Roman numerals at the ends of names of metals to distinguish the charges on monatomic cations. It is really the oxidation number that is in parentheses. This nomenclature system is called the *Stock system*. For monatomic ions, the oxidation number is equal to the charge. For other cations, again the oxidation number is used in the name. For example, Hg_2^{2+} is named mercury(I) ion. Its charge is $2+$; the oxidation number of each atom is $+1$. Oxidation numbers are also used for other cations, such as dioxovanadium(V) ion, VO_2^+. The prefix *oxo-* stands for oxygen. Oxidation numbers can be used with nonmetal-nonmetal compounds, as in sulfur(VI) oxide for SO_3, but the older system using prefixes (Table 6-2) is still used more often.

EXAMPLE 14.13.   Name the following according to the Stock system: (*a*) $NiCl_2$, (*b*) UO_2SO_4, and (*c*) P_2O_5.

Ans.   (*a*) Nickel(II) chloride   (*b*) dioxouranium(VI) sulfate   (*c*) phosphorus(V) oxide

14.5. BALANCING OXIDATION-REDUCTION EQUATIONS

In every reaction in which the oxidation number of an element (or more than one) in one reactant goes up, an element (or more than one) in some reactant must go down in oxidation number. An increase in oxidation number is called an *oxidation*. A decrease in oxidation number is called a *reduction*. The term *redox* (the first letters of *red*uction and *ox*idation) is often used as a synonym for *oxidation-reduction*. The total change in oxidation number (change in each atom times number of atoms) must be the same in the oxidation as in the reduction, because the number of electrons whose "control" is transferred from one species must be the same as the number transferred to the other. The species that causes another to be reduced is called the *reducing agent*; in the process, it is oxidized. The species that causes the oxidation is called the *oxidizing agent*; in the process, it is reduced.

EXAMPLE 14.14.   (*a*) In drying dishes, a dish towel could be termed a drying agent, and the dish a wetting agent. What happens to the towel and to the dish? (*b*) In its reaction with $CrCl_2$, Cl_2 is the oxidizing agent and $CrCl_2$ is the reducing agent. What happens to Cl_2 and to Cr^{2+}?

Ans.   (*a*)   The towel, the drying agent, gets wet. The dish, the wetting agent, gets dry.

(*b*) $2\,CrCl_2 + Cl_2 \longrightarrow 2\,CrCl_3$

The oxidizing agent, Cl_2, is reduced to Cl^-. The oxidation number goes from 0 to -1. The reducing agent, Cr^{2+}, is oxidized. Its oxidation number goes from $+2$ to $+3$. Just as the water must go somewhere in part *a*, the electrons must go somewhere in part *b*.

One of the most important uses of oxidation numbers is in balancing redox (oxidation-reduction) equations. These equations can get very complicated, and a systematic method of balancing them is essential. There are many such methods, however, and each textbook seems to use its own. There are many similarities among the methods, and the following discussion will help no matter what method your instructor and your textbook use.

It is of critical importance in this section to keep in mind the difference between oxidation number and charge. You balance charge one way and changes in oxidation number another way.

There are two essentially different methods to balance redox reactions—the oxidation number change method and the ion-electron method. The first is perhaps easier, and the second is somewhat more useful, especially for electrochemical reactions (Sec. 14.6).

Oxidation Number Change

The total of the oxidation numbers gained in a reaction must equal the total of the oxidation numbers lost, since the numbers of electrons "gained" and "lost" must be equal. Therefore, we can balance the species in which the elements that are oxidized and reduced appear by the changes in oxidation numbers. We use the numbers of atoms of each of these elements that will give us equal numbers of electrons gained and lost. If necessary, first balance the number of atoms of the element oxidized and/or the number of atoms of the element reduced. Finally, we balance the rest of the species by inspection, as we did in Chap. 8.

EXAMPLE 14.15.   Balance the following equation:

$$? \, HCl + ? \, HNO_3 + ? \, CrCl_2 \longrightarrow ? \, CrCl_3 + ? \, NO + ? \, H_2O$$

Ans.   By inspecting the oxidation states of all the elements, we find that the Cr goes from $+2$ to $+3$ and the N goes from $+5$ to $+2$. They are the only elements undergoing change in oxidation number.

$$(+2 \longrightarrow +3) = 1$$

$$? \, HCl + ? \, HNO_3 + ? \, CrCl_2 \longrightarrow ? \, CrCl_3 + ? \, NO + ? \, H_2O$$

$$(+5 \longrightarrow +2) = -3$$

To balance the oxidation numbers lost and gained, we need three $CrCl_2$ and three $CrCl_3$ for each N atom reduced:

$$? \, HCl + 1 \, HNO_3 + 3 \, CrCl_2 \longrightarrow 3 \, CrCl_3 + 1 \, NO + ? \, H_2O$$

It is now easy to balance the HCl by balancing the Cl atoms and to balance the H_2O by balancing the O atoms:

$$3 \, HCl + 1 \, HNO_3 + 3 \, CrCl_2 \longrightarrow 3 \, CrCl_3 + 1 \, NO + 2 \, H_2O$$

or

$$3 \, HCl + HNO_3 + 3 \, CrCl_2 \longrightarrow 3 \, CrCl_3 + NO + 2 \, H_2O$$

We check to see that there are now four H atoms on each side, and the equation is balanced.

EXAMPLE 14.16.   Balance the following equation:

$$HCl + K_2CrO_4 + H_2C_2O_4 \longrightarrow CO_2 + CrCl_3 + KCl + H_2O$$

Ans.   Here, Cr is reduced and C is oxidized. Before attempting to balance the oxidization numbers gained and lost, we have to balance the number of carbon atoms. Before the carbon atoms are balanced, we don't know whether to work with one carbon (as in the CO_2) or two (as in the $H_2C_2O_4$).

$$HCl + K_2CrO_4 + H_2C_2O_4 \longrightarrow 2 \, CO_2 + CrCl_3 + KCl + H_2O$$

Now proceed as before:

$$(+6 \longrightarrow +3) = -3$$

$$HCl + K_2CrO_4 + H_2C_2O_4 \longrightarrow 2 \, CO_2 + CrCl_3 + KCl + H_2O$$

$$2(+3 \longrightarrow +4) = +2$$

$$HCl + 2 \, K_2CrO_4 + 3 \, H_2C_2O_4 \longrightarrow 6 \, CO_2 + 2 \, CrCl_3 + KCl + H_2O$$

Balance H_2O from the number of O atoms, KCl by the number of K atoms, and finally HCl by the number of Cl or H atoms. Check.

$$10\,HCl + 2\,K_2CrO_4 + 3\,H_2C_2O_4 \longrightarrow 6\,CO_2 + 2\,CrCl_3 + 4\,KCl + 8\,H_2O$$

It is often easier to balance redox equations in net ionic form than in overall form.

EXAMPLE 14.17.   Balance:

$$H^+ + CrO_4{}^{2-} + H_2C_2O_4 \longrightarrow CO_2 + Cr^{3+} + H_2O$$

Ans.   First balance the C atoms. Then balance the elements changing oxidation state. Then balance the rest of the atoms by inspection.

$$10\,H^+ + 2\,CrO_4{}^{2-} + 3\,H_2C_2O_4 \longrightarrow 6\,CO_2 + 2\,Cr^{3+} + 8\,H_2O$$

The net charge (6+) and the numbers of atoms of each element are equal on the two sides.

Ion-Electron, Half-Reaction Method

In the ion-electron method of balancing redox equations, an equation for the oxidation half-reaction and one for the reduction half-reaction are written and balanced separately. Only when each of these is complete and balanced are the two combined into one complete equation for the reaction as a whole. It is worthwhile to balance the half-reactions separately since the two half-reactions can be carried out in separate vessels if they are suitably connected electrically. (See Sec. 14.6.) In general, net ionic equations are used in this process; certainly some ions are required in each half-reaction. In the equations for the two half-reactions, electrons appear explicitly; in the equation for the complete reaction—the combination of the two half-reactions—no electrons are included.

During the balancing of redox equations by the ion-electron method, species may be added to one side of the equation or the other. These species are present in the solution, and their inclusion in the equation indicates that they also react. Water is a prime example. When it reacts or is produced, the chemist is not likely to notice that there is less or more water present. In general, all the formulas for the atoms that change oxidation number will be given to you in equations to balance. Sometimes, the formula of a compound or ion of oxygen will be omitted, but rarely will the formula for a compound or ion of some other element not be provided. If a formula is not given, see if you can figure out what that formula might be by considering the possible oxidation states (Sec. 14.3).

EXAMPLE 14.18.   Guess a formula for the oxygen-containing product in the following redox reaction:

$$H_2O_2 + I^- \longrightarrow I_2 +$$

Ans.   Since the iodine is oxidized (from -1 to 0), the oxygen must be reduced. It starts out in the -1 oxidation state; it must be reduced to the -2 oxidation state. Water or OH^- is the probable product.

$$H_2O_2 + 2\,I^- \longrightarrow I_2 + 2\,OH^-$$

There are many methods of balancing redox equations by the half-reaction method. One such method is presented here. You should do steps 1 through 5 for one half-reaction and then those same steps for the other half-reaction before proceeding to the rest of the steps.

1. Identify the element(s) oxidized and the element(s) reduced. Start a separate half-reaction for each of these.

2. Balance these elements.

3. Balance the change in oxidation number by adding electrons to the side with the higher total of oxidation numbers. That is, add electrons on the left for a reduction half-reaction and on the right for an oxidation half-reaction. One way to remember on which side to add the electrons is the following mnemonic:

Loss of Electrons is Oxidation (LEO).

Gain of Electrons is Reduction (GER). "LEO the Lion says 'GER'."

4. In acid solution, balance the net charge with hydrogen ions, H^+.

5. Balance the hydrogen and oxygen atoms with water.

6. Multiply every item in one or both equations by small integers, if necessary, so that the number of electrons is the same in each. The same small integer is used throughout each half-reaction, and is different from that used in the other half-reaction. Then add the two half-reactions.

7. Cancel all species that appear on both sides of the equation. All the electrons must cancel out in this step, and often some hydrogen ions and water molecules also cancel.

8. Check to see that atoms of all the elements are balanced and that the net charge is the same on both sides of the equation.

Note that all the added atoms in steps 4 and 5 have the same oxidation number as the atoms already in the equation. The atoms changing oxidation number have already been balanced in steps 1 and 2.

EXAMPLE 14.19. Balance the following equation by the ion-electron, half-reaction method:

$$Cr^{2+} + H^+ + NO_3^- \longrightarrow Cr^{3+} + NO + H_2O$$

Ans. Step 1: $Cr^{2+} \longrightarrow Cr^{3+}$ $NO_3^- \longrightarrow NO$

Step 2: Cr already balanced.   N already balanced.

Step 3: $Cr^{2+} \longrightarrow Cr^{3+} + e^-$ $3\,e^- + NO_3^- \longrightarrow NO$

Step 4: $Cr^{2+} \longrightarrow Cr^{3+} + e^-$ $4\,H^+ + 3\,e^- + NO_3^- \longrightarrow NO$

Step 5: $Cr^{2+} \longrightarrow Cr^{3+} + e^-$ $4\,H^+ + 3\,e^- + NO_3^- \longrightarrow NO + 2\,H_2O$

Step 6: Multiplying by 3 :

$$3\,Cr^{2+} \longrightarrow 3\,Cr^{3+} + 3\,e^-$$

Adding:

$$3\,Cr^{2+} + 4\,H^+ + 3\,e^- + NO_3^- \longrightarrow NO + 2\,H_2O + 3\,Cr^{3+} + 3\,e^-$$

Step 7: $3\,Cr^{2+} + 4\,H^+ + NO_3^- \longrightarrow NO + 2\,H_2O + 3\,Cr^{3+}$

Step 8: The equation is balanced. There are three Cr atoms, four H atoms, one N atom, and three O atoms on each side. The net charge on each side is 9+.

EXAMPLE 14.20. Complete and balance the following equation in acid solution:

$$Cr_2O_7^{2-} + Cl_2 \longrightarrow ClO_4^- + Cr^{3+}$$

Ans. Step 1: $Cr_2O_7^{2-} \longrightarrow Cr^{3+}$ $Cl_2 \longrightarrow ClO_4^-$

Step 2: $Cr_2O_7^{2-} \longrightarrow 2\,Cr^{3+}$ $Cl_2 \longrightarrow 2\,ClO_4^-$

Step 3:   2 atoms are reduced 3 units each   2 atoms are oxidized 7 units each

$$6\,e^- + Cr_2O_7^{2-} \longrightarrow 2\,Cr^{3+}$$ $Cl_2 \longrightarrow 2\,ClO_4^- + 14\,e^-$

Step 4: $14\,H^+ + 6\,e^- + Cr_2O_7^{2-} \longrightarrow 2\,Cr^{3+}$ $Cl_2 \longrightarrow 2\,ClO_4^- + 14\,e^- + 16\,H^+$

Step 5: $14\,H^+ + 6\,e^- + Cr_2O_7^{2-} \longrightarrow 2\,Cr^{3+} + 7\,H_2O$ $8\,H_2O + Cl_2 \longrightarrow 2\,ClO_4^- + 14\,e^- + 16\,H^+$

Step 6: $7\,[14\,H^+ + 6\,e^- + Cr_2O_7^{2-} \longrightarrow 2\,Cr^{3+} + 7\,H_2O]$ $3\,[8\,H_2O + Cl_2 \longrightarrow 2\,ClO_4^- + 14\,e^- + 16\,H^+]$

$98\,H^+ + 42\,e^- + 7\,Cr_2O_7^{2-} \longrightarrow 14\,Cr^{3+} + 49\,H_2O$ $24\,H_2O + 3\,Cl_2 \longrightarrow 6\,ClO_4^- + 42\,e^- + 48\,H^+$

Step 7: $24\,H_2O + 3\,Cl_2 + 98\,H^+ + 42\,e^- + 7\,Cr_2O_7^{2-} \longrightarrow 14\,Cr^{3+} + 49\,H_2O + 6\,ClO_4^- + 42\,e^- + 48\,H^+$

$3\,Cl_2 + 50\,H^+ + 7\,Cr_2O_7^{2-} \longrightarrow 14\,Cr^{3+} + 25\,H_2O + 6\,ClO_4^-$

Step 8: The equation is balanced. There are 6 Cl atoms, 50 H atoms, 14 Cr atoms, and 49 O atoms on each side, as well as a net 36+ charge on each side.

If a reaction is carried out in basic solution, the same process may be followed. After all the other steps have been completed, any H^+ present can be neutralized by adding OH^- ions to each side, creating water and excess OH^- ions. The water created may be combined with any water already on that side or may cancel any water on the other side.

EXAMPLE 14.21. Complete and balance the following equation in basic solution:

$$Fe(OH)_2 + CrO_4{}^{2-} \longrightarrow Fe(OH)_3 + Cr(OH)_3$$

Ans. Step 1: $\qquad Fe(OH)_2 \longrightarrow Fe(OH)_3 \qquad\qquad\qquad\qquad CrO_4{}^{2-} \longrightarrow Cr(OH)_3$

Step 2: Already done. $\qquad\qquad\qquad\qquad\qquad\qquad$ Already done

Step 3: $\qquad Fe(OH)_2 \longrightarrow Fe(OH)_3 + e^- \qquad\qquad 3\,e^- + CrO_4{}^{2-} \longrightarrow Cr(OH)_3$

Step 4: $\qquad Fe(OH)_2 \longrightarrow Fe(OH)_3 + e^- + H^+ \qquad 5\,H^+\;3\,e^- + CrO_4{}^{2-} \longrightarrow Cr(OH)_3$

Step 5: $\quad H_2O + Fe(OH)_2 \longrightarrow Fe(OH)_3 + e^- + H^+ \qquad 5\,H^+ + 3\,e^- + CrO_4{}^{2-} \longrightarrow Cr(OH)_3 + H_2O$

Step 6: $\;3\,H_2O + 3\,Fe(OH)_2 \longrightarrow 3\,Fe(OH)_3 + 3\,e^- + 3\,H^+$

$$3\,H_2O + 3\,Fe(OH)_2 + 5\,H^+ + 3\,e^- + CrO_4{}^{2-} \longrightarrow 3\,Fe(OH)_3 + 3\,e^- + 3\,H^+ + Cr(OH)_3 + H_2O$$

$$2\,H_2O + 3\,Fe(OH)_2 + 2\,H^+ + CrO_4{}^{2-} \longrightarrow 3\,Fe(OH)_3 + Cr(OH)_3$$

To eliminate the H^+, which cannot exist in basic solution, add 2 OH^- to each side, forming 2 H_2O on the left:

$$2\,H_2O + 3\,Fe(OH)_2 + 2\,H_2O + CrO_4{}^{2-} \longrightarrow 3\,Fe(OH)_3 + Cr(OH)_3 + 2\,OH^-$$

Finally,

$$4\,H_2O + 3\,Fe(OH)_2 + CrO_4{}^{2-} \longrightarrow 3\,Fe(OH)_3 + Cr(OH)_3 + 2\,OH^-$$

14.6. ELECTROCHEMISTRY

The interaction of electricity with matter was introduced in Chap. 8, where the electrical decomposition of a melted salt was used to prepare active elements from their compounds. An illustration of an electrolysis was shown in Fig. 5-2. Chemical reactions occur at the two electrodes. The electrode at which oxidation occurs is called the *anode*; the one at which reduction takes place is called the *cathode*. Electricity passes through a *circuit* under the influence of a *potential* or *voltage*, the driving force of the movement of charge. There are two different types of interaction of electricity and matter, as follows:

Electrolysis: Electric current causes chemical reaction.

Galvanic cell action: Chemical reaction causes electric current, as in the use of a battery.

Electrolysis

The following are the requirements for electrolysis:

1. Ions (There must be charged particles to carry current. It might not be the ions that react, however.)
2. Liquid, either a pure liquid or a solution, so that the ions can migrate
3. Source of potential (In a galvanic cell, the chemical reaction is the source of potential, but not in an electrolysis cell.)
4. Mobile ions, complete circuit (including wires to carry electrons), and electrodes (at which the current changes from the flow of electrons to the movement of ions or vice versa)

If you electrolyze a solution containing a compound of a very active metal and/or a very active nonmetal, the water (or other solvent) might be electrolyzed instead of the ion. For example, if you electrolyze molten sodium chloride, you get the free elements:

$$2\,NaCl(l) \xrightarrow{\text{electricity}} 2\,Na(l) + Cl_2(g)$$

However, if you electrolyze a dilute aqueous solution of NaCl, the water is decomposed. (The NaCl is necessary to conduct the current, but neither Na^+ nor Cl^- reacts at the electrodes.)

$$2\,H_2O \xrightarrow[\text{NaCl(dilute)}]{\text{electricity}} 2\,H_2 + O_2$$

If you electrolyze a concentrated solution of NaCl instead, H_2 is produced at the cathode and Cl_2 is produced at the anode:

$$2\,e^- + 2\,H_2O \xrightarrow[\text{NaCl}]{\text{electricity}} H_2 + 2\,OH^-$$

$$2\,Cl^- \xrightarrow[\text{NaCl}]{\text{electricity}} Cl_2 + 2\,e^-$$

It is obvious that the reaction conditions are very important to the products.

Electrolysis is used in a wide variety of ways. Three examples follow: (1) Electrolysis cells are used to produce very active elements in their elemental form. The aluminum industry is based on the electrolytic reduction of aluminum oxide, for example. (2) Electrolysis may be used to electroplate objects. A thin layer of metal, such as silver, can be deposited on other metals, such as steel, by electrodeposition (Fig. 14-2). (3) Electrolysis is also used to purify metals, such as copper. Copper is thus made suitable to conduct electricity. The anode is made out of the impure material; the cathode is made from a thin piece of pure copper. Under carefully controlled conditions, copper goes into solution at the anode, but less active metals, notably silver and gold, fall to the bottom of the container. The copper ion deposits on the cathode, but more active metals stay in solution. Thus very pure copper is produced. The pure copper turns out to be less expensive than the impure copper, which is not too surprising when you think about it. (Which would you expect to be more expensive, pure copper or a copper-silver-gold mixture?)

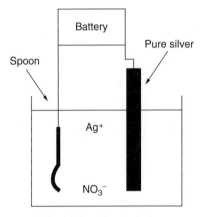

Fig. 14-2.   Electroplating

Galvanic Cells

When you place a piece of zinc metal into a solution of $CuSO_4$, you expect a chemical reaction because the more active zinc displaces the less active copper from its compound (Sec. 8.3). This is an oxidation-reduction reaction, involving transfer of electrons from zinc to copper.

$$Zn \longrightarrow Zn^{2+} + 2\,e^-$$
$$Cu^{2+} + 2\,e^- \longrightarrow Cu$$

It is possible to carry out these same half-reactions in different places if we connect them suitably. We must deliver the electrons from Zn to Cu^{2+}, and we must have a complete circuit. The apparatus is shown in Fig. 14-3. A galvanic cell with this particular combination of reactants is called a *Daniell cell*. The pieces of zinc and copper serve as electrodes, at which chemical reaction takes place. It is at the electrodes that the electron current is changed to an ion current or vice versa. The salt bridge is necessary to complete the circuit. If it were not there, the buildup of charge in each beaker (positive in the left, negative in the right) would stop the reaction extremely quickly (less than 1 s). The same chemical reactions are taking place in this apparatus as would take place if we dipped zinc metal in $CuSO_4$ solution, but the zinc half-reaction is taking place in the left beaker and the copper half-reaction is taking place in the right beaker. Electrons flow from left to right in the wire, and they could be made to do electrical work, such as lighting a small bulb. To keep the beakers from acquiring a charge, cations flow through the salt bridge toward the right and anions flow to the left. The salt bridge is filled with a solution of an unreacting salt, such as KNO_3. The oxidation-reduction reaction provides the potential to produce the current in a complete circuit.

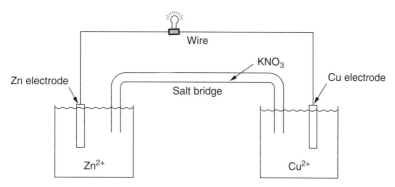

Fig. 14-3. Daniell cell

One such combination of anode and cathode is called a *cell*. Theoretically, any spontaneous oxidation-reduction reaction can be made to produce a galvanic cell. A combination of cells is a *battery*.

We are used to the convenience of cells (more usually in the form of batteries) to power flashlights, cars, portable radios, and other devices requiring electric power. Many different kinds of cells are available, each with some advantages and some disadvantages. The desirable features of a practical cell include low cost, light weight, relatively constant potential, rechargeability, and long life. Not all the desirable features can be incorporated into each cell.

The lead storage cell (six of which make up the lead storage battery commonly used in automobiles) is discussed as an example of a practical cell. The cell, diagrammed in Fig. 14-4, consists of a lead electrode and a lead(IV) oxide electrode immersed in relatively concentrated H_2SO_4 in a single container. When the cell delivers power (when it is used), the electrodes react as follows:

$$Pb + H_2SO_4 \longrightarrow PbSO_4(s) + 2\,H^+ + 2\,e^-$$

$$2\,H^+ + 2\,e^- + PbO_2 + H_2SO_4 \longrightarrow PbSO_4(s) + 2\,H_2O$$

The solid $PbSO_4$ formed in each reaction adheres to the electrode. The H_2SO_4 becomes diluted in two ways: some of it is used up, and some water is formed. The electrons going from anode to cathode do the actual work for which the electric power source is employed, such as starting the car. Both electrodes are placed in the same solution; no salt bridge is needed. This is possible because both the oxidizing agent and the reducing agent, as well as the products of the oxidation and reduction, are solids, so they cannot migrate to the other electrode and react directly.

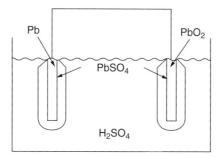

Fig. 14-4. Lead storage cell

The lead storage cell may be recharged by forcing electrons back the other way. Each electrode reaction is reversed, and Pb, PbO_2, and H_2SO_4 are produced again. In a car, the recharge reaction takes place each time the car is driven and the alternator changes some of the mechanical energy of the engine to electric energy and distributes it to the battery, and much less often when the battery is recharged at a gasoline station.

Solved Problems

INTRODUCTION

14.1. (a) What is the formula of a compound of two ions: X^+ and Y^{2-}? (b) What is the formula of a covalent compound of two elements: "W" with an oxidation state of $+1$ and "Z" with an oxidation state of -2?

> *Ans.*   We treat the oxidation states in part b just like the charges in part a. In this manner, we can predict formulas for ionic and covalent compounds. (a) X_2Y (b) W_2Z.

ASSIGNING OXIDATION NUMBERS

14.2. Show that rules 2 and 3 (Sec. 14.2) are corollaries of rule 1.

> *Ans.*   Rule 2: Uncombined elements have zero charges, and so the oxidation numbers must add up to zero. Since all the atoms are the same, all the oxidation numbers must be the same—0. Rule 3: For monatomic ions, the oxidation numbers of all the atoms add up to the charge on the ion. Since there is only one atom (it is monatomic), the oxidation number of that atom must add up to the charge on the ion; that is, it is equal to the charge on the ion.

14.3. Draw an electron dot diagram for H_2O_2. Assign an oxidation number to oxygen on this basis. Compare this number with that assigned by rule 6 (Sec. 14.2).

> *Ans.*
>
> $$H:\overset{..}{\underset{..}{O}}:\overset{..}{\underset{..}{O}}:H$$
>
> | | |
> |---|---|
> | Free atom | 6 |
> | − Controlled | − 7 |
> | Oxidation number | − 1 |
>
> The electrons shared between the oxygen atoms are counted one for each atom. Peroxide oxygen is assigned an oxidation state of -1 by rule 6 also.

14.4. What is the sum of the oxidation numbers of all the *atoms* in the following compounds or ions? (a) PO_4^{3-}, (b) VO_2^+, (c) ClO_2^-, (d) $Cr_2O_7^{2-}$, (e) $SiCl_4$, and (f) NaCl.

> *Ans.*   The sum equals the charge on the species in each case: (a) -3 (b) $+1$ (c) -1 (d) -2 (e) 0   (f) 0

14.5. Determine the oxidation numbers for the underlined elements: $(\underline{V}O_2)_3\underline{P}O_4$.

> *Ans.*   We recognize the phosphate ion, PO_4^{3-}. Each VO_2 ion therefore must have a 1+ charge. The oxidation numbers are $+5$ for P and $+5$ for V.

14.6. What is the oxidation number of chlorine in each of the following? (a) Cl_2O_3, (b) ClO_4^-, and (c) ClF_5.

> *Ans.*   (a) $+3$ (b) $+7$ (c) $+5$

14.7. Determine the oxidation number for the underlined element: (a) $\underline{P}OCl_3$, (b) $H\underline{N}O_2$, (c) $Na_2\underline{S}O_4$, (d) $\underline{P}Cl_5$, and (e) $\underline{N}_2O_3$.

> *Ans.*   (a) $+5$ (b) $+3$ (c) $+6$ (d) $+5$ (e) $+3$

14.8. Determine the oxidation number for the underlined element: (a) $\underline{Cl}O^-$, (b) $\underline{P}O_4^{3-}$, (c) $\underline{S}O_4^{2-}$, and (d) $\underline{V}O^{2+}$.

> *Ans.*   (a) $+1$ (b) $+5$ (c) $+6$ (d) $+4$

14.9. What is the oxidation number of Si in $Si_6O_{18}^{12-}$?

> *Ans.*
> $$6x + 18(-2) = -12 \qquad x = +4$$

14.10. What oxyacid of nitrogen can be prepared by adding water to N_2O_5? *Hint:* Both compounds have nitrogen in the same oxidation state.

Ans. HNO_3

$$H_2O + N_2O_5 \longrightarrow 2\,HNO_3$$

PERIODIC RELATIONSHIPS OF OXIDATION NUMBERS

14.11. Predict the formulas of two compounds of each of the following pairs of elements: (*a*) S and O, (*b*) Cl and O, (*c*) P and S, (*d*) P and F, (*e*) I and F, and (*f*) S and F.

Ans.   The first element in each part is given in its highest oxidation state and in an oxidation state 2 less than the highest. The second element is in its minimum oxidation state.   (*a*) SO_3 and SO_2 (*b*) Cl_2O_7 and Cl_2O_5 (*c*) P_2S_5 and P_2S_3 (*d*) PF_5 and PF_3 (*e*) IF_7 and IF_5 (*f*) SF_6 and SF_4

14.12. Predict the formulas of four fluorides of iodine.

Ans. IF_7, IF_5, IF_3 and IF. The oxidation states of iodine in these compounds correspond to the maximum oxidation state for a group VII element and to states 2, 4, and 6 lower. (See Fig. 14-1.)

14.13. Write the formulas for two monatomic ions for each of the following metals: (*a*) Co, (*b*) Tl, (*c*) Sn, and (*d*) Cu.

Ans.   (*a*) Co^{3+} and Co^{2+} (the oxidation states of transition metals very in steps of one.)   (*b*) Tl^{3+} and Tl^+ (the maximum oxidation state of a group III element and the state 2 less than the maximum.)   (*c*) Sn^{4+} and Sn^{2+} (the maximum oxidation state of a group IV element and the state 2 less than the maximum.)   (*d*) Cu^+ and Cu^{2+} (the maximum oxidation state for the coinage metals is greater than the group number.)

OXIDATION NUMBERS IN INORGANIC NOMENCLATURE

14.14. Name NO_2 and N_2O_4, using the Stock system. Explain why the older system using prefixes is still useful.

Ans.   Both compounds have nitrogen in the +4 oxidation state, so if we call NO_2 nitrogen(IV) oxide, what do we call N_2O_4? We actually use the older system for N_2O_4—dinitrogen tetroxide.

BALANCING OXIDATION-REDUCTION EQUATIONS

14.15. Why is it possible to add H^+ and/or H_2O to an equation for a reaction carried out in aqueous acid solution when none visibly appears or disappears?

Ans.   The H_2O and H^+ are present in excess in the solution. Therefore, they can react or be produced without the change being noticed.

14.16. How many electrons are involved in a reaction of one atom with a change of oxidation number from (*a*) +2 to −3 and (*b*) +5 to −2?

Ans.   (*a*) $2 - (-3) = 5$ 5 electrons are involved   (*b*) $5 - (-2) = 7$ 7 electrons are involved

14.17. Identify (*a*) the oxidizing agent, (*b*) the reducing agent, (*c*) the element oxidized, and (*d*) the element reduced in the following reaction:

$$8\,H^+ + CrO_4{}^{2-} + 3\,Co^{2+} \longrightarrow Cr^{3+} + 3\,Co^{3+} + 4\,H_2O$$

Ans.   (*a*) $CrO_4{}^{2-}$ (*b*) Co^{2+} (*c*) Co   (*d*) Cr.
An element in the reducing agent is oxidized; an element in the oxidizing agent is reduced.

14.18. Balance the equation for the reduction of HNO_3 to NH_4NO_3 by Mn by the oxidation number change method. Add other compounds as needed.

Ans. $HNO_3 + Mn \longrightarrow NH_4NO_3 + Mn(NO_3)_2$

The manganese goes up two oxidation numbers, and some of the nitrogen atoms are reduced from $+5$ to -3:

$$(0 \longrightarrow +2) = +2$$

$$HNO_3 + Mn \longrightarrow NH_4NO_3 + Mn(NO_3)_2$$

$$(+5 \longrightarrow -3) = -8$$

It takes four Mn atoms per N atom reduced:

$$HNO_3 + 4\,Mn \longrightarrow 1\,NH_4NO_3 + 4\,Mn(NO_3)_2$$

There are other N atoms that were not reduced, but still are present in the nitrate ions. Additional HNO_3 is needed to provide for these.

$$10\,HNO_3 + 4\,Mn \longrightarrow NH_4NO_3 + 4\,Mn(NO_3)_2$$

Water is also produced, needed to balance both the H and O atoms:

$$10\,HNO_3 + 4\,Mn \longrightarrow NH_4NO_3 + 4\,Mn(NO_3)_2 + 3\,H_2O$$

The equation is now balanced, having 10 H atoms, 10 N atoms, 30 O atoms, and 4 Mn atoms on each side.

ELECTROCHEMISTRY

14.19. Even if sodium metal were produced by the electrolysis of aqueous NaCl, what would happen to the sodium produced in the water?

Ans.   The sodium is so active that it would react immediately with the water. No elementary sodium should ever be expected from a water solution.

14.20. Explain why Al cannot be produced from its salts in aqueous solution.

Ans.   Al is too active; the water will be reduced instead.

14.21. Can a Daniell cell be recharged?

Ans.   No. If the Daniell cell were to be recharged, the Cu^{2+} ions would get into the zinc half-cell through the salt bridge. There, they would react directly with the zinc electrode, and the cell would be destroyed.

14.22. Explain why a Daniell cell cannot be placed in a single container, like the lead storage cell.

Ans.   The copper(II) ions would migrate to the zinc electrode and be reduced to copper metal directly. The zinc electrode would become copper-plated, and the cell would not function.

14.23. Write a balanced chemical equation for (*a*) the direct reaction of zinc with copper(II) sulfate and (*b*) the overall reaction in a Daniell cell containing $ZnSO_4$ and $CuSO_4$.

Ans.   (*a*) and (*b*) $Zn + CuSO_4 \longrightarrow ZnSO_4 + Cu$

14.24. The electrolysis of brine (concentrated NaCl solution) produces hydrogen at the cathode and chlorine at the anode. Write a net ionic equation for each half-reaction and the total reaction. What other chemical is produced in this commercially important process?

Ans.
$$2\,H_2O + 2\,e^- \longrightarrow H_2 + 2\,OH^-$$
$$2\,Cl^- \longrightarrow Cl_2 + 2\,e^-$$
$$2\,Cl^- + 2\,H_2O \longrightarrow Cl_2 + H_2 + 2\,OH^-$$

The other product is NaOH.

14.25. Must two oxidation-reduction half-reactions be carried out (*a*) in the same location and (*b*) at the same time?

Ans.   (*a*) No, they may be in different locations, as in electrochemical processes   (*b*) Yes. (The electrons cannot appear from nowhere or go nowhere.)

Supplementary Problems

14.26. Explain why we were able to use the charges on the monatomic cations in names in Chap. 6 instead of the required oxidation numbers.

Ans. For monatomic ions, the charge is equal to the oxidation number.

14.27. Determine the oxidation number of oxygen in (*a*) Na_2O_2, (*b*) RbO_2, and (*c*) OF_2.

Ans. (*a*) -1 (*b*) $-\frac{1}{2}$ ($+1 + 2x = 0$, therefore $x = -\frac{1}{2}$) (*c*) $+2$

14.28. Since the reactions in Problem 14.23 are the same, and the Daniell cell produces electric energy, what kind of energy does the direct reaction produce?

Ans. It produces more heat energy than the electrochemical reaction produces.

14.29. Complete and balance the following redox equations:

(*a*) $MnO_4^- + Hg_2^{2+} \longrightarrow Hg^{2+} + Mn^{2+}$ 　　　　 (*d*) $P_4 + OH^- \longrightarrow PH_3 + HPO_3^{2-}$

(*b*) $I^- + H_2O_2 \longrightarrow I_2 + H_2O$ 　　　　　　　 (*e*) $H_2C_2O_4 + MnO_4^- \longrightarrow Mn^{2+} + CO_2$

(*c*) $Br_2 + OH^- \longrightarrow Br^- + BrO_3^-$ 　　　　　　 (*f*) $Cr(OH)_2 + H_2O_2 \longrightarrow Cr(OH)_3$

Ans. (*a*) $16\,H^+ + 2\,MnO_4^- + 5\,Hg_2^{2+} \longrightarrow 10\,Hg_8^{2+} + 2\,Mn^{2+} + 8\,H_2O$

(*b*) $2\,H^+ + 2\,I^- + H_2O_2 \longrightarrow I_2 + 2\,H_2O$

(*c*) $3\,Br_2 + 6\,OH^- \longrightarrow 5\,Br^- + BrO_3^- + 3\,H_2O$

(*d*) $2\,H_2O + P_4 + 4\,OH^- \longrightarrow 2\,PH_3 + 2\,HPO_3^{2-}$

(*e*) $6\,H^+ + 5\,H_2C_2O_4 + 2\,MnO_4^- \longrightarrow 2\,Mn^{2+} + 10\,CO_2 + 8\,H_2O$

(*f*) $2\,Cr(OH)_2 + H_2O_2 \longrightarrow 2\,Cr(OH)_3$

14.30. Which of the equations in Problem 14.29 represent reactions in basic solution? How can you tell?

Ans. Reactions (*c*), (*d*), and (*f*) occur in basic solution. The presence of OH^- ions shows immediately that the solution is basic. (The presence of NH_3 also indicates basic solution; in acid solution, this base would react to form NH_4^+.) The other reactions are not in base; H^+ is present. Also, if they were in base, the metal ions would react to form insoluble hydroxides and the simple ions would not be present.

14.31. Complete and balance the following equations:

(*a*) $Zn + H^+ + NO_3^- \longrightarrow NH_4^+ + Zn^{2+}$ 　　　　 (*b*) $Zn + HNO_3 \longrightarrow NH_4NO_3 + Zn(NO_3)_2$

(*c*) How are these equations related? Which is easier to balance?

Ans. (*a*) 　　　　　　　 $NO_3^- \longrightarrow NH_4^+$ 　　　　　 $Zn \longrightarrow Zn^{2+} + 2\,e^-$

$8\,e^- + NO_3^- \longrightarrow NH_4^+$

$10\,H^+ + 8\,e^- + NO_3^- \longrightarrow NH_4^+$

$10\,H^+ + 8\,e^- + NO_3^- \longrightarrow NH_4^+ + 3\,H_2O$ 　　　 $4\,Zn \longrightarrow 4\,Zn^{2+} + 8\,e^-$

$$10\,H^+ + 4\,Zn + NO_3^- \longrightarrow NH_4^+ + 3\,H_2O + 4\,Zn^{2+}$$

(*b*) Either add $9\,NO_3^-$ to each side of the equation in part *a*:

$$10\,HNO_3 + 4\,Zn \longrightarrow NH_4NO_3 + 3\,H_2O + 4\,Zn(NO_3)_2$$

or

$2\,HNO_3 \longrightarrow NH_4NO_3$ 　　　　　　 $Zn \longrightarrow Zn(NO_3)_2 + 2\,e^-$

$8\,e^- + 2\,HNO_3 \longrightarrow NH_4NO_3$ 　　　　 $2\,HNO_3 + Zn \longrightarrow Zn(NO_3)_2 + 2\,e^-$

$8\,H^+ + 8\,e^- + 2\,HNO_3 \longrightarrow NH_4NO_3$ 　　 $2\,HNO_3 + Zn \longrightarrow Zn(NO_3)_2 + 2\,e^- + 2\,H^+$

$8\,H^+ + 8\,e^- + 2\,HNO_3 \longrightarrow NH_4NO_3 + 3\,H_2O$ 　 $8\,HNO_3 + 4\,Zn \longrightarrow 4\,Zn(NO_3)_2 + 8\,e^- + 8\,H^+$

$$10\,HNO_3 + 4\,Zn \longrightarrow 4\,Zn(NO_3)_2 + NH_4NO_3 + 3\,H_2O$$

(c) Part *a* is the net ionic equation for part *b*. It is easier to balance part *a*. To balance part *b*, either amend the balanced net ionic equation of part *a*, or do part *b*, starting from scratch. (There must always be at least one type of ion represented in a balanced half-reaction equation.)

14.32. Balance the following equation by the oxidation number change method:

$$NH_3 + O_2 \longrightarrow NO + H_2O$$

Ans.   The O_2 is reduced to the -2 oxidation state; both products contain the reduction product.

$$(-3 \rightarrow +2) = +5$$
$$NH_3 + O_2 \longrightarrow NO + H_2O$$
$$2(0 \rightarrow -2) = -4$$

We know that five O_2 molecules are required, but we still do not know from this calculation how many of the oxygen atoms go to NO or H_2O.

$$4\,NH_3 + 5\,O_2 \longrightarrow 4\,NO + H_2O$$

Finish by inspection:

$$4\,NH_3 + 5\,O_2 \longrightarrow 4\,NO + 6\,H_2O$$

14.33. Complete and balance the following equations:

(a) $H_2O_2 + Sn^{2+} \longrightarrow Sn^{4+} + H_2O$

(b) $Br^- + BrO_3^- \longrightarrow Br_2 + H_2O$

(c) $Ce^{4+} + SO_3^{2-} \longrightarrow Ce^{3+} + SO_4^{2-}$

(d) $Zn + OH^- \longrightarrow Zn(OH)_4^{2-} + H_2$

(e) $Cu_2O + H^+ \longrightarrow Cu + Cu^{2+} + H_2O$

(f) $H_2SO_4(conc) + Zn \longrightarrow H_2S + Zn^{2+}$

(g) $HNO_3 + Pb(NO_3)_2 \longrightarrow Pb(NO_3)_4 + NO$

(h) $H^+ + NO_3^- + Sn^{2+} \longrightarrow NO + Sn^{4+}$

(i) $SO_4^{2-} + I^- + H^+ \longrightarrow SO_2 + I_2 + H_2O$

(j) $Cr^{3+} + MnO_4^- \longrightarrow Cr_2O_7^{2-} + Mn^{2+}$

(k) $Ag^+ + S_2O_3^{2-} \longrightarrow S_4O_6^{2-} + Ag$

Ans.   (a)
$$\overset{+2}{H_2O_2 + Sn^{2+}} \longrightarrow Sn^{4+} + 2\,H_2O$$
$$2(-1 \rightarrow -2) = -2$$
$$2\,H^+ + H_2O_2 + Sn^{2+} \longrightarrow Sn^{4+} + 2\,H_2O$$

(b) $6\,H^+ + 5\,Br^- + BrO_3^- \longrightarrow 3\,Br_2 + 3\,H_2O$

(c) $e^- + Ce^{4+} \longrightarrow Ce^{3+}$
$$\underline{H_2O + SO_3^{2-} \longrightarrow SO_4^{2-} + 2\,e^- + 2\,H^+}$$
$$2\,Ce^{4+} + H_2O + SO_3^{2-} \longrightarrow SO_4^{2-} + 2\,H^+ + 2\,Ce^{3+}$$

(d) $4\,OH^- + Zn \longrightarrow Zn(OH)_4^{2-} + 2\,e^-$
$$\underline{2\,e^- + 2\,H_2O \longrightarrow 2\,OH^- + H_2}$$
$$2\,H_2O + 2\,OH^- + Zn \longrightarrow Zn(OH)_4^{2-} + H_2$$

(e) $Cu_2O + 2\,H^+ \longrightarrow Cu + Cu^{2+} + H_2O$

Cu^+ is not stable in aqueous solution. Cu(I) is stable in solid compounds like Cu_2O, but when that reacts with an acid, the Cu^+ disproportionates—reacts with itself—to produce a lower and a higher oxidation state.

(f)
$$8\,e^- + 8\,H^+ + H_2SO_4 \longrightarrow H_2S + 4\,H_2O$$
$$\underline{Zn \longrightarrow Zn^{2+} + 2\,e^-}$$
$$4\,Zn + 8\,H^+ + H_2SO_4 \longrightarrow H_2S + 4\,H_2O + 4\,Zn^{2+}$$

$(+2 \rightarrow +4) = +2$

(g) $HNO_3 + Pb(NO_3)_2 \longrightarrow Pb(NO_3)_4 + NO$

$(+5 \rightarrow +2) = -3$

$6\,HNO_3 + 2\,HNO_3 + 3\,Pb(NO_3)_2 \longrightarrow 3\,Pb(NO_3)_4 + 2\,NO + 4\,H_2O$

(for extra
nitrate ions)

$8\,HNO_3 + 3\,Pb(NO_3)_2 \longrightarrow 3\,Pb(NO_3)_4 + 2\,NO + 4\,H_2O$

(h) $8\,H^+ + 2\,NO_3^- + 3\,Sn^{2+} \longrightarrow 2\,NO + 3\,Sn^{4+} + 4\,H_2O$

(i) $SO_4^{2-} + 2\,I^- + 4\,H^+ \longrightarrow SO_2 + I_2 + 2\,H_2O$

(j) $10\,Cr^{3+} + 6\,MnO_4^- + 11\,H_2O \longrightarrow 5\,Cr_2O_7^{2-} + 6\,Mn^{2+} + 22\,H^+$

(k) $2\,Ag^+ + 2\,S_2O_3^{2-} \longrightarrow S_4O_6^{2-} + 2\,Ag$

14.34. Consider the following part of an equation:

$$NO_2 + MnO_4^- \longrightarrow Mn^{2+} +$$

(a) If one half-reaction is a reduction, what must the other half-reaction be? (b) To what oxidation state can the nitrogen be changed? (c) Complete and balance the equation.

Ans.   (a)   Since one half-reaction is a reduction, the other half-reacttion must be an oxidation.

(b)   The maximum oxidation state for nitrogen is $+5$, because nitrogen is in periodic group V. Since it starts out in oxidation number $+4$, it must be oxidized to $+5$.

(c)
$$NO_2 + MnO_4^- \longrightarrow Mn^{2+} + NO_3^-$$
$$5\,NO_2 + MnO_4^- \longrightarrow Mn^{2+} + 5\,NO_3^-$$
$$H_2O + 5\,NO_2 + MnO_4^- \longrightarrow Mn^{2+} + 5\,NO_3^- + 2\,H^+$$

14.35. What is the oxidation number of sulfur in $S_2O_8^{2-}$, the peroxydisulfate ion?

Ans.   If you calculate the oxidation number assuming that the oxygen atoms are normal oxide ions, you get an answer of $+7$, which is greater than the maximum oxidation number for sulfur. That must mean that one of the pairs of oxygen atoms is a peroxide, and thus the sulfur must be in its highest oxidation number, $+6$.

14.36. Which of the following reactions (indicated by unbalanced equations) occur in acid solution and which occur in basic solution?

(a) $HNO_3 + Cu \longrightarrow NO + Cu^{2+}$

(b) $CrO_4^{2-} + Fe(OH)_2 \longrightarrow Cr(OH)_3 + Fe(OH)_3$

(c) $N_2H_4 + I^- \longrightarrow NH_4^+ + I_2$

Ans.   (a) Acid solution (HNO_3 would not be present in base.)   (b) Basic solution (The hydroxides would not be present in acid.)   (c) Acid solution (NH_3 rather than NH_4^+ would be present in base.)

14.37. Complete and balance the following equations:

(a) $Cl^- + MnO_2 + H^+ \longrightarrow Mn^{2+} + H_2O + Cl_2$ (e) $V^{2+} + H_3AsO_4 \longrightarrow HAsO_2 + VO^{2+}$

(b) $ClO^- \longrightarrow Cl^- + ClO_3^-$ (f) $Hg_2^{2+} + CN^- \longrightarrow C_2N_2 + Hg$

(c) $Pb + PbO_2 + SO_4^{2-} \longrightarrow PbSO_4 + H_2O$ (g) $VO^{2+} + AsO_4^{3-} \longrightarrow AsO_2^- + VO_2^+$

(d) $I_3^- + Co(CN)_6^{4-} \longrightarrow Co(CN)_6^{3-} + I^-$

Ans.   (a) $2\,Cl^- + MnO_2 + 4\,H^+ \longrightarrow Mn^{2+} + 2\,H_2O + Cl_2$

(b) $3\,ClO^- \longrightarrow 2\,Cl^- + ClO_3^-$

(c) $4\,H^+ + Pb + PbO_2 + 2\,SO_4{}^{2-} \longrightarrow 2\,PbSO_4 + 2\,H_2O$

(d) $I_3{}^- + 2\,Co(CN)_6{}^{4-} \longrightarrow 2\,Co(CN)_6{}^{3-} + 3\,I^-$

(e) $V^{2+} + H_3AsO_4 \longrightarrow HAsO_2 + VO^{2+} + H_2O$

(f) $Hg_2{}^{2+} + 2\,CN^- \longrightarrow C_2N_2 + 2\,Hg$

(g) $2\,VO^{2+} + AsO_4{}^{3-} \longrightarrow AsO_2{}^- + 2\,VO_2{}^+$

14.38. The oxidizing ability of H_2SO_4 depends on its concentration. Which element is reduced by reaction of Zn on H_2SO_4 in each of the following reactions?

$$Zn + H_2SO_4(\text{dilute}) \longrightarrow ZnSO_4 + H_2$$

$$4\,Zn + 5\,H_2SO_4(\text{conc}) \longrightarrow H_2S + 4\,H_2O + 4\,ZnSO_4$$

Ans.   In the first reaction, hydrogen is reduced. In the second reaction, sulfur is reduced.

14.39. Complete and balance the following equation:

$$Sb_2S_3 + HNO_3 \longrightarrow H_3SbO_4 + SO_2 + NO$$

Ans. $3\,Sb_2S_3 + 22\,HNO_3 \longrightarrow 6\,H_3SbO_4 + 9\,SO_2 + 22\,NO + 2\,H_2O$

14.40. What is the maximum oxidation state of fluorine in any compound?

Ans.   The only oxidation state of fluorine in a compound is -1; it is the most electronegative element. (It always has control of both shared electrons, except in the element F_2.)

14.41. Calculate the oxidation number of carbon in (a) CH_2O and (b) CH_2F_2.

Ans. (a) 0 (b) 0

14.42. What is the more likely formula for bismuth in the $+5$ oxidation state—Bi^{5+} or $BiO_3{}^-$?

Ans. $BiO_3{}^-$. There are no 5+ monatomic ions.

14.43. Explain why direct current (dc) rather than alternating current (ac) is used for electrolysis. Why is direct current used in cars?

Ans.   In direct current, the electrons flow in the same direction all the time. In alternating current, the electrons flow one way for a short period of time (typically $\frac{1}{60}$s), and then they flow the other way. To get any electrolysis that is not immediately undone, direct current is required. Direct current is also used in cars because cells generate direct current.

CHAPTER 15

Solutions

15.1. QUALITATIVE CONCENTRATION TERMS

Solutions are mixtures, and therefore do not have definite compositions. For example, in a glass of water it is possible to dissolve one teaspoonful of sugar or two or three or more. However, for most solutions there is a limit to how much *solute* will dissolve in a given quantity of *solvent* at a given temperature. The maximum concentration of solute that will dissolve in given quantity of solvent is called the *solubility* of the solute. Solubility depends on temperature. Most solids dissolve in liquids more at high temperatures than at low temperatures, while gases dissolve in cold liquids better than in hot liquids.

A solution in which the concentration of the solute is equal to the solubility is called a *saturated* solution. If the concentration is lower, the solution is said to be *unsaturated*. It is also possible to prepare a *supersaturated* solution, an unstable solution containing a greater concentration of solute than is present in a saturated solution. Such a solution deposits the excess solute if a crystal of the solute is added to it. It is prepared by dissolving solute at one temperature and carefully changing the temperature to a point where the solution is unstable.

EXAMPLE 15.1. A solution at 0°C contains 119 g of sodium acetate per 100 g of water. If more sodium acetate is added, it appears not to dissolve, and no sodium acetate appears to crystallize from solution either. Describe the following solutions as saturated, unsaturated, or supersaturated. (*a*) 105 g sodium acetate in 100 g water at 0°C. (*b*) 142 g sodium acetate in 100 g water at 0°C. (*c*) 1.19 g sodium acetate in 1.00 g water at 0°C.

Ans.   (*a*) The solution is unsaturated; more solute could dissolve.   (*b*) The solution is supersaturated; only 119 g is stable in water at this temperature.   (*c*) The solution is saturated; the *concentration* is the same as that given in the statement of the problem.

EXAMPLE 15.2. How can the solution described in Example 15.1*b* be prepared?

Ans.   The 142 g of sodium acetate is mixed with 100 g of water and heated to nearly 100°C (where 170 g of solute would dissolve). The mixture is stirred until solution is complete, and then the solution is cooled until it gets to 0°C. The solute would crystallize if it could, but this particular solute has difficulty doing so, and thus a supersaturated solution is formed. Adding a crystal of solid sodium acetate allows the excess solute to crystallize around the solid added, and the excess solute precipitates out of the solution, leaving the solution saturated.

15.2. MOLALITY

Molarity (Chap. 11) is defined in terms of the volume of a solution. Since the volume is temperature-dependent, so is the molarity of the solution. Two units of concentration that are independent of temperature are introduced in this chapter. *Molality* is defined as the number of moles of solute per kilogram of solvent in

219

a solution. The symbol for molality is m. Its unit is *molal*. Carefully note the differences between molality and molarity:

1. Molality is defined in terms of kilograms, not liters.
2. Molality is defined in terms of solvent, not solution.
3. The symbol for molality is a lowercase m, not capital M.

Take great care not to confuse molality and molarity.

EXAMPLE 15.3. Calculate the molality of a solution prepared by adding 0.300 mol of solute to 225 g of water.

Ans.
$$\frac{0.300 \text{ mol}}{0.225 \text{ kg}} = 1.33 \, m$$

EXAMPLE 15.4. What mass of H_2O is required to make a 2.50 m solution with 0.511 mol NaCl?

Ans.
$$0.511 \text{ mol NaCl} \left(\frac{1.00 \text{ kg } H_2O}{2.50 \text{ mol NaCl}} \right) = 0.204 \text{ kg } H_2O$$

EXAMPLE 15.5. What mass of 1.25 m CH_3OH solution in water can be prepared with 50.0 g of CH_3OH?

Ans.
$$50.0 \text{ g } CH_3OH \left(\frac{1 \text{ mol } CH_3OH}{32.0 \text{ g } CH_3OH} \right) \left(\frac{1.00 \text{ kg solvent}}{1.25 \text{ mol } CH_3OH} \right) = 1.25 \text{ kg solvent}$$

$$50.0 \text{ g } CH_3OH + 1250 \text{ g } H_2O = 1300 \text{ g solution}$$

Note the importance of keeping track of what materials you are considering, in addition to their units.

EXAMPLE 15.6. Calculate the molarity of a 1.85 m solution of NaCl in water if the density of the solution is 1.04 g/mL.

Ans.   Since neither concentration depends on the quantity of solution under consideration, let us work with a solution containing 1.000 kg H_2O. Then we have

$$1.85 \text{ mol NaCl} \left(\frac{58.5 \text{ g NaCl}}{1 \text{ mol NaCl}} \right) = 108 \text{ g NaCl}$$

The total mass of the solution is therefore 1108 g, and its volume is

$$1108 \text{ g} \left(\frac{1 \text{ mL}}{1.04 \text{ g}} \right) = 1070 \text{ mL} = 1.07 \text{ L}$$

The molarity is

$$\frac{1.85 \text{ mol NaCl}}{1.07 \text{ L}} = 1.73 \, M$$

15.3. MOLE FRACTION

The other temperature-independent concentration unit introduced in this chapter is the mole fraction. The *mole fraction* of a substance in a solution is the ratio of the number of moles of that substance to the total number of moles in the solution. The symbol for mole fraction of A is usually X_A, although some texts use the symbol N_A. Thus, for a solution containing x mol of A, y mol of B, and z mol of C, the mole fraction of A is

$$X_A = \frac{x \text{ mol A}}{x \text{ mol A} + y \text{ mol B} + z \text{ mol C}}$$

EXAMPLE 15.7. What is the mole fraction of CH_3OH in a solution of 12.0 g CH_3OH and 25.0 g H_2O?

Ans.
$$12.0 \text{ g } CH_3OH \left(\frac{1 \text{ mol } CH_3OH}{32.0 \text{ g } CH_3OH} \right) = 0.375 \text{ mol } CH_3OH$$

$$25.0 \text{ g H}_2\text{O}\left(\frac{1 \text{ mol H}_2\text{O}}{18.0 \text{ g H}_2\text{O}}\right) = 1.39 \text{ mol H}_2\text{O}$$

$$X_{CH_3OH} = \frac{0.375 \text{ mol CH}_3\text{OH}}{0.375 \text{ mol CH}_3\text{OH} + 1.39 \text{ mol H}_2\text{O}} = 0.212$$

Since the mole fraction is a ratio of moles (of one substance) to moles (total), the units cancel and mole fraction has no units.

EXAMPLE 15.8.   Show that the total of both mole fractions in a solution of two compounds is equal to 1.

Ans.
$$X_A = \frac{x \text{ mol A}}{x \text{ mol A} + y \text{ mol B}} \qquad X_B = \frac{y \text{ mol B}}{x \text{ mol A} + y \text{ mol B}}$$

$$X_A + X_B = \frac{x \text{ mol A}}{x \text{ mol A} + y \text{ mol B}} + \frac{y \text{ mol B}}{x \text{ mol A} + y \text{ mol B}} = \frac{x \text{ mol A} + y \text{ mol B}}{x \text{ mol A} + y \text{ mol B}} = 1$$

15.4.   EQUIVALENTS

Equivalents are measures of the quantity of a substance present, analogous to moles. The *equivalent* is defined in terms of a chemical reaction. It is defined in one of two different ways, depending on whether an oxidation-reduction reaction or an acid-base reaction is under discussion. For an oxidation-reduction reaction, *one equivalent* is the quantity of a substance that will react with or yield 1 mol of *electrons*. For an acid-base reaction, *one equivalent* is the quantity of a substance that will react with or yield 1 mol of *hydrogen ions or hydroxide ions*. Note that the equivalent is defined in terms of a reaction, not merely in terms of the formula of a compound. Thus, the same mass of the same compound undergoing different reactions can correspond to different numbers of equivalents. The ability to determine the number of equivalents per mole is the key to calculations in this section.

EXAMPLE 15.9.   How many equivalents are there in 82.0 g of H_2SO_3, in each of the following reactions?

(*a*) $H_2SO_3 + NaOH \longrightarrow NaHSO_3 + H_2O$ (*b*) $H_2SO_3 + 2 NaOH \longrightarrow Na_2SO_3 + 2 H_2O$

Ans.   Since 82.0 g of H_2SO_3 is 1.00 mol of H_2SO_3, we will concentrate on that quantity of H_2SO_3. Both of these reactions are acid-base reactions, and so we define the number of equivalents of H_2SO_3 in terms of the number of moles of hydroxide ion with which it reacts. In the first equation, 1 mol of H_2SO_3 reacts with 1 mol of OH^-. By definition, that quantity of H_2SO_3 is 1 equiv. For that equation, 1 equiv of H_2SO_3 is equal to 1 mol of H_2SO_3. In the second equation, 1 mol of H_2SO_3 reacts with 2 mol OH^-. By definition, that quantity of H_2SO_3 is equal to 2 equiv. Thus, in that equation, 2 equiv of H_2SO_3 is 1 mol of H_2SO_3.

(*a*)   1 equiv = 1 mol   (*b*)   2 equiv = 1 mol

We can use these equalities as factors to change moles to equivalents or equivalents to moles, especially to calculate normalities (Sec. 15.5) or equivalent masses (Sec. 15.6).

EXAMPLE 15.10.   How many equivalents are there in 82.0 g of H_2SO_3 in the following reaction?

$$6 H^+ + H_2SO_3 + 3 Zn \longrightarrow H_2S + 3 Zn^{2+} + 3 H_2O$$

Ans.   This reaction is a redox reaction (Chap. 14), and so we define the number of equivalents of H_2SO_3 in terms of the number of moles of electrons with which it reacts. Since no electrons appear explicitly in an overall equation, we will write the half-reaction in which the H_2SO_3 appears:

$$6 H^+ + H_2SO_3 + 6 e^- \longrightarrow H_2S + 3 H_2O$$

It is now apparent that 1 mol of H_2SO_3 reacts with 6 mol e^-, and by definition 6 equiv of H_2SO_3 react with 6 mol e^-. Thus, 6 equiv equal 1 mol in this reaction. Since 82.0 g H_2SO_3 is 1 mol, there are 6 equiv in 82.0 g H_2SO_3.

Some instructors and texts ask for the number of equivalents per mole of an acid or base without specifying a particular reaction. In that case, merely assume that the substance undergoes an acid-base reaction as completely as possible. State that assumption in your answers on examinations.

EXAMPLE 15.11. What is the number of equivalents in 4.00 mol H_2SO_3?

Ans. Assuming that the H_2SO_3 will react with a base to replace both hydrogen atoms, we have reaction (*b*) from Example 15.9, and there are 2 equiv per mole. In 4.00 mol H_2SO_3,

$$4.00 \text{ mol } H_2SO_3 \left(\frac{2 \text{ equiv}}{1 \text{ mol}} \right) = 8.00 \text{ equiv}$$

The major use of equivalents stems from its definition. Once you define the number of equivalents in a certain mass of a substance, you do not need to write the equation for its reaction. That equation has already been used in defining the number of equivalents. Thus, a chemist can calculate the number of equivalents in a certain mass of substance, and technicians can subsequently use that definition without knowing the details of the reaction.

> *One equivalent of one substance in a reaction always reacts with one equivalent of each of the other substances in that reaction.*

EXAMPLE 15.12. How many equivalents of NaOH does the 82.0 g of H_2SO_3 react with in reaction (*b*) of Example 15.9?

Ans. Since there are 2 equiv of H_2SO_3, they must react with 2 equiv of NaOH. Checking, we see that 2 equiv of NaOH liberate 2 mol of OH^- and also react with 2 mol H^+, and thus there are 2 equiv by definition also.

15.5. NORMALITY

Analogous to molarity, *normality* is defined as the number of *equivalents* of solute per liter of solution. Its unit is *normal* with the symbol N. Thus the normality of a solution may be 3.0 normal, denoted 3.0 N.

EXAMPLE 15.13. What is the normality of a solution containing 1.75 equiv in 2.50 L of solution?

Ans.
$$\frac{1.75 \text{ equiv}}{2.50 \text{ L}} = 0.700 \text{ N}$$

The normality of the solution is 0.700 normal, or, stated another way, the solution is 0.700 normal.

Normality is some integral multiple (1, 2, 3, ...) of molarity, since there are always some integral number of equivalents per mole.

EXAMPLE 15.14. What is the normality of 2.25 M H_2SO_3 in each of the following reactions?

(*a*) $H_2SO_3 + NaOH \longrightarrow NaHSO_3 + H_2O$ (*b*) $H_2SO_3 + 2 NaOH \longrightarrow Na_2SO_3 + 2 H_2O$

Ans. We found in Example 15.9 that there were 1 equiv per mol in the first reaction and 2 equiv per mol in the second. We use these factors to solve for normality:

(*a*) $2.25 \text{ } M = \dfrac{2.25 \text{ mol}}{1 \text{ L}} \left(\dfrac{1 \text{ equiv}}{1 \text{ mol}} \right) = \dfrac{2.25 \text{ equiv}}{1 \text{ L}} = 2.25 \text{ N}$

(*b*) $2.25 \text{ } M = \dfrac{2.25 \text{ mol}}{1 \text{ L}} \left(\dfrac{2 \text{ equiv}}{1 \text{ mol}} \right) = \dfrac{4.50 \text{ equiv}}{1 \text{ L}} = 4.50 \text{ N}$

EXAMPLE 15.15. What is the molarity of 1.25 N H_2SO_3 in each of the following reactions?

(*a*) $H_2SO_3 + NaOH \longrightarrow NaHSO_3 + H_2O$ (*c*) $6 H^+ + H_2SO_3 + 6 e^- \longrightarrow H_2S + 3 H_2O$
(*b*) $H_2SO_3 + 2 NaOH \longrightarrow Na_2SO_3 + 2 H_2O$

Ans. Again we use the factors found in Sec. 15.4:

(*a*) $1.25 \text{ } N = \dfrac{1.25 \text{ equiv}}{1 \text{ L}} \left(\dfrac{1 \text{ mol}}{1 \text{ equiv}} \right) = \dfrac{1.25 \text{ mol}}{1 \text{ L}} = 1.25 \text{ M}$

(*b*) $1.25 \text{ } N = \dfrac{1.25 \text{ equiv}}{1 \text{ L}} \left(\dfrac{1 \text{ mol}}{2 \text{ equiv}} \right) = \dfrac{0.625 \text{ mol}}{1 \text{ L}} = 0.625 \text{ M}$

(c) $1.25\ N = \dfrac{1.25\ \text{equiv}}{1\ \text{L}}\left(\dfrac{1\ \text{mol}}{6\ \text{equiv}}\right) = \dfrac{0.208\ \text{mol}}{1\ \text{L}} = 0.208\ M$

We can do concentration problems with normality just as we did them with molarity (Chap. 11). The quantities are expressed in equivalents, of course.

EXAMPLE 15.16.   How many equivalents are present in 3.0 L of 2.0 N solution?

Ans.
$$3.0\ \text{L}\left(\dfrac{2.0\ \text{equiv}}{1\ \text{L}}\right) = 6.0\ \text{equiv}$$

EXAMPLE 15.17.   How many liters of 2.00 N H_3PO_4 does it take to hold 1.80 equiv?

Ans.
$$1.80\ \text{equiv}\left(\dfrac{1\ \text{L}}{2.00\ \text{equiv}}\right) = 0.900\ \text{L}$$

EXAMPLE 15.18.   What is the final concentration of HCl if 2.0 L of 2.2 N HCl and 1.5 L of 3.4 N HCl are mixed and diluted to 5.0 L?

Ans.
$$(2.0\ \text{L})(2.2\ N) = 4.4\ \text{equiv}$$
$$(1.5\ \text{L})(3.4\ N) = 5.1\ \text{equiv}$$
$$\text{total} = 9.5\ \text{equiv}$$
$$\dfrac{9.5\ \text{equiv}}{5.0\ \text{L}} = 1.9\ N$$

Equivalents are especially useful in dealing with stoichiometry problems in solution. Since 1 equiv of one thing reacts with 1 equiv of any other thing in the reaction, it is also true that the volume times the normality of the first thing is equal to the volume times the normality of the second.

EXAMPLE 15.19.   What volume of 2.00 N NaOH is required to neutralize 25.00 mL of 2.70 N H_2SO_4?

Ans.
$$N_1 V_1 = N_2 V_2$$
$$V_2 = \dfrac{N_1 V_1}{N_2} = \dfrac{(2.70\ N\ H_2SO_4)(25.00\ \text{mL})}{2.00\ N\ \text{NaOH}} = 33.8\ \text{mL}$$

Note that less volume of H_2SO_4 is required than of NaOH because its normality is greater.

15.6.   EQUIVALENT MASS

The *equivalent mass* of a substance is the mass in grams of 1 equiv of the substance. The equivalent mass is often a useful property to characterize a substance.

EXAMPLE 15.20.   A new solid acid was prepared in a laboratory; its molar mass was not known. It was titrated with standard base, and the number of moles of base was calculated. Without knowing the formula of the acid, can you tell how many moles of the acid were present in a certain mass of acid? Can you tell how many equivalents of acid were present?

Ans.   Without knowing the formula, you cannot tell the number of moles, but you can tell the number of equivalents. For example, HX and H_2X_2 would both neutralize the same number of moles of NaOH per gram of acid. Suppose X had a molar mass of 35 g/mol. HX would have a molar mass of 36 g/mL; H_2X_2 would have a molar mass of 72 g/mol. To neutralize 1.00 mol NaOH would take 1.00 mol HX or 0.500 mol H_2X_2:

$$HX + NaOH \longrightarrow NaX + H_2O$$

or
$$H_2X_2 + 2\,NaOH \longrightarrow Na_2X_2 + 2\,H_2O$$

One mole of HX is 36 g; 0.500 mol of H_2X_2 is also 36 g. You cannot tell from the mass which is the formula of the acid. In either case, the equivalent mass of the acid is 36 g per equivalent. The equivalent mass of H_2X_2 is given by

$$\dfrac{72\ \text{g}}{1\ \text{mol}}\left(\dfrac{1\ \text{mol}}{2\ \text{equiv}}\right) = \dfrac{36\ \text{g}}{1\ \text{equiv}}$$

To convert from molar mass to equivalent mass, use the same factors as were introduced in Sec. 15.4.

EXAMPLE 15.21. Calculate the equivalent mass of $H_2C_2O_4$ in its complete neutralization by NaOH.

Ans.
$$H_2C_2O_4 + 2\,OH^- \longrightarrow C_2O_4{}^{2-} + 2\,H_2O$$
$$\frac{90.0\ g}{1\ mol}\left(\frac{1\ mol}{2\ equiv}\right) = \frac{45.0\ g}{1\ equiv}$$

EXAMPLE 15.22. Calculate the molar mass of an acid with three replaceable hydrogen ions and an equivalent mass of 31.2 g/equiv, assuming complete neutralization.

Ans.
$$\frac{31.2\ g}{1\ equiv}\left(\frac{3\ equiv}{1\ mol}\right) = \frac{93.6\ g}{1\ mol}$$

Solved Problems

QUALITATIVE CONCENTRATION TERMS

15.1. Describe each of the solutions indicated as saturated, unsaturated, or supersaturated. (*a*) More solute is added to a solution of that solute, and the additional solute all dissolves. Describe the original solution. (*b*) More solute is added to a solution of that solute, and the additional solute does not all dissolve. Describe the final solution. (*c*) A solution is left standing, and some of the solvent evaporates. After a time, some solute crystallizes out. Describe the final solution. (*d*) A hot saturated solution is cooled slowly, and no solid crystallizes out. The solute is a solid that is more soluble when hot than when cold. Describe the cold solution. (*e*) A hot solution is cooled slowly, and after a time some solid crystallizes out. Describe the cold solution.

    Ans. (*a*) The original solution was unsaturated, since it was possible to dissolve more solute at that temperature. (*b*) The final solution is saturated; it is holding all the solute that it can hold stably at that temperature, so not all the excess solute dissolves. (*c*) The final solution is saturated; it has solid solute in contact with the solution, and that solute does not dissolve. The solution must be holding as much as it can at this temperature. (*d*) The solution is supersaturated; it is holding more solute than is stable at the cold temperature. (*e*) The solution is saturated.

15.2. $Ce_2(SO_4)_3$ is unusual because it is a solid that dissolves in water better at low temperatures than at high temperatures. State how you might attempt to make a supersaturated solution of this compound in water at 50°C.

    Ans. Dissolve as much as possible in water at 0°C. Then warm the solution carefully to 50°C. (Unless it has been done before, there is no guarantee that any particular compound will produce a supersaturated solution, but this is the way to try.)

15.3. The solubility of ethyl alcohol in water is said to be infinite. What does that mean?

    Ans. It means that the alcohol is completely soluble in water no matter how much alcohol or how little water is present.

MOLALITY

15.4. Calculate the molality of each of the following solutions: (*a*) 2.00 g C_2H_5OH in 30.0 g H_2O and (*b*) 3.00 g NaCl in 20.0 g H_2O.

    Ans. (*a*) $2.00\ g\ C_2H_5OH\left(\dfrac{1\ mol\ C_2H_5OH}{46.0\ g\ C_2H_5OH}\right) = 0.0435\ mol\ C_2H_5OH$ $\dfrac{0.0435\ mol\ C_2H_5OH}{0.0300\ kg\ H_2O} = 1.45\ m$

         (*b*) $3.00\ g\ NaCl\left(\dfrac{1\ mol\ NaCl}{58.5\ g\ NaCl}\right) = 0.0513\ mol\ NaCl$ $\dfrac{0.0513\ mol\ NaCl}{0.0200\ kg\ H_2O} = 2.57\ m\ NaCl$

15.5. How many moles of solute are there in a 1.50 m solution containing 2.22 kg of solvent?

    Ans. $2.22\ kg\ solvent\left(\dfrac{1.50\ mol\ solute}{1\ kg\ solvent}\right) = 3.33\ mol\ solute$

15.6. Calculate the molality of a solution prepared by adding 30.0 g of water to a 2.00 m solution containing 50.0 g of water.

Ans.
$$0.0500 \text{ kg } H_2O\left(\frac{2.00 \text{ mol solute}}{1 \text{ kg } H_2O}\right) = 0.100 \text{ mol solute}$$

$$\frac{0.100 \text{ mol solute}}{0.0800 \text{ kg } H_2O} = 1.25 \text{ } m$$

15.7. What mass of solvent is required to make a 2.00 m solution with 3.18 mol of solute?

Ans.
$$3.18 \text{ mol solute}\left(\frac{1 \text{ kg solvent}}{2.00 \text{ mol solute}}\right) = 1.59 \text{ kg solvent}$$

MOLE FRACTION

15.8. What are the units of mole fraction? Which component is the solvent?

Ans. Mole fraction has no units. It is defined as one number of moles divided by another, and the units cancel. None of the components is defined as the solvent in dealing with mole fractions.

15.9. Explain why $X_A = 1.11$ cannot be correct.

Ans. The total of all mole fractions in a solution is 1, and no single mole fraction can exceed 1.

15.10. Calculate the molality of C_2H_5OH in a solution with mole fraction 0.300 of C_2H_5OH in water.

Ans. Assume 1.00 mol total, which contains:

$$0.300 \text{ mol } C_2H_5OH$$

$$0.700 \text{ mol } H_2O\left(\frac{18.0 \text{ g } H_2O}{1 \text{ mol } H_2O}\right) = 12.6 \text{ g } H_2O$$

$$\frac{0.300 \text{ mol } C_2H_5OH}{0.0126 \text{ kg } H_2O} = 23.8 \text{ } m$$

15.11. Calculate the molality and the mole fraction of the first compound in each of the following solutions: (*a*) 1.00 mol CH_2O and 1.00 mol H_2O, (*b*) 0.150 mol CH_3OH and 50.0 g H_2O, and (*c*) 20.0 g C_2H_5OH and 50.0 g H_2O.

Ans. (*a*)
$$1.00 \text{ mol } H_2O\left(\frac{18.0 \text{ g } H_2O}{1 \text{ mol } H_2O}\right) = 18.0 \text{ g } H_2O = 0.0180 \text{ kg } H_2O$$

$$\text{Molality} = \frac{1.00 \text{ mol } CH_2O}{0.0180 \text{ kg } H_2O} = 55.6 \text{ } m$$

$$X_{CH_2O} = \frac{1.00 \text{ mol } CH_2O}{1.00 \text{ mol } CH_2O + 1.00 \text{ mol } H_2O} = 0.500$$

(*b*)
$$\text{Molality} = \frac{0.150 \text{ mol } CH_3OH}{0.0500 \text{ kg } H_2O} = 3.00 \text{ } m$$

$$50.0 \text{ g } H_2O\left(\frac{1 \text{ mol } H_2O}{18.0 \text{ g } H_2O}\right) = 2.78 \text{ mol } H_2O$$

$$X_{CH_3OH} = \frac{0.150 \text{ mol } CH_3OH}{2.93 \text{ mol total}} = 0.0512$$

(*c*)
$$20.0 \text{ g } C_2H_5OH\left(\frac{1 \text{ mol } C_2H_5OH}{46.0 \text{ g } C_2H_5OH}\right) = 0.435 \text{ mol } C_2H_5OH$$

$$\frac{0.435 \text{ mol}}{0.0500 \text{ kg}} = 8.70 \text{ } m$$

From part *b* we have 2.78 mol H_2O.

$$X_{C_2H_5OH} = \frac{0.435 \text{ mol}}{0.435 \text{ mol} + 2.78 \text{ mol}} = 0.135$$

15.12. Calculate the mole fraction of the first component in each of the following solutions: (a) 15.0 g $C_6H_{12}O_6$ in 55.0 g H_2O and (b) 0.200 m solution of $C_6H_{12}O_6$ in water.

Ans.   (a)

$$15.0 \text{ g } C_6H_{12}O_6\left(\frac{1 \text{ mol } C_6H_{12}O_6}{180 \text{ g } C_6H_{12}O_6}\right) = 0.0833 \text{ mol } C_6H_{12}O_6$$

$$55.0 \text{ g } H_2O\left(\frac{1 \text{ mol } H_2O}{18.0 \text{ g } H_2O}\right) = 3.06 \text{ mol } H_2O$$

$$X_{C_6H_{12}O_6} = \frac{0.0833 \text{ mol}}{3.06 \text{ mol} + 0.0833 \text{ mol}} = 0.0265$$

(b)   In 1.00 kg H_2O there are 0.200 mol $C_6H_{12}O_6$ and

$$1000 \text{ g } H_2O\left(\frac{1 \text{ mol } H_2O}{18.0 \text{ g } H_2O}\right) = 55.6 \text{ mol } H_2O$$

The mole fraction is

$$X_{C_6H_{12}O_6} = \frac{0.200 \text{ mol}}{55.8 \text{ mol total}} = 0.00358$$

EQUIVALENTS

15.13. An equivalent is defined as the amount of a substance that reacts with or produces 1 mol of what?

Ans.   In a redox reaction, electrons. In an acid-base reaction, hydrogen ions or hydroxide ions.

15.14. What is the number of equivalents per mole of HCl?

Ans.   Assuming complete neutralization (or even oxidation of the Cl^- to Cl_2), we can see that 1 mol HCl will react with 1 mol NaOH (or 1 mol e^-); thus, 1 equiv = 1 mol. (In its reactions to produce oxyanions of chlorine (ClO^-, ClO_2^-, etc.), there is more than 1 equivalent per mole.)

15.15. What numbers of equivalents per mole are possible for H_3PO_4 in its acid-base reactions?

Ans.   1, 2, or 3, depending on how complete its reaction with base is.

15.16. How many equivalents of NaOH react with the 82.0 g of H_2SO_3 in part a of Example 15.9?

Ans.   1 equiv. (One OH^- can react with 1 mol H^+; 1 equiv of H_2SO_3 is involved.)

15.17. How many equivalents are there per mole of the first reactant in each of the following equations?

(a) $Zn \longrightarrow Zn^{2+} + 2e^-$

(b) $H_2SO_4 + NaOH \longrightarrow NaHSO_4 + H_2O$

(c) $2\,HCl + Ba(OH)_2 \longrightarrow BaCl_2 + 2\,H_2O$

(d) $MnO_4^- + 5\,e^- + 8\,H^+ \longrightarrow Mn^{2+} + 4\,H_2O$

(e) $H_2SO_4 + 2\,H^+ + Cu \longrightarrow$
     $Cu^{2+} + SO_2 + 2\,H_2O$

(f) $NaHCO_3 + HCl \longrightarrow NaCl + CO_2 + H_2O$

(g) $NaH_2PO_4 + 2\,NaOH \longrightarrow Na_3PO_4 + 2\,H_2O$

(h) $2\,H_2SO_4 + Cu \longrightarrow CuSO_4 + SO_2 + 2\,H_2O$

(i) $NaHCO_3 + NaOH \longrightarrow Na_2CO_3 + H_2O$

Ans.   (a) 2 (b) 1 (c) 1 (d) 5 (e) 2 (f) 1 (g) 2 (h) 2 [same as (e) for the one H_2SO_4 molecule that is reduced] (i) 1

NORMALITY

15.18. If a bottle is labeled 3.109 N H_3PO_4, which reaction should be assumed for the definition of its normality?

Ans. $H_3PO_4 + 3\,NaOH \longrightarrow Na_3PO_4 + 3\,H_2O$ (complete neutralization)

15.19. Explain why normality is the number of milliequivalents per milliliter.

Ans. By definition, normality is the number of equivalents per liter. Then

$$\frac{x \text{ equiv}}{1 \text{ L}} \left(\frac{1000 \text{ mequiv}}{1 \text{ equiv}} \right) \left(\frac{1 \text{ L}}{1000 \text{ mL}} \right) = \frac{x \text{ mequiv}}{1 \text{ mL}}$$

15.20. Explain why volume times normality of one reagent is equal to volume times normality of each of the other reagents in the reaction.

Ans. Volume times normality is the number of equivalents, and by definition, the number of equivalents of one reagent is the same as the number of equivalents of any other reagent in a given reaction.

15.21. What volume of 1.50 *N* H_2SO_4 is completely neutralized by 35.8 mL of (*a*) 2.50 *N* NaOH, (*b*) 2.50 *M* NaOH, (*c*) 2.50 *M* $Ba(OH)_2$, and (*d*) 2.50 *N* $Ba(OH)_2$?

Ans. (*a*)   The number of equivalents of NaOH is 0.0358 L × 2.50 *N* = 0.0895 equiv. That also is the number of equivalents of H_2SO_4.

$$0.0895 \text{ equiv} \left(\frac{1 \text{ L}}{1.50 \text{ equiv}} \right) = 0.0597 \text{ L} = 59.7 \text{ mL}$$

(*b*)   Since 2.50 *M* NaOH is 2.50 *N* NaOH, the answer is the same as that in part *a*.

(*c*) $2.50 \ M \ Ba(OH)_2 = \dfrac{2.50 \text{ mol } Ba(OH)_2}{1 \text{ L}} \left(\dfrac{2 \text{ equiv}}{1 \text{ mol}} \right) = 5.00 \ N \ Ba(OH)_2$

$$5.00 \ N (0.0358 \text{ L}) = 0.179 \text{ equiv}$$

The quantity of $Ba(OH)_2$ is 0.179 equiv. The volume of H_2SO_4 is therefore

$$0.179 \text{ equiv } H_2SO_4 \left(\frac{1 \text{ L}}{1.50 \text{ equiv}} \right) = 0.119 \text{ L} = 119 \text{ mL}$$

(*d*)   2.50 *N* $Ba(OH)_2$ has the same neutralizing ability as 2.50 *N* NaOH, and the answer is the same as that in part *a*.

15.22. Calculate the number of milliequivalents per milliliter in 6.11 *N* H_2SO_4.

Ans. $6.11 \ N = \dfrac{6.11 \text{ equiv}}{1 \text{ L}} \left(\dfrac{1000 \text{ mequiv}}{1 \text{ equiv}} \right) \left(\dfrac{1 \text{ L}}{1000 \text{ mL}} \right) = \dfrac{6.11 \text{ mequiv}}{1 \text{ mL}}$

Normality may be defined as the number of milliequivalents per milliliter as well as the number of equivalents per liter.

15.23. What volume of 2.25 *N* H_3PO_4 will 14.7 equivalents of a base neutralize completely? Explain why you did not need to know the formula of the base to answer this question.

Ans. 14.7 equiv of base reacts with 14.7 equiv of acid, no matter what the base.

$$14.7 \text{ equiv } H_3PO_4 \left(\frac{1 \text{ L}}{2.25 \text{ equiv}} \right) = 6.53 \text{ L}$$

(The identity of the base was used previously to determine the number of equivalents of base; it is not necessary to know its identity now.)

15.24. A bottle is marked 1.00 *N* H_2SO_4. What is its probable molarity?

Ans. Assuming that the normality is for complete neutralization, the molarity is given by

$$\frac{1.00 \text{ equiv}}{1 \text{ L}} \left(\frac{1 \text{ mol}}{2 \text{ equiv}} \right) = 0.500 \ M$$

15.25. A bottle is marked 1.00 *N* H_2SO_4 for production of $NaHSO_4$. What is its molarity?

Ans. $H_2SO_4 + NaOH \longrightarrow NaHSO_4 + H_2O$

$$\frac{1.00 \text{ equiv}}{1 \text{ L}} \left(\frac{1 \text{ mol}}{1 \text{ equiv}} \right) = 1.00 \ M$$

EQUIVALENT MASS

15.26. If 2.500 g of a solid acid is neutralized by 31.31 mL of 1.022 N NaOH, what is its equivalent mass?

Ans.

$$31.31 \text{ mL} \left(\frac{1.022 \text{ mequiv}}{1 \text{ mL}} \right) = 32.00 \text{ mequiv}$$

$$\frac{2500 \text{ mg}}{32.00 \text{ mequiv}} = 78.13 \text{ mg/mequiv} = 78.13 \text{ g/equiv}$$

15.27. It takes 39.27 mL of a base to titrate 2.400 g of potassium hydrogen phthalate ($KHC_8H_4O_4$). What is the normality of the base?

Ans.

$$2.400 \text{ g} \left(\frac{1 \text{ equiv}}{204.2 \text{ g}} \right) = 0.011\ 75 \text{ equiv } KHC_8H_4O_4$$

$$\frac{0.011\ 75 \text{ equiv base}}{0.039\ 27 \text{ L}} = 0.2993\ N$$

Supplementary Problems

15.28. Calculate the number of moles of CH_3OH in 4.00 L of 3.000 m solution if the density of the solution is 0.920 g/mL.

Ans.

$$4.00 \text{ L} \left(\frac{0.920 \text{ kg}}{1 \text{ L}} \right) = 3.68 \text{ kg solution}$$

Per kilogram of water:

$$3.000 \text{ mol } CH_3OH \left(\frac{32.04 \text{ g}}{1 \text{ mol}} \right) = 96.12 \text{ g } CH_3OH = 0.096\ 12 \text{ kg } CH_3OH$$

$$1.000 \text{ kg } H_2O + 0.096\ 12 \text{ kg } CH_3OH = 1.096 \text{ kg solution}$$

For the entire solution:

$$3.68 \text{ kg solution} \left(\frac{1 \text{ kg } H_2O}{1.096 \text{ kg solution}} \right) \left(\frac{3.00 \text{ mol } CH_3OH}{1 \text{ kg } H_2O} \right) = 10.1 \text{ mol } CH_3OH$$

15.29. What is the total of the mole fractions of a three-component system?

Ans. 1.00.

15.30. The solubility of $Ba(OH)_2 \cdot 8H_2O$ in water at 15°C is 5.6 g per 100.0 g water. What is the molality of a saturated solution at 15°C?

Ans.

$$5.6 \text{ g } Ba(OH)_2 \cdot 8H_2O \left[\frac{1 \text{ mol } Ba(OH)_2 \cdot 8H_2O}{315 \text{ g } Ba(OH)_2 \cdot 8H_2O} \right] = 0.018 \text{ mol } Ba(OH)_2 \cdot 8\ H_2O$$

$$\frac{0.018 \text{ mol}}{0.1000 \text{ kg}} = 0.18\ m$$

15.31. (*a*) What is the concentration of NaOH if 48.19 mL of 1.750 N H_2SO_4 neutralizes 25.00 mL of the base?

(*b*) What is the concentration of NaOH if 48.19 mL of 1.750 M H_2SO_4 neutralizes 25.00 mL of the base?

Ans. (*a*)

$$N_1 V_1 = N_2 V_2$$

$$N_2 = \frac{N_1 V_1}{V_2} = \frac{(1.750\ N)(48.19 \text{ mL})}{25.00 \text{ mL}} = 3.373\ N$$

(*b*) The number of millimoles of acid is given by

$$48.19 \text{ mL} \left(\frac{1.750 \text{ mmol}}{1 \text{ mL}} \right) = 84.33 \text{ mmol } H_2SO_4$$

The number of millimoles of base is determined from the 2 : 1 ratio in the balanced chemical equation:

$$84.33 \text{ mmol H}_2\text{SO}_4 \left(\frac{2 \text{ mmol NaOH}}{1 \text{ mmol H}_2\text{SO}_4} \right) = 168.7 \text{ mmol NaOH}$$

$$\frac{168.7 \text{ mmol NaOH}}{25.00 \text{ mL NaOH}} = 6.748 \; M \text{ NaOH} = 6.748 \; N \text{ NaOH}$$

15.32. Calculate the equivalent mass of each of the following acids toward complete neutralization. (a) H_3PO_3, (b) H_2SO_4, (c) HBr, (d) $HC_2H_3O_2$, and (e) $H_2C_2O_4$.

Ans. (a) 27.3 g/equiv (b) 49.0 g/equiv (c) 80.9 g/equiv (d) 60.0 g/equiv (e) 45.0 g/equiv.

15.33. A sample of 5.00 g of a solid acid was treated with 50.00 mL of 2.500 N NaOH, which dissolved it completely (by reacting with it). There was enough excess NaOH to require 9.13 mL of 1.000 N HCl to neutralize the excess base. What is the equivalent mass of the acid?

Ans. The numbers of equivalents of base and HCl are as follows:

$$50.00 \text{ mL} \left(\frac{2.500 \text{ mequiv}}{1 \text{ mL}} \right) = 125.0 \text{ mequiv NaOH}$$

$$9.13 \text{ mL} \left(\frac{1.000 \text{ mequiv}}{1 \text{ mL}} \right) = 9.13 \text{ mequiv HCl}$$

The difference is the number of milliequivalents of the solid acid:

$$125.0 \text{ mequiv} - 9.13 \text{ mequiv} = 115.9 \text{ mequiv}$$

The equivalent mass is

$$\frac{5000 \text{ mg}}{115.9 \text{ mequiv}} = \frac{43.1 \text{ mg}}{1 \text{ mequiv}} = \frac{43.1 \text{ g}}{1 \text{ equiv}}$$

15.34. Write an equation for the half-reaction in which $H_2C_2O_4$ is oxidized to CO_2. What is the equivalent mass of $H_2C_2O_4$?

Ans.
$$H_2C_2O_4 \longrightarrow 2 \, CO_2 + 2 \, H^+ + 2 \, e^-$$

$$\frac{90.0 \text{ g}}{1 \text{ mol}} \left(\frac{1 \text{ mol}}{2 \text{ equiv}} \right) = \frac{45.0 \text{ g}}{1 \text{ equiv}}$$

15.35. What is the equivalent mass of $H_2C_2O_4$ in the following reaction?

$$6 \, H^+ + 2 \, MnO_4^- + 5 \, H_2C_2O_4 \longrightarrow 10 \, CO_2 + 8 \, H_2O + 2 \, Mn^{2+}$$

Ans. The half-reaction and thus the equivalent mass are given in the prior problem. The 1 mol of acid liberates 2 mol of electrons, and so there are 2 equiv of acid per mole of acid. If we wrote the half-reaction for the equation given, we would find that there was 10 mol of electrons involved in reducing 5 mol of acid, and so again we get

$$10 \text{ equiv/5 mol} = 2 \text{ equiv/mol}$$

Still another method to balance the equation is by the oxidation state change method:

$$H_2C_2O_4 \longrightarrow 2 \, CO_2$$
$$2(3 \rightarrow 4) = +2$$

The change in oxidation state is 2 per molecule of $H_2C_2O_4$, and so there are 2 equiv per mole.

CHAPTER 16

Rates and Equilibrium

16.1. INTRODUCTION

We have learned (Chap. 8) that some reactions occur under one set of conditions while an opposite reaction occurs under another set of conditions. For example, we learned that sodium and chlorine combine when treated with each other, but that molten NaCl decomposes when electrolyzed:

$$2\,Na + Cl_2 \longrightarrow 2\,NaCl$$

$$2\,NaCl(molten) \xrightarrow{\text{electricity}} 2\,Na + Cl_2$$

However, some sets of substances can undergo both a forward and a reverse reaction under the same set of conditions. This circumstances leads to a state called *chemical equilibrium*. Before we take up equilibrium, however, we have to learn about the factors that affect the rate of a chemical reaction.

16.2. RATES OF CHEMICAL REACTION

Some chemical reactions proceed very slowly, others with explosive speed, and still others somewhere in between. The "dissolving" of underground limestone deposits by water containing carbon dioxide to form caverns is an example of a slow reaction; it can take centuries. The explosion of trinitrotoluene (TNT) is an example of a very rapid reaction.

The *rate of a reaction* is defined as the change in concentration of any of its reactants or products per unit time. There are six factors that affect the rate of a reaction:

1. *The nature of the reactants.* Carbon tetrachloride (CCl_4) does not burn in oxygen, but methane (CH_4) burns very well indeed. In fact, CCl_4 used to be used in fire extinguishers, while CH_4 is the major component of natural gas. This factor is least controllable by the chemist, and so is of least interest here.

2. *Temperature.* In general, the higher the temperature of a system, the faster the chemical reaction will proceed. A rough rule of thumb is that a 10°C rise in temperature will approximately double the rate of a reaction.

3. *The presence of a catalyst.* A *catalyst* is a substance that can accelerate (or slow down) a chemical reaction without undergoing a permanent change in its own composition. For example, the decomposition of $KClO_3$ by heat is accelerated by the presence of a small quantity of MnO_2. After the reaction, the $KClO_3$ has been changed to KCl and O_2, but the MnO_2 is still MnO_2.

4. *The concentration of the reactants.* In general, the higher the concentration of the reactants, the faster the reaction.

5. *The pressure of gaseous reactants.* In general, the higher the pressure of gaseous reactants, the faster the reaction. This factor is merely a corollary of factor 4, since the higher pressure is in effect a higher concentration (Chap. 12).

6. *State of subdivision.* The smaller the pieces of a solid reactant—the smaller the state of subdivision—the faster the reaction. Wood shavings burn faster than solid wood, for example, because they have greater surface area in contact with the oxygen with which they are combining (for a given mass of wood). In a sense, this is also a corollary of factor 4.

The *collision theory* is presented to explain the factors that affect reaction rates. The theory considers the molecules undergoing reaction to explain the observed phenomena. The theory postulates that in order for a reaction to occur, molecules must collide with one another with sufficient energy to break chemical bonds in the reactants. A very energetic and highly unstable species is formed, called an *activated complex*. Not every collision between reacting molecules, even those with sufficient energy, produces products. The molecules might be oriented in the wrong directions to produce products, or the activated complex may break up to re-form the reactants instead of forming the products. But the huge majority of collisions do not have enough energy to cause bond breakage in the first place.

The minimum energy that may cause a reaction to occur is called the *activation energy*, designated E_a. An everyday analogy to activation energy is a golfer whose ball has landed in a deep bunker near the green (Fig. 16-1). It does not matter if the green is above the bunker or below, the golfer must give the ball sufficient energy to get over the hill that separates the ball and the green. If the ball is hit with too little energy, it will merely return to its original level in the bunker. Similarly, if molecular collisions are not energetic enough, the molecules will merely return to their original states even if temporarily they have been somewhat deformed.

Fig. 16-1.   Activation energy

EXAMPLE 16.1.   If the activation energy of a certain reaction is 15 kJ/mol and the overall reaction process produces 25 kJ/mol of energy in going from reactants to products, how much energy is given off when the activated complex is converted to products?

Ans.   That process produces 40 kJ/mol:

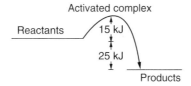

The collision theory allows us to explain the factors that affect the reaction rate. There is a wide range of energies among molecules in any sample, and generally only the most energetic molecules can undergo reaction. An increase in temperature increases the number of molecules that have sufficient energy to react (the activation energy); a rise in temperature of $10°C$ about doubles the number of molecules with that energy. An increase in concentration or pressure causes the molecules to collide more often; with more collisions, more effective collisions are expected. The state of subdivision of a solid affects the reaction rate because the more surface area there is, the more collisions there are between the fluid molecules and the solid surface. A catalyst works by reducing the activation energy, making an easier path for the reactants to get to products. Since more reactant molecules have this (lower) activation energy, the reaction goes faster.

16.3.   CHEMICAL EQUILIBRIUM

Many chemical reactions convert practically all the reactant(s) (at least the limiting quantity) to products under a given set of conditions. These reactions are said to *go to completion*. In other reactions, as the products are formed, they in turn react to form the original reactants again. This situation—two opposing reactions occurring at the same time—leads to formation of some products but the reactants are not completely converted to products. A state in which two exactly opposite reactions are occurring at the same rate is called *chemical equilibrium*. (In fact, all chemical reactions are equilibrium reactions, at least theoretically.) For example, nitrogen and hydrogen gases react with each other at 500°C and high pressure to form ammonia; under the same conditions, ammonia decomposes to produce hydrogen and nitrogen:

$$3\,H_2 + N_2 \longrightarrow 2\,NH_3$$
$$2\,NH_3 \longrightarrow 3\,H_2 + N_2$$

To save effort, we often write these two exactly opposite equations as one, with double arrows:

$$3\,H_2 + N_2 \rightleftarrows 2\,NH_3 \qquad \text{or} \qquad 2\,NH_3 \rightleftarrows 3\,H_2 + N_2$$

We call the reagents on the right of the chemical equation as it is written the *products* and those on the left the *reactants*, despite the fact that we can write the equation with either set of reagents on either side.

With the reaction just above, if you start with a mixture of nitrogen and hydrogen and allow it to come to 500°C at 200-atm pressure, some nitrogen and hydrogen combine to form ammonia. If you heat ammonia to 500°C at 200-atm pressure, some of it decomposes to nitrogen and hydrogen. Both reactions can occur in the same vessel at the same time.

What happens when we first place hydrogen and nitrogen in a container at 500°C and allow them to react? At first, there is no ammonia present, and so the only reaction that occurs is the combination of the two elements. As time passes, there is less and less nitrogen and hydrogen, and the combination reaction therefore slows down (factor 4 or 5, Sec. 16.2). Meanwhile, the concentration of ammonia is building up, and the decomposition rate of the ammonia therefore increases. There comes a time when both the combination reaction and the decomposition reaction occur at the same rate. When that happens, the concentration of ammonia will not change any more. The reaction apparently stops. However, in reality both the combination reaction and the decomposition reaction continue to occur with no net change taking place. A state of *equilibrium* has been achieved.

Le Châtelier's Principle

If we change the conditions on a system at equilibrium such as the N_2, H_2, NH_3 system at equilibrium, for example by changing the temperature, we can get some further net reaction. Soon, however, the system will achieve a new equilibrium at the new set of conditions.

Le Châtelier's principle states that if a stress is applied to a system *at equilibrium*, the equilibrium will shift in a tendency to reduce that stress. A *stress* is something done *to* the system (not *by* the equilibrium reaction). The stresses that we consider are change of concentration(s), change of temperature, change of pressure, and addition of a catalyst. Let us consider the effect on a typical equilibrium by each of these stresses.

The Effect of Concentration

An increase in concentration of one of the reactants or products of the equilibrium will cause the equilibrium to shift to try to reduce that concentration increase.

EXAMPLE 16.2.   How will addition of hydrogen gas affect the following equilibrium system?

$$3\,H_2 + N_2 \rightleftarrows 2\,NH_3$$

Ans.   Addition of hydrogen will at first increase the concentration of hydrogen. The equilibrium will therefore shift to reduce some of that increased concentration; it will shift right. That is, some of the added hydrogen will react with some of the nitrogen originally present to produce more ammonia. Note especially that the hydrogen concentration will be above the original hydrogen concentration but below the concentration it would have if no shift had taken place.

EXAMPLE 16.3. Suppose that, under a certain set of conditions, a mixture of nitrogen, hydrogen, and ammonia is at equilibrium. The concentration of hydrogen is 0.0350 mol/L; that of nitrogen is 0.0100 mol/L, and that of ammonia is 0.0250 mol/L. Now 0.0003 mol of hydrogen is added to 1.00 L of the mixture. What is the widest possible range of hydrogen concentration in the new equilibrium?

Ans.   Before addition of the extra hydrogen, its concentration was 0.0350 mol/L. After addition of the extra hydrogen, but before any equilibrium shift could take place, there is 0.0353 mol/L. The shift of the equilibrium uses up some but not all of the added hydrogen, and so the final hydrogen concentration must be above 0.0350 mol/L and below 0.0353 mol/L. Some nitrogen has been used up and its final concentration must be less than its original concentration, 0.0100 mol/L. Some additional ammonia has been formed, and so its final concentration has been increased over 0.0250 mol/L. Notice that Le Châtelier's principle does not tell us how much of a shift there will be, but only qualitatively in which direction a shift will occur.

Temperature Change

To consider the effect of temperature change, let us rewrite one of the preceeding equations including the heat involved:

$$3\,H_2 + N_2 \rightleftarrows 2\,NH_3 + \text{heat}$$

How do we get the temperature of the system to rise? By adding heat. When we add heat, this equilibrium system reacts to reduce that stress, that is, to use up some of the added heat. It can use up heat in the reverse reaction, the decomposition of ammonia to hydrogen and nitrogen. When the substances written as products of the reaction (on the right side of the equation) react to produce more reactants (on the left side of the equation), we say that the reaction has *shifted to the left*. When the opposite process occurs, we say that the equilibrium has *shifted to the right*. Thus, raising the temperature on this system already at equilibrium causes a shift to the left; some of the ammonia decomposes without being replaced.

EXAMPLE 16.4. What happens to the above system at equilibrium if the temperature is lowered?

Ans.   We lower the temperature of a system by removing heat. The equilibrium tries to minimize that change by restoring some heat; it can do that by shifting to the right. Thus, more nitrogen and hydrogen are converted to ammonia.

The Effect of Pressure

Pressure affects gases in a system much more than it affects liquid or solids. We will investigate the same ammonia, hydrogen, nitrogen system discussed before. If the system is at equilibrium, what will an increase in pressure by the chemist do to the equilibrium? The system will shift to try to reduce the stress, as required by Le Châtelier's principle. How can this system reduce its own pressure? By reducing the total number of moles present. It can shift to the right to produce 2 mol of gas for every 4 mol used up:

$$3\,H_2 + N_2 \rightleftarrows 2\,NH_3$$

Of course, it need not shift much. For example, if 0.00030 mol H_2 reacts with 0.00010 mol N_2 to produce 0.00020 mol NH_3, the total number of moles will have been reduced (by 0.00020 mol), and the pressure will therefore have been reduced.

EXAMPLE 16.5. What will be the effect of increased pressure caused by decreasing the volume on the following system at equilibrium?

$$2\,NH_3 \rightleftarrows 3\,H_2 + N_2$$

Ans.   Again some nitrogen and hydrogen will be converted to ammonia. This time, since the equation is written in the reverse of the way it was written above, the equilibrium will shift to the left. Of course, the same physical effect is produced. More ammonia is formed. But the answer in terms of the direction of the shift is different since the equation is written "backward."

EXAMPLE 16.6. What effect would an increase in volume have on the following system at equilibrium at 500°C, where all the reagents are gases?

$$H_2 + I_2 \rightleftarrows 2\,HI$$

Ans.   The increase in volume would decrease the pressure on each of the gases (Chap. 12). The equilibrium would not shift, however, because there are equal numbers of moles of gases on the two sides. Neither possible shift would cause an increase in pressure.

Presence of a Catalyst

Addition of a catalyst to a system at equilibrium will not cause any change in the position of the equilibrium; it will shift neither left nor right. (Addition of the catalyst speeds up both the forward reaction and the reverse reaction equally.)

16.4. EQUILIBRIUM CONSTANTS

Although Le Châtelier's principle does not tell us how much an equilibrium will be shifted, there is a way to determine the position of an equilibrium once data have been determined for the equilibrium experimentally. At a given temperature the ratio of concentrations of products to reactants, each raised to a suitable power, is constant for a given equilibrium reaction. The letters A, B, C, and D are used here to stand for general chemical species. Thus, for a chemical reaction in general,

$$a\text{A} + b\text{B} \rightleftarrows c\text{C} + d\text{D}$$

The following ratio always has the same value at a given temperature:

$$K = \frac{[\text{C}]^c[\text{D}]^d}{[\text{A}]^a[\text{B}]^b}$$

Here the square brackets indicate the concentration of the chemical species within the bracket. That is, [A] means the concentration of A, and so forth. $[\text{A}]^a$ means the concentration of A raised to the power a, where a is the value of the coefficient of A in the balanced equation for the chemical equilibrium. The value of the ratio of concentration terms is symbolized by the letter K, called the *equilibrium constant*. For example, for the reaction of hydrogen and nitrogen referred to in Sec. 16.3,

$$3\,H_2 + N_2 \rightleftarrows 2\,NH_3$$

the ratio is

$$K = \frac{[NH_3]^2}{[H_2]^3[N_2]}$$

The exponents 2 and 3 are the coefficients of ammonia and hydrogen, respectively, in the balanced equation.

Please note carefully the following points about the *equilibrium constant expression*:

1. This is a mathematical equation. Its values are numbers, involving the concentrations of the chemicals and their coefficients.
2. Each concentration is raised to the power given by the coefficient in the chemical equation.
3. The concentrations of the products of the reaction are written in the numerator of the equilibrium constant expression; the concentrations of the reactants are in the denominator.
4. The terms are multiplied together, not added.
5. Each equilibrium constant expression is associated with a particular chemical reaction written in a given direction.

EXAMPLE 16.7. Write the equilibrium constant expression for the reaction

$$2\,NH_3 \rightleftarrows 3\,H_2 + N_2$$

Ans.   For this equation, the terms involving the concentrations of the elements are in the numerator, because they are the products:

$$K = \frac{[H_2]^3[N_2]}{[NH_3]^2}$$

There are a great many types of equilibrium problems. We will start with the easiest and work up to harder ones.

EXAMPLE 16.8.   Calculate the value of the equilibrium constant for the following reaction if at equilibrium the concentration of A is 1.60 M, that of B is 0.800 M, that of C is 0.240 M, and that of D is 0.760 M.

$$A + B \rightleftharpoons C + D$$

Ans.   Since all the coefficients in the balanced equation are equal to 1, the equilibrium constant expression is

$$K = \frac{[C][D]}{[A][B]}$$

We merely substitute the equilibrium concentrations into this equation to determine the value of the equilibrium constant:

$$K = \frac{(0.240\ M)(0.760\ M)}{(1.60\ M)(0.800\ M)} = 0.143$$

One way to make Example 16.8 harder is by giving numbers of moles and a volume instead of concentrations at equilibrium. Since the equilibrium constant is defined in terms of concentrations, we must first convert the numbers of moles and volume to concentrations. Note especially that the volume of all the reactants is the same, since they are all in the same system.

EXAMPLE 16.9.   Calculate the value of the equilibrium constant if at equilibrium there are 0.400 mol A, 0.200 mol B, 0.0600 mol C, and 0.190 mol D in 0.250 L of solution.

$$A + B \rightleftharpoons C + D$$

Ans.   Since numbers of moles and a volume are given, it is easy to calculate the equilibrium concentrations of the species. In this case, [A] = 1.60 M, [B] = 0.800 M, [C] = 0.240 M, and [D] = 0.760 M, which are exactly the concentrations given in the prior example. Thus, this problem has exactly the same answer.

It is somewhat more difficult to determine the value of the equilibrium constant if some initial concentrations instead of equilibrium concentrations are given.

EXAMPLE 16.10.   Calculate the value of the equilibrium constant in the following reaction if 1.50 mol of A and 2.50 mol of B are placed in 1.00 L of solution and allowed to come to equilibrium. The equilibrium concentration of C is found to be 0.45 M.

$$A + B \rightleftharpoons C + D$$

Ans.   To determine the *equilibrium* concentrations of all the reactants and products, we must deduce the changes that have occurred. We can assume by the wording of the problem that no C or D was added by the chemist, that some A and B have been used up, and that some D has also been produced. It is perhaps easiest to tabulate the various concentrations. We use the chemical equation as our table headings, and we enter the values that we know now:

	A	+	B	$\rightleftharpoons$	C	+	D
Initial concentrations	1.50		2.50		0.00		0.00
Changes produced by reaction							
Equilibrium concentrations					0.45		

We deduce that to produce 0.45 M C it takes 0.45 M A and 0.45 M B. Moreover, we know that 0.45 M D was also produced. *The magnitudes of the values in the second row of the table—the changes produced by the reaction—are always in the same ratio as the coefficients in the balanced chemical equation.*

	A	+	B	$\rightleftharpoons$	C	+	D
Initial concentrations	1.50		2.50		0.00		0.00
Changes produced by reaction	−0.45		−0.45		+0.45		+0.45
Equilibrium concentrations					0.45		

Note that the concentrations of reactants are reduced. We then merely add the columns to find the rest of the equilibrium concentrations:

	A	+	B	$\rightleftharpoons$	C	+	D
Initial concentrations	1.50	2.50		0.00	0.00
Changes produced by reaction	−0.45	−0.45		+0.45	+0.45
Equilibrium concentrations	1.05	2.05		0.45	0.45

Now that we have calculated the equilibrium concentrations, we can substitute these values into the equilibrium constant expression:

$$K = \frac{[C][D]}{[A][B]} = \frac{(0.45)(0.45)}{(1.05)(2.05)} = 0.094$$

When the chemical equation is more complex, the equilibrium constant expression is also more complex and the deductions about the equilibrium concentrations of reactants and products are more involved, too.

EXAMPLE 16.11.   Calculate the value of the equilibrium constant for the following reaction if 1.50 mol of A and 2.50 mol of B are placed in 1.00 L of solution and allowed to come to equilibrium. The equilibrium concentration of C is found to be 0.45 M.

$$2\,A + B \rightleftharpoons C + D$$

Ans.   The equilibrium constant expression for this equation is

$$K = \frac{[C][D]}{[A]^2[B]}$$

	2 A	+	B	$\rightleftharpoons$	C	+	D
Initial concentrations	1.50	2.50		0.00	0.00
Changes produced by reaction					
Equilibrium concentrations				0.45	

The changes brought about by the chemical reaction are a little different in this case. Twice as many moles per liter of A are used up as moles per liter of C are produced. Note that the magnitudes in the middle row of this table and the coefficients in the balanced chemical equation are in the same ratio.

	2 A	+	B	$\rightleftharpoons$	C	+	D
Initial concentrations	1.50	2.50		0.00	0.00
Changes produced by reaction	−0.90	−0.45		+0.45	+0.45
Equilibrium concentrations				0.45	

Adding the columns gives

	2 A	+	B	$\rightleftharpoons$	C	+	D
Initial concentrations	1.50	2.50		0.00	0.00
Changes produced by reaction	−0.90	−0.45		+0.45	+0.45
Equilibrium concentrations	0.60	2.05		0.45	0.45

The equilibrium values are substituted into the equilibrium constant expression:

$$K = \frac{[C][D]}{[A]^2[B]} = \frac{(0.45)(0.45)}{(0.60)^2(2.05)} = 0.27$$

The next type of problem gives the initial concentrations of the reactants (and products) plus the value of the equilibrium constant and requires calculation of one or more equilibrium concentrations. We use algebraic quantities (such as x) to represent at least one equilibrium concentration, and we solve for the others if necessary in terms of that quantity.

EXAMPLE 16.12. For the reaction $\qquad\qquad\qquad\qquad\qquad\qquad$ $A + B \rightleftarrows C + D$

1.25 mol of A and 1.50 mol of B are placed in a 1.00-L vessel and allowed to come to equilibrium. The value of the equilibrium constant is 0.0010. Calculate the concentration of C at equilibrium.

Ans.

	A	+ B	$\rightleftarrows$ C	+ D
Initial concentrations	1.25	1.50	0.00	0.00
Changes produced by reaction				
Equilibrium concentrations			x	

The changes due to the chemical reaction are as easy to determine as before, and therefore so are the equilibrium concentrations:

	A	+ B	$\rightleftarrows$ C	+ D
Initial concentrations	1.25	1.50	0.00	0.00
Changes produced by reaction	$-x$	$-x$	x	x
Equilibrium concentrations	$1.25 - x$	$1.50 - x$	x	x

We could substitute these equilibrium concentrations in the equilibrium constant expression and solve by using the quadratic equation (Appendix). However, it is more convenient to attempt to approximate the equilibrium concentrations by neglecting a small quantity (x) *when added to or subtracted from* a larger quantity (such as 1.25 or 1.50). We do not neglect small quantities unless they are added to or subtracted from larger quantities! Thus, we approximate the equilibrium concentrations as $[A] \cong 1.25\ M$ and $[B] \cong 1.50\ M$. The equilibrium constant expression is thus

$$K = \frac{[C][D]}{[A][B]} = \frac{(x)(x)}{(1.25)(1.50)} = \frac{x^2}{1.88} = 0.0010$$
$$x = 0.043\ M$$

Next we must check whether our approximation was valid. Is it true that $1.25 - 0.043 \cong 1.25$ and $1.50 - 0.043 \cong 1.50$? Considering the limits of accuracy of the equilibrium constant expression, results within 5% accuracy are considered valid for the general chemistry course. These results are thus valid. The equilibrium concentrations are

$$[A] = 1.25 - 0.043 = 1.21\ M$$
$$[B] = 1.50 - 0.043 = 1.46\ M$$
$$[C] = [D] = 0.043\ M$$

If the approximation had caused an error of 10% or more, you would not be able to use it. You would have to solve by a more rigorous method, such as the quadratic equation or using an electronic calculator to solve for the unknown value. (Ask your instructor if you are responsible for the quadratic equation method.)

EXAMPLE 16.13. Repeat the prior example, but with an equilibrium constant value of 0.100.

Ans.   The problem is done in exactly the same manner until you find that the value of x is 0.434 M.

$$\frac{x^2}{1.88} = 0.100$$
$$x = 0.434$$

When that value is subtracted from 1.25 or 1.50, the answer is not nearly the original 1.25 or 1.50. The approximation is not valid. Thus, we must use the quadratic equation.

$$K = \frac{(x)^2}{(1.25 - x)(1.50 - x)} = 0.100$$
$$x^2 = (0.100)(1.25 - x)(1.50 - x) = 0.100(x^2 - 2.75x + 1.88)$$
$$0.900x^2 + 0.275x - 0.188 = 0$$
$$x = \frac{-b \pm \sqrt{(b^2 - 4ac)}}{2a} = \frac{-0.275 \pm \sqrt{(0.275)^2 - 4(0.900)(-0.188)}}{2(0.900)}$$

Two solutions for x, one positive and one negative, can be obtained from this equation, but only one will have physical meaning. There cannot be any negative concentrations.

$$x = 0.329 \ M = [C] = [D]$$

The concentrations of A and B are then

$$[A] = 1.25 \ - 0.329 = 0.92 \ M$$

$$[B] = 1.50 \ - 0.329 = 1.17 \ M$$

The value from the equilibrium constant expression is therefore equal to the value of K given in the problem:

$$K = \frac{(0.329)^2}{(0.92)(1.17)} = 0.10$$

Solved Problems

RATES OF CHEMICAL REACTION

16.1. Does sugar dissolve faster in hot coffee or in iced coffee? Explain.

Ans.   In hot coffee. The higher the temperature, the faster the process.

16.2. Does lump sugar or granular sugar dissolve faster in water, all other factors being equal? Explain.

Ans.   Granular sugar dissolves faster; it has more surface area in contact with the liquid.

16.3. Does a glowing lump of charcoal react faster in air or in pure oxygen? Explain.

Ans.   It will react faster in pure oxygen; the concentration of oxygen is greater under those conditions.

16.4. At the same temperature, which sample of a gas has more molecules per unit volume—one at high pressure or one at low pressure? Which sample would react faster with a solid substance, all other factors being equal?

Ans.   The high-pressure gas has more molecules per unit volume and therefore reacts faster.

16.5. The activation energy of a certain reaction is 15 kJ/mol. The reaction is exothermic, yielding 19 kJ/mol. What is the activation energy of the reverse reaction?

Ans.   34 kJ/mol

16.6. Using the equation for the average kinetic energy of the molecules in a gaseous sample (Chap. 13) $\overline{KE} = 1.5kT$, show that a 10° rise in temperature from 273 K to 283 K does not double the average kinetic energy of the molecules.

Ans. $$\overline{KE}_{283} = 1.5k(283 \ \text{K})$$

$$\overline{KE}_{273} = 1.5k(273 \ \text{K})$$

Dividing gives

$$\frac{\overline{KE}_{283}}{\overline{KE}_{273}} = \frac{283}{273} = 1.04$$

The average energy increases about 4%; it does not double. Note that what doubles with a 10°C increase in temperature is the number of molecules with sufficiently high energy to react.

CHEMICAL EQUILIBRIUM

16.7. Write an "equation" for the addition of heat to a water-ice mixture at 0°C to produce more liquid water at 0°C. Which way does the equilibrium shift when you try to lower the temperature?

Ans. $$H_2O(s) + \text{heat} \rightleftharpoons H_2O(l)$$

The equilibrium shifts toward ice as you remove heat in an attempt to lower the temperature. (However, the temperature does not change until all the water freezes and the equilibrium system is destroyed.)

16.8. What does Le Châtelier's principle state about the effect of an addition of NH_3 on a mixture of N_2 and H_2 before it attains equilibrium?

Ans.   Nothing. Le Châtelier's principle applies only to a system already at equilibrium.

16.9. What effect would an increase in volume have on the following system at equilibrium at 500°C?

$$2\,C(s) + O_2(g) \rightleftarrows 2\,CO(g)$$

Ans.   The equilibrium would shift to the right. The increase in volume would decrease the pressure of each of the gases, but not of the carbon, which is a solid. The number of moles of gas would be increased by the shift to the right.

16.10. Hydrogen gas is added to an equilibrium system of hydrogen, nitrogen, and ammonia. The equilibrium shifts to reduce the stress of the added hydrogen. Will there be more, less, or the same concentration of hydrogen present at the new equilibrium compared with the old equilibrium?

Ans. $$N_2 + 3\,H_2 \rightleftarrows 2\,NH_3 + heat$$

There will be a greater concentration of hydrogen and of ammonia at the new equilibrium, and a lower concentration of nitrogen. The concentration of hydrogen is not as great as the original concentration plus that which would have resulted from the addition of more hydrogen, however. Some of the total has been used up in the equilibrium shift.

16.11. For the reaction: $heat + A + B \rightleftarrows C$, does a rise in temperature increase or decrease the rate of (*a*) the forward reaction? (*b*) the reverse reaction? (*c*) Which effect is greater?

Ans.   (*a*) and (*b*) Increase. (An increase in temperature increases all rates.)   (*c*) The added heat shifts the equilibrium to the right. That means that the forward reaction was speeded up more than the reverse reaction was speeded up.

16.12. The balanced chemical equation

$$3\,H_2 + N_2 \rightleftarrows 2\,NH_3$$

means which one(s) of the following? (*a*) One can place H_2 and N_2 in a reaction vessel only in the ratio 3 mol to 1 mol. (*b*) If one places NH_3 in a reaction vessel, one cannot put any N_2 and/or H_2 in it. (*c*) If one puts 3 mol of H_2 and 1 mol of N_2 into a reaction vessel, it will produce 2 mol of NH_3. (*d*) For every 1 mol of N_2 that reacts, 3 mol of H_2 will also react and 2 mol of NH_3 will be produced. (*e*) For every 2 mol of NH_3 that decomposes, 3 mol of H_2 and 1 mol of N_2 are produced.

Ans.   Only parts *d* and *e* are correct. The balanced equation governs the **reacting** ratios only. It cannot determine how much of any chemical may be placed in a vessel—(*a*) and (*b*)—or if a reaction will go to completion—(*c*).

EQUILIBRIUM CONSTANTS

16.13. Write equilibrium constant expressions for the following equations. Tell how they are related.

(*a*) $N_2O_4 \rightleftarrows NO_2 + NO_2$ (*b*) $N_2O_4 \rightleftarrows 2\,NO_2$

Ans.   (*a*) $K = \dfrac{[NO_2][NO_2]}{[N_2O_4]}$ (*b*) $K = \dfrac{[NO_2]^2}{[N_2O_4]}$

The equilibrium constant expressions are the same because the chemical equations are the same. It is easy to see why the coefficients of the chemical equation are used as exponents in the equilibrium constant expression by writing out the equation and expression as in part *a*.

16.14. Write equilibrium constant expressions for the following equations. Tell how these expressions are related.

    (a) $N_2O_4 \rightleftharpoons 2\,NO_2$ (b) $2\,NO_2 \rightleftharpoons N_2O_4$ (c) $NO_2 \rightleftharpoons \frac{1}{2}\,N_2O_4$

    Ans.   (a) $K = \dfrac{[NO_2]^2}{[N_2O_4]}$ (b) $K = \dfrac{[N_2O_4]}{[NO_2]^2}$ (c) $K = \dfrac{[N_2O_4]^{1/2}}{[NO_2]}$

        The K in part b is the reciprocal of that in part a. The K in part c is the square root of that in part b.

16.15. Write an equilibrium constant expression for each of the following:

    (a) $SO_2 + \frac{1}{2}\,O_2 \rightleftharpoons SO_3$ (b) $2\,NO + \frac{1}{2}\,O_2 \rightleftharpoons N_2O_3$

    (c) $CH_3C(OH){=}CHCOCH_3 \rightleftharpoons CH_3COCH_2COCH_3$

    Ans.   (a) $K = \dfrac{[SO_3]}{[SO_2][O_2]^{1/2}}$ (b) $K = \dfrac{[N_2O_3]}{[NO]^2[O_2]^{1/2}}$ (c) $K = \dfrac{[CH_3COCH_2COCH_3]}{[CH_3C(OH){=}CHCOCH_3]}$

16.16. Write an equilibrium constant expression for each of the following:

    (a) $N_2 + 3\,H_2 \rightleftharpoons 2\,NH_3$ (d) $PCl_5(g) \rightleftharpoons PCl_3(g) + Cl_2(g)$

    (b) $H_2 + Br_2(g) \rightleftharpoons 2\,HBr(g)$ (e) $2\,NO_2 \rightleftharpoons 2\,NO + O_2$

    (c) $SO_2Cl_2 \rightleftharpoons SO_2 + Cl_2$

    Ans.   (a) $K = \dfrac{[NH_3]^2}{[N_2][H_2]^3}$ (c) $K = \dfrac{[SO_2][Cl_2]}{[SO_2Cl_2]}$ (e) $K = \dfrac{[NO]^2[O_2]}{[NO_2]^2}$

               (b) $K = \dfrac{[HBr]^2}{[H_2][Br_2]}$ (d) $K = \dfrac{[PCl_3][Cl_2]}{[PCl_5]}$

16.17. $X + Y \rightleftharpoons Z$

A mixture of 1.20 mol of X, 2.10 mol of Y, and 0.950 mol of Z is found at equilibrium in a 1.00-L vessel. (*a*) Calculate K. (*b*) If the same mixture had been found in a 2.00-L reaction mixture, would the value of K have been the same? Explain.

    Ans.   (a) $K = \dfrac{[Z]}{[X][Y]} = \dfrac{0.950}{(1.20)(2.10)} = 0.377$ (b) $K = \dfrac{0.475}{(0.600)(1.05)} = 0.754$

        The value is not the same. The value of K is related to the *concentrations* of the reagents, not to their numbers of moles.

16.18. $A + B \rightleftharpoons C + D$

After 1.00 mol of A and 1.80 mol of B were placed in a 1.00-L vessel and allowed to achieve equilibrium, 0.30 mol of C was found. In order to determine the value of the equilibrium constant, determine (*a*) the quantity of C produced, (*b*) the quantity of D produced, (*c*) the quantities of A and B used up, (*d*) the quantity of A remaining at equilibrium, by considering the initial quantity and the quantity used up, (*e*) the quantity of B remaining at equilibrium, and (*f*) the value of the equilibrium constant.

    Ans.   (*a*), (*b*), and (*c*) 0.30 mol each.   (*d*) $1.00 - 0.30 = 0.70$ mol   (*e*) $1.80 - 0.30 = 1.50$ mol

        (*f*) $K = \dfrac{[C][D]}{[A][B]} = \dfrac{(0.30)^2}{(0.70)(1.50)} = 8.6 \times 10^{-2}$

16.19. $2\,A + B \rightleftharpoons C + 2\,D$

If 0.90 mol of A and 1.60 mol of B are placed in a 1.00-L vessel and allowed to achieve equilibrium, 0.35 mol of C is found. Use a table such as shown in Example 16.10 to determine the value of the equilibrium constant.

Ans.

	2 A	+	B	⇌	C	+	2 D
Initial	0.90		1.60		0		0
Change	−0.70		−0.35		+0.35		+0.70
Equilibrium	0.20		1.25		0.35		0.70

$$K = \frac{[C][D]^2}{[A]^2[B]} = \frac{(0.35)(0.70)^2}{(0.20)^2(1.25)} = 3.4$$

16.20. In which line of the table used to calculate the equilibrium concentrations are the values in the ratio of the coefficients of the balanced chemical equation? Is this true in other lines?

Ans.   The terms in the second line are in that ratio, since it describes the changes made by the reaction. The terms in the other lines are not generally in that ratio.

16.21. For the following reaction, $K = 1.0 \times 10^{-7}$:

$$W + Q \rightleftharpoons R + Z$$

If 1.0 mol of W and 2.5 mol of Q are placed in a 1.0-L vessel and allowed to come to equilibrium, calculate the equilibrium concentration of Z, using the following steps: (*a*) If the equilibrium concentration of Z is equal to *x*, how much Z was produced by the chemical reaction? (*b*) How much R was produced by the chemical reaction? (*c*) How much W and Q were used up by the reaction? (*d*) How much W is left at equilibrium? (*e*) How much Q is left at equilibrium? (*f*) With the value of the equilibrium constant given, will *x* (equal to the Z concentration at equilibrium) be significant when subtracted from 1.0? (*g*) Approximately what concentrations of W and Q will be present at equilibrium? (*h*) What is the value of *x*? (*i*) What is the concentration of R at equilibrium? (*j*) Is the answer to part *f* justified?

Ans.   (*a*) *x*. (There was none originally, so all that is there must have come from the reaction.)   (*b*) *x*. (The Z and R are produced in equal mole quantities, and they are in the same volume.)   (*c*) *x* each. (The reacting ratio of W to Q to R to Z is 1: 1: 1: 1.)   (*d*) $1.0 - x$ (the original concentration minus the concentration used up).   (*e*) $2.5 - x$ (the original concentration minus the concentration used up).   (*f*) No. (We will try it first and later see that this is true.)   (*g*) 1.0 mol/L and 2.5 mol/L, respectively.

(*h*) $K = \dfrac{[R][Z]}{[W][Q]} = \dfrac{x^2}{(1.0)(2.5)} = 1.0 \times 10^{-7}$ $x^2 = 2.5 \times 10^{-7}$ $x = 5.0 \times 10^{-4} = [Z]$

(*i*) $[R] = x = 5.0 \times 10^{-4}$ mol/L

(*j*)   Yes. $1.0 - (5.0 \times 10^{-4}) = 1.0$

16.22. For the following reaction of gases, $K = 1.0 \times 10^{-11}$:

$$W + 2 Q \rightleftharpoons 3 R + Z$$

Determine the concentration of Z at equilibrium by repeating each of the steps in the prior problem. Calculate the equilibrium concentration of R if 1.0 mol of W and 2.5 mol of Q are placed in a 1.0-L vessel and allowed to attain equilibrium.

Ans.   (*a*) *x*   (*b*) 3*x*   (*c*) W: *x*, Q: 2*x*   (*d*) 1.0−*x*   (*e*) 2.5−2*x*   (*f*) no   (*g*) 1.0 mol/L and 2.5 mol/L, respectively

(*h*) $K = \dfrac{[R]^3[Z]}{[W][Q]^2} = \dfrac{(3x)^3(x)}{(1.0)(2.5)^2} = 4.3x^4 = 1.0 \times 10^{-11}$ $x = 1.2 \times 10^{-3} = [Z]$

(*i*) $[R] = 3x = 3.6 \times 10^{-3}$ mol/L

(*j*)   Yes

Supplementary Problems

16.23. Compare the function of a catalyst in a chemical reaction with that of a marriage broker—a matchmaker.

Ans.   They both facilitate an effect without permanent change in themselves.

16.24. Identify or explain each of the following terms: (*a*) equilibrium, (*b*) rate of reaction, (*c*) catalyst, (*d*) completion, (*e*) Le Châtelier's principle, (*f*) stress, (*g*) shift, (*h*) shift to the right or left, (*i*) equilibrium constant, and (*j*) equilibrium constant expression.

Ans.   (*a*) Equilibrium is a state in which two exactly opposite processes occur at equal rates. No apparent change takes place at equilibrium.   (*b*) Rate of reaction is the number of moles per liter of reactant that reacts per unit time.   (*c*) A catalyst is a substance that alters the rate of a chemical reaction without undergoing permanent change in its own composition.   (*d*) A reaction goes to completion when one or more of the reactants is entirely used up. In contrast, a reaction at equilibrium has some of each reactant formed again from products, and no reactant or product concentration goes to zero.   (*e*) Le Châtelier's principle states that when a stress is applied to a system at equilibrium, the equilibrium tends to shift to reduce the stress.   (*f*) Stress is a change of conditions imposed on a system at equilibrium.   (*g*) Shift is a change in the concentrations of reactants and products as a result of a stress.   (*h*) Shift to the right is production of more products by using up reactants; shift to the left is exactly the opposite.   (*i*) An equilibrium constant is a set value of the ratio of concentrations of products to concentrations of reactants, each raised to the appropriate power and multiplied together.   (*j*) An equilibrium constant expression is the equation that relates the equilibrium constant to the concentration ratio.

16.25. Nitrogen and sulfur oxides undergo the following rapid reactions in oxygen. Determine the overall reaction. What is the role of NO?

$$2\,NO + O_2 \longrightarrow 2\,NO_2$$
$$NO_2 + SO_2 \longrightarrow NO + SO_3$$

Ans.   Combining the first reaction with twice the second yields the following reaction:

$$2\,SO_2 + O_2 \longrightarrow 2\,SO_3$$

That reaction will proceed only slowly in the absence of the NO and NO_2. The NO is a catalyst. It speeds up the conversion of SO_2 to SO_3 and is not changed permanently in the process.

16.26. State, if possible, what happens to the position of the equilibrium in each of the following cases.

$$N_2 + 3\,H_2 \rightleftharpoons 2\,NH_3 + heat$$

(*a*) Heat is added and ammonia is added. (*b*) Nitrogen is added and the volume is reduced. (*c*) Ammonia is taken out and a catalyst is added. (*d*) Nitrogen and ammonia are both added.

Ans.   (*a*) Addition of ammonia and addition of heat will each cause the equilibrium to shift to the left; addition of both will also cause shift to the left.   (*b*) Each stress—addition of nitrogen and increase in pressure (by reduction of the volume)—causes a shift to the right, and so the equilibrium will shift to the right when both stresses are applied.   (*c*) Addition of a catalyst will cause no shift, but removal of ammonia will cause a shift to the right, and so these stresses together will cause a shift to the right.   (*d*) You cannot tell, because addition of the nitrogen would cause a shift to the right and addition of ammonia would cause a shift to the left. Since no data are given and Le Châtelier's principle is only qualitative, no conclusion is possible.

16.27. What is the best set of temperature and pressure conditions for the Haber process—the industrial process to convert hydrogen and nitrogen to ammonia?

Ans.   To get the equilibrium to shift as much as possible toward ammonia, you should use a high pressure and a low temperature. However, the lower the temperature, the slower the reaction. Therefore, a high pressure, 200 atm, and an intermediate temperature, 500°C, are actually used in the industrial process. A catalyst to speed up the reaction is also used.

16.28. Consider a change in total volume of the equilibria described. In each case, calculate *K* for a 1.0-L total volume and again for a 2.0-L total volume, and explain in each case why there is or is not a difference.

(a) In the reaction B $\rightleftarrows$ C + D, 0.16 mol of B, 0.10 mol of C, and 0.10 mol of D are found in a 1.0-L reaction mixture at equilibrium. Calculate the value of K. If the same mixture were found in a 2.0-L reaction mixture at equilibrium, would K be different?

(b) In the reaction 2 B $\rightleftarrows$ C + D, 0.16 mol of B, 0.10 mol of C, and 0.10 mol of D are found in a 1.0-L reaction mixture at equilibrium. Calculate the value of K. If the same mixture were found in a 2.0-L reaction mixture at equilibrium, would K be different? Compare this answer to that in part a, and explain.

Ans.   (a)   In the two reactions we have

$$K_1 = \frac{(0.10)(0.10)}{0.16} = 0.063 \qquad K_2 = \frac{(0.050)(0.050)}{0.080} = 0.031$$

and the values are different.

(b)   In the two reactions we have

$$K_1 = \frac{(0.10)(0.10)}{(0.16)^2} = 0.39 \qquad K_2 = \frac{(0.050)(0.050)}{(0.080)^2} = 0.39$$

The values are the same. In part b, 2.0 L appears twice in the numerator and twice in the denominator, so that it cancels out. In part a, it appears only once in the denominator and twice in the numerator, and so the constant in the 2.0-L volume would be half as large as that in the 1.0-L volume. Note that in part a, the same mixture would *not* appear in the 2.0-L mixture at the same temperature, because the value of K does not change with volume.

16.29. Calculate the units of K in Problem 16.14b and c. Can you tell from the units which·of the equations is referred to?

Ans.   The unit for part b is L/mol; that for part c is $(L/mol)^{1/2}$. If a value is given in L/mol for the reaction, the equation in part b is the one to be used. If the units are the square root of those, the equation of part c is to be used.

16.30. What is the sequence of putting values in the table in the solution of Problem 16.19?

Ans.   The superscript numbers before the table entries give the sequence.

	2 A +		B		$\rightleftarrows$	C		+	2 D	
Initial	[1]	0.90	[1]	1.60		[1]	0	[1]	0	
Change	[3]	−0.70	[3]	−0.35		[2]	+0.35	[3]	+0.70	
Equilibrium	[4]	0.20	[4]	1.25		[1]	0.35	[4]	0.70	

16.31. For the reaction A + B $\rightleftarrows$ X + heat, $K = 1.0 \times 10^{-4}$, would an increase in temperature raise or lower the value of K?

Ans.   According to Le Châtelier's principle, raising the temperature would shift this equilibrium to the left. That means that there would be less X and more A and B present at the new equilibrium temperature. The value of the equilibrium constant at that temperature would therefore be lower than the one at the original temperature.

$$K = \frac{[X]}{[A][B]}$$

16.32. Calculate the value of the equilibrium constant at a certain temperature for the following reaction if there are present at equilibrium 0.10 mol of N_2, 0.070 mol of O_2, and 1.4×10^{-3} mol of NO_2 in 2.0 L.

$$N_2 + 2 O_2 \rightleftarrows 2 NO_2$$

Ans.

$$[N_2] = 0.10 \text{ mol}/2.0 \text{ L} = 0.050 \text{ mol/L}$$

$$[O_2] = 0.070 \text{ mol}/2.0 \text{ L} = 0.035 \text{ mol/L}$$

$$[NO_2] = 0.0014 \text{ mol}/2.0 \text{ L} = 0.000\,70 \text{ mol/L}$$

$$K = \frac{[NO_2]^2}{[N_2][O_2]^2} = \frac{(0.000\,70)^2}{(0.050)(0.035)^2} = 8.0 \times 10^{-3}$$

16.33. For the reaction $N_2O_4 \rightleftarrows 2\,NO_2$ calculate the value of the equilibrium constant if initially 1.00 mol of NO_2 and 1.00 mol of N_2O_4 are placed in a 1.00-L vessel and at equilibrium 0.75 mol of N_2O_4 is left in the vessel.

Ans.

	$N_2O_4 \rightleftarrows$	$2\,NO_2$
Initial	1.00	1.00
Change	-0.25	$+0.50$
Equilibrium	0.75	1.50

$$K = \frac{[NO_2]^2}{[N_2O_4]} = \frac{(1.50)^2}{0.75} = 3.0$$

16.34. In a 3.00-L vessel at 488°C, H_2 and $I_2(g)$ react to form HI with $K = 50$. If 0.150 mol of each reactant is used, how many moles of I_2 remain at equilibrium?

Ans. The initial concentration of each reagent is 0.150 mol/3.00 L = 0.0500 mol/L. Let the equilibrium concentration of HI be $2x$ for convenience.

	H_2	$+\ I_2$	$\rightleftarrows\ 2HI$
Initial	0.0500	0.0500	0
Change	$-x$	$-x$	$+2x$
Equilibrium	$0.0500 - x$	$0.0500 - x$	$2x$

$$K = \frac{[HI]^2}{[H_2][I_2]} = \frac{(2x)^2}{(0.0500 - x)^2} = 50$$

Taking the square root of each side of the last equation yields

$$\frac{2x}{0.0500 - x} = 7.1$$

$$2x = 7.1(0.0500) - 7.1x$$

$$9.1x = 7.1(0.0500)$$

$$x = 0.039$$

$$[I_2] = 0.0500 - 0.039 = 0.011\ \text{mol/L}$$

In 3.00 L, there is then 0.033 mol I_2.

16.35. Calculate the number of moles of Cl_2 produced at equilibrium in a 20.0-L vessel when 2.00 mol of PCl_5 is heated to 250°C. $K = 0.041$ mol/L.

$$PCl_5 \rightleftarrows PCl_3 + Cl_2$$

Ans.

$$K = 0.041 = \frac{[PCl_3][Cl_2]}{[PCl_5]} = \frac{x^2}{0.100 - x}$$

$$x^2 = 0.0041 - 0.041x$$

$$x^2 + 0.041x - 0.0041 = 0$$

$$x = \frac{-0.041 + \sqrt{(0.041)^2 + 4(0.0041)}}{2} = 0.047$$

$$\text{moles } Cl_2 = (0.047\ \text{mol/L})(20.0\ \text{L}) = 0.94\ \text{mol}$$

16.36. Calculate the equilibrium CO_2 concentration if 0.200 mol of CO and 0.200 mol of H_2O are placed in a 1.00-L vessel and allowed to come to equilibrium at 975°C.

$$CO + H_2O(g) \rightleftarrows CO_2 + H_2 \qquad K = 0.64$$

Ans.

$$K = \frac{[CO_2][H_2]}{[CO][H_2O]} = \frac{x^2}{(0.200-x)^2} = 0.64$$

Taking the square root of this expression yields

$$\frac{x}{0.200-x} = 0.80$$

$$x = 0.16 - 0.80x$$

$$1.80x = 0.16$$

$$x = 0.089 \text{ mol/L}$$

16.37. Would you expect reactions between molecules to proceed faster than certain reactions between ions in which no covalent bonds are to be broken?

Ans.   No, in general they are much slower.

16.38. What effect would halving the volume have on the HI pressure in the following system at equilibrium at 500°C?

$$H_2 + I_2 \rightleftharpoons 2\,HI$$

Ans.   At 500°C, all these reagents are gases, so there are equal numbers of moles of gas on the two sides, and the equilibrium would not shift. The only change in pressure is due to Boyle's law (Chap. 12). The decrease in volume would double the pressure of each of the gases.

CHAPTER 17

Acid-Base Theory

17.1. INTRODUCTION

Thus far, we have used the *Arrhenius theory* of acids and bases (Secs. 6.4 and 8.3), in which acids are defined as hydrogen-containing compounds that react with bases. Bases are compounds containing OH^- ions or that form OH^- ions when they react with water. Bases react with acids to form salts and water. Metallic hydroxides and ammonia are the most familiar bases to us.

The *Brønsted-Lowry theory* expands the definition of acids and bases to allow us to explain much more of solution chemistry. For example, the Brønsted-Lowry theory allows us to explain why a solution of ammonium nitrate tests acidic and a solution of potassium acetate tests basic. Most of the substances that we consider acids in the Arrhenius theory are also acids in the Brønsted-Lowry theory, and the same is true of bases. In both theories, *strong acids* are those that react completely with water to form ions. *Weak acids* ionize only slightly. We can now explain this partial ionization as an equilibrium reaction of the weak acid, the ions, and the water. A similar statement can be made about weak bases:

$$HC_2H_3O_2 + H_2O \rightleftharpoons C_2H_3O_2^- + H_3O^+$$
$$NH_3 + H_2O \rightleftharpoons NH_4^+ + OH^-$$

17.2. THE BRØNSTED-LOWRY THEORY

In the Brønsted-Lowry theory, (often called the Brønsted theory for short), an *acid* is defined as a substance that donates a proton to another substance. In this sense, a *proton* is a hydrogen atom that has lost its electron; it has nothing to do with the protons in the nuclei of other atoms. (The nuclei of ^2H are also considered protons for this purpose; they are hydrogen ions.) A *base* is a substance that accepts a proton from another substance. The reaction of an acid and a base produces another acid and base. The following reaction is thus an acid-base reaction according to Brønsted:

$$\underset{\text{acid}}{HC_2H_3O_2} + \underset{\text{base}}{H_2O} \rightleftharpoons \underset{\text{base}}{C_2H_3O_2^-} + \underset{\text{acid}}{H_3O^+}$$

The $HC_2H_3O_2$ is an acid because it denotes its proton to the H_2O to form $C_2H_3O_2^-$ and H_3O^+. The H_2O is a base because it accepts that proton. But this is an equilibrium reaction, and $C_2H_3O_2^-$ reacts with H_3O^+ to form $HC_2H_3O_2$ and H_2O. The $C_2H_3O_2^-$ is a base because it accepts the proton from H_3O^+; the H_3O^+ is an acid because it donates a proton. H_3O^+ is called the *hydronium ion*. It is the combination of a proton and a water molecule, and is the species that we have been abbreviating H^+ thus far in this book. (H^+ is not stable; it does not have the configuration of a noble gas; see Sec. 5.4.)

The acid on the left of this equation is related to the base on the right; they are said to be *conjugates* of each other. The $HC_2H_3O_2$ is the conjugate acid of the base $C_2H_3O_2^-$. Similarly, H_2O is the conjugate base of H_3O^+. *Conjugate differ in each case by H^+.*

EXAMPLE 17.1.   Write an equilibrium equation for the reaction of NH_3 and H_2O, and label each of the conjugate acids and bases.

Ans.

$$\overset{\displaystyle\overbrace{}^{\text{Conjugates}}}{\underset{\displaystyle\underbrace{}_{\text{Conjugates}}}{\underset{\text{base}}{NH_3} + \underset{\text{acid}}{H_2O} \rightleftharpoons \underset{\text{acid}}{NH_4^+} + \underset{\text{base}}{OH^-}}}$$

The NH_3 is a base because it accepts a proton from water, which is therefore an acid. The NH_4^+ is an acid because it can donate a proton to OH^-, a base.

We have now labeled water as both an acid and a base. It is useful to think of water as either, because it really has no more properties of one than the other. Water is sometimes referred to as *amphiprotic*. It reacts as an acid in the presence of bases, and it reacts as a base in the presence of acids.

Various acids have different strengths. Some acids are strong; that is, they react with water completely to form their conjugate bases. Other acids are weak, and they form conjugate bases that are stronger than the conjugate bases of strong acids. In fact, the stronger the acid, the weaker its conjugate base. Some acids are so extremely weak that they do not donate protons at all at ordinary temperatures. In this text we refer to these conjugates as feeble. We can classify the acids and the related conjugate bases as follows:

Conjugate Acid	Conjugate Base
Strong	Feeble
Weak	Weak
Feeble	Strong

The same reasoning applies to bases that are molecules and their conjugate acids (that are ions).

Note especially that weak acids **do not** have strong conjugate bases, as stated in some texts. For example, acetic acid is weak, and its conjugate base, the acetate ion, is certainly not strong. It is even weaker as a base than acetic acid is as an acid.

EXAMPLE 17.2.   Classify the following acids and bases according to their strength: HNO_3, $HC_2H_3O_2$, KOH, and NH_3.

Ans. HNO_3 is a strong acid; $HC_2H_3O_2$ is a weak acid; KOH is a strong base; NH_3 is a weak base.

EXAMPLE 17.3.   Classify the conjugates of the species in the prior example according to their strength.

Ans. NO_3^- is a feeble base; $C_2H_3O_2^-$ is a weak base; K^+ is a feeble acid; NH_4^+ is a weak acid.

This classification indicates that NO_3^- and K^+ have no tendency to react with water to form their conjugates. The $C_2H_3O_2^-$ does react with water to some extent to form $HC_2H_3O_2$ and OH^-. The NH_4^+ reacts with water to a small extent to form H_3O^+ and NH_3. Consider the following equation:

$$\underset{\text{acid}}{HNO_3} + \underset{\text{base}}{H_2O} \rightleftharpoons \underset{\text{base}}{NO_3^-} + \underset{\text{acid}}{H_3O^+}$$

Since HNO_3 reacts with water 100%, NO_3^- does not react with H_3O^+ at all. If the nitrate ion cannot take a proton from the hydronium ion, it certainly cannot take the proton from water, which is a much weaker acid than the hydronium ion is.

EXAMPLE 17.4.   What is the difference between the reaction of $HC_2H_3O_2$ with H_2O and with OH^-?

Ans.   The first reaction takes place to a slight extent; $HC_2H_3O_2$ is a weak acid. The second reaction goes almost 100%. Even weak acids react almost completely with OH^-.

The acidity of a solution is determined by the hydronium ion concentration of the solution. The greater the $[H_3O^+]$, the more acidic the solution; the lower the $[H_3O^+]$, the more basic the solution. Other substances, for example, OH^-, affect the acidity of a solution by affecting the concentration of H_3O^+. The presence in water of OH^- in greater concentration than H_3O^+ makes the solution basic. If the relative concentrations are reversed, the solution is acidic.

EXAMPLE 17.5.   Explain why $KC_2H_3O_2$ tests basic in water solution.

Ans.   The K^+ does not react with water at all. The $C_2H_3O_2^-$ reacts with water to a slight extent:

$$C_2H_3O_2^- + H_2O \rightleftharpoons HC_2H_3O_2 + OH^-$$

The excess OH^- makes the solution basic.

17.3.   ACID-BASE EQUILIBRIUM

Equilibrium constants can be written for the ionization of weak acids and weak bases, just as for any other equilibria. For the equation

$$HC_2H_3O_2 + H_2O \rightleftharpoons C_2H_3O_2^- + H_3O^+$$

we would originally (Chap. 16) write

$$K = \frac{[C_2H_3O_2^-][H_3O^+]}{[HC_2H_3O_2][H_2O]}$$

However, in dilute aqueous solution, the concentration of H_2O is practically constant, and its concentration is conventionally built into the value of the equilibrium constant. The new constant, variously called K_a or K_i for acids (K_b or K_i bases), does not have the water concentration term in the denominator:

$$K_a = \frac{[C_2H_3O_2^-][H_3O^+]}{[HC_2H_3O_2]}$$

EXAMPLE 17.6.   Calculate the value of K_a for $HC_2H_3O_2$ in 0.150 M solution if the H_3O^+ concentration of the solution is found to be 1.64×10^{-3} M.

Ans.

	$HC_2H_3O_2$	+	$H_2O \rightleftharpoons$	$C_2H_3O_2^-$	+	H_3O^+
Initial	0.150			0		0
Change						
Equilibrium						1.64×10^{-3}

	$HC_2H_3O_2$	+	$H_2O \rightleftharpoons$	$C_2H_3O_2^-$	+	H_3O^+
Initial	0.150			0		0
Change	-1.64×10^{-3}			1.64×10^{-3}		1.64×10^{-3}
Equilibrium						1.64×10^{-3}

	$HC_2H_3O_2$	+	$H_2O \rightleftharpoons$	$C_2H_3O_2^-$	+	H_3O^+
Initial	0.150			0		0
Change	-1.64×10^{-3}			1.64×10^{-3}		1.64×10^{-3}
Equilibrium	0.148			1.64×10^{-3}		1.64×10^{-3}

The value of the equilibrium constant is given by

$$K_a = \frac{[C_2H_3O_2^-][H_3O^+]}{[HC_2H_3O_2]} = \frac{(1.64 \times 10^{-3})(1.64 \times 10^{-3})}{0.148} = 1.82 \times 10^{-5}$$

EXAMPLE 17.7.   Calculate the hydronium ion concentration of a 0.250 M solution of acetic acid, using the equilibrium constant of Example 17.6.

Ans.   In this example, we will use x for the unknown concentration of hydronium ions. We will solve in terms of x.

$$HC_2H_3O_2 \;+\; H_2O \rightleftharpoons C_2H_3O_2^- + H_3O^+$$

Initial	0.250	0	0
Change	$-x$	x	x
Equilibrium	$0.250 - x$	x	x

$$K_a = \frac{[C_2H_3O_2^-][H_3O^+]}{[HC_2H_3O_2]} = \frac{(x)(x)}{0.250 - x} = 1.82 \times 10^{-5}$$

Since the acid is weak, we can solve this equation most easily by assuming that x is small enough to neglect when *added to* or *subtracted from* a greater concentration. Therefore,

$$K_a = \frac{(x)(x)}{0.250} = 1.82 \times 10^{-5}$$

$$x^2 = 4.55 \times 10^{-6}$$

$$x = 2.13 \times 10^{-3} = [H_3O^+]$$

In cases where x is too large to neglect from the concentration from which it is subtracted, a more exact method is required. The quadratic formula is used (Appendix).

$$K_a = \frac{(x)(x)}{0.250 - x} = 1.82 \times 10^{-5}$$

$$x^2 = 1.82 \times 10^{-5}(0.250 - x) = 4.55 \times 10^{-6} - 1.82 \times 10^{-5}x$$

$$x^2 + 1.82 \times 10^{-5}x - 4.55 \times 10^{-6} = 0$$

$$x = \frac{-1.82 \times 10^{-5} + \sqrt{(1.82 \times 10^{-5})^2 - 4(-4.55 \times 10^{-6})}}{2}$$

$$x = 2.12 \times 10^{-3}$$

In this case, the approximate solution gives almost the same answer as the exact solution. In general, you should use the approximate method and check your answer to see that it is reasonable; only if it is not should you use the quadratic equation.

17.4.   AUTOIONIZATION OF WATER

Since water is defined as both an acid and a base (Sec. 17.2), it is not surprising to find that water can react with itself, even though only to a very limited extent, in a reaction called *autoionization*:

$$\underset{\text{acid}}{H_2O} + \underset{\text{base}}{H_2O} \rightleftharpoons \underset{\text{acid}}{H_3O^+} + \underset{\text{base}}{OH^-}$$

An equilibrium constant for this reaction, called K_w, does not have terms for the concentration of water; otherwise it is like the other equilibrium constants considered so far.

$$K_w = [H_3O^+][OH^-]$$

The value for this constant in dilute aqueous solution at 25°C is 1.0×10^{-14}. Thus, water ionizes very little when it is pure, and even less in acidic or basic solution. You must remember this value.

The equation for K_w indicates that there is always some H_3O^+ and always some OH^- in any aqueous solution. Their concentrations are inversely proportional. A solution is acidic if the H_3O^+ concentration exceeds the OH^- concentration; it is neutral if the two concentrations are equal; and it is basic if the OH^- concentration exceeds the H_3O^+ concentration:

$$[H_3O^+] > [OH^-] \quad \text{acidic}$$
$$[H_3O^+] = [OH^-] \quad \text{neutral}$$
$$[H_3O^+] < [OH^-] \quad \text{basic}$$

EXAMPLE 17.8.   Calculate the hydronium ion concentration in pure water at 25°C.

Ans.
$$2\,H_2O \rightleftharpoons H_3O^+ + OH^-$$
$$K_w = [H_3O^+][OH^-] = 1.0 \times 10^{-14}$$

Both ions must be equal in concentration since they were produced in equal molar quantities by the autoionization reaction. Let the concentration of each be equal to x.

$$x^2 = 1.0 \times 10^{-14}$$
$$x = 1.0 \times 10^{-7} = [H_3O^+] = [OH^-]$$

EXAMPLE 17.9.   Calculate the hydronium ion concentration in 0.0010 M NaOH.

Ans.
$$K_w = [H_3O^+][OH^-] = 1.0 \times 10^{-14}$$
$$[OH^-] = 0.0010\,M$$
$$[H_3O^+](0.0010) = 1.0 \times 10^{-14}$$
$$[H_3O^+] = 1.0 \times 10^{-11}$$

The OH^- from the NaOH has caused the autoionization reaction of water to shift to the left, yielding only a tiny fraction of the H_3O^+ that is present in pure water.

17.5.   THE pH SCALE

The pH scale was invented to reduce the necessity for using exponential numbers to report acidity. The pH is defined as

$$pH = -\log[H_3O^+]$$

EXAMPLE 17.10.   Calculate the pH of (*a*) 0.010 M H_3O^+ solution and (*b*) 0.010 M HCl solution.

Ans.   (*a*)
$$[H_3O^+] = 0.010 = 1.0 \times 10^{-2}$$
$$pH = -\log[H_3O^+] = -\log(1.0 \times 10^{-2}) = 2.00$$

(*b*)   Since HCl is a strong acid, it reacts completely with water to give 0.010 M H_3O^+ solution. Thus, the pH is the same as that in part *a*.

EXAMPLE 17.11.   Calculate the pH of (*a*) 0.0010 M NaOH and (*b*) 0.0010 M $Ba(OH)_2$.

Ans.   (*a*)
$$[OH^-] = 0.0010 = 1.0 \times 10^{-3}$$
$$[H_3O^+] = 1.0 \times 10^{-11} \quad (\textit{See Example } 17.9)$$
$$pH = -\log(1.0 \times 10^{-11}) = +11.00$$

(*b*) $[OH^-] = 0.0020\,M$. (There are two OH^- ions per formula unit.)
$$[OH^-] = 0.0020 = 2.0 \times 10^{-3}$$
$$[H_3O^+] = (1.0 \times 10^{-14})/(2.0 \times 10^{-3}) = 5.0 \times 10^{-12}$$
$$pH = 11.30$$

Special note for electronic calculator users: To determine the negative of the logarithm of a quantity, on some calculators, enter the quantity, press the $\boxed{\text{LOG}}$ key, then press the change-sign key, $\boxed{+/-}$. On other calculators, the first two steps are done in the opposite order. See your owner's manual. Do not use the minus key. (See the Appendix.)

EXAMPLE 17.12. What is the difference in calculating the pH of a solution of a weak acid and that of a strong acid?

Ans. For a strong acid, the H_3O^+ concentration can be determined directly from the concentration of the acid. For a weak acid, the H_3O^+ concentration must be determined first from an equilibrium constant calculation (Sec. 17.3); then the pH is calculated.

EXAMPLE 17.13. Calculate the pH of (a) 0.10 *M* HCl and (b) 0.10 *M* $HC_2H_3O_2$ ($K_a = 1.8 \times 10^{-5}$).

Ans. (a)
$$[H_3O^+] = 0.10 = 1.0 \times 10^{-1}$$
$$pH = 1.00$$

(b)
$$HC_2H_3O_2 + H_2O \rightleftharpoons H_3O^+ + C_2H_3O_2^-$$

Initial	0.10	0	0
Change	$-x$	x	x
Equilibrium	$0.10 - x$	x	x

$$\frac{x^2}{0.10} = 1.8 \times 10^{-5}$$
$$x = 1.3 \times 10^{-3}$$
$$pH = 2.89$$

From the way pH is defined and the value of K_w, we can deduce that solutions with pH = 7 are neutral, those with pH less than 7 are acidic, and those with pH greater than 7 are basic.

$$[H_3O^+] > [OH^-] \quad \text{acidic} \quad pH < 7$$
$$[H_3O^+] = [OH^-] \quad \text{neutral} \quad pH = 7$$
$$[H_3O^+] < [OH^-] \quad \text{basic} \quad pH > 7$$

17.6. BUFFER SOLUTIONS

A *buffer solution* is a solution of a weak acid and its conjugate base or a weak base and its conjugate acid. The main property of a buffer solution is its resistance to changes in its pH despite the addition of small quantities of strong acid or strong base. The student must know the following three things about buffer solutions:

1. How they are prepared.
2. What they do.
3. How they do it.

A buffer solution may be prepared by the addition of a weak acid to a salt of that acid or addition of a weak base to a salt of that base. For example, a solution of acetic acid and sodium acetate is a buffer solution. The weak acid ($HC_2H_3O_2$) and its conjugate base ($C_2H_3O_2^-$, from the sodium acetate) constitute a buffer solution. There are other ways to make such a combination of weak acid plus conjugate base (Problem 17.26).

A buffer solution resists change in its acidity. For example, a certain solution of acetic acid and sodium acetate has a pH of 4.0. When a small quantity of NaOH is added, the pH goes up to 4.2. If that quantity of NaOH had been added to the same volume of an unbuffered solution of HCl at pH 4, the pH would have gone up to a value as high as 12 or 13.

The buffer solution works on the basis of Le Châtelier's principle. Consider the equation for the reaction of acetic acid with water:

$$\underset{\text{excess}}{HC_2H_3O_2} + \underset{\substack{\text{huge} \\ \text{excess}}}{H_2O} \rightleftharpoons \underset{\text{excess}}{C_2H_3O_2^-} + \underset{\substack{\text{limited} \\ \text{quantity}}}{H_3O^+}$$

The solution of $HC_2H_3O_2$ and $C_2H_3O_2^-$ in H_2O results in the relative quantities of each of the species in the equation as shown under the equation. If H_3O^+ is added to the equilibrium system, the equilibrium shifts left to use up some of the added H_3O^+. If the acetate ion were not present to take up the added H_3O^+, the pH would drop. Since the acetate ion reacts with much of the added H_3O^+, there is little increase in H_3O^+ and little drop in pH. If OH^- is added to the solution, it reacts with H_3O^+ present. But the removal of that H_3O^+ is a stress, which causes this equilibrium to shift to the right, replacing much of the H_3O^+ removed by the OH^-. The pH does not rise nearly as much in the buffered solution as it would have in an unbuffered solution.

Calculations may be made as to how much the pH is changed by the addition of a strong acid or base. You should first determine how much of each conjugate would be present if that strong acid or base reacted completely with the weak acid or conjugate base originally present. Use the results as the initial values of concentrations for the equilibrium calculations.

EXAMPLE 17.14. A solution containing 0.250 mol $HC_2H_3O_2$ and 0.150 mol $NaC_2H_3O_2$ in 1.00 L is treated with 0.010 mol NaOH. Assume that there is no change in the volume of the solution. (*a*) What was the original pH of the solution? (*b*) What is the pH of the solution after the addition of the NaOH? $K_a = 1.8 \times 10^{-5}$.

Ans. (*a*)

$$HC_2H_3O_2 + H_2O \rightleftharpoons C_2H_3O_2^- + H_3O^+$$

$$K_a = \frac{[C_2H_3O_2^-][H_3O^+]}{[HC_2H_3O_2]} = 1.8 \times 10^{-5}$$

$$HC_2H_3O_2 + H_2O \rightleftharpoons H_3O^+ + C_2H_3O_2^-$$

Initial	0.250	0	0.150
Change	$-x$	x	x
Equilibrium	$0.250 - x$	x	$0.150 + x$

Assuming x to be negligible when added to or subtracted from larger quantities, we find

$$K_a = \frac{(0.150)x}{0.250} = 1.8 \times 10^{-5}$$

$$x = 3.0 \times 10^{-5}$$

The assumption was valid.

$$pH = 4.52$$

(*b*) We assume that the 0.010 mol of NaOH reacted with 0.010 mol of $HC_2H_3O_2$ to produce 0.010 mol more of $C_2H_3O_2^-$ (Sec. 10.3). That gives us

$$HC_2H_3O_2 + H_2O \rightleftharpoons H_3O^+ + C_2H_3O_2^-$$

Initial	0.240	0	0.160
Change	$-x$	x	x
Equilibrium	$0.240 - x$	x	$0.160 + x$

$$K_a = \frac{(0.160)x}{0.240} = 1.8 \times 10^{-5}$$

$$x = 2.7 \times 10^{-5}$$

$$pH = 4.57$$

The pH has risen from 4.52 to 4.57 by the addition of 0.010 mol NaOH. (That much NaOH would have raised the pH of 1.0 L of an unbuffered solution of HCl originally at the same pH to a final pH value of 12.00.)

Solved Problems

THE BRØNSTED-LOWRY THEORY

17.1. Explain the difference between the strength of an acid and its concentration.

Ans. *Strength* refers to the extent the acid will ionize in water. *Concentration* is a measure of the number of moles of the acid in a certain volume of solution.

17.2. Draw an electron dot diagram for the hydronium ion, and explain why it is expected to be more stable than the hydrogen ion, H^+.

Ans.
$$H:\overset{\cdot\cdot}{\underset{\cdot\cdot}{O}}:H^{+}$$
$$\overset{|}{H}$$

Since each hydrogen atom has two electrons and the oxygen atom has eight electrons, the octet rule is satisfied. In H^+, the atom would have no electrons, which is not the electronic configuration of a noble gas.

17.3. Is NH_4Cl solution in water acidic or basic? Explain.

Ans. It is acidic. Cl^- does not react with water. The NH_4^+ reacts somewhat to form H_3O^+, making the solution somewhat acidic.

$$NH_4^+ + H_2O \rightleftharpoons NH_3 + H_3O^+$$
$$\text{acid} \quad\quad \text{base} \quad\quad\quad \text{base} \quad\quad \text{acid}$$

Another way to explain this effect is to say that NH_4^+ is a weak conjugate acid and Cl^- does not react. The stronger of these makes the solution slightly acidic.

17.4. How is an Arrhenius salt defined in the Brønsted-Lowry system?

Ans. In the Brønsted-Lowry system, an Arrhenius salt is defined as a combination of two conjugates—the conjugate acid of the original base and the conjugate base of the original acid. For example, the salt NH_4ClO_2 is a combination of the conjugate acid (NH_4^+) of NH_3 and the conjugate base (ClO_2^-) of $HClO_2$.

17.5. H_3BO_3 is a much weaker acid than is $HC_2H_3O_2$. The conjugate of which one is the stronger base?

Ans. $H_2BO_3^-$ is a stronger base than $C_2H_3O_2^-$ since it is the conjugate of the weaker acid.

17.6. (*a*) In a 0.150 *M* solution of HA (a weak acid), 1.00% of the HA is ionized. What are the actual concentrations of HA, H_3O^+, and A^- in a solution prepared by adding 0.150 mol HA to sufficient water to make 1.00 L of solution? (*b*) What are the actual concentrations of HA, H_3O^+, and A^- in a solution prepared by adding 0.150 mol A^- (from NaA) plus 0.150 mol H_3O^+ (from HCl) to sufficient water to make 1.00 L of solution?

Ans. (*a*) HA ionizes 1.00% to yield:

$$[HA] = 0.149\ M \quad\quad (99.0\%\ \text{of}\ 0.150\ M)$$
$$[H_3O^+] = 0.0015\ M \quad\quad (1.00\%\ \text{of}\ 0.150\ M)$$
$$[A^-] = 0.0015\ M$$

(*b*) The H_3O^+ reacts with the A^- to give HA and H_2O:

$$H_3O^+ + A^- \rightleftharpoons HA + H_2O$$

This reaction is exactly the reverse of the ionization reaction of HA, so that if the ionization reaction goes 1.00%, this reaction goes 99.0%. Thus, after this reaction there is the same concentration of each of these species.

17.7. Write the net ionic equation for the reaction of HCl(aq) and $NaC_2H_3O_2$. What relationship is there between this equation and the equation for the ionization of acetic acid?

Ans. HCl(aq) is really a solution of H_3O^+ and Cl^-:

$$HCl + H_2O \longrightarrow H_3O^+ + Cl^-$$

The net ionic equation of interest is then

$$H_3O^+ + C_2H_3O_2^- \longrightarrow HC_2H_3O_2 + H_2O$$

This equation is the reverse of the ionization equation for acetic acid.

ACID-BASE EQUILIBRIUM

17.8. The $[H_3O^+]$ in 0.150 M benzoic acid ($HC_7H_5O_2$) is 3.08×10^{-3} M. Calculate K_a.

Ans.
$$HC_7H_5O_2 + H_2O \rightleftarrows H_3O^+ + C_7H_5O_2^-$$

$$[H_3O^+] = [C_7H_5O_2^-] = 3.08 \times 10^{-3}$$

$$[HC_7H_5O_2] = 0.150 - (3.08 \times 10^{-3}) = 0.147$$

$$K_a = \frac{[H_3O^+][C_7H_5O_2^-]}{[HC_7H_5O_2]} = \frac{(3.08 \times 10^{-3})^2}{0.147} = 6.45 \times 10^{-5}$$

17.9. For formic acid, $HCHO_2$, K_a is 1.80×10^{-4}. Calculate $[H_3O^+]$ in 0.0150 M $HCHO_2$.

Ans.
$$HCHO_2 + H_2O \rightleftarrows H_3O^+ + CHO_2^-$$

Initial	0.0150	0	0
Change	$-x$	x	x
Equilibrium	$0.0150 - x$	x	x

If we neglect x when it is subtracted from 0.0150, then

$$K_a = \frac{[H_3O^+][CHO_2^-]}{[HCHO_2]} = \frac{(x)(x)}{0.0150} = 1.80 \times 10^{-4}$$

$$x^2 = 2.70 \times 10^{-6}$$

$$x = 1.64 \times 10^{-3}$$

This value of x is 10.9% of the concentration from which it was subtracted. The approximation is wrong. The quadratic formula must be used:

$$K_a = \frac{(x)(x)}{0.0150 - x} = 1.80 \times 10^{-4}$$

$$x^2 = -1.80 \times 10^{-4}x + 2.70 \times 10^{-6}$$

$$x^2 + 1.80 \times 10^{-4}x - 2.70 \times 10^{-6} = 0$$

$$x = \frac{-b \pm \sqrt{b^2 - 4ac}}{2a}$$

$$= \frac{-1.80 \times 10^{-4} + \sqrt{(1.80 \times 10^{-4})^2 - 4(-2.70 \times 10^{-6})}}{2}$$

$$x = 1.56 \times 10^{-3} = [H_3O^+]$$

Check: $\quad \dfrac{(1.56 \times 10^{-3})^2}{0.0150 - (1.56 \times 10^{-3})} = 1.8 \times 10^{-4}$

17.10. What is the CN^- concentration in 0.030 M HCN? $K_a = 6.2 \times 10^{-10}$.

Ans.
$$HCN \quad + \quad H_2O \rightleftarrows CN^- + H_3O^+$$

Initial	0.030	0	0
Change	$-x$	x	x
Equilibrium	$0.030 - x$	x	x

$$K_a = \frac{[CN^-][H_3O^+]}{[HCN]} = \frac{x^2}{0.030} = 6.2 \times 10^{-10}$$

$$x = 4.3 \times 10^{-6} = [CN^-]$$

17.11. What concentration of acetic acid, $HC_2H_3O_2$, is needed to give a hydronium ion concentration of 4.0×10^{-3} M? $K_a = 1.8 \times 10^{-5}$.

Ans.

$$HC_2H_3O_2 \quad + \quad H_2O \rightleftharpoons H_3O^+ \ + \ C_2H_3O_2{}^-$$

Initial	x		0	0
Change	-0.0040		0.0040	0.0040
Equilibrium	$x - 0.0040 \cong x$		0.0040	0.0040

$$K_a = \frac{[H_3O^+][C_2H_3O_2{}^-]}{[HC_2H_3O_2]} = \frac{(0.0040)^2}{x} = 1.8 \times 10^{-5}$$

$$x = 0.89 \ M$$

AUTOIONIZATION OF WATER

17.12. According to Le Châtelier's principle, what does the H_3O^+ concentration from the ionization of an acid (strong or weak) do to the ionization of water?

Ans.

$$2 \ H_2O \rightleftharpoons H_3O^+ + OH^-$$

The presence of H_3O^+ from an acid will repress this water ionization. The water will ionize less than it does in neutral solution, and the H_3O^+ generated by the water will be negligible in all but the most dilute acid. The OH^- generated by the water will be equally small, but since it is the only OH^- present, it still has to be considered.

17.13. Calculate the hydronium ion concentration in 0.50 M NaCl.

Ans.   Since both Na^+ and Cl^- are feeble, neither reacts with water at all. Thus, the hydronium ion concentration in this solution is the same as that in pure water, $1.0 \times 10^{-7} M$.

17.14. What does a 0.20 M HNO_3 solution contain?

Ans.   It contains 0.20 M H_3O^+, 0.20 M $NO_3{}^-$, H_2O, and a slight concentration of OH^- from the ionization of the water.

17.15. What does a 0.20 M $HC_2H_3O_2$ solution contain?

Ans.   It contains nearly 0.20 M $HC_2H_3O_2$, a little H_3O^+ and $C_2H_3O_2{}^-$, H_2O, and a slight concentration of OH^- from the ionization of the water.

THE pH SCALE

17.16. Write an equilibrium constant expression for each of the following:

    (*a*) $HC_2H_3O_2 + H_2O \rightleftharpoons H_3O^+ + C_2H_3O_2{}^-$ (*b*) $2 \ H_2O \rightleftharpoons H_3O^+ + OH^-$

    Ans.   (*a*) $K_a = \dfrac{[H_3O^+][C_2H_3O_2{}^-]}{[HC_2H_3O_2]}$ (*b*) $K_w = [H_3O^+][OH^-]$

17.17. Calculate the pH of each of the following solutions: (*a*) 0.0100 M HCl, (*b*) 0.0100 M NaOH, (*c*) 0.0100 M $Ba(OH)_2$, and (*d*) 0.0100 M NaCl.

    Ans.   (*a*) 2.000 (*b*) 12.000 (*c*) 12.301 (*d*) 7.000

17.18. Calculate the pH of 0.100 M NH_3. $K_b = 1.8 \times 10^{-5}$

Ans.

$$NH_3 + H_2O \rightleftharpoons NH_4^+ + OH^-$$

Initial	0.100	0	0
Change	$-x$	x	x
Equilibrium	$0.100 - x$	x	x

$$K_b = \frac{[NH_4^+][OH^-]}{[NH_3]} = \frac{(x)(x)}{0.100} = 1.8 \times 10^{-5}$$

$$x^2 = 1.8 \times 10^{-6}$$

$$x = 1.3 \times 10^{-3} = [OH^-]$$

$$[H_3O^+] = \frac{K_w}{[OH^-]} = \frac{1.0 \times 10^{-14}}{1.3 \times 10^{-3}} = 7.5 \times 10^{-12}$$

$$pH = 11.13$$

17.19. Calculate the pH, where applicable, of every solution mentioned in Problem 17.8 through 17.14.

Ans. Merely take the logarithm of the *hydronium ion concentration* and change the sign: Problem 17.8, 2.511; Problem 17.9, 2.807; Problem 17.10, 5.37; Problem 17.11, 0.05; Problem 17.13, 7.00; and Problem 17.14, 0.70.

BUFFER SOLUTIONS

17.20. Is it true that when one places a weak acid and its conjugate base in the same solution, they react with each other? That is, do they both appear on the same side of the equation?

Ans. They do not react with each other in that way. The conjugate acts as a stress as defined by Le Châtelier's principle, and does not allow the weak acid to ionize as much as it would if the conjugate were not present.

17.21. According to Le Châtelier's principle, what is the effect of NH_4Cl on a solution of NH_3?

Ans. The added NH_4^+ will cause the ionization of NH_3 to be repressed. The base will not ionize as much as if the ammonium ion were not present.

17.22. One way to get acetic acid and sodium acetate into the same solution is to add them both to water. State another way to get both these reagents in a solution.

Ans. One can add acetic acid and neutralize some of it with sodium hydroxide (Sec. 10.3). The excess acetic acid remains in the solution, and the quantity that reacts with the sodium hydroxide will form sodium acetate. Thus, both acetic acid and sodium acetate will be present in the solution without any sodium acetate having been added directly. See the next problem.

17.23. State another way that a solution can be made containing acetic acid and sodium acetate, besides the way indicated in Problem 17.22.

Ans. One can add sodium acetate and hydrochloric acid. The reverse of the ionization reaction occurs, yielding acetic acid:

$$C_2H_3O_2^- + H_3O^+ \rightleftharpoons HC_2H_3O_2 + H_2O$$

If a smaller number of moles of hydrochloric acid is added than moles of sodium acetate present, some of the sodium acetate will be in excess (Sec. 10.3), and will be present in the solution with the acetic acid formed in this reaction.

17.24. Calculate the hydronium ion concentration of a 0.250 M acetic acid solution also containing 0.190 M sodium acetate. $K_a = 1.81 \times 10^{-5}$.

Ans.

$$HC_2H_3O_2 + H_2O \rightleftharpoons H_3O^+ + C_2H_3O_2^-$$

Initial	0.250	0	0.190
Change	$-x$	x	x
Equilibrium	$0.250 - x$	x	$0.190 + x$

$$K_a = \frac{[C_2H_3O_2^-][H_3O^+]}{[HC_2H_3O_2]} = \frac{(x)(0.190 + x)}{0.250 - x} = 1.81 \times 10^{-5}$$

$$K_a \cong \frac{(x)(0.190)}{0.250} \cong 1.81 \times 10^{-5}$$

$$x = 2.38 \times 10^{-5} = [H_3O^+]$$

17.25. A solution contains 0.20 mol of HA (a weak acid) and 0.15 mol of NaA (a salt of the acid). Another solution contains 0.20 mol of HA to which 0.15 mol of NaA is added. The final volume of each solution is 1.0 L. What difference, if any, is there between the two solutions?

Ans.   There is no difference between the two solutions. The wording of the problems is different, but the final result is the same.

17.26. What are the concentrations in 1.00 L of solution after 0.100 mol of NaOH is added to 0.200 mol of $HC_2H_3O_2$ but before any equilibrium reaction is considered? Of what use would these be to calculate the pH of this solution?

Ans.   The balanced equation is

$$NaOH + HC_2H_3O_2 \longrightarrow NaC_2H_3O_2 + H_2O$$

The limiting quantity is NaOH, and so 0.100 mol NaOH reacts with 0.100 mol $HC_2H_3O_2$ to produce 0.100 mol $NaC_2H_3O_2$ + 0.100 mol H_2O. There is 0.100 mol excess $HC_2H_3O_2$ left in the solution. So far, this problem is exactly the same as Problem 10.70. Now we can start the equilibrium calculations with these initial quantities. We have a buffer solution problem.

17.27. Calculate the pH of a solution containing 0.250 M NH_3 plus 0.200 M NH_4Cl. $K_b = 1.8 \times 10^{-5}$.

Ans.

$$NH_3 \quad + \quad H_2O \rightleftharpoons NH_4^+ \quad + \quad OH^-$$

Initial	0.250	0.200	0
Change	$-x$	x	x
Equilibrium	$0.250 - x$	$0.200 + x$	x

$$K_b = \frac{[NH_4^+][OH^-]}{[NH_3]} = \frac{(x)(0.200)}{0.250} = 1.8 \times 10^{-5}$$

$$x = 2.25 \times 10^{-5} = [OH^-]$$

$$[H_3O^+] = \frac{K_w}{[OH^-]} = \frac{1.0 \times 10^{-14}}{2.25 \times 10^{-5}} = 4.4 \times 10^{-10}$$

$$pH = 9.35$$

Supplementary Problems

17.28. Identify each of the following terms: (*a*) hydronium ion, (*b*) Brønsted-Lowry theory, (*c*) proton (Brønsted sense), (*d*) acid (Brønsted sense), (*e*) base (Brønsted sense), (*f*) conjugate, (*g*) strong acid or base, (*h*) acid dissociation constant, (*i*) base dissociation constant, (*j*) autoionization, (*k*) pH, and (*l*) K_w.

Ans. (*a*) The hydronium ion is H_3O^+, which is a proton added to a water molecule. (*b*) The Brønsted-Lowry theory defines acids as proton donors and bases as proton acceptors. (*c*) A proton in the Brønsted sense is a hydrogen nucleus—a hydrogen ion. (*d*) An acid is a proton donor. (*e*) A base in the Brønsted sense is a proton acceptor. (*f*) A conjugate is the product of a proton loss or gain from a Brønsted acid or base. (*g*) An acid or a base that reacts completely with water to form H_3O^+ or OH^- ions, respectively. The common strong acids are HCl, $HClO_3$, $HClO_4$, HBr, HI, HNO_3, and H_2SO_4 (first proton). All soluble hydroxides are strong bases. (*h*) The equilibrium constant for the reaction of a weak acid with water. (*i*) The equilibrium constant for the reaction of a weak base with water. (*j*) An acid-base reaction of a substance with itself, for example, $2\,H_2O \rightleftharpoons H_3O^+ + OH^-$. (*k*) $pH = -\log[H_3O^+]$. (*l*) $K_w = [H_3O^+][OH^-]$.

17.29. What is the effect of NH_4Cl on a solution of NH_3?

Ans. The added NH_4^+ will cause the ionization of NH_3 to be repressed. The NH_3 will not ionize as much as if the ammonium ion were not present. This is the same as Problem 17.21, whether or not Le Châtelier's principle is mentioned.

17.30. What chemicals are left in solution after 0.100 mol of NaOH is added to 0.200 mol of NH_4Cl?

Ans. The balanced equation is

$$NaOH + NH_4Cl \longrightarrow NH_3 + NaCl + H_2O$$

The limiting quantity is NaOH, so that 0.100 mol NaOH reacts with 0.100 mol NH_4Cl to produce 0.100 mol NH_3 + 0.100 mol NaCl + 0.100 mol H_2O. There is also 0.100 mol excess NH_4Cl left in the solution. This is a buffer solution of NH_3 plus NH_4^+ in which the Na^+ and Cl^- are inert.

17.31. Fluoroacetic acid, $HC_2H_2FO_2$, has a K_a of 2.6×10^{-3}. What concentration of the acid is required to get $[H_3O^+] = 1.5 \times 10^{-3}$?

Ans.
$$HC_2H_2FO_2 + H_2O \rightleftharpoons C_2H_2FO_2^- + H_3O^+$$

$$K_a = \frac{[C_2H_2FO_2^-][H_3O^+]}{[HC_2H_2FO_2]} = 2.6 \times 10^{-3}$$

$$HC_2H_2FO_2 + H_2O \rightleftharpoons C_2H_2FO_2^- + H_3O^+$$

Initial	x	0	0
Change	-0.0015	0.0015	0.0015
Equilibrium	$x - 0.0015$	0.0015	0.0015

If you try to neglect 0.0015 with respect to x, you get a foolish answer. This problem must be solved exactly.

$$K_a = \frac{(0.0015)^2}{x - 0.0015} = 2.6 \times 10^{-3}$$

$$2.25 \times 10^{-6} = (2.6 \times 10^{-3})x - 3.9 \times 10^{-6}$$

$$x = 2.4 \times 10^{-3} = [HC_2H_2FO_2]$$

17.32. For the reaction $W + X \rightleftharpoons Y + Z$

(*a*) Calculate the value of the equilibrium constant if 0.100 mol of W and 55.40 mol of X were placed in a 1.00-L vessel and allowed to come to equilibrium, at which point 0.010 mol of Z was present. (*b*) Using the value of the equilibrium constant calculated in (*a*), determine the concentration of Y at equilibrium if 0.200 mol of W and 55.40 mol of X were placed in a 1.00-L vessel and allowed to come to equilibrium. (*c*) Calculate the ratio of the concentration of W at equilibrium to that initially present in part *b*. Calculate the same ratio for X. (*d*) Calculate the value of

$$K' = \frac{[Y][Z]}{[W]}$$

from the data of part a. (e) Using the equation of part d and the data of part b, calculate the concentration of Y at equilibrium. (f) Explain why the answer to part e is the same as that to part b.

Ans.   (a)

$$K = \frac{[Y][Z]}{[W][X]} = \frac{(0.010)^2}{(0.090)(55.39)} = 2.0 \times 10^{-5}$$

(b)

$$\frac{x^2}{(0.200)(55.40)} = 2.0 \times 10^{-5}$$

$$x^2 = 2.2 \times 10^{-4}$$

$$x = 1.5 \times 10^{-2}$$

(c) $\dfrac{0.185}{0.200} = 0.925$ $\dfrac{55.38}{55.40} = 0.9996$ (practically no change for X)

(d)

$$K' = \frac{(0.010)^2}{0.090} = 1.1 \times 10^{-3}$$

(e)

$$\frac{x^2}{0.200} = 1.1 \times 10^{-3}$$

$$x^2 = 2.2 \times 10^{-4}$$

$$x = 1.5 \times 10^{-2}$$

(f) The concentration of X is essentially constant throughout the reaction (see part c). The equilibrium constant of part a is amended to include that constant concentration; the new constant is K'. That constant is just as effective in calculating the equilibrium concentration of Y as is the original constant K. This is the same effect as using K_a, not K and [H$_2$O], for weak acid and weak base equilibrium calculations.

17.33. (a) What ions are present in a solution of sodium acetate? (b) According to Le Châtelier's principle, what effect would the addition of sodium acetate have on a solution of 0.250 M acetic acid? (c) Compare the hydronium ion concentrations of Example 17.7 and Problem 17.24 to check your answer to part b.

Ans.   (a) Na$^+$ and C$_2$H$_3$O$_2^-$ (b) The acetate ion should repress the ionization of the acetic acid, so that there would be less hydronium ion concentration in the presence of the sodium acetate.   (c) The hydronium ion concentration dropped from 2.12×10^{-3} to 2.38×10^{-5} as a result of the addition of 0.190 M sodium acetate.

17.34. Consider the ionization of the general acid HA:

$$\text{HA} + \text{H}_2\text{O} \rightleftharpoons \text{H}_3\text{O}^+ + \text{A}^-$$

Calculate the hydronium ion concentration of a 0.100 M solution of acid for (a) $K_a = 2.0 \times 10^{-10}$ and (b) $K_a = 8.0 \times 10^{-2}$

Ans.   (a)   Let [H$_3$O$^+$] $= x$

$$K_a = \frac{[\text{H}_3\text{O}^+][\text{A}^-]}{[\text{HA}]} = 2.0 \times 10^{-10}$$

$$\frac{x^2}{0.10 - x} = 2.0 \times 10^{-10}$$

$$x^2 = 2.0 \times 10^{-11}$$

$$x = 4.5 \times 10^{-6}$$

(b)   Let $[H_3O^+] = x$.

$$K_a = \frac{[H_3O^+][A^-]}{[HA]} = 8.0 \times 10^{-2}$$

$$\frac{x^2}{0.10 - x} = 8.0 \times 10^{-2}$$

$$x^2 = 8.0 \times 10^{-2}(0.10 - x)$$

Here the x cannot be neglected when subtracted from 0.10, because the equilibrium constant is too big. The problem is solved by using the quadratic formula (Appendix).

$$x = 5.8 \times 10^{-2}$$

CHAPTER 18

Organic Chemistry

18.1. INTRODUCTION

Historically, the term *organic chemistry* has been associated with the study of compounds obtained from plants and animals. However, about 175 years ago it was found that typical organic compounds could be prepared in the laboratory without the use of any materials derived directly from living organisms. Indeed, today great quantities of synthetic materials, having properties as desirable as, or more desirable than, those of natural products, are produced commercially. These materials include fibers, perfumes, medicines, paints, pigments, rubber, and building materials.

In modern terms, an organic compound is one that contains at least one carbon-to-carbon and/or carbon-to-hydrogen bond. (Urea, thiourea, and a few other compounds that do not fit this description are considered to be organic compounds.) In addition to carbon and hydrogen, the elements that are most likely to be present in organic compounds are oxygen, nitrogen, phosphorus, sulfur, and the halogens (fluorine, chlorine, bromine, and iodine). With just these few elements, literally millions of organic compounds are known, and thousands of new compounds are synthesized every year.

That an entire branch of chemistry can be based on such a relatively small number of elements can be attributed to the fact that carbon atoms have the ability to link together to form long chains, rings, and a variety of combinations of branched chains and fused rings.

In this chapter, some basic concepts of organic chemistry will be described. The objectives of the discussions will be to emphasize the systematic relationships that exist in simple cases. Extension of the concepts presented will be left to more advanced texts.

18.2. BONDING IN ORGANIC COMPOUNDS

The elements that are commonly part of organic compounds are all located in the upper right corner of the periodic table. They are all nonmetals. The bonds between atoms of these elements are essentially covalent. (Some organic molecules may form ions; nevertheless, the bonds *within* each organic ion are covalent. For example, the salt sodium acetate consists of sodium ions, Na^+, and acetate ions, $C_2H_3O_2^-$. Despite its charge, the bonds *within* the acetate ion are all covalent.)

The covalent bonding in organic compounds can be described by means of the electron dot notation (Chap. 5). The carbon atoms has four electrons in its outermost shell:

In order to complete its octet, each carbon atom must share a total of four electron pairs. The *order* of a bond is the number of electron pairs shared in that bond. The total number of shared pairs is called the *total bond order* of an atom. Thus, carbon must have a total bond order of 4 (except in CO). A *single bond* is a sharing of one pair of electrons; a *double bond*, two; and a *triple bond*, three. Therefore, in organic compounds, each carbon atom forms four single bonds, a double bond and two single bonds, a triple bond and a single bond, or two double bonds. As shown in the table below, each of these possibilities corresponds to a total bond order of 4.

Number and Types of Bonds	Total Bond Order
Four single bonds	$4 \times 1 = 4$
One double bond and two single bonds	$(1 \times 2) + (2 \times 1) = 4$
One triple bond and one single bond	$(1 \times 3) + (1 \times 1) = 4$
Two double bonds	$2 \times 2 = 4$

A hydrogen atom has only one electron in its outermost shell, and can accommodate a maximum of two electrons in its outermost shell. Hence, in any molecule, each hydrogen atom can form only one bond—a single bond. The oxygen atom, with six electrons in its outermost shell, can complete its octet by forming either two single bonds or one double bond, for a total bond order of 2 (except in CO). The total bond orders of the other elements usually found in organic compounds can be deduced in a similar manner. The results are given in Table 18-1.

Table 18-1   Total Bond Orders in Organic Compounds

Element	Symbol	Total Bond Order	Periodic Group Number
Carbon	C	4	4
Nitrogen	N	3	5
Phosphorus	P	3	5
Oxygen	O	2	6
Sulfur	S	2	6
Halogen	X	1	7
Hydrogen	H	1	1

With the information given in Table 18-1, it is possible to write an electron dot structure for many organic molecules.

EXAMPLE 18.1.   Write an electron dot structure for each of the following molecular formulas: (*a*) CH_4, (*b*) CH_4O, (*c*) CH_5N, (*d*) CH_2O, and (*e*) C_2F_2.

Ans.   (*a*) H:C:H with H above and H below   (*b*) H:C:O:H with H above and H below   (*c*) H:C:N:H with H above and H H below   (*d*) H:C::O: (*e*) :F:C:::C:F:

Each atom in this example is characterized by a total bond order corresponding to that listed in Table 18-1.

18.3.   STRUCTURAL, CONDENSED, AND LINE FORMULAS

Electron dot formulas are useful for deducing the structures of organic molecules, but it is more convenient to use simpler representations—structural or graphical formulas—in which a line is used to denote a shared pair of electrons. Because each pair of electrons shared between two atoms is equivalent to a bond order of 1, each shared pair can be represented by a line between the symbols of the elements. Unshared electrons on the atoms are often not shown in this kind of representation. The resulting representations of molecules are called *structural formulas* or *graphical formulas*. The structural formulas for compounds (*a*) to (*e*) described in Example 18.1 may be written as follows:

$$
\begin{array}{ccc}
& \text{H} & \text{H} & \text{H} \\
& | & | & | \\
(a)\quad \text{H—C—H} & (b)\quad \text{H—C—O—H} & (c)\quad \text{H—C—N—H} \\
& | & | & |\ \ | \\
& \text{H} & \text{H} & \text{H}\ \ \text{H}
\end{array}
$$

$$
\begin{array}{cc}
& \text{H} \\
& | \\
(d)\quad \text{H—C=O} & \qquad (e)\quad \text{F—C}\equiv\text{C—F}
\end{array}
$$

As the number of carbon atoms per molecule increases, the structures are written with significantly less effort using *condensed formulas* rather than structural formulas. Condensed formulas are like structural formulas except that in a condensed formula, the hydrogen atoms are written to the right of the larger atom to which they are attached. (Since it can form only one bond, a hydrogen atom cannot be attached to another hydrogen atom and still be part of the organic molecule.) For example, a molecule with a carbon atom connected (1) to a nitrogen atom bonded to two hydrogen atoms and also (2) to three other carbon atoms, each bonded to three hydrogen atoms, may be represented as

$$
\begin{array}{c}
\text{CH}_3 \\
| \\
\text{CH}_3\text{—C—NH}_2 \\
| \\
\text{CH}_3
\end{array}
$$

To see the efficiency of this representation, write out both this condensed formula and the structural formula for this compound and compare the effort that it takes.

For even greater convenience in representing the structures of organic compounds, particularly in printed material, *line formulas* are used, so-called because they are printed on one line. In line formulas, the symbol for each carbon atom is written on a line adjacent to the symbols for the other elements to which it is bonded. Line formulas show the general sequence in which the carbon atoms are attached, but to interpret them properly, the permitted total bond orders for all the respective atoms must be kept in mind. Again referring to compounds (*a*) to (*e*) described in Example 18.1, the line formulas are as follows:

(*a*) CH_4 (*b*) CH_3OH (*c*) CH_3NH_2 (*d*) CH_2O (*e*) $CF\equiv CF$ or $FC\equiv CF$

If the permitted total bond orders of the respective atoms are remembered, it is apparent that the line formula CH_4 cannot represent such structures as

$$
\begin{array}{c}
\text{H—C—H—H} \qquad \text{(incorrect)} \\
| \\
\text{H}
\end{array}
$$

which has a total order of 2 for one of the hydrogen atoms and a total bond order of only 3 for the carbon atom. Similarly, the formula CH_2O cannot represent the structure H—C—O—H or H—C=O—H, because in either of these cases, the total bond order of the carbon atom is less than 4 and in H—C=O—H the total bond order of the oxygen is greater than 2. Accordingly, line formulas must be interpreted in terms of the permitted total bond orders.

EXAMPLE 18.2. Write structural and condensed formulas for the molecules represented by the following formulas: (*a*) C_2H_4 and (*b*) CH_3COCH_3.

Ans.   (*a*)   Since the hydrogen atoms can have only a total bond order of 1, the two carbon atoms must be linked together. In order for each carbon atom to have a total bond order of 4, the two carbon atoms must be linked to each other by a double bond and also be bonded to two hydrogen atoms each.

$$H-C=C-H \qquad CH_2=CH_2$$

(b) The line formula CH_3COCH_3 implies that two of the three carbon atoms each has three hydrogen atoms attached. This permits them to form one additional single bond, to the middle carbon atom. The middle carbon atom, with two single bonds to carbon atoms, must complete its total bond order of 4 with a double bond to the oxygen atom.

$$H-C-C-C-H \qquad CH_3-C-CH_3$$

18.4. HYDROCARBONS

Compounds consisting of only carbon and hydrogen have the simplest compositions of all organic compounds. These compounds are called *hydrocarbons*. It is possible to classify the hydrocarbons into series, based on the characteristic structures of the molecules in each series. The four most fundamental series are known as the (1) alkane series, (2) alkene series, (3) alkyne series, and (4) aromatic series. There are many subdivisions of each series, and it is also possible to have molecules that could be classified as belonging to more than one series.

Saturated Hydrocarbons

The *alkane series* is also called the *saturated hydrocarbon series* because the molecules of this class have carbon atoms connected by single bonds only, and therefore have the maximum number of hydrogen atoms possible for the number of carbon atoms. These substances may be represented by the general formula C_nH_{2n+2}, and molecules of successive members of the series differ from each other by only a CH_2 unit. The line formulas and names of the first 10 unbranched members of the series, given in Table 18-2, should be memorized because these names form the basis for naming many other organic compounds. Note that the first parts of the names of the later members listed are the Greek or Latin prefixes that denote the number of carbon atoms. Also note that the characteristic ending of each name is -*ane*. Names of other organic compounds are derived from these names by dropping the -*ane* ending and adding other endings. At room temperature, the first four members of this series are gases; the remainder of those listed in Table 18-2 are liquids. Members of the series having more than 13 carbon atoms are solids at room temperature.

Table 18-2   The Simplest Unbranched Alkanes

Number of C Atoms	Molecular Formula	Line Formula	Name
1	CH_4	CH_4	Methane
2	C_2H_6	CH_3CH_3	Ethane
3	C_3H_8	$CH_3CH_2CH_3$	Propane
4	C_4H_{10}	$CH_3CH_2CH_2CH_3$	Butane
5	C_5H_{12}	$CH_3CH_2CH_2CH_2CH_3$	Pentane
6	C_6H_{14}	$CH_3CH_2CH_2CH_2CH_2CH_3$	Hexane
7	C_7H_{16}	$CH_3CH_2CH_2CH_2CH_2CH_2CH_3$	Heptane
8	C_8H_{18}	$CH_3CH_2CH_2CH_2CH_2CH_2CH_2CH_3$	Octane
9	C_9H_{20}	$CH_3CH_2CH_2CH_2CH_2CH_2CH_2CH_2CH_3$	Nonane
10	$C_{10}H_{22}$	$CH_3CH_2CH_2CH_2CH_2CH_2CH_2CH_2CH_2CH_3$	Decane

Compounds in the alkane series are chemically rather inert. Aside from burning in air or oxygen to produce carbon dioxide and water (or carbon monoxide and water), the most characteristic reaction they undergo is reaction with halogen molecules. The latter reaction is initiated by light. Using pentane as a typical alkane hydrocarbon and bromine as a typical halogen, these reactions are represented by the following equations:

$$C_5H_{12} + 8\,O_2 \longrightarrow 5\,CO_2 + 6\,H_2O$$
$$C_5H_{12} + Br_2 \xrightarrow{\text{light}} C_5H_{11}Br + HBr$$

Because of their limited reactivity, the saturated hydrocarbons are also called the *paraffins*. This term is derived from the Latin words meaning "little affinity."

Unsaturated Hydrocarbons

The *alkene series* of hydrocarbons is characterized by having one double bond in the carbon chain of each molecule. The characteristic formula for members of the series is C_nH_{2n}. Since there must be at least two carbon atoms present to have a carbon-to-carbon double bond, the first member of this series is ethene, C_2H_4, also known as ethylene. Propene (propylene), C_3H_6, and butene (butylene), C_4H_8, are the next two members of the series. Note that the systematic names of these compounds denote the number of carbon atoms in the chain with the name derived from that of the alkane having the same number of carbon atoms (Table 18-2). Note also that the characteristic ending *-ene* is part of the name of each of these compounds.

Owing to the presence of the double bond, the alkenes are said to be *unsaturated* and are more reactive than the alkanes. The formula for propene, for example, is $CH_2{=}CHCH_3$. The alkenes may react with hydrogen gas in the presence of a catalyst to produce the corresponding alkane; they may react with halogens or with hydrogen halides at relatively low temperatures to form compounds containing only single bonds. These possibilities are illustrated in the following equations in which ethene is used as a typical alkene:

$$CH_2{=}CH_2 + H_2 \xrightarrow{\text{catalyst}} CH_3CH_3$$
$$CH_2{=}CH_2 + Br_2 \longrightarrow CH_2BrCH_2Br$$
$$CH_2{=}CH_2 + HBr \longrightarrow CH_3CH_2Br$$

The *alkyne series* of hydrocarbons is characterized by having molecules with one triple bond each. They have the general formula C_nH_{2n-2} and the name ending *-yne*. Like other unsaturated hydrocarbons, the alkynes are quite reactive. Ethyne is commonly known as *acetylene*. It is the most important member of the series commercially, being widely used as a fuel in acetylene torches and also as a raw material in the manufacture of synthetic rubber and other industrial chemicals.

The *aromatic hydrocarbons* are characterized by molecules containing six-member rings of carbon atoms with each carbon atom attached to a maximum of one hydrogen atom. The simplest member of the series is benzene, C_6H_6. Using the total bond order rules discussed above, the structural formula of benzene can be written as follows:

Such a molecule, containing alternating single and double bonds, would be expected to be quite reactive. Actually, benzene is quite unreactive, and its chemical properties resemble those of the alkanes much more than those of the unsaturated hydrocarbons. For example, the characteristic reaction of benzene with halogens resembles that of the reaction of the alkanes:

$$C_6H_6 + Br_2 \xrightarrow[\text{Fe}]{} C_6H_5Br + HBr$$

Unlike the alkanes, however, the reaction of benzene with the halogens is catalyzed by iron. The relative lack of reactivity in aromatic hydrocarbons is attributed to *delocalized double bonds*. That is, the second pair of electrons in each of the three possible carbon-to-carbon double bonds is shared by all six carbon atoms rather than by any two specific carbon atoms. Two ways of writing structural formulas which indicate this type of bonding in the benzene molecules are as follows:

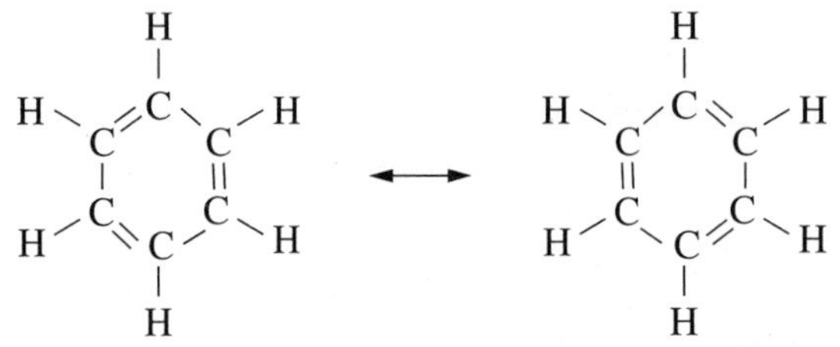

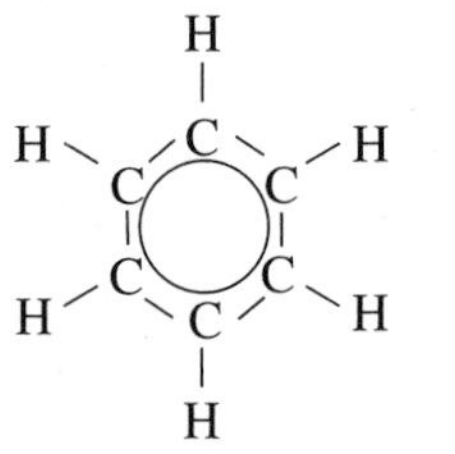

Because of its great stability, the six-member ring of carbon atoms persists in most reactions. For simplicity, the ring is sometimes represented as a hexagon, each corner of which is assumed to be occupied by a carbon atom with a hydrogen atom attached (unless some other atom is explicitly indicated at that point). The delocalized electrons are indicated by a circle within the hexagon. The following representations illustrate these rules:

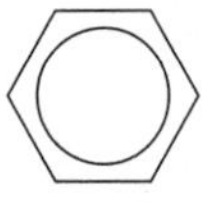

benzene, C_6H_6 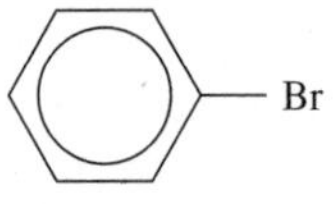   bromobenzene, C_6H_5Br

Other aromatic hydrocarbons include naphthalene, $C_{10}H_8$, and anthracene, $C_{14}H_{10}$, whose structures are as follows:

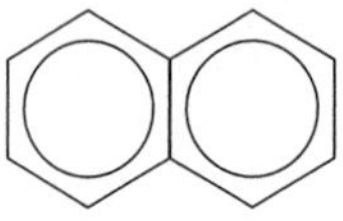

naphthalene 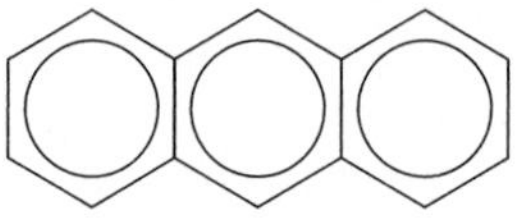   anthracene

Derivatives of aromatic hydrocarbons may have chains of carbon atoms substituted on the aromatic ring. Examples are toluene, $C_6H_5CH_3$, and styrene, $C_6H_5CH\!=\!CH_2$:

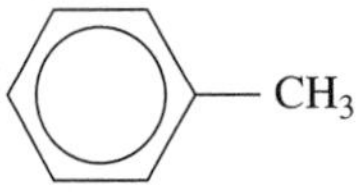

toluene   styrene
(methyl benzene)

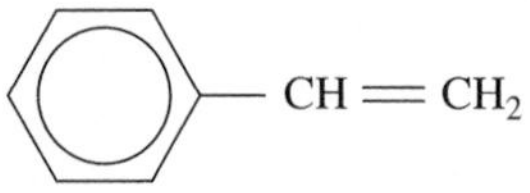

18.5. ISOMERISM

The ability of a carbon atom to link to more than two other carbon atoms makes it possible for two or more compounds to have the same molecular formula but different structures. Sets of compounds related in this way are called *isomers* of each other. For example, there are two different compounds having the molecular formula C_4H_{10}. Their condensed formulas are as follows:

$$CH_3-CH_2-CH_2-CH_3$$

butane
(normal boiling point, 1°C)

$$CH_3-\underset{\underset{CH_3}{|}}{CH}-CH_3$$

methylpropane
(normal boiling point, −10°C)

These are two distinctly different compounds, with different chemical and physical properties; for example, they have different boiling points.

Note that there are only two possible isomers having the molecular formula C_4H_{10}. Condensed formulas written as

$$
\begin{array}{ccc}
CH_2 - CH_2 & & CH_3 - CH_2 \\
| \qquad | & \text{and} & | \\
CH_3 \quad CH_3 & & CH_2 - CH_3
\end{array}
$$

are both butane. They have a continuous chain of four carbon atoms and are completely saturated with hydrogen atoms; they have no carbon branches.

EXAMPLE 18.3.   Write the line formulas (*a*) for butane and (*b*) for methylpropane, both C_4H_{10}.

Ans.   (*a*) $CH_3CH_2CH_2CH_3$
                           butane

          (*b*) $CH_3CH(CH_3)CH_3$ or $CH_3CH(CH_3)_2$ or $CH(CH_3)_3$ or $(CH_3)_3CH$
                                                    methylpropane

Similarly, three isomers of pentane, C_5H_{12}, exist. As the number of carbon atoms in a molecule increases, the number of possible isomers increases markedly. Theoretically, for the formula $C_{20}H_{42}$ there are 366 319 possible isomers. Other types of isomerism are possible in molecules that contain atoms other than carbon and hydrogen atoms and in molecules with double bonds or triple bonds in the chain of carbon atoms.

Compounds called *cycloalkanes*, having molecules with no double bonds but having a cyclic or ring structure, are isomeric with alkenes whose molecules contain the same number of carbon atoms. For example, cyclopentane and 2-pentene have the same molecular formula, C_5H_{10}, but have completely different structures:

$$
\begin{array}{cc}
\begin{array}{c}
CH_2 - CH_2 \\
/ \qquad \quad \backslash \\
CH_2 \qquad CH_2 \\
\backslash \qquad / \\
CH_2
\end{array}
&
\qquad CH_3 - CH = CH - CH_2 - CH_3
\end{array}
$$

Cyclopentane has the low chemical reactivity which is typical of saturated hydrocarbons, while 2-pentene is much more reactive. Similarly, ring structures each containing a double bond, called cycloalkenes, can be shown to be isomeric with alkynes.

Literally thousands of isomers can exist. Even relatively simple molecules can have many isomers. Thus, the phenomenon of isomerism accounts in part for the enormous number and variety of compounds of carbon.

18.6.   RADICALS AND FUNCTIONAL GROUPS

The millions of organic compounds other than hydrocarbons can be regarded as *derivatives* of hydrocarbons, where one (or more) of the hydrogen atoms on the parent molecule is replaced by another kind of atom or group of atoms. For example, if one hydrogen atom in a molecule of methane, CH_4, is replaced by an —OH group, the resulting compound is methanol, also called methyl alcohol, CH_3OH. (This replacement is often not easy to perform in the laboratory and is meant in the context used here as a mental exercise.) In most cases, the compound is named in a manner that designates the hydrocarbon parent from which it was derived. Thus, the word *methyl* is derived from the word *methane*. The hydrocarbon part of the molecule is often called the *radical*. To name a radical, change the ending of the parent hydrocarbon name from -*ane* to -*yl*. The names of some common radicals are listed in Table 18-3. Note that the radical derived from benzene, C_6H_5—, is called the *phenyl* radical. Some other radicals are also given names that are not derived from the names of the parent hydrocarbons, but these other cases will not be discussed here. Since, in many reactions, the hydrocarbon part of the organic compound is not changed and does not affect the nature of the reaction, it is useful to generalize many reactions by using the symbol R— to denote any radical. Thus, the compounds CH_3Cl, CH_3CH_2Cl, $CH_3CH_2CH_2Cl$, and so forth, all of which undergo similar chemical reactions, can be represented by the formula RCl. If the radical is derived

Table 18-3 Some Common Radicals

Parent Hydrocarbon		Radical		
Name	Line Formula	Name	Line Formula	Structural Formula
Methane	CH_4	Methyl	CH_3-	$H-\overset{\overset{\displaystyle H}{\mid}}{\underset{\underset{\displaystyle H}{\mid}}{C}}-$
Ethane	CH_3CH_3	Ethyl	CH_3CH_2-	$H-\overset{\overset{\displaystyle H}{\mid}}{\underset{\underset{\displaystyle H}{\mid}}{C}}-\overset{\overset{\displaystyle H}{\mid}}{\underset{\underset{\displaystyle H}{\mid}}{C}}-$
Propane	$CH_3CH_2CH_3$	Propyl	$CH_3CH_2CH_2-$	$H-\overset{\overset{\displaystyle H}{\mid}}{\underset{\underset{\displaystyle H}{\mid}}{C}}-\overset{\overset{\displaystyle H}{\mid}}{\underset{\underset{\displaystyle H}{\mid}}{C}}-\overset{\overset{\displaystyle H}{\mid}}{\underset{\underset{\displaystyle H}{\mid}}{C}}-$
Benzene	C_6H_6	Phenyl	C_6H_5-	

from an alkane, the radical is called an *alkyl radical*; and if it is derived from an aromatic hydrocarbon, it is called an *aryl radical*.

The names of alkanes with radical branches use the name from Table 18-2 for the longest continuous chain of carbon atoms. The radicals are named using the -*yl* ending presented in Table 18-3, and their positions along the carbon chain are denoted by a number. For example,

$$CH_3-CH_2-\underset{\underset{\displaystyle CH_3}{\mid}}{CH}-CH_2-CH_2-CH_3$$

is called 3-methylhexane. The longest continuous carbon chain is six carbon atoms long, therefore it is a hexane. The single-carbon branch is the methyl group, which is on the third carbon atom from the left end of the chain. *We actually start counting from the end of the chain that gives the lower number*. In this case, the branch is on the fourth carbon atom from the right, so the smaller number is used. *Be sure to use the longest continuous chain of carbon atoms; they are not always presented on a horizontal line.*

If there are two or more branches, each is named. If there are two or more identical branches, the prefixes from Table 6-2 are used. Thus the following names:

$$CH_3-\underset{\underset{\displaystyle CH_3}{\mid}}{CH}-\underset{\underset{\displaystyle CH_2-CH_3}{\mid}}{CH}-CH_2-CH_2-CH_3 \qquad CH_3-\underset{\underset{\displaystyle CH_3}{\mid}}{CH}-\underset{\underset{\displaystyle CH_3}{\mid}}{CH}-CH_2-CH_2-CH_3$$

2-methyl-3-ethylhexane   2,3-dimethylhexane

When a hydrogen atom of an alkane or aromatic hydrocarbon molecule is replaced by an atom of another element or group of atoms containing another element, the hydrocarbon-like part of the molecule is relatively inert, like these hydrocarbons themselves. Therefore, the resulting compound will have properties characteristic of the substituting group. Specific groups of atoms responsible for the characteristic properties of the compound are called *functional groups*. For the most part, organic compounds can be classified according to the

Table 18-4   Formulas for Functional Groups

Type	Characteristic Functional Group	Example
Alcohol	$-OH$	$R-OH$ or ROH
Ether	$-O-$	$R-O-R'$ or ROR'
Aldehyde	$-\overset{\displaystyle }{\underset{\displaystyle H}{C}}=O$	$R-\overset{\displaystyle }{\underset{\displaystyle H}{C}}=O$ or RCHO
Ketone	$-\underset{\displaystyle O}{\overset{\displaystyle \|}{C}}-$	$R-\underset{\displaystyle O}{\overset{\displaystyle \|}{C}}-R'$ or RCOR'
Acid	$-\underset{\displaystyle O}{\overset{\displaystyle \|}{C}}-OH$	$R-\underset{\displaystyle O}{\overset{\displaystyle \|}{C}}-OH$ or RCO_2H
Ester	$-\underset{\displaystyle O}{\overset{\displaystyle \|}{C}}-O-$	$R-\underset{\displaystyle O}{\overset{\displaystyle \|}{C}}-OR'$ or RCO_2R'
Amine	$-NH_2,\quad \diagdown NH,\quad \diagdown N \diagdown$	$R-NH_2$, RNHR', RR'R''N
Amide	$-\underset{\displaystyle O}{\overset{\displaystyle \|}{C}}-N\diagup\diagdown$	$R-\underset{\displaystyle O}{\overset{\displaystyle \|}{C}}-\underset{\displaystyle R''}{\overset{\displaystyle }{N}}-R'$ or RCONRR'

The radicals labeled R' or R'' may be the same as or different from the radicals labeled R in the same compounds.

functional group they contain. The most important classes of such compounds include (1) alcohols, (2) ethers, (3) aldehydes, (4) ketones, (5) acids, (6) amines, (7) esters, and (8) amides. The general formulas for these classes are given in Table 18-4, where the symbol for a radical, R, is written for the hydrocarbon part(s) of the molecule.

Molecules of some compounds contain more than one functional group, and there may be even more than one kind of functional group in each molecule. The purpose of this discussion is merely to describe some of the possible compounds. Therefore, the methods of preparation of the various classes of compounds will not be given here for every class, nor will more than a few of their properties be described.

18.7. ALCOHOLS

Compounds containing the functional group —OH are called *alcohols*. The —OH group is covalently bonded to a carbon atom in an alcohol molecule, and the molecules do *not* ionize in water solution to give OH⁻ ions. On the contrary, they react with metallic sodium to liberate hydrogen in a reaction analogous to that of sodium with water.

$$2\,HOH + 2\,Na \longrightarrow H_2 + 2\,NaOH$$

$$2\,ROH + 2\,Na \longrightarrow H_2 + 2\,NaOR$$

The simplest alcohol is methanol, CH_3OH, also called methyl alcohol in a less systematic system of naming. Methanol is also known as wood alcohol. Ethanol, CH_3CH_2OH, also known as ethyl alcohol or grain alcohol, is the active constituent of intoxicating beverages. Other alcohols of importance are included in Table 18-5. Note that the systematic names of alcohols characteristically end in *-ol*.

Table 18-5   Common Alcohols

Systematic Name	Formula	Common Name	Familiar Source or Use
Methanol	CH_3OH	Methyl alcohol	Wood alcohol
Ethanol	CH_3CH_2OH	Ethyl alcohol	Grain alcohol
1-Propanol	$CH_3CH_2CH_2OH$	Propyl alcohol	Rubbing alcohol
2-Propanol	$CH_3CHOHCH_3$	Isopropyl alcohol	Rubbing alcohol
1,2-Ethanediol	CH_2OHCH_2OH	Ethylene glycol	Antifreeze
1,2,3-Propanetriol	$CH_2OHCHOHCH_2OH$	Glycerine	Animal fat

When the number of carbon atoms in an alcohol molecule is greater than 2, several isomers are possible, depending on the location of the —OH group as well as on the nature of the carbon chain. For example, the condensed formulas of four isomeric alcohols with the formula C_4H_9OH are given in Table 18-6.

Table 18-6   Isomeric Alcohols

Systematic Name	Formula	Common Name
1-Butanol	$CH_3 — CH_2 — CH_2 — CH_2 — OH$	Butyl alcohol (primary)
2-Butanol	$CH_3 — CH_2 — \underset{\underset{OH}{\mid}}{CH} — CH_3$	Secondary butyl alcohol
2-Methyl-1-propanol	$CH_3 — \underset{\underset{CH_3}{\mid}}{CH} — CH_2 — OH$	Isobutyl alcohol
2-Methyl-2-propanol	$CH_3 — \overset{\overset{OH}{\mid}}{\underset{\underset{CH_3}{\mid}}{C}} — CH_3$	Tertiary butyl alcohol

18.8.   ETHERS

A chemical reaction that removes a molecule of water from two alcohol molecules results in the formation of an *ether*.

$$CH_3CH_2OH + HOCH_2CH_3 \longrightarrow CH_3CH_2OCH_2CH_3 + H_2O$$

This reaction is run by heating the alcohol in the presence of concentrated sulfuric acid, a good dehydrating agent. Diethyl ether, $CH_3CH_2OCH_2CH_3$, was formerly used as an anesthetic in surgery, but it caused undesirable side effects in many patients and was replaced by other types of agents.

The two radicals of the ether molecule need not be identical. A mixed ether is named after both radicals.

EXAMPLE 18.4.   Name the following ethers: (*a*) $CH_3OCH_2CH_2CH_2CH_3$, (*b*) $C_6H_5OCH_2CH_3$, and (*c*) CH_3OCH_3.

Ans.   (*a*) Methyl butyl ether   (*b*) Phenyl ethyl ether   (*c*) Dimethyl ether

18.9. ALDEHYDES AND KETONES

Aldehydes are produced by the mild oxidation of primary alcohols—alcohols with the —OH group on an end carbon atom.

$$CH_3—CH_2—OH \xrightarrow[\text{mild oxidation}]{} CH_3—\overset{\overset{\textstyle O}{\textstyle \|}}{C}H$$

ethanol or ethyl alcohol   ethanal or acetaldehyde

The mild oxidation of a secondary alcohol (with the —OH group on a carbon atom connected to two other carbon atoms) produces a *ketone*.

$$CH_3—\underset{\underset{\textstyle OH}{\textstyle |}}{C}H—CH_3 \xrightarrow[\text{oxidation}]{} CH_3—\overset{\overset{\textstyle }{\textstyle \|}}{\underset{\underset{\textstyle O}{\textstyle }}{C}}—CH_3$$

2-propanol or isopropyl alcohol   propanone or acetone

Each of these groups is characterized by having a *carbonyl group*, $\diagdown C{=}O$, but the aldehyde has the $\diagdown C{=}O$ group on one end of the carbon chain, and the ketone has the $\diagdown C{=}O$ group on a carbon other than one on the end. Acetone is a familar solvent for varnishes and lacquers. As such, it is used as a nail polish remover. The systematic name ending for *al*dehydes is *-al*; that for ket*ones* is *-one*.

18.10. ACIDS AND ESTERS

Acids can be produced by the oxidation of aldehydes or primary alcohols. For example, the souring of wine results from the oxidation of the ethyl alcohol in wine to acetic acid:

$$CH_3CH_2OH \xrightarrow[\text{oxidation}]{} CH_3—\overset{\overset{\textstyle }{\textstyle \|}}{\underset{\underset{\textstyle O}{\textstyle }}{C}}—O—H$$

ethanol   ethanoic acid or acetic acid

A solution of acetic acid formed in this manner is familar as vinegar. The name ending for organic acids is *—oic acid*, so the systematic name for acetic acid is ethanoic acid.

Acids react with alcohols to produce *esters*:

$$CH_3CO_2H + HOCH_2CH_3 \longrightarrow CH_3CO_2CH_2CH_3 + H_2O$$

acetic acid   ethyl alcohol   ethyl acetate   water
(an ester)

Many simple esters have a pleasant, fruity odor. Indeed, the odors of many fruits and oils are due to esters. Esters are named by combining the radical name of the alcohol with that of the negative ion of the acid. The ending *-ate* replaces the *-oic acid* of the parent acid. The ester in the equation above is an example.

EXAMPLE 18.5. Name the following esters:

$$(a)\ C_4H_9OC\overset{\|}{\underset{O}{}}CH_3 \qquad \text{and} \qquad (b)\ C_6H_5OC\overset{\|}{\underset{O}{}}H$$

Ans. (*a*) Butyl acetate or butyl ethanoate   (*b*) Phenyl formate or phenyl methanoate

The formation of an ester from an alcohol and an acid is an equilibrium reaction. The reverse reaction can be promoted by removing the acid from the reaction mixture, for example, by treating it with NaOH. Animal fats are converted to soaps (fatty acid salts) and glycerine (a trialcohol) in this manner.

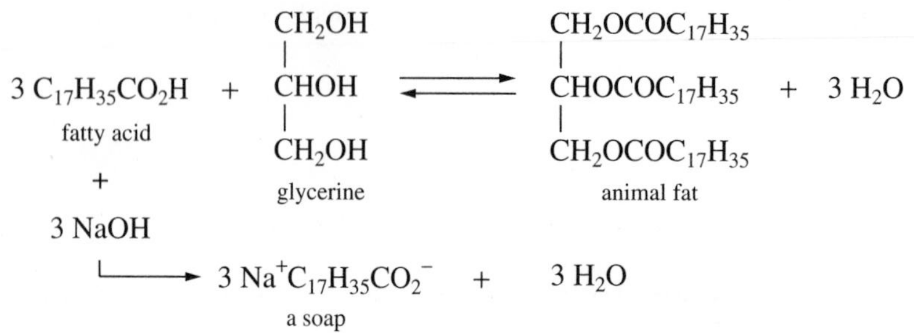

$$3\ C_{17}H_{35}CO_2H\ +\ \begin{matrix}CH_2OH\\|\\CHOH\\|\\CH_2OH\end{matrix}\ \rightleftharpoons\ \begin{matrix}CH_2OCOC_{17}H_{35}\\|\\CHOCOC_{17}H_{35}\\|\\CH_2OCOC_{17}H_{35}\end{matrix}\ +\ 3\ H_2O$$

fatty acid

+

3 NaOH

glycerine   animal fat

$$\longrightarrow\ 3\ Na^+C_{17}H_{35}CO_2^-\ +\ 3\ H_2O$$

a soap

Nitroglycerine, a powerful explosive, is an ester of the inorganic acid HNO_3 and glycerol (glycerine).

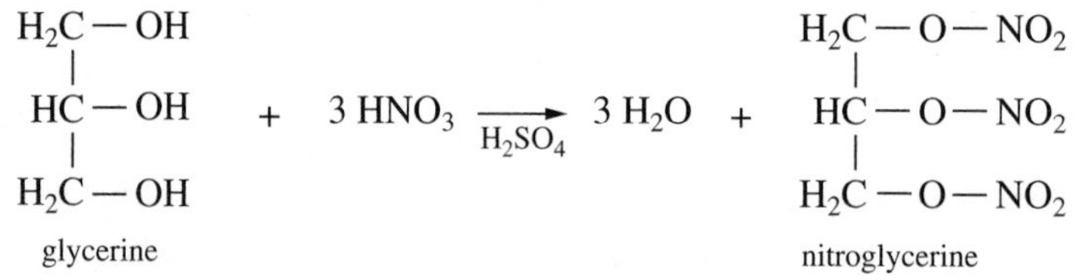

$$\begin{matrix}H_2C-OH\\|\\HC-OH\\|\\H_2C-OH\end{matrix}\ +\ 3\ HNO_3\ \xrightarrow{H_2SO_4}\ 3\ H_2O\ +\ \begin{matrix}H_2C-O-NO_2\\|\\HC-O-NO_2\\|\\H_2C-O-NO_2\end{matrix}$$

glycerine   nitroglycerine

18.11. AMINES

Amines can be considered derivatives of ammonia, NH_3, in which one or more hydrogen atoms have been replaced by organic radicals. For example, replacing one hydrogen atom on the nitrogen atom results in a *primary amine*, RNH_2. A *secondary amine* has a formula of the type R_2NH, and a *tertiary amine* has the formula R_3N. Like ammonia, amines react as Brønsted bases:

$$RNH_2 + H_2O \rightleftharpoons RNH_3^+ + OH^-$$

Aromatic amines are of considerable importance commercially. The simplest aromatic amine, aniline, $C_6H_5NH_2$, is used in the production of various dyes and chemicals for color photography.

18.12. AMIDES

This class of compounds is formed by the reaction of organic acids with amines. The functional group and the general formula are

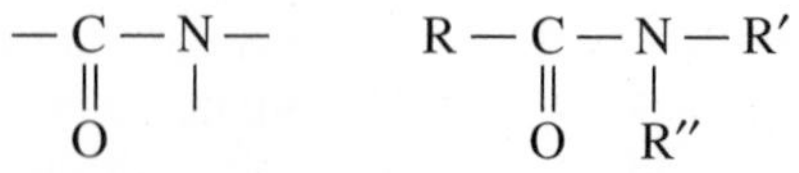

$$\begin{matrix}-C-N-\\\|\|\ \ |\\O\end{matrix}\qquad\begin{matrix}R-C-N-R'\\\|\|\ \ |\\O\ \ \ R''\end{matrix}$$

Any or all of the R groups can be the same or different, and indeed may even be a hydrogen atom. The simplest amide is formamide, $HCONH_2$. The name of each member of the group starts with any radicals attached to the nitrogen atom. Then comes the name of the organic acid, with *-oic acid* replaced by the word amide.

$$CH_3-\underset{\underset{O}{\|\|}}{C}-OH\ +\ HN(CH_3)_2\ \longrightarrow\ CH_3-\underset{\underset{O}{\|\|}}{C}-N(CH_3)_2\ +\ H_2O$$

acetic acid or
ethanoic acid

dimethylamine

dimethyl acetamide or
dimethyl ethanamide

Amino acids form proteins in the human body with amide linkages, and the same linkages hold together the building blocks of nylon.

Solved Problems

INTRODUCTION

18.1. Name one type of reaction that practically all organic compounds undergo.

Ans.   Combustion

BONDING IN ORGANIC COMPOUNDS

18.2. Write an electron dot diagram for C_2H_5Br.

Ans.

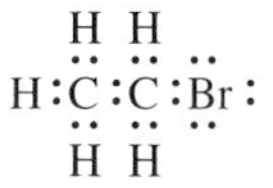

$$\begin{array}{ccc} H & H & \\ H\!:\!\overset{..}{\underset{..}{C}}\!:\!\overset{..}{\underset{..}{C}}\!:\!\overset{..}{\underset{..}{Br}}\!: \\ H & H & \end{array}$$

18.3. What is the total bond order of sulfur in CH_3SH?

Ans.   2

18.4. What is the total bond order of carbon in (*a*) CO_2, (*b*) CH_4, (*c*) $H_2C{=}O$, and (*d*) CO?

Ans.   (*a*) 4   (*b*) 4   (*c*) 4   (*d*) 3

18.5. Draw electron dot diagrams for C_3H_4 and C_3H_8.

Ans.

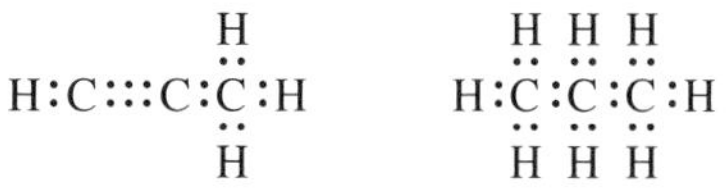

$$\begin{array}{ccc} & H & \\ H\!:\!C\!:\!:\!:\!C\!:\!\overset{..}{\underset{..}{C}}\!:\!H \\ & H & \end{array} \qquad \begin{array}{ccc} H & H & H \\ H\!:\!\overset{..}{\underset{..}{C}}\!:\!\overset{..}{\underset{..}{C}}\!:\!\overset{..}{\underset{..}{C}}\!:\!H \\ H & H & H \end{array}$$

STRUCTURAL, CONDENSED, AND LINE FORMULAS

18.6. Draw electron dot diagrams and structural formulas for molecules with the following molecular formulas: (*a*) CH_3N and (*b*) HCN.

Ans.   (*a*) $\begin{array}{c} H\!:\!C\!:\!:\!N\!:\!H \\ \overset{..}{H} \end{array}$ $\begin{array}{c} H-C{=}N-H \\ | \\ H \end{array}$ (*b*) $H\!:\!C\!:\!:\!:\!N\!:$ $H-C{\equiv}N\!:$

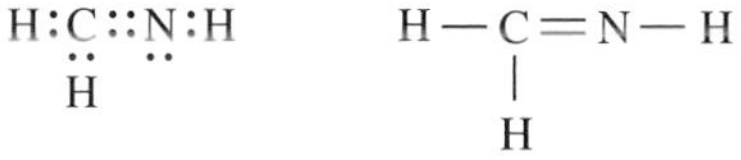

18.7. Write the condensed formula for $P(CH_3)_3$. What is the total bond order of phosphorus in this compound?

Ans. $\begin{array}{c} CH_3-P-CH_3 \\ | \\ CH_3 \end{array}$ The total bond order of P is 3

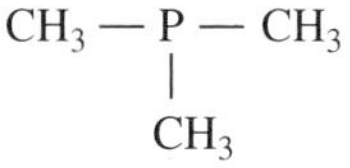

18.8. Describe the difference, if any, between the compound(s) represented by the formulas CH_3OH and $HOCH_3$.

Ans.   There is no difference.

18.9. Write condensed formulas for (*a*) $CH_3CH_2OCH_2CH_3$ and (*b*) $(CH_3CH_2)_2NH$.

Ans.   (*a*) $CH_3{-}CH_2{-}O{-}CH_2{-}CH_3$ (*b*) $CH_3{-}CH_2{-}NH{-}CH_2{-}CH_3$

18.10. Explain the meaning in line formulas of a pair of parentheses with no subscript behind it, such as in $CH_3CH_2CH(CH_3)C_3H_7$.

Ans.   The parentheses designate a branch, or side chain.

18.11. Draw as many representations as are introduced in this chapter for (*a*) butane and (*b*) benzene.

Ans. (*a*) C_4H_{10} $CH_3CH_2CH_2CH_3$ $CH_3 — CH_2 — CH_2 — CH_3$

$$\begin{array}{ccccccccc}
 & H & & H & & H & & H & \\
 & | & & | & & | & & | & \\
H- & C & - & C & - & C & - & C & -H \\
 & | & & | & & | & & | & \\
 & H & & H & & H & & H &
\end{array}$$

(*b*)

C_6H_6

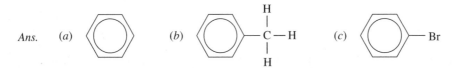

HYDROCARBONS

18.12. (*a*) Explain why alkenes have two fewer hydrogen atoms per molecule than alkanes with the same number of carbon atoms. (*b*) Explain why cycloalkanes have two fewer hydrogen atoms per molecule than alkanes with the same number of carbon atoms. (*c*) Explain why the cycloalkanes react more like alkanes than alkenes, despite the similarity in molecular formula of alkenes and cycloalkanes.

Ans. (*a*) The additional pair of electrons in the double bond between two carbon atoms leaves two fewer electrons to bond to hydrogen atoms. (*b*) The extra bond between the two "end" carbon atoms leaves two fewer electrons to bond to hydrogen atoms. (*c*) There are no multiple bonds in alkanes and cycloalkanes.

18.13. Write structural formulas for (*a*) benzene, (*b*) toluene (methyl benzene), and (*c*) bromobenzene.

Ans. (*a*) ⬡   (*b*) ⬡ $— \overset{\displaystyle H}{\underset{\displaystyle H}{\overset{|}{\underset{|}{C}}}} — H$ (*c*) ⬡ $—$ Br

18.14. (*a*) List the common names of the first three members of the alkene series and (*b*) the first member of the alkyne series.

Ans. (*a*) Ethylene, propylene, and butylene   (*b*) Acetylene

ISOMERISM

18.15. Write formulas for all possible isomers of butadiene which contain only one multiple bond and no rings. Butadiene is $CH_2 = CH — CH = CH_2$.

Ans. $HC \equiv C — CH_2 — CH_3$ and $CH_3 — C \equiv C — CH_3$

18.16. Write condensed formulas for the three isomers of pentane.

Ans. $CH_3 — CH_2 — CH_2 — CH_2 — CH_3$ $CH_3 — \underset{\displaystyle CH_3}{\overset{|}{CH}} — CH_2 — CH_3$ $CH_3 — \overset{\displaystyle CH_3}{\underset{\displaystyle CH_3}{\overset{|}{\underset{|}{C}}}} — CH_3$

18.17. Condensed formulas for C_3H_8 and C_2H_4 are fairly easy to write. Explain why they would be harder to write for C_4H_{10} and C_3H_6.

    Ans. These compounds can each form isomers (having the same molecular formulas). C_4H_{10} can occur as an unbranched compound and a branched-chain compound. C_3H_6 can occur with a double bond or in a ring.

18.18. Explain why there can be no two-carbon branch in an alkane that contains a four-carbon longest continuous chain.

    Ans. A branch cannot start at the end of a carbon chain, and so it must be from one of the two center carbon atoms in the parent compound. If a two-carbon chain is added there, the two-carbon chain becomes part of the longest chain, and a one-carbon side chain is left.

18.19. Draw condensed formulas to represent the two isomers of butene which contain no branches.

    Ans. $CH_3 - CH_2 - CH = CH_2$ $CH_3 - CH = CH - CH_3$

18.20. Which ones of the following are identical to each other?

    Ans. (*a*) and (*c*) are identical because the double bonds are really delocalized, (*b*) and (*d*) have the Br atoms on different carbon atoms.

18.21. Write condensed formulas for the four isomers of octane each of which contains three one-carbon branches.

RADICALS AND FUNCTIONAL GROUPS

18.22. Is it true that isomers can occur only with hydrocarbons?

    Ans. No. Compounds with functional groups can form isomers even more extensively than can hydrocarbons (given the same number of carbon atoms).

18.23. List the pairs of organic functional groups that can be isomeric with each other if the numbers of carbon atoms in their compounds are the same.

    Ans. Alcohols and ethers; aldehydes and ketones; acids and esters.

18.24. Identify the radical(s) and the functional group in each of the following molecules: (a) CH_3OH, (b) CH_3CO_2H, (c) CH_3CHO, and (d) $CH_3CH_2OCH_3$.

Ans.

	Radical(s)	Functional Group
(a)	CH_3-	$-OH$
(b)	CH_3-	$-CO_2H$
(c)	CH_3-	$-CHO$
(d)	CH_3CH_2- and $-CH_3$	$-O-$

ALCOHOLS

18.25. Explain why an alcohol, ROH, is not a base.

Ans.   The OH group is not ionic.

18.26. Write the formula of another alcohol isomeric with $CH_3CH_2CH_2OH$.

Ans. $CH_3CHOHCH_3$ (The $-OH$ group can be attached to the middle carbon atom.)

ETHERS

18.27. Write line formulas for (a) dipropyl ether, (b) ethyl phenyl ether, and (c) methyl butyl ether.

Ans.   (a) $CH_3CH_2CH_2OCH_2CH_2CH_3$ (b) $CH_3CH_2OC_6H_5$ (c) $CH_3OCH_2CH_2CH_2CH_3$

18.28. Write the line formula for an ether that is an isomer of (a) CH_3CH_2OH and (b) $CH_3CH_2CH_2OH$.

Ans.   (a) CH_3OCH_3 (b) $CH_3CH_2OCH_3$

18.29. Write the formulas of all alcohols isomeric with $CH_3CH_2OCH_2CH_3$.

Ans. $CH_3CH_2CH_2CH_2OH$ and $CH_3CH_2CHOHCH_3$, $(CH_3)_2CHCH_2OH$, and $(CH_3)_2COHCH_3$.

ALDEHYDES AND KETONES

18.30. Explain why the simplest ketone has three carbon atoms.

Ans.   A ketone has a doubly bonded oxygen atom on a carbon that is *not* an end carbon atom. The smallest carbon chain that has a carbon not on an end has three carbon atoms.

18.31. Write balanced chemical equations for the catalytic reduction by hydrogen gas to the corresponding alcohol for (a) CH_3CH_2CHO and (b) CH_3COCH_3.

Ans.   (a) $CH_3CH_2CHO + H_2 \xrightarrow{\text{catalyst}} CH_3CH_2CH_2OH$

(b) $CH_3COCH_3 + H_2 \xrightarrow{\text{catalyst}} CH_3CHOHCH_3$
In each case, the hydrogen molecule adds across the $C{=}O$ double bond.

ACIDS AND ESTERS

18.32. Contrast the following two formulas for acetic acid: $HC_2H_3O_2$ and CH_3CO_2H. Explain the advantages of each. Which hydrogen atom is lost upon ionization of acetic acid in water?

Ans.   The first formula is easily identified as an acid, with H written first. It is that hydrogen atom which ionizes in water. The second formula shows the bonding. The hydrogen atom attached to the oxygen atom ionizes in water.

18.33. Explain which hydrogen atom on ethanoic acid, CH_3CO_2H, has acidic properties.

    Ans.   The hydrogen atom attached to the oxygen atom is ionizable.

18.34. Since hexanoic acid is $C_5H_{11}CO_2H$, what is the formula for pentanoic acid?

    Ans. $C_4H_9CO_2H$

18.35. The mild oxidation and moderate oxidation of CH_3CH_2OH are described in the text. What are the products of the vigorous oxidation of CH_3CH_2OH with excess oxygen gas?

    Ans. CO_2 and H_2O

18.36. Write balanced chemical equations for the reaction of CH_3CO_2H with (*a*) CH_3OH, (*b*) CH_3CH_2OH, (*c*) $CH_3CH_2CH_2OH$, (*d*) $CH_3CH_2CH_2CH_2OH$, and (*e*) explain the utility of the symbol R.

    Ans.   (*a*) $CH_3OH + HOCOCH_3 \longrightarrow CH_3OCOCH_3 + H_2O$

               (*b*) $CH_3CH_2OH + HOCOCH_3 \longrightarrow CH_3CH_2OCOCH_3 + H_2O$

               (*c*) $CH_3CH_2CH_2OH + HOCOCH_3 \longrightarrow CH_3CH_2CH_2OCOCH_3 + H_2O$

               (*d*) $CH_3CH_2CH_2CH_2OH + HOCOCH_3 \longrightarrow CH_3CH_2CH_2CH_2OCOCH_3 + H_2O$

               (*e*)   Each of these reactions corresponds to the equation below. The fact that all four of these and many more can be represented by this single equation makes R a useful representation.

$$ROH + HOCOCH_3 \longrightarrow ROCOCH_3 + H_2O$$

18.37. Ethyl acetate is reduced by hydrogen in the presence of a catalyst to yield one organic compound that contains oxygen but is not an ether. What is the compound?

    Ans. CH_3CH_2OH

AMINES

18.38. Write the formulas for (*a*) ammonium chloride, (*b*) methyl ammonium chloride, (*c*) dimethyl ammonium chloride, (*d*) trimethyl ammonium chloride, and (*e*) tetramethyl ammonium chloride.

    Ans.   (*a*) NH_4Cl.
               (*b*) CH_3NH_3Cl. (One H atom has been replaced by a CH_3 group.)
               (*c*) $(CH_3)_2NH_2Cl$. (Two H atoms have been replaced by CH_3 groups.)
               (*d*) $(CH_3)_3NHCl$. (Three H atoms have been replaced by CH_3 groups.)
               (*e*) $(CH_3)_4NCl$. (All four H atoms have been replaced by CH_3 groups.)

18.39. Write balanced chemical equations for the reaction with H_3O^+ (excess) of aqueous (*a*) NH_3, (*b*) CH_3NH_2, and (*c*) $NH_2CH_2CH_2NH_2$.

    Ans.   (*a*) $H_3O^+ + NH_3 \longrightarrow NH_4^+ + H_2O$

               (*b*) $H_3O^+ + CH_3NH_2 \longrightarrow CH_3NH_3^+ + H_2O$

               (*c*) $2 H_3O^+ + NH_2CH_2CH_2NH_2 \longrightarrow {}^+NH_3CH_2CH_2NH_3^+ + 2 H_2O$

AMIDES

18.40. What amide is the product of the reaction of CH_3CO_2H and $CH_3CH_2NH_2$?

    Ans. $CH_3CONHCH_2CH_3$

18.41. Write line formulas for all amides that are isomers of dimethylmethanamide.

    Ans.   Dimethylmethanamide is $HCON(CH_3)_2$. The possible isomers are $CH_3CONHCH_3$ and $CH_3CH_2CONH_2$.

Supplementary Problems

18.42. Define or identify each of the following terms: (*a*) organic chemistry, (*b*) total bond order, (*c*) condensed formula, (*d*) structural formula, (*e*) line formula, (*f*) hydrocarbon, (*g*) alkane, (*h*) alkene, (*i*) alkyne, (*j*) aromatic hydrocarbon, (*k*) saturated, (*l*) delocalized double bond, (*m*) isomerism, (*n*) cycloalkane, (*o*) radical, (*p*) functional group, (*q*) alcohol, (*r*) ether, (*s*) aldehyde, (*t*) ketone, (*u*) carbonyl group, and (*v*) ester.

  Ans.   See the Glossary.

18.43. What is the total bond order of oxygen in all aldehydes, alcohols, ketones, acids, amides, and ethers?

  Ans.   2

18.44. What is the smallest number of carbon atoms possible in a molecule of (*a*) a ketone, (*b*) an aldehyde, and (*c*) an aromatic hydrocarbon?

  Ans.   (*a*) 3 (A molecule cannot have a "middle" carbon atom unless it has at least three carbon atoms.)   (*b*) 1
             (*c*) 6

18.45. In which classes of compounds can the R group(s) be a hydrogen atom?

  Ans.   Acids, amides, amines (two maximum), and esters (R on the oxygen atom cannot be H),

18.46. (*a*) Which classes of compounds contain the element grouping $\diagdown C = O$? (*b*) In which is it the entire functional group?

  Ans.   (*a*) Aldehydes, ketones, acids, amides, and esters   (*b*) Aldehydes and ketones

18.47. Which of the following is apt to be the stronger acid, FCOOH or HCOOH? Explain.

  Ans.   The electrons in the OH bond are attracted away from the hydrogen atom more by the F atom in FCOOH than by the H atom in HCOOH, since F is more electronegative than H. The H atom on the O atom in FCOOH is therefore easier to remove; that is, it is more acidic.

18.48. Urea and thiourea have formulas NH_2CONH_2 and NH_2CSNH_2. (*a*) Explain why they are considered to be organic compounds. (*b*) Explain why they do not fit the general definition given in Sec. 18.1.

  Ans.   (*a*) Urea is a product of animal metabolism, and thiourea is its sulfur analog.   (*b*) Both hydrogen atoms of their "parent," formaldehyde, H_2CO, have been replaced with amino groups, and there are no C—C or C—H bonds left.

18.49. In the formula $CH_3CH_2OCOCH_3$, which oxygen atom is double-bonded?

  Ans.   The double-bonded oxygen is conventionally written after the carbon atom to which it is attached, so the second oxygen is double-bonded to the carbon. The condensed formula is

$$CH_3 - CH_2 - O - \underset{\underset{O}{\|}}{C} - CH_3$$

18.50. Using toothpicks and marshmallows or gumdrops, make a model of butane. The angles between the toothpicks must be 109.5°. Show that the following represent the same compound:

$$\begin{matrix} C-C \\ | \quad | \\ C-C \end{matrix} \qquad \begin{matrix} \diagup C \diagdown \\ C \qquad C \end{matrix} \qquad \begin{matrix} C \\ | \\ C-C-C \end{matrix} \qquad \begin{matrix} C-C \\ | \\ C-C \end{matrix}$$

18.51. Give the name of the class of organic compound represented by each of the following: (*a*) CH_3COCH_3, (*b*) $CH_3OCH_2CH_3$, (*c*) $CH_3CO_2CH_3$, (*d*) CH_3CH_3, (*e*) CH_3CO_2H, (*f*) CH_3CHO, (*g*) CH_3CH_2OH, (*h*) $CH_2{=}CHCH_3$, and (*i*) CH_3NH_2.

  Ans.   (*a*) Ketone   (*b*) Ether   (*c*) Ester   (*d*) Alkane   (*e*) Acid   (*f*) Aldehyde   (*g*) Alcohol   (*h*) Alkene   (*i*) Amine

18.52. Burning of which compound, CH_3CH_2OH or CH_3CH_3, is apt to provide more heat? Explain why you made your choice.

Ans. CH_3CH_3. The CH_3CH_2OH is already partially oxidized.

18.53. Calculate the concentration of hydroxide ion in 1.00 L of a solution containing 0.100 mol of $CH_3CH_2NH_2$ ($K_b = 1.6 \times 10^{-11}$).

Ans.

$$RNH_2 + H_2O \rightleftharpoons RNH_3^+ + OH^-$$

Let $x = [OH^-]$.

$$\frac{x^2}{0.100} = 1.6 \times 10^{-11}$$

$$x = 1.3 \times 10^{-6} = [OH^-]$$

18.54. What functional group is most likely to be found in a polyester shirt?

Ans.   Ester $-\underset{\underset{O}{\overset{\|}{}}}{C} - O -$

18.55. The formation of which of the functional groups described in this chapter results in the simultaneous formation of water?

Ans.   Ether, ester, and amide

18.56. Ethyl acetate is reduced by hydrogen in the presence of a catalyst to yield one oxygen-containing organic compound, but propyl acetate yields two. What are the two compounds?

Ans. CH_3CH_2OH and $CH_3CH_2CH_2OH$

18.57. In this question consider organic molecules with no rings or carbon-to-carbon double or triple bonds. (*a*) Show that ethers can be isomers of alcohols but not of aldehydes. (*b*) Show that aldehydes and ketones can be isomers of each other, but not of acids or alcohols. (*c*) Write the condensed formula for all alcohols isomeric with diethyl ether.

Ans.   (*a*)   These alcohols have molecular formulas $C_nH_{2n+2}O$. Ethers can be thought of as $C_nH_{2n+1}OC_{n'}H_{2n'+1}$ or $C_{n+n'}H_{2(n+n')+2}O$. If $n + n' = n''$, the general formula of the ether would be $C_{n''}H_{2n''+2}O$. Since the n in the general formula of the alcohol and the n'' in the general formula of the ether are arbitrary, they could be the same. The ether can be an isomer of an alcohol.

   (*b*)   Both correspond to $C_nH_{2n}O$

   (*c*) $CH_3 - CH_2 - CH_2 - CH_2 - OH$ $CH_3 - CH_2 - \underset{\underset{OH}{\overset{\|}{|}}}{CH} - CH_3$

   $CH_3 - \underset{\underset{CH_3}{|}}{CH} - CH_2 - OH$ $CH_3 - \underset{\underset{CH_3}{|}}{\overset{\overset{OH}{|}}{C}} - CH_3$

CHAPTER 19

Nuclear Reactions

19.1. INTRODUCTION

In all the processes discussed so far, the nucleus of every atom remained unchanged. In this chapter, the effect of changes in the nucleus will be considered. Reactions involving changes in the nuclei are called *nuclear reactions*. They involve great quantities of energy, and this energy is referred to as *atomic energy* or more precisely as *nuclear energy*. In general, nuclear reactions may be either spontaneous or induced. We will consider them in that order.

Nuclear reactions differ from ordinary chemical reactions in the following ways:

1. Most often, atomic numbers change.
2. Although there is no change in the total of the mass numbers, the quantity of matter does change significantly. Some matter is changed to energy.
3. The atom reacts independently of other atoms to which it might be bonded.
4. Reactions are those of specific isotopes rather than the naturally occurring mixtures of isotopes.
5. The quantities usually used in calculations are atoms rather than moles of atoms.

19.2. NATURAL RADIOACTIVITY

The nuclei of the atoms of some isotopes are inherently unstable, and they disintegrate over time to yield other nuclei (Sec. 3.5.). The process is called *radioactive decay*, and the decay of each nucleus is called an *event*. There is nothing that humans can do about this type of radioactivity; as long as these isotopes exist, the nuclei will decompose. Depending on the isotope involved, some number of existing atoms decompose over a period of time (Sec. 19.3). Three main types of small particles are emitted from the nuclei during natural radioactive decay; they are named after the first three letters of the Greek alphabet. Their names and properties are listed in Table 19-1. When an *alpha* or *beta particle* is emitted from the nucleus, the identity of the element is changed because the atomic number is changed,

A stream of alpha particles is sometimes called an *alpha ray*. A stream of beta particles is called a *beta ray*. A *gamma ray* is composed of a stream of gamma particles.

We can denote the charge and mass number of these small particles as we denoted atomic numbers and mass numbers in Sec. 3.5.

Table 19-1   Products of Natural Radioactivity

Symbol	Name	Mass Number	Charge	Identity
α	Alpha particle	4	2+	Helium nucleus
β	Beta particle	0	1−	High-energy electron
γ	Gamma ray	0	0	High-energy particle of light

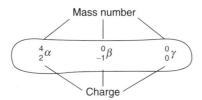

The superscripts refer to the mass numbers of the particles; the subscripts refer to their charges.

Nuclear equations are written with both the total charge and the total of the mass numbers unchanged from reactants to products. That is, the total of the subscripts of the reactants equals the total of the subscripts of the products, and the total of the superscripts of the reactants equals the total of the superscripts of the products. The subscripts of isotopes may be omitted because the symbol of the element gives the atomic number.

EXAMPLE 19.1.   Show that the mass number and the total charge are both conserved in the natural disintegration of $^{238}_{92}$U:

$$^{238}U \longrightarrow ^{234}Th + \alpha$$

Ans.   The equation may be rewritten including all atomic numbers and mass numbers:

$$^{238}_{92}U \longrightarrow ^{234}_{90}Th + ^{4}_{2}He \qquad \text{or} \qquad ^{238}_{92}U \longrightarrow ^{234}_{90}Th + ^{4}_{2}\alpha$$

Adding the $234 + 4$ superscripts of the products gives the superscript of the reactant. Adding the $90 + 2$ subscripts of the products gives the subscript of the reactant. The nuclear equation is balanced.

EXAMPLE 19.2.   Complete the following nuclear equation:

$$^{233}Pa \longrightarrow ^{233}U + ?$$

Ans.   Inserting the proper subscript and superscript indicates that the product is a particle with 1− charge and 0 mass number:

$$^{233}_{91}Pa \longrightarrow ^{233}_{92}U + ^{0}_{-1}?$$

The missing particle is a beta particle (Table 19-1).

$$^{233}_{91}Pa \longrightarrow ^{233}_{92}U + ^{0}_{-1}\beta$$

Note that a beta particle has been emitted from the nucleus. This change has been accompanied by the increase in the number of protons by 1 and a decrease in the number of neutrons by 1. In effect, a neutron has been converted to a proton and an electron, and the electron has been ejected from the nucleus.

The emission of a gamma particle causes no change in the charge or mass number of the original particle. (It does cause a change in the internal energy of the nucleus, however.) For example,

$$^{119}_{50}Sn \longrightarrow ^{119}_{50}Sn + ^{0}_{0}\gamma$$

The same $^{119}_{50}$Sn isotope is produced, but it has a lower energy after the emission of the gamma particle.

19.3. HALF-LIFE

Not all the nuclei of a given sample of a radioactive isotope disintegrate at the same time, but do so over a period of time. The number of radioactive disintegrations per unit time that occur in a given sample of a naturally radioactive isotope is directly proportional to the quantity of that isotope present. The more nuclei present, the more will disintegrate per second (or per year, etc.).

EXAMPLE 19.3. Sample A of $^{235}_{92}$U has a mass of 0.500 kg; sample B of the same isotope has a mass of 1.00 kg. Compare the rate of decay (the number of disintegrations per second) in the two samples.

Ans.   There will be twice the number of disintegrations per second in sample B (the 1.00-kg sample) as in sample A, because there were twice as many $^{235}_{92}$U atoms there to start. The number of disintegrations per gram per second is a constant, because both samples are the same isotope—$^{235}_{92}$U.

After a certain period of time, sample B of Example 19.3 will have disintegrated so much that there will be only 0.500 kg of $^{235}_{92}$U left. (Products of its decay will also be present.) How would the decay rate of this 0.500 kg of $^{235}_{92}$U compare with that of 0.500 kg of $^{235}_{92}$U originally present in sample A of Example 19.3? The rates should be the same, since each contains 0.500 kg of $^{235}_{92}$U. That means that sample B is disintegrating at a slower rate as its number of $^{235}_{92}$U atoms decreases. Half of this sample will disintegrate in a time equal to the time it took one-half the original sample B to disintegrate. The period in which half a naturally radioactive sample disintegrates is called its *half-life* because that is the time required for half of any given sample of the isotope to disintegrate (Fig. 19-1). The half-lives of several isotopes are given in Table 19-2.

Table 19-2   Half-Lives of Some Nuclei

Isotope	Half-Life	Radiation[*]
^{238}U	4.5×10^9 years	Alpha
^{235}U	7.1×10^8 years	Alpha
^{237}Np	2.2×10^6 years	Alpha
^{14}C	5730 years	Beta
^{90}Sr	19.9 years	Beta
^{3}H	12.3 years	Beta
^{140}Ba	12.5 days	Beta
^{131}I	8.0 days	Beta
^{15}O	118 s	Beta
^{94}Kr	1.4 s	Beta

If 0.500 kg takes 1 half-life to undergo the process shown for sample A, it also should take the same 1 half-life to undergo the process shown for sample B, after sample B gets down to 0.500 kg.

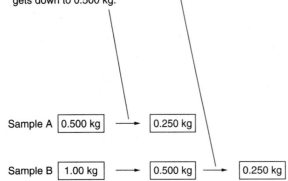

Sample A   0.500 kg   →   0.250 kg

Sample B   1.00 kg   →   0.500 kg   →   0.250 kg

Fig. 19-1.   A half-life example

[*] In most of these processes, gamma radiation is also emitted.

EXAMPLE 19.4. A certain isotope has a half-life of 3.00 years. How much of a 8.00-g sample of this isotope will remain after 12.0 years?

Ans.   After the first 3.00 years, half the 8.00-g sample will still be the same isotope. After 3.00 years more, half of the 4.00 g remaining will still be the original isotope. That is, 2.00 g will remain. After the third 3.00-year period, half of this 2.00 g will remain, or 1.00 g, of the original isotope will have its original identity, and after the fourth 3.00 years, half of that sample, or 0.500 g, will remain.

EXAMPLE 19.5. A certain isotope has a half-life of 3.00 years. How much of an 8.00-g sample of this isotope will have decomposed after 12.0 years?

Ans.   We calculated in Example 19.4 that 0.500 g would remain. That means that 8.00 g − 0.500 g = 7.50 g of the isotope will have disintegrated. (Most of the 7.50 g will produce some other isotope.) Notice the difference in the wording of

Examples 19.4 and 19.5. How much *remains* is determined from the half-life; how much *decomposes* is determined from the original quantity and how much remains.

Half-life problems may be solved with the equation

$$\ln\frac{N_0}{N} = \left(\frac{0.693}{t_{1/2}}\right)t$$

where N is the number of radioactive atoms of the original isotope remaining at time t, N_0 is the number of those atoms at the beginning of the process, $t_{1/2}$ is the half-life, and ln is the natural logarithm. The electronic calculator gives natural logarithms and antilogarithms at the touch of a key.

EXAMPLE 19.6. Calculate the time required for a radioactive sample to lose one-fourth of the atoms of its parent isotope. The half-life is 32.0 s.

Ans. If one-fourth of the atoms are lost, three-fourths remain. Therefore,

$$\ln\frac{1}{3/4} = \left(\frac{0.693}{32.0\text{ s}}\right)t$$

$$t = \frac{(32.0\text{ s})(0.288)}{0.693} = 13.3\text{ s}$$

In 13.3 s, one-fourth of the atoms will disintegrate. Note that this interval of time is less than the half-life, in which one-half the atoms would disintegrate, which is reasonable.

19.4. RADIOACTIVE SERIES

When an isotope disintegrates spontaneously, the products include one of the small particles from Table 19-1 and a large nucleus. The new nucleus is not necessarily stable; it might itself decompose spontaneously, yielding another small particle and still another nucleus. A disintegrating nucleus is called a *parent nucleus*, and the product is called the *daughter nucleus*. Such disintegrations can continue until a stable nucleus is produced. It turns out that four series of nuclei are generated from the disintegration of four different naturally occurring isotopes with high mass numbers. Since each nucleus can produce only an alpha particle, a beta particle, or a gamma particle, it turns out that the mass number of the daughter nucleus can be different from that of the parent nucleus by 4 units or 0 unit. If an alpha particle is emitted, the daughter nucleus will be 4 units smaller; if a beta or gamma particle is produced, the daughter nucleus will have the same mass number as the parent nucleus. Thus, the mass numbers of all the members of a given series can differ from each other by some multiple of 4 units.

EXAMPLE 19.7. When a ^{238}U nucleus disintegrates, the following series of alpha and beta particles is emitted: alpha, beta, beta, alpha, alpha, alpha, alpha, beta, alpha, beta, beta, beta, alpha. (Since emission of gamma particles accompanies practically every disintegration and since gamma particles do not change the atomic number or mass number of an isotope, they are not listed.) Show that each isotope produced has a mass number that differs from 238 by some multiple of 4.

Ans. Each alpha particle loss lowers the mass number of the product nucleus by 4. Each beta particle loss lowers the mass number by 0. The mass numbers thus go down as follows:

$$238 \xrightarrow{\alpha} 234 \xrightarrow{\beta} 234 \xrightarrow{\beta} 234 \xrightarrow{\alpha} 230 \xrightarrow{\alpha} 226 \xrightarrow{\alpha} 222 \xrightarrow{\alpha} 218$$
$$\downarrow \alpha$$
$$206 \xleftarrow{\alpha} 210 \xleftarrow{\beta} 210 \xleftarrow{\beta} 210 \xleftarrow{\beta} 210 \xleftarrow{\alpha} 214 \xleftarrow{\beta} 214$$

There are four such series of naturally occurring isotopes. The series in which all mass numbers are evenly divisible by 4 is called the *4n series*. The series with mass numbers 1 more than the corresponding *4n* series members is called the *4n + 1 series*. Similarly, there are a *4n + 2* series and a *4n + 3* series. Since the mass numbers change by 4 or 0, no member of any series can produce a product in a different series.

19.5. NUCLEAR FISSION AND FUSION

Nuclear fission refers to splitting a (large) nucleus into two smaller ones, plus one or more tiny particles listed in Table 19-3. *Nuclear fusion* refers to the combination of small nuclei to make a larger one. Both types of processes are included in the term *artificial transmutation*.

Table 19-3   Nuclear Projectiles and Products*

Name	Symbol	Identity	Nuclear Rest Mass (amu)
Proton	^{1}H or p	Hydrogen nucleus	1.00728
Deuteron	^{2}H or d	Heavy hydrogen nucleus	2.0135
Tritium	^{3}H	Tritium nucleus	3.01550
Helium-3	^{3}He	Light helium nucleus	3.01493
Neutron	^{1}n or n	Free neutron	1.008665
Alpha	α	Helium nucleus	4.001503
Beta	β	High-energy electron	0.00054858
Gamma	γ	High-energy light particle	0.0
Positron	$_+\beta$	Positive electron	0.00054858

* Larger projectiles are identified by their regular isotopic symbols, such as $^{12}_{6}$C.

Transmutation means converting one element to another (by changing the nucleus). The first artificial transmutation was the bombardment of $^{14}_{7}$N by alpha particles in 1919 by Lord Rutherford.

$$^{14}_{7}\text{N} + {}^{4}_{2}\text{He} \longrightarrow {}^{17}_{8}\text{O} + {}^{1}_{1}\text{H}$$

The alpha particles could be obtained from a natural decay process. At present, a variety of particles can be used to bombard nuclei (Table 19-3), some of which are raised to high energies in "atom smashing" machines. Again, nuclear equations are written in which the net charge and the total of the mass numbers on one side must be the same as their counterparts on the other side.

EXAMPLE 19.8.   What small particle(s) must be produced with the other products of the reaction of a neutron with a $^{235}_{92}$U nucleus by the following reaction?

$$^{235}_{92}\text{U} + {}^{1}_{0}\text{n} \longrightarrow {}^{90}_{38}\text{Sr} + {}^{143}_{54}\text{Xe} + ?$$

Ans.   In order to get the subscripts and the superscripts in the equation to balance, the reaction must produce three neutrons:

$$^{235}_{92}\text{U} + {}^{1}_{0}\text{n} \longrightarrow {}^{90}_{38}\text{Sr} + {}^{143}_{54}\text{Xe} + 3\,{}^{1}_{0}\text{n}$$

This reaction is an example of a nuclear *chain reaction*, in which the products of the reaction cause more of the same reaction to proceed. The three neutrons can, if they do not escape from the sample first, cause three more such reactions. The nine neutrons produced from these reactions can cause nine more such reactions, and so forth. Soon, a huge number of nuclei are converted, and simultaneously a small amount of matter is converted to a great deal of energy. Atomic bombs and nuclear energy plants both run on this principle.

EXAMPLE 19.9.   If each neutron in a certain nuclear reaction can produce three new neutrons, and each reaction takes 1 s, how many neutrons can be produced theoretically in the 15th second?

Ans.   Assuming that no neutrons escaped, the number of neutrons produced during the 15th second is

$$3^{15} = 14\,348\,907$$

The number produced in the 60th second is 4.24×10^{28}.

These nuclear reactions actually take place in much less than 1 s each, and the number of reactions can exceed 10^{28} within much less than 1 min. Since the energy of each "event" is relatively great, a large amount of energy is available.

Such nuclear reactions are controllable by keeping the sample size small so that most of the neutrons escape from the sample instead of causing further reactions. The smallest mass of sample that can cause a sustained nuclear reaction, called a *chain reaction*, is called the *critical mass*. Another way to control the nuclear reaction is to insert control rods into the nuclear fuel. The rods absorb some of the neutrons and prevent a runaway reaction.

When a positron is emitted from a nucleus, it can combine with an electron to produce energy. Show that the following equations, when combined, yield exactly the number of electrons required for the product nucleus.

$$^{22}\text{Na} \longrightarrow {}^{22}\text{Ne} + {}_{+1}\beta$$

$$e^- + {}_{+1}\beta \longrightarrow \text{energy}$$

One of the 11 electrons outside the Na nucleus could be annihilated in the second reaction, leaving 10 electrons for the Ne nucleus.

19.6. NUCLEAR ENERGY

Nuclear energy in almost inconceivable quantities can be obtained from nuclear fission and fusion reactions according to Einstein's famous equation

$$E = mc^2$$

The E in this equation is the energy of the process. The m is the mass of the matter that is converted to energy—the *change* in rest mass. Note well that it is *not* the total mass of the reactant nucleus, but only the mass of the matter that is converted to energy. Sometimes the equation is written as

$$E = (\Delta m)c^2$$

The c in the equation is the velocity of light, 3.00×10^8 m/s. The constant c^2 is so large that conversion of a very tiny quantity of matter produces a huge quantity of energy.

EXAMPLE 19.10.   Calculate the amount of energy produced when 1.00 g of matter is converted to energy. (*Note:* More than 1.00 g of isotope is used in this reaction.) $1 \text{ J} = 1 \text{ kg} \cdot \text{m}^2/\text{s}^2$

Ans. $E = (\Delta m)c^2 = (1.00 \times 10^{-3} \text{ kg})(3.00 \times 10^8 \text{ m/s})^2 = 9.00 \times 10^{13} \text{ J} = 9.00 \times 10^{10} \text{ kJ}$

Ninety billion kilojoules of energy is produced by the conversion of 1 g of matter to energy! The tremendous quantities of energy available in the atomic bomb and the hydrogen bomb stem from the large value of the constant c^2 in Einstein's equation. Conversion of a tiny portion of matter yields a huge quantity of energy. Nuclear plants also rely on this type of energy to produce electricity commercially.

Nuclear fusion reactions involve combinations of nuclei. The fusion reaction of the hydrogen bomb involves the fusing of deuterium, ${}_1^2\text{H}$, in lithium deuteride, Li^2H:

$$_1^2\text{H} + {}_1^2\text{H} \longrightarrow {}_2^3\text{He} + {}_0^1\text{n}$$

$$_1^2\text{H} + {}_1^2\text{H} \longrightarrow {}_1^3\text{H} + {}_1^1\text{H}$$

The ${}^3\text{H}$ produced (along with that produced from the fission of ${}^6\text{Li}$) can react further, yielding even greater energy per event.

$$_1^3\text{H} + {}_1^2\text{H} \longrightarrow {}_2^4\text{He} + {}_0^1\text{n}$$

These fusion processes must be started at extremely high temperatures—on the order of tens of millions of degrees Celsius—which are achieved on earth by fission reactions. That is, the hydrogen bomb is triggered by an atomic bomb. Nuclei have to get very close for a fusion reaction to occur, and the strong repulsive force between two positively charged nuclei tends to keep them apart. Very high temperatures give the nuclei enough kinetic energy to overcome this repulsion. No such problem exists with fission, since the neutron projectile has no charge and can easily get close to the nuclear target. (A report was made in 1989 of a "cold fusion" reaction—a fusion reaction at ordinary temperatures with relatively little energy input—but that report has not yet been confirmed.) The stars get their energy from fusion reactions at extremely high temperatures.

The mass of matter at rest is referred to as its *rest mass*. When matter is put into motion, its mass increases corresponding to its increased energy. The extra mass is given by

$$E = mc^2$$

When a nuclear event takes place, some rest mass is converted to extra mass of the product particles because of their high speed or to the mass of photons of light. While the total mass is conserved in the process, some rest mass (i.e., some matter) is converted to energy.

Nuclear *binding energy* is the energy equivalent (in $E = mc^2$) of the difference between the mass of the nucleus of an atom and the sum of the masses of its uncombined protons and neutrons. For example, the mass of a $_2^4$He nucleus is 4.0015 amu. The mass of a free proton is 1.00728 amu, and that of a free neutron is 1.00866 amu. The free particles exceed the nucleus in mass by

$$2(1.00728 \text{ amu}) + 2(1.00866 \text{ amu}) - 4.0015 \text{ amu} = 0.0304 \text{ amu}$$

This mass has an energy equivalent of 4.54×10^{-12} J for each He nucleus. You would have to put in that much energy into the combined nucleus to get the free particles; that is why that energy is called the binding energy.

The difference in binding energies of the reactants and products of a nuclear reaction can be used to calculate the energy which the reaction will provide.

EXAMPLE 19.11. The mass of a ^{7}Li nucleus is 7.0154 amu. Using this value and those given above, calculate the energy given off in the reaction of 1 mol of ^{7}Li:

$$^7\text{Li} + {}^1\text{H} \longrightarrow 2\,{}^4\text{He}$$

Ans. A mole each of the reactant nuclei has a mass of 7.0154 g + 1.00728 g = 8.0227 g. The product has a mass of 2(4.0015 g) = 8.0030 g. The difference in mass, 0.0197 g, has an energy equivalent to

$$E = mc^2 = (1.97 \times 10^{-5} \text{ kg})(3.00 \times 10^8 \text{ m/s})^2 = 1.77 \times 10^{12} \text{ J}$$

More than 1 billion kilojoules of energy is produced from a mole of ^{7}Li nuclei plus a mole of ^{1}H nuclei. (Burning a mole of carbon in oxygen yields 3.93×10^5 J.)

Solved Problems

INTRODUCTION

19.1. What is the difference between C and ^{12}C?

    Ans. The first symbol stands for the naturally occurring mixture of isotopes of carbon; the second stands for only one isotope—the most common one.

19.2. The alchemists of the middle ages spent years and years trying to convert base metals such as lead to gold. Is such a change possible? Why did they not succeed?

    Ans. Base metals can be changed to gold in nuclear reactions, but not in ordinary chemical reactions, since both are elements. (The process costs much more than the gold is worth, however.) The alchemists never succeeded because they could not perform nuclear reactions. (Indeed, they never dreamed of their existence.)

NATURAL RADIOACTIVITY

19.3. In a nuclear equation, the subscripts are often omitted, but the superscripts are not. Where can you look to find the subscripts? Why can you *not* look there for the superscripts?

    Ans. You can look at a periodic table for the atomic numbers, which are the subscripts. You cannot look at the periodic table for the superscripts, because mass numbers are not generally there. (Mass numbers for the few elements that do not occur naturally are provided in parentheses in most periodic tables.) The atomic masses that are given can give a clue to the mass number of the most abundant isotope in many cases, but not all. For example, the atomic mass of Br is nearest 80, but the two isotopes that represent the naturally occurring mixture have mass numbers 79 and 81.

19.4.   What is the difference, if any, between α, $^4_2\alpha$, and $^4_2He^{2+}$?

   Ans.   There is no difference. All are representations of an alpha particle (also called a helium nucleus).

19.5.   Complete the following nuclear equations:

   (a) $^{211}_{82}Pb \longrightarrow \, ^{211}_{83}Bi + ?$ (c) $^{220}_{86}Rn \longrightarrow ? + \, ^4_2\alpha$

   (b) $^{223}_{87}Fr \longrightarrow \, ^{223}_{88}Ra + ?$ (d) $^{213}_{83}Bi \longrightarrow \, ^{213}_{84}Po + ?$

   Ans.   (a) $^{211}_{82}Pb \longrightarrow \, ^{211}_{83}Bi + \, ^{\,0}_{-1}\beta$ (c) $^{220}_{86}Rn \longrightarrow \, ^{216}_{84}Po + \, ^4_2\alpha$

         (b) $^{223}_{87}Fr \longrightarrow \, ^{223}_{88}Ra + \, ^{\,0}_{-1}\beta$ (d) $^{213}_{83}Bi \longrightarrow \, ^{213}_{84}Po + \, ^{\,0}_{-1}\beta$

19.6.   Complete the following nuclear equations:

   (a) $^{225}_{89}Ac \longrightarrow \, ^{221}_{87}Fr + ?$ (c) $? \longrightarrow \, ^{209}_{83}Bi + \, ^{\,0}_{-1}\beta$

   (b) $^{214}_{83}Bi \longrightarrow \, ^{214}_{84}Po + ?$ (d) $^{232}_{90}Th \longrightarrow \, ^{228}_{88}Ra + ?$

   Ans.   (a) $^{225}_{89}Ac \longrightarrow \, ^{221}_{87}Fr + \, ^4_2\alpha$ (c) $^{209}_{82}Pb \longrightarrow \, ^{209}_{83}Bi + \, ^{\,0}_{-1}\beta$

         (b) $^{214}_{83}Bi \longrightarrow \, ^{214}_{84}Po + \, ^{\,0}_{-1}\beta$ (d) $^{232}_{90}Th \longrightarrow \, ^{228}_{88}Ra + \, ^4_2\alpha$

19.7.   Complete the following nuclear equations:

   (a) $? \longrightarrow \, ^{127}_{52}Te + \, ^0\gamma$ (b) $^{237}Np \longrightarrow \, ^{233}Pa + ?$ (c) $^{22}Na \longrightarrow ? + \, ^{\,0}_{+1}\beta$ (positron)

   Ans.   (a) $^{127}_{52}Te \longrightarrow \, ^{127}_{52}Te + \, ^0_0\gamma$ (b) $^{237}_{93}Np \longrightarrow \, ^{233}_{91}Pa + \, ^4_2\alpha$ (c) $^{22}_{11}Na \longrightarrow \, ^{22}_{10}Ne + \, ^{\,0}_{+1}\beta$

19.8.   Consider the equation

$$^{238}_{92}U \longrightarrow \, ^{234}_{90}Th + \, ^4_2He$$

   Does this equation refer to the nuclei of these elements or to atoms of the elements as a whole?

   Ans.   Nuclear equations are written to describe nuclear changes. However, in this equation, since the correct number of electrons is available for the products, the equation can also be regarded as describing the reactions of complete atoms as well as just their nuclei.

HALF-LIFE

19.9.   A certain isotope has a half-life of 1.24 years. How much of a 1.12-g sample of this isotope will remain after 3.72 years?

   Ans.   After the first 1.24 years, half the 1.12-g sample will still be the same isotope. After 1.24 years more, half of the 0.560 g remaining will still be the original isotope. That is 0.280 g will remain. After the third 1.24 years, 0.140 g remain.

19.10. One-eighth of a certain sample of a radioactive isotope is present 30.0 min after its original weighing. How much will be present after 10.0 min more?

   Ans.   The 30.0 min represents three half-lives, since the number of atoms is reduced to one-eighth the original quantity. (One-half disintegrated in the first half-life, one-half of those left disintegrated in the second, and half of that disintegrated in the third, leaving one-eighth of the original number at the end of 30.0 min.) The half-life is therefore 10.0 min, and the sample is reduced to one-sixteenth of its original quantity after 10.0 min more. That is, half of the one-eighth number of atoms remains after one more half-life.

19.11. Draw a graph of the mass of the radioactive atoms left in the decomposition of a 1200-g sample of a radioactive isotope with a half-life of 10.0 h. Extend the graph to allow readings up to 50 h. Use the vertical axis for mass and the horizontal axis for time.

Ans.

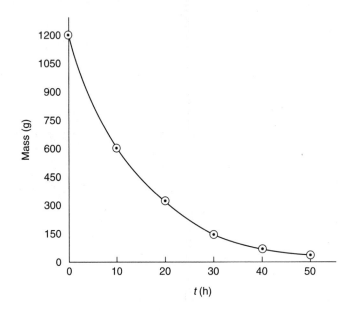

19.12. From the graph of Problem 19.11, estimate how many grams of the isotope will remain after 15.0 h.

   Ans.   About 425 g.

RADIOACTIVE SERIES

19.13. Match the end product and the parent of each of the four radioactive series without consulting any reference tables or other data.

Parents	End Products
^{232}Th	^{206}Pb
^{235}U	^{207}Pb
^{237}Np	^{208}Pb
^{238}U	^{209}Bi

   Ans. ^{232}Th & ^{208}Pb; ^{235}U & ^{207}Pb; ^{237}Np & ^{209}Bi; ^{238}U & ^{206}Pb
   In each case, the final product must differ from the original parent by some multiple of 4 mass numbers. For example, the ^{208}Pb differs in mass number from ^{232}Th by $24 = 4 \times 6$. There must have been six alpha particles emitted in this decay series, with a reduction of four mass numbers each. (The beta and gamma particles emitted do not affect the mass number.)

NUCLEAR FISSION AND FUSION

19.14. What is the difference between (*a*) the mass of an ^{1}H nucleus and the mass of an ^{1}H atom and (*b*) the mass of an ^{1}H atom and the mass number of ^{1}H?

   Ans.   (*a*) The difference is the mass of the electron.   (*b*) The actual mass of an atom is nonintegral. (Calculations involving mass, such as those using $E = mc^2$, should use the actual mass.) The mass number is an integer, equal to the number of protons plus neutrons in the nucleus. In this case, the mass number is 1.

19.15. Complete the following nuclear equations:

(a) $^{14}_{7}\text{N} + ^{4}_{2}\alpha \longrightarrow ? + ^{1}_{1}\text{H}$

(c) $^{235}_{92}\text{U} + ^{1}_{0}\text{n} \longrightarrow ^{140}_{56}\text{Ba} + ^{94}_{36}\text{Kr} + ? ^{1}_{0}\text{n}$

(b) $^{27}_{13}\text{Al} + ^{4}_{2}\alpha \longrightarrow ? + ^{1}_{0}\text{n}$

(d) $^{235}_{92}\text{U} + ^{1}_{0}\text{n} \longrightarrow ^{90}_{38}\text{Sr} + ? + 3 ^{1}_{0}\text{n}$

Ans.   (a) $^{14}_{7}\text{N} + ^{4}_{2}\alpha \longrightarrow ^{17}_{8}\text{O} + ^{1}_{1}\text{H}$

(c) $^{235}_{92}\text{U} + ^{1}_{0}\text{n} \longrightarrow ^{140}_{56}\text{Ba} + ^{94}_{36}\text{Kr} + 2 ^{1}_{0}\text{n}$

(b) $^{27}_{13}\text{Al} + ^{4}_{2}\alpha \longrightarrow ^{30}_{15}\text{P} + ^{1}_{0}\text{n}$

(d) $^{235}_{92}\text{U} + ^{1}_{0}\text{n} \longrightarrow ^{90}_{38}\text{Sr} + ^{143}_{54}\text{Xe} + 3 ^{1}_{0}\text{n}$

NUCLEAR ENERGY

19.16. Calculate the energy of the reaction of a positron with an electron.

$$_{+}\beta + e^{-} \longrightarrow \text{energy}$$

Ans.   Data from Table 19-3 are used. The sum of the masses of the reactants is converted to energy with Einstein's equation.

$$\Delta m = 2(0.00054858 \text{ amu}) = 0.0010972 \text{ amu}$$

This mass is changed to kilograms to calculate the energy in joules:

$$0.0010972 \text{ amu}\left(\frac{1 \text{ g}}{6.022 \times 10^{23} \text{ amu}}\right)\left(\frac{1 \text{ kg}}{1000 \text{ g}}\right) = 1.822 \times 10^{-30} \text{ kg}$$

The energy is given by

$$E = mc^2 = (1.822 \times 10^{-30} \text{ kg})(3.000 \times 10^8 \text{ m/s})^2 = 1.640 \times 10^{-13} \text{ J}$$

19.17. Calculate the energy of the reaction that the free neutron undergoes if it does not encounter a nucleus to react with.

$$\text{n} \longrightarrow \text{p} + e^{-} + \text{energy}$$

Ans.   The data from Table 19-3 are used. The sum of the masses of the products minus the mass of the neutron, converted to energy with Einstein's equation, yields the energy produced.

$$\Delta m = [1.008665 - (1.00728 + 0.00054858)] \text{ amu} = 0.00084 \text{ amu}$$

$$0.00084 \text{ amu}\left(\frac{1 \text{ g}}{6.022 \times 10^{23} \text{ amu}}\right)\left(\frac{1 \text{ kg}}{1000 \text{ g}}\right) = 1.4 \times 10^{-30} \text{ kg}$$

The energy is given by

$$E = mc^2 = (1.4 \times 10^{-30} \text{ kg})(3.00 \times 10^8 \text{ m/s})^2 = 1.3 \times 10^{-13} \text{ J}$$

19.18. Calculate the energy of the reaction of one atom ^{14}C to yield ^{14}N and a beta particle. The *atomic* masses are $^{14}\text{C} = 14.003241$ amu and $^{14}\text{N} = 14.003074$ amu.

Ans.   The difference in masses of products and reactants for the nuclear equation

$$^{14}\text{C} \longrightarrow ^{14}\text{N} + \beta$$

is

$$\Delta m = m_{\text{N nucleus}} + m_{\text{beta}} - m_{\text{C nucleus}}$$

That is equal to

$$(m_{\text{N atom}} - 7m_{\text{e}}) + m_{\text{beta}} - (m_{\text{C atom}} - 6m_{\text{e}})$$

The mass of the seven electrons cancels out, and the difference in mass between the reactants and products in the nuclear equation is equal to the difference in masses of the atoms.

$$\Delta m = 14.003241 \text{ amu} - 14.003074 \text{ amu} = 0.000167 \text{ amu}$$

$$0.000167 \text{ amu}\left(\frac{1 \text{ g}}{6.022 \times 10^{23} \text{ amu}}\right)\left(\frac{1 \text{ kg}}{1000 \text{ g}}\right) = 2.77 \times 10^{-31} \text{ kg}$$

The energy is given by

$$E = mc^2 = (2.77 \times 10^{-31} \text{kg})(3.00 \times 10^8 \text{ m/s})^2 = 2.49 \times 10^{-14} \text{ J}$$

Supplementary Problems

19.19. Without consulting any references, text, or tables, add the correct symbol corresponding to each pair of subscript and superscript. (Example: $^{12}_{6}$? is $^{12}_{6}$C)   (a) $^{0}_{+1}$?, (b) $^{0}_{0}$?, (c) $^{1}_{1}$?, (d) $^{1}_{0}$?, (e) $^{4}_{2}$?, (f) $^{0}_{-1}$?, and (g) $^{3}_{1}$?.

    Ans.   (a) $^{0}_{+1}\beta$ (b) $^{0}_{0}\gamma$ (c) $^{1}_{1}$p or $^{1}_{1}$H   (d) $^{1}_{0}$n   (e) $^{4}_{2}\alpha$ or $^{4}_{2}$He   (f) $^{0}_{-1}\beta$ or $^{0}_{-1}$e   (g) $^{3}_{1}$H

19.20. What is the difference between β and $_{+}\beta$?

    Ans. β is a high-energy electron. $_{+}\beta$ is a positron—a particle with the same properties as the electron except for the sign of its charge, which is positive.

19.21. (a) How many alpha particles are emitted in the radioactive decay series starting with $^{235}_{92}$U and ending with $^{207}_{82}$Pb? (b) How many beta particles are emitted? (c) Can you tell the order of these emissions without consulting reference data? (d) Can you tell how many gamma particles are emitted?

    Ans.   (a) The mass number changes by 28 units in this series, corresponding to seven alpha particles ($7 \times 4 = 28$). (b) The seven alpha particles emitted would have reduced the atomic number by 14 units if no beta particles had been emitted; since the atomic number is reduced by only 10 units, four ($14 - 10$) beta particles are also emitted.   (c) It is impossible to tell from these data alone what the order of emissions is.   (d) It is impossible to tell the number of gamma particles emitted, since emission of a gamma particle does not change either the mass number or the atomic number.

19.22. What is the difference between radioactive decay processes and other types of nuclear reactions?

    Ans.   Other types of reactions require a small particle to react with a nucleus to produce a nuclear reaction; radioactive decay processes are spontaneous with only one nucleus as a reactant.

19.23. In a certain type of nuclear reaction, one neutron is a projectile (a reactant) and two neutrons are produced. Assume that each process takes 1 s. If every product neutron causes another event, how many neutrons will be produced (and not be used up again) (a) in 4 s and (b) in 10 s?

    Ans.

$$\text{n} \xrightarrow[\text{1 s}]{} 2\,\text{n} \xrightarrow[\text{1 s}]{} 4\,\text{n} \xrightarrow[\text{1 s}]{} 8\,\text{n} \xrightarrow[\text{1 s}]{} 16\,\text{n}$$

    (a) In 4 s, $2^{4} = 16$ neutrons are produced.   (b) In 10 s, $2^{10} = 1024$ neutrons will be produced.

19.24. In a certain type of nuclear reaction, one neutron is a projectile (a reactant) and two neutrons are produced. Assume that each process takes 1 s. Suppose that half of all the product neutrons cause another event each, and the other half escape from the sample. How many neutrons will be produced in the third second?

    Ans.   Two.

    The reaction continues with the same number of neutrons being produced as started the reaction in the first place.

19.25. Can the half-life of an isotope be affected by changing the compound it is in?

    Ans.   No. The chemical environment has a negligible effect on the nuclear properties of the atom.

19.26. How many beta particles are emitted in the decomposition series from $^{238}_{92}$U to $^{206}_{82}$Pb?

    Ans.   The number of alpha particles, calculated from the loss of mass number, is 8, because the mass number was lowered by 32. The number of beta particles is equal to twice the number of alpha particles minus the difference in atomic numbers of the two isotopes.

$$(2 \times 8) - 10 = 6$$

    Six beta particles are emitted. (See Example 19.7 and Problem 19.21.)

19.27. If 1.00 mol of C is burned to CO_2 in an ordinary chemical reaction, 393×10^3 J of energy is liberated. (*a*) If 12.0 g of C could be totally converted to energy, how much energy could be liberated? (*b*) If 0.00100% of the mass could be converted to energy, how much energy could be liberated?

    Ans.   (*a*) $E = mc^2 = (12.0 \times 10^{-3} \text{ kg})(3.00 \times 10^8 \text{ m/s})^2 = 1.08 \times 10^{15} \text{ kg·m}^2/\text{s}^2 = 1.08 \times 10^{15}$ J

            (*b*) $0.0000100(1.08 \times 10^{15} \text{ J}) = 1.08 \times 10^{10}$ J

                Over 25 000 times more energy would be liberated by converting 0.00100% of the carbon to energy than by burning all of it chemically.

Scientific Calculations

This appendix introduces two mathematics topics important for chemistry students: scientific algebra and electronic calculator mathematics. The scientific algebra section (Sec. A.1) presents the relationships between scientific algebra and ordinary algebra. The two topics are much more similar than different; however, since you already know ordinary algebra, the differences are emphasized here. The calculator math section (Sec. A.2) discusses points with which students most often have trouble. This section is not intended to replace the instruction booklet that comes with a calculator, but to emphasize the points in that booklet that are most important to science students.

For more practice with the concepts in this appendix, you might recalculate the answers to some of the examples the text.

A.1.   SCIENTIFIC ALGEBRA

Designation of Variables

To solve an algebraic equation such as

$$5x + 25 = 165$$

we first isolate the term containing the unknown ($5x$) by addition or subtraction on each side of the equation of any terms not containing the unknown. In this case, we subtract 25 from each side:

$$5x + 25 - 25 = 165 - 25$$
$$5x = 140$$

We then isolate the variable by multiplication or division. In this case, we divide by 5:

$$\frac{5x}{5} = \frac{140}{5}$$
$$x = 28$$

If values are given for some variables, for example, for the equation

$$x = \frac{y}{z}$$

if $y = 14$ and $z = 5.0$, then we simply substitute the given values and solve:

$$x = \frac{14}{5.0} = 2.8$$

In chemistry and other sciences, such equations are used continually. Since density is defined as mass divided by volume, we could use the equation

$$x = \frac{y}{z}$$

with y representing the mass and z representing the volume, to solve for x, the density. However, we find it easier to use letters (or combinations of letters) that remind us of the quantities they represent. Thus, we write

$$d = \frac{m}{V}$$

with m representing the mass, V representing the volume, and d representing the density. In this way, we do not have to keep looking at the statement of the problem to see what each variable represents. However, we solve this equation in exactly the same way as the equation in x, y, and z.

Chemists need to represent so many different kinds of quantities that the same letter may have to represent more than one quantity. The necessity for duplication is lessened in the following ways:

Method	*Example*
Using *italic* letters for variables and roman (regular) letters for units	m for mass and m for meter
Using capital and lowercase (small) letters to mean different things	V for volume and v for velocity
Using subscripts to differentiate values of the same type	V_1 for the first volume, V_2 for the second, and so on
Using Greek letters	π (pi) for osmotic pressure
Using combinations of letters	MM for molar mass

Each such symbol is handled just as an ordinary algebraic variable, such as x or y.

EXAMPLE A.1. Solve each of the following equations for the first variable, assuming that the second and third variables are equal to 15 and 3.0, respectively. For example, in part a, M is the first variable, n is the second, and V is the third. So n is set equal to 15, and V is set equal to 3.0, allowing you to solve for M.

(a) $M = n/V$

(b) $n = m/\text{MM}$ (where MM is a single variable)

(c) $v = \lambda \nu$ (λ and ν are the Greek letters *lambda* and *nu*.)

(d) $P_1 V_1 = P_2(2.0)$ (P_1 and P_2 represent different pressures.)

Ans.   (a) $M = 15/3.0 = 5.0$ (c) $v = (15)(3.0) = 45$

          (b) $n = 5.0$ (d) $P_1 = (3.0)(2.0)/15 = 0.40$

EXAMPLE A.2. Solve the following equation, in which each letter stands for a different quantity, for R:

$$PV = nRT$$

Ans.   Dividing both sides of the equation by n and T yields

$$R = \frac{PV}{nT}$$

EXAMPLE A.3. Solve the following equation for F in terms of t:

$$\frac{t}{F - 32.0} = \frac{5}{9}$$

Ans.   Inverting each side of the equation yields

$$\frac{F - 32.0}{t} = \frac{9}{5}$$

Simplifying gives $F = \frac{9}{5}t + 32.0$

EXAMPLE A.4.   Using the equation in Example A.3, find the value of F if $t = 25.0$.

Ans. $F = \frac{9}{5}t + 32.0 = \frac{9}{5}(25.0) + 32.0 = 77.0$

UNITS

Perhaps the biggest difference between ordinary algebra and scientific algebra is that scientific measurements (and most other measurements) are always expressed with *units*. Like variables, units have standard symbols. The units are part of the measurements and very often help you determine what operation to perform.

Units are often multiplied or divided, but never added or subtracted. (The associated quantities may be added or subtracted, but the units are not.) For example, if we add the lengths of two ropes, each of which measures 4.00 yards (Fig. A-1*a*), the final answer includes just the unit *yards* (abbreviated yd). Two units of distance are multiplied to get area, and three units of distance are multiplied to get the volume of a rectangular solid (such as a box). For example, to get the area of a carpet, we multiply its length in yards by its width in yards. The result has the unit square yards (Fig. A-1 *b*):

$$\text{Yard} \times \text{yard} = \text{yard}^2$$

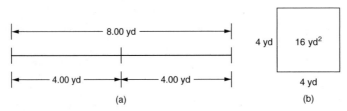

(a)   (b)

Fig. A-1.   Addition and multiplication of lengths

   (*a*) When two (or more) lengths are added, the result is a length, and the unit is a unit of length, such as yard.
   (*b*) When two lengths are multiplied, the result is an area, and the unit is the square of the unit of length, such as square yards.

Be careful to distinguish between similarly worded phrases, such as 2.00 *yards, squared* and 2.00 *square yards* (Fig. A-2).

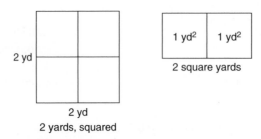

Fig. A-2.   An important difference in wording

   Knowing the difference between such phrases as 2 *yards, squared* and 2 *square yards* is important. Multiplying 2 yards by 2 yards gives 2 yards, squared, which is equivalent to 4 square yards, or four blocks with sides 1 yard long, as you can see. In contrast, 2 square yards is two blocks, each having sides measuring 1 yard.

EXAMPLE A.5.   What is the unit of the volume of a cubic box whose edge measures 2.00 ft?

Ans.   A cube has the same length along each of its edges, so the volume is

$$V = (2.00 \text{ ft})^3 = 8.00 \text{ ft}^3$$

The unit is: ft × ft × ft = ft³

The unit of a quantity may be treated as an algebraic variable. For example, how many liters of soda are purchased if someone buys a 1.00-L bottle of soda plus three 2.00-L bottles of soda?

$$1.00 \text{ L} + 3(2.00 \text{ L}) = 7.00 \text{ L}$$

The same result would have been obtained if L were an algebraic variable instead of a unit. In dollars, how much will the 7.00 L of soda cost if the average price is 89 cents per liter?

$$7.00 \text{ L} \left(\frac{89 \text{ cents}}{1 \text{ L}} \right) = 623 \text{ cents} \left(\frac{1 \text{ dollar}}{100 \text{ cents}} \right) = 6.23 \text{ dollars}$$

EXAMPLE A.6.   What is the unit of the price of ground meat at the supermarket?

Ans.   The price is given in dollars per pound.

If two or more quantities representing the same type of measurement—for example, a distance—are multiplied, they are usually expressed in the same unit. For example, to calculate the area of a rug that is 9.0 ft wide and 4.0 yd long, we express the length and the width in the same unit before they are multiplied. The width in yards is

$$9.0 \text{ ft} \left(\frac{1 \text{ yd}}{3 \text{ ft}} \right) = 3.0 \text{ yd}$$

The area is

$$(4.0 \text{ yd})(3.0 \text{ yd}) = 12 \text{ yd}^2$$

If we had multiplied the original measurements without first converting one to the unit of the other, we would have obtained an incomprehensible set of units:

$$(9.0 \text{ ft})(4.0 \text{ yd}) = 36 \text{ ft·yd}$$

EXAMPLE A.7.   What is the cost of 8.00 ounces (oz) of hamburger if the store charges $2.98 per pound?

Ans.   *Do not* simply multiply:

$$8.00 \text{ oz} \left(\frac{2.98 \text{ dollars}}{1 \text{ lb}} \right) = \frac{23.84 \text{ oz·dollar}}{\text{lb}}$$

Instead, first convert one of the quantities to a unit that matches that of the other quantity:

$$8.00 \text{ oz} \left(\frac{1 \text{ lb}}{16 \text{ oz}} \right) = 0.500 \text{ lb}$$

$$0.500 \text{ lb} \left(\frac{2.98 \text{ dollars}}{1 \text{ lb}} \right) = 1.49 \text{ dollars}$$

The same principles apply to the metric units used in science.

EXAMPLE A.8.   A car accelerates from 10.0 mi/h to 40.0 mi/h in 10.0 s. What is the acceleration of the car? (Acceleration is the change in velocity per unit of time.)

Ans.   The change in velocity is

$$40.0 \text{ mi/h} - 10.0 \text{ mi/h} = 30.0 \text{ mi/h}$$

$$\frac{30.0 \text{ mi/h}}{10.0 \text{ s}} = \frac{3.00 \text{ mi/h}}{\text{s}}$$

The acceleration is 3.00 mi/h per second. This is an example of one of the few times when two different units are used for the same quantity (time) in one value (the acceleration).

QUADRATIC EQUATIONS

A quadratic equation is an equation of the form

$$ax^2 + bx + c = 0$$

Two solutions are given by the equation

$$x = \frac{-b \pm \sqrt{b^2 - 4ac}}{2a}$$

This equation giving the values of x is known as the *quadratic formula*. Two answers are given by this equation (depending on whether the plus or minus sign is used), but often, only one of them has any physical significance.

EXAMPLE A.9. Determine the values of a, b, and c in each of the following equations (after it is put in the form $ax^2 + bx + c = 0$). Then calculate two values for x in each case.

(a) $x^2 - x - 12 = 0$ (b) $x^2 + 3x = 10$

Ans.   (a)   Here $a = 1$, $b = -1$, and $c = -12$.

$$x = \frac{-(-1) \pm \sqrt{(-1)^2 - 4(1)(-12)}}{2(1)}$$

Using the plus sign before the square root yields

$$x = \frac{+1 + \sqrt{1 + 48}}{2} = 4$$

Using the minus sign before the square root yields

$$x = \frac{1 - \sqrt{49}}{2} = -3$$

The two values for x are 4 and -3. Check:

$$(4)^2 - (4) - 12 = 0$$

$$(-3)^2 - (-3) - 12 = 0$$

(b)   First, rearrange the equation into the form

$$ax^2 + bx + c = 0$$

In this case, subtracting 10 from each side yields

$$x^2 + 3x - 10 = 0$$

Thus, $a = 1$, $b = 3$, and $c = -10$. The two values of x are

$$x = \frac{-3 + \sqrt{9 + 40}}{2} = \frac{-3 + \sqrt{49}}{2} = 2$$

and

$$x = \frac{-3 - \sqrt{49}}{2} = -5$$

Conversion to Integral Ratios

It is sometimes necessary to convert a ratio of decimal fraction numbers to integral ratios (Chaps. 3 and 8.) (Note that you cannot round a number more than about 1%.) The steps necessary to perform this operation follow, with an example given at the right.

Table A-1   Certain Decimal Fractions and Their Common Fraction Equivalents

Decimal Part of Number	Common Fraction Equivalent	Multiply by:	Example
0.5	$\frac{1}{2}$	2	$\frac{1.5 \times 2}{1 \times 2} = \frac{3}{2}$
0.333	$\frac{1}{3}$	3	$\frac{1.333 \times 3}{1 \times 3} = \frac{4}{3}$
0.667	$\frac{2}{3}$	3	$\frac{4.667 \times 3}{1 \times 3} = \frac{14}{3}$
0.25	$\frac{1}{4}$	4	$\frac{2.25 \times 4}{1 \times 4} = \frac{9}{4}$
0.75	$\frac{3}{4}$	4	$\frac{1.75 \times 4}{1 \times 4} = \frac{7}{4}$
0.20	$\frac{1}{5}$	5	$\frac{1.2 \times 5}{1 \times 5} = \frac{6}{5}$

	Step	*Example*
1.	Divide the larger of the numbers by the smaller to get a new ratio of equal value with a denominator equal to 1.	$\frac{6.324}{4.744} = \frac{1.333}{1}$
2.	Recognize the *fractional* part of the numerator of the new fraction as a common fraction. (The whole-number part does not matter.) Table A-1, lists decimal fractions and their equivalents.	0.333 is the decimal equivalent of $\frac{1}{3}$
3.	Multiply both numerator and denominator of the calculated ratio in step 1 by the denominator of the common fraction of step 2.	$\frac{1.333 \times 3}{1 \times 3} = \frac{4}{3}$

A.2.   CALCULATOR MATHEMATICS

There is such a variety of models of electronic calculators available that it is impossible to describe them all. This subsection describes many simple calculators. Read your owner's manual to determine the exact steps to use to solve the various types of problems presented here.

A chemistry student needs to have and know how to operate a scientific calculator capable of handling exponential numbers. A huge variety of features are available on calculators, but any calculator with exponential capability should be sufficient for this and other introductory chemistry courses. These calculators generally have the other function keys necessary for these courses. Once you have obtained such a calculator, practice doing calculations with it so that you will not have to think about how to use the calculator while you should be thinking about how to solve the chemistry problems.

Some calculators have more than twice as many functions as function keys. Each key stands for two different operations, one typically labeled *on* the key and the other *above* the key. To use the second function (above the key), you press a special key first and then the function key. This special key is labeled INV (for inverse) on some calculators, 2ndF (for second function) on others, and something else on still others. On a few calculators, the MODE key gives several keys a third function. Most scientific calculators have a memory in which you can store one or more values and later recall the value.

Some of the examples in this section involve very simple calculations. The idea is to learn to use the calculator for operations that you can easily do in your head; if you make a mistake with the calculator, you will recognize it immediately. After you practice with simple calculations, you will have the confidence to do more complicated types.

Precedence Rules

In algebra, when more than one operation is indicated in a calculation, the operations are done in a prescribed order. The order in which they are performed is called the *precedence*, or *priority*, of the operations. The order of the common algebraic operations is given in Table A-2. If operations having the same precedence are used, they are performed as they appear from left to right (except for exponentiation and unary minus, which are done right to left). For example, if a calculation involves a multiplication and an addition, the multiplication should be done first, since it has a higher precedence. In each of the following calculations, the multiplication should be done before the addition:

$$x = 2 + 4 \times 6$$
$$x = 4 \times 6 + 2$$

The answer in each case is 26. Try each of these calculations on your calculator to make sure that it does the operations in the correct order automatically.

Table A-2   Order of Precedence of Common Operations

	Calculator	Algebra
Highest	Parentheses	Parentheses
	Exponentiation (root) or unary minus*	Exponentiation (root) or unary minus*
	Multiplication or division	Multiplication
		Division
Lowest	Addition or subtraction	Addition or subtraction

*Unary minus makes a single value negative, such as in the number −2.

If parentheses are used in an equation, all calculations within the parentheses are to be done *before* the result is used for the rest of the calculations. For example,

$$y = (2 + 4) \times 6$$

means that the addition (within the parentheses) is to be done first, before the other operation (multiplication). *The parentheses override the normal precedence rules.* We might say that parentheses have the highest precedence.

When you are using a calculator, some operation may be waiting for its turn to be done. For example, when $2 + 4 \times 6$ is being entered, the addition will not be done when the multiplication key is pressed. It will await the final equals key, when first the multiplication and then the addition will be carried out. If you want the addition to be done first, you may press the equals key right after entering the four. If you want the calculator to do operations in an order different from that determined by the precedence rules, you may insert parentheses (if they are provided on your calculator) or press the equals $\boxed{=}$ key to finalize all calculations so far before you continue with others.

EXAMPLE A.10.   What result will be shown on the calculator for each of the following sequences of keystrokes?

(a) $\boxed{5}\,\boxed{\times}\,\boxed{3}\,\boxed{+}\,\boxed{4}\,\boxed{=}$ (c) $\boxed{5}\,\boxed{\times}\,\boxed{(}\,\boxed{3}\,\boxed{+}\,\boxed{4}\,\boxed{)}\,\boxed{=}$

(b) $\boxed{(}\,\boxed{5}\,\boxed{\times}\,\boxed{3}\,\boxed{)}\,\boxed{+}\,\boxed{4}\,\boxed{=}$ (d) $\boxed{(}\,\boxed{5}\,\boxed{\times}\,\boxed{3}\,\boxed{+}\,\boxed{4}\,\boxed{)}\,\boxed{=}$

Ans.   (a) 19   (b) 19   (c) 35   (d) 19

EXAMPLE A.11.   What result will be shown on the calculator for each of the following sequences of keystrokes?

(a) $\boxed{5}\,\boxed{+}\,\boxed{3}\,\boxed{\times}\,\boxed{4}\,\boxed{=}$ (b) $\boxed{5}\,\boxed{+}\,\boxed{3}\,\boxed{=}\,\boxed{\times}\,\boxed{4}\,\boxed{=}$

Ans.   (a) 17   (b) 32

The rules of precedence followed by the calculator are exactly the same as those for algebra and arithmetic, except that in algebra and arithmetic, multiplication is done before division (multiplication has a higher precedence than division). Algebraically, ab/cd means that the product of a and b is divided by the product of c and d. On the calculator, since multiplication and division have equal precedence, if the keys are pressed in the order

$$a \boxed{\times} b \boxed{\div} c \boxed{\times} d \boxed{=}$$

the quotient ab/c is *multiplied* by d. To get the correct algebraic result for ab/cd, the *division* key should be pressed before the value of d is entered, or parentheses should be used around the denominator.

EXAMPLE A.12. What result will be obtained from pressing the following keys?

$$\boxed{3}\,\boxed{6}\,\boxed{\div}\,\boxed{9}\,\boxed{\times}\,\boxed{2}\,\boxed{=}$$

Ans.   The result is 8; first, 36 is divided by 9 and then the result (4) is multiplied by 2. If you want 36 divided by both 9 and 2, use either of the following keystroke sequences:

$$\boxed{3}\,\boxed{6}\,\boxed{\div}\,\boxed{9}\,\boxed{\div}\,\boxed{2}\,\boxed{=}$$

or

$$\boxed{3}\,\boxed{6}\,\boxed{\div}\,\boxed{(}\,\boxed{9}\,\boxed{\times}\,\boxed{2}\,\boxed{)}\,\boxed{=}$$

EXAMPLE A.13. Solve:

(a) $\left(\dfrac{9.3}{6.0}\right) \times 2.0$ (b) $\dfrac{9.3}{6.0 \times 2.0}$

Ans.   (a)   The answer is 3.1. The keystrokes are

$$\boxed{9}\,\boxed{.}\,\boxed{3}\,\boxed{\div}\,\boxed{6}\,\boxed{.}\,\boxed{0}\,\boxed{\times}\,\boxed{2}\,\boxed{.}\,\boxed{0}\,\boxed{=}$$

(b)   The answer is 0.775. The keystrokes used in part *a* do not carry out the calculations required for this part. The correct keystroke sequence is

$$\boxed{9}\,\boxed{.}\,\boxed{3}\,\boxed{\div}\,\boxed{6}\,\boxed{.}\,\boxed{0}\,\boxed{\div}\,\boxed{2}\,\boxed{.}\,\boxed{0}\,\boxed{=}$$

or

$$\boxed{9}\,\boxed{.}\,\boxed{3}\,\boxed{\div}\,\boxed{(}\,\boxed{6}\,\boxed{.}\,\boxed{0}\,\boxed{\times}\,\boxed{2}\,\boxed{.}\,\boxed{0}\,\boxed{)}\,\boxed{=}$$

Either *divide* by *each number* in the denominator, or use parentheses around the two numbers so that their product will be divided into the numerator.

Division

In algebra, division is represented in any of the following ways, all of which mean the same thing:

$$\frac{a}{b} \qquad a/b \qquad a \div b$$

Note that any *operation* in the numerator or in the denominator of a built-up fraction (a fraction written on two lines), no matter what its precedence, is done *before* the division indicated by the fraction bar. For example, to simplify the expression

$$\frac{a+b}{c-d}$$

the sum $a+b$ is divided by the difference $c-d$. The addition and subtraction, despite being lower in precedence, are done before the division. In the other forms of representation, this expression is written as

$$(a+b)/(c-d) \qquad \text{or} \qquad (a+b) \div (c-d)$$

where the parentheses are required to signify that the addition and subtraction are to be done first. A calculator has only one operation key for division, $\boxed{\div}$. When you use your calculator, be careful to indicate what is divided by what when more than two variables are involved.

EXAMPLE A.14.   Write the sequence of keystrokes required to perform each of the following calculations, where a, b, and c represent numbers to be entered:

(a) $\dfrac{a+b}{c+d}$ (b) $a + b/c + d$ (c) $a + b/(c+d)$

Ans.   (a) $\boxed{(}\ \boxed{a}\ \boxed{+}\ \boxed{b}\ \boxed{)}\ \boxed{\div}\ \boxed{(}\ \boxed{c}\ \boxed{+}\ \boxed{d}\ \boxed{)}\ \boxed{=}$ (c) $\boxed{a}\ \boxed{+}\ \boxed{b}\ \boxed{\div}\ \boxed{(}\ \boxed{c}\ \boxed{+}\ \boxed{d}\ \boxed{)}\ \boxed{=}$

   (b) $\boxed{a}\ \boxed{+}\ \boxed{b}\ \boxed{\div}\ \boxed{c}\ \boxed{+}\ \boxed{d}\ \boxed{=}$

The Change-Sign Key

If we want to enter a negative number, the number is entered first, and then its sign is changed with the change-sign $\boxed{+/-}$ key, not the subtraction $\boxed{-}$ key. The change-sign key changes the value on the display from positive to negative or vice versa.

EXAMPLE A.15.   Write the sequence of keystrokes required to calculate the product of 6 and -3.

Ans. $\boxed{6}\ \boxed{\times}\ \boxed{3}\ \boxed{+/-}\ \boxed{=}$

   The change-sign key converts 3 to -3 before the two numbers are multiplied.

EXAMPLE A.16.   What result will be displayed after the following sequence of keystrokes?

                                            $\boxed{7}\ \boxed{+/-}\ \boxed{+/-}$

Ans.   7 (The first $\boxed{+/-}$ changes the value to -7, and then the second $\boxed{+/-}$ changes it back to $+7$.)

EXAMPLE A.17.   (a) What result will be displayed after the following sequence of keystrokes?

                                            $\boxed{3}\ \boxed{+/-}\ \boxed{x^2}\ \boxed{=}$

   (b) What keystrokes are needed to calculate the value of -3^2?

Ans.   (a)   The $\boxed{+/-}$ key changes 3 to -3 (minus 3). The squaring key $\boxed{x^2}$ squares the quantity in the display (-3), yielding $+9$.

   (b)   For the algebraic quantity -3^2, the operations are done right to left; that is, the squaring is done first, and the final answer is -9. To make the operations on the calculator follow the algebraic rules, enter 3 press $\boxed{x^2}$, and then press $\boxed{+/-}$.

Exponential Numbers

When displaying a number in exponential notation, most calculators show the coefficient followed by a space or a minus sign and then two digits giving the value of the exponent. For example, the following numbers represent 2.27×10^3 and 2.27×10^{-3}, respectively:

$$2.27 \quad\ \ 03$$
$$2.27 - \ \underset{\smile}{0}3$$
$$\underset{\text{Coefficient}}{} \quad \underset{\text{Exponent}}{}$$

Note that the base (10) is not shown explicitly on most calculators. If it is shown, interpreting the values of numbers in exponential notation is slightly easier.

To enter a number in exponential form, press the keys corresponding to the coefficient, then press either the $\boxed{\text{EE}}$ key or the $\boxed{\text{EXP}}$ key (you will have one or the other on your calculator), and finally press the keys corresponding to the exponent. For example, to enter 4×10^5 into the calculator, press

$$\boxed{4}\ \boxed{\text{EE}}\ \boxed{5} \quad \text{or} \quad \boxed{4}\ \boxed{\text{EXP}}\ \boxed{5}$$

Do not press the times $\boxed{\times}$ *key or the* $\boxed{1}$ *and* $\boxed{0}$ *keys when entering an exponential number!* The $\boxed{\text{EE}}$ or $\boxed{\text{EXP}}$ key stands for "times 10 to the power." For simplicity, we will use EXP to mean either EXP or EE from this point on.

If the *coefficient* is negative, press the $\boxed{+/-}$ key *before* the $\boxed{\text{EXP}}$ key. If the *exponent* is negative, press the $\boxed{+/-}$ key *after* the $\boxed{\text{EXP}}$ key.

EXAMPLE A.18.   What keys should be pressed to enter the number -4.44×10^{-5}?

Ans.

$$\boxed{4}\ \boxed{.}\ \boxed{4}\ \boxed{4}\ \boxed{+/-}\ \boxed{\text{EXP}}\ \boxed{5}\ \boxed{+/-}$$

or

$$\boxed{4}\ \boxed{.}\ \boxed{4}\ \boxed{4}\ \boxed{+/-}\ \boxed{\text{EXP}}\ \boxed{+/-}\ \boxed{5}$$

EXAMPLE A.19.   What value will be displayed if you enter the following sequence of keystrokes?

$$\boxed{4}\ \boxed{.}\ \boxed{4}\ \boxed{4}\ \boxed{\times}\ \boxed{1}\ \boxed{0}\ \boxed{\text{EXP}}\ \boxed{5}\ \boxed{=}$$

Ans.   These keystrokes perform the calculation

$$4.44 \times (10 \times 10^5) =$$

The resulting value will be 4.44×10^6 (which might be displayed in floating-point format as $4\,440\,000$). These keystrokes instruct the calculator to multiply 4.44 by 10×10^5, which yields a value 10 times larger than was intended if you wanted to enter 4.44×10^5.

EXAMPLE A.20.   A student presses the following sequence of keys: to get the value of the quotient

$$\frac{4 \times 10^7}{4 \times 10^7}$$

$$\boxed{4}\ \boxed{\times}\ \boxed{1}\ \boxed{0}\ \boxed{\text{EXP}}\ \boxed{7}\ \boxed{\div}\ \boxed{4}\ \boxed{\times}\ \boxed{1}\ \boxed{0}\ \boxed{\text{EXP}}\ \boxed{7}\ \boxed{=}$$

What value is displayed on the calculator as a result?

Ans.   The result is 1×10^{16}. The calculator divides 4×10^8 by 4, then multiplies that answer by 10 times 10^7. (See the precedence rules in Table A-1.) This answer is wrong because any number divided by itself should give an answer of 1. (You should always check to see if your answer is reasonable.)

EXAMPLE A.21.   What keystrokes should the student have used to get the correct result in Example A.20?

Ans.

$$\boxed{4}\ \boxed{\text{EXP}}\ \boxed{7}\ \boxed{\div}\ \boxed{4}\ \boxed{\text{EXP}}\ \boxed{7}\ \boxed{=}$$

The precedence rules are not invoked for this sequence of keystrokes, since only one operation, division, is done.

Some calculators display answers in decimal notation unless they are programmed to display them in scientific notation. If a number is too large to fit on the display, such a calculator will use scientific notation automatically. To get a display in scientific notation for a reasonably sized decimal number, press the $\boxed{\text{SCI}}$ key or an equivalent key, if available. (See your instruction booklet.) If automatic conversion is not available on your calculator, you can multiply the decimal value by 1×10^{10} (if the number is greater than 1) or 1×10^{-10} (if the number is less than 1), which forces the display automatically into scientific notation. Then you mentally subtract or add 10 to the resulting exponent.

The Reciprocal Key

The reciprocal of a number is 1 divided by the number. It has the same number of significant digits as the number itself. For example, the reciprocal of 5.00 is 0.200. A number times its reciprocal is equal to 1.

On the calculator, the reciprocal $\boxed{1/x}$ key takes the reciprocal of whatever value is in the display. To get the reciprocal of a/b, enter the value of a, press the division $\boxed{\div}$ key, enter the value of b, press the equals $\boxed{=}$ key, and finally press the reciprocal $\boxed{1/x}$ key. Alternatively, enter the value of b, press the division key, enter the value of a, and press the equals key. The reciprocal of a/b is b/a.

The reciprocal key is especially useful if you have a calculated value in the display that you want to use as a denominator. For example, if you want to calculate $a/(b + c)$ and you already have the value of $b + c$ in the display, you divide by a, press the equals key, and then press the reciprocal key to get the answer. Alternatively, with the value of $b + c$ in the display, you can press the reciprocal key and then *multiply* that value by a.

$$\frac{a}{b + c} = \left(\frac{1}{b + c}\right)a$$

EXAMPLE A.22.   The value equal to $1.77 + 1.52 = 3.29$ is in the display of your calculator. What keystrokes should you use to calculate the value of the following expression?

$$\frac{3.08}{1.77 + 1.52}$$

Ans. $\boxed{1/x}\ \boxed{\times}\ \boxed{3}\ \boxed{.}\ \boxed{0}\ \boxed{8}\ \boxed{=}$ or $\boxed{\div}\ \boxed{3}\ \boxed{.}\ \boxed{0}\ \boxed{8}\ \boxed{=}\ \boxed{1/x}$

The display should read $0.93617\cdots$.

Logarithms and Antilogarithms

Determining the logarithm of a number or what number a certain value is the logarithm of (the antilogarithm of the number) is sometimes necessary. The logarithm $\boxed{\log}$ key takes the *common logarithm of the value in the display*. The $\boxed{10^x}$ key, or the $\boxed{INV}$ key or the $\boxed{2ndF}$ key followed by the $\boxed{\log}$ key, takes the *antilogarithm of the value in the display*. That is, this sequence gives the number whose logarithm was in the display.

The $\boxed{\ln}$ key takes the natural logarithm of the value in the display. The $\boxed{e^x}$ key or the $\boxed{INV}$ key or the $\boxed{2ndF}$ key followed by the $\boxed{\ln}$ key, yields the natural antilogarithm of the value in the display.

EXAMPLE A.23.   What sequence of keystrokes is required to determine (*a*) the logarithm of 1.15 and (*b*) the antilogarithm of 1.15? Reminder: Some calculators require log to be entered before the number. Check the owner's manual.

Ans.   (*a*) $\boxed{1}\ \boxed{.}\ \boxed{1}\ \boxed{5}\ \boxed{\log}$

(*b*) $\boxed{1}\ \boxed{.}\ \boxed{1}\ \boxed{5}\ \boxed{10^x}$ or $\boxed{1}\ \boxed{.}\ \boxed{1}\ \boxed{5}\ \boxed{2ndF}\ \boxed{\log}$ or $\boxed{1}\ \boxed{.}\ \boxed{1}\ \boxed{5}\ \boxed{INV}\ \boxed{\log}$

EXAMPLE A.24.   What sequence of keystrokes is required to determine (*a*) the natural logarithm of 5.1/5.43 and (*b*) the natural antilogarithm of 5.1/5.43?

Ans.   (*a*) $\boxed{5}\ \boxed{.}\ \boxed{1}\ \boxed{\div}\ \boxed{5}\ \boxed{.}\ \boxed{4}\ \boxed{3}\ \boxed{=}\ \boxed{\ln}$

(*b*) $\boxed{5}\ \boxed{.}\ \boxed{1}\ \boxed{\div}\ \boxed{5}\ \boxed{.}\ \boxed{4}\ \boxed{3}\ \boxed{=}\ \boxed{2ndF}\ \boxed{\ln}$

In part *a*, be sure to press $\boxed{=}$ before $\boxed{\ln}$ so that you take the natural logarithm of the quotient, not that of the denominator. In part *b*, press $\boxed{=}$ before 2ndF (or $\boxed{INV}$).

EXAMPLE A.25.   Write the sequence of keystrokes required to solve for x, and calculate the value of x in each case:

(*a*) $x = \log(2.22/1.03)$ (*b*) $x = 2.22/\log 1.03$

Ans. (*a*) $\boxed{2}\ \boxed{.}\ \boxed{2}\ \boxed{2}\ \boxed{\div}\ \boxed{1}\ \boxed{.}\ \boxed{0}\ \boxed{3}\ \boxed{=}\ \boxed{\log}$ (*b*) $\boxed{2}\ \boxed{.}\ \boxed{2}\ \boxed{2}\ \boxed{\div}\ \boxed{1}\ \boxed{.}\ \boxed{0}\ \boxed{3}\ \boxed{\log}\ \boxed{=}$

$\qquad\qquad\qquad x = 0.334$ $\qquad\qquad\qquad x = 173$

Note the similarity of the keystroke sequences but the great difference in answers.

Significant Figures

An electronic calculator gives its answers with as many digits as are available on the display unless the last digits are zeros to the right of the decimal point. The calculator has no regard for the rules of significant figures (Sec. 2.5). *You* must apply the rules when reporting the answer. For example, the reciprocal of 9.00 is really 0.111, but the calculator displays something like 0.111 111 11. Similarly, dividing 8.08 by 2.02 should yield 4.00, but the calculator displays 4. You must report only the three significant 1s in the first example and must add the two significant 0s in the second example.

Summary

Note especially the following points that are important for your success:

1. Variables, constants, and units are represented with standard symbols in scientific mathematics (for example, *d* for density). Be sure to learn and use the standard symbols.

2. Units may be treated as algebraic quantities in calculations.

3. Be sure to distinguish between similar symbols for variables and units. Variables are often printed in *italics*. For example, mass is symbolized by *m*, and meter is represented by m. Capitalization can be crucial: *v* represents velocity and *V* stands for volume.

4. Algebraic and calculator operations must be done in the proper order (according to the rules of precedence). (See Table A-1.) Note that multiplication and division are treated somewhat differently on the calculator than in algebra.

5. Operations of equal precedence are done left to right except for exponentiation and unary minus, which are done right to left.

Solved Problems

SCIENTIFIC ALGEBRA

A.1. Simplify each of the following:

(*a*) $3x = 42$ (*b*) $\dfrac{27y}{3x}$ (*c*) $\dfrac{1}{x} = 6.00$ (*d*) $14\left(\dfrac{x}{42}\right)$ (*e*) $23(x-16) = 0$

(*f*) $(x+9)-(2x+1)$

Ans. Be sure to distinguish between equations and expressions when simplifying.

(*a*) $x = 14$ (*b*) $9y/x$ (*c*) $x = 0.167$ (*d*) $0.33x$ (*e*) $x = 16$ (*f*) $-x+8$

A.2. Using your calculator, find the value of *x* from

$$x = (a+b)-(c+d)$$

where $a = -3, b = 5, c = -6$, and $d = 10$. Repeat this calculation twice, once using the parentheses keys and once without using them.

Ans. $x = -2$

A.3. Simplify each of the following, if possible:

(*a*) $\dfrac{V_1 T_2}{V_1 T_2}$ (*b*) $\dfrac{P_1 V_1}{P_2 V_2}$ (*c*) $\dfrac{P_2}{3T_1} = \dfrac{2P_2}{T_2}$

Ans.   (*a*)   1

          (*b*)   The expression cannot be simplified.

          (*c*) $T_2 = 6T_1$

A.4.   What is the difference between (5.0 cm^3) and $(5.0 \text{ cm})^3$?

    Ans.   The exponent 3 *outside* the parentheses indicates that the 5.0 is to be cubed, as well as the unit centimeter. The exponent inside the parentheses acts on the unit only.

A.5.   In each of the following sets, are the expressions equal?

    (*a*) $\left(\dfrac{1}{b+c}\right)a$ $\dfrac{a}{b+c}$ (*c*) $\dfrac{a}{b+c}$ $\dfrac{a}{b} + \dfrac{a}{c}$ $\dfrac{a}{b} + \dfrac{1}{c}$

    (*b*) $\dfrac{a}{bc}$ $\dfrac{1}{b} \times \dfrac{a}{c}$ $\dfrac{a}{b} \times \dfrac{1}{c}$

    Ans.   (*a*)   Both expressions are equal.

          (*b*)   All three expressions are equal.

          (*c*)   None of the expressions is equal to any of the others.

A.6.   Are the following expressions equal to each other?

    (*a*) $a(b + c) - d(e - f)$ $ab + ac - de - df$

    (*b*) $(a + b)(c - d) - (e + f)(g - h)$ $ac + bc - ad - bd - eg + eh - fg + fh$

    Ans.   (*a*)   They are not equal: $-d$ times $-f$ is equal to $+df$, not $-df$.

          (*b*)   They are equal.

A.7.   Evaluate each of the following pairs of expressions, using $a = 4$ and $b = 3$. In each case, determine if the two expressions are equal.

    (*a*) $\dfrac{1}{a} + \dfrac{1}{b}$ $\dfrac{1}{a+b}$ (*b*) $\dfrac{1}{a} - \dfrac{1}{b}$ $\dfrac{1}{a-b}$ (*c*) $\dfrac{1}{a} \times \dfrac{1}{b}$ $\dfrac{1}{ab}$

    Ans.   (*a*)   Not equal: $0.583 \neq 0.143$

          (*b*)   Not equal: $-0.083 \neq 1$

          (*c*)   Equal: $0.0833 = 0.0833$

A.8.   What is the unit of the price of gasoline?

    Ans.   Dollars/gallon or dollars/liter

A.9.   What units are obtained when a mass in grams is divided by a volume in (*a*) milliliters (mL), (*b*) cubic centimeters (cm^3), and (*c*) cubic meters (m^3)?

    Ans.   (*a*) g/mL   (*b*) g/cm^3 (*c*) g/m^3

A.10.   What units are obtained when a volume in milliliters is divided into a mass in (*a*) kilograms, (*b*) grams, and (*c*) milligrams?

    Ans.   (*a*) kg/mL   (*b*) g/mL   (*c*) mg/mL

A.11. Simplify each of the following:

    (a) 23.50 dollars $\left(\dfrac{1 \text{ lb}}{3.50 \text{ dollars}} \right)$

    (b) 350.0 mi $\left(\dfrac{1 \text{ gal}}{31.4 \text{ mi}} \right)$

    Ans.   (a) 6.71 lb (b) 11.1 gal

A.12. What *unit* does the answer have in each of the following:

    (a) 2 cm is subtracted from 1 cm   (c) 2 cm is multiplied by 1 cm^2

    (b) 2 cm is multiplied by 1 cm   (d) 2 cm^3 is divided by 1 cm

    Ans.   (a) cm   (b) cm^2 (c) cm^3 (d) cm^2

A.13. (a) Solve $M = n/V$ for n (d) Solve $d = m/V$ for V

      (b) Solve $\pi V = nRT$ for T (e) Solve $\dfrac{P_1}{T_1} = \dfrac{P_2}{T_2}$ for P_2

      (c) Solve $PV = nRT$ for R

    Ans.   (a) $n = MV$ (d) $V = m/d$

               (b) $T = \dfrac{\pi V}{nR}$ (e) $P_2 = \dfrac{P_1 T_2}{T_1}$

               (c) $R = \dfrac{PV}{nT}$

A.14. (a) Solve $d = m/V$ for m (e) Solve $P_1 V_1 = P_2 V_2$ for V_1

      (b) Solve $PV = nRT$ for V (f) Solve $\pi V = nRT$ for n

      (c) Solve $M = n/V$ for V (g) Solve $\Delta t = k_b m$ for m

      (d) Solve $\dfrac{P_1}{T_1} = \dfrac{P_2}{T_2}$ for T_1 (h) Solve $E = hv$ for h

    Ans.   (a) $m = dV$ (e) $V_1 = P_2 V_2 / P_1$

               (b) $V = \dfrac{nRT}{P}$ (f) $n = \pi V/(RT)$

               (c) $V = n/M$ (g) $m = \Delta t/k_b$

                                      (h) $h = E/v$

               (d) $T_1 = \dfrac{P_1 T_2}{P_2}$

A.15. What *unit* does the answer have in each of the following?
    (a) 2 g is divided by 1 cm^3
    (b) 2 g is divided by 1 cm and also by 2 cm^2

    Ans.   (a) g/cm^3 (b) g/cm^3

A.16. What *unit* does the result have when each of the following expressions is simplified?

    (a) 4.00 ft $\left(\dfrac{12 \text{ in.}}{1 \text{ ft}} \right) \left(\dfrac{4.95 \text{ dollars}}{1 \text{ in.}} \right)$ (c) $(6.22 \text{ cm}^2) \times (1.37 \text{ cm})$

                                              (d) $(6.22 \text{ cm}^2)/(1.37 \text{ cm})$

    (b) $(12.0 \text{ cm}^3)/(2.00 \text{ cm})$

    Ans.   (a) Dollars   (b) cm^2 (c) cm^3 (d) cm

A.17. Solve each of the following equations for the indicated variable, given the other values listed.

Equation	Solve for:	Given	Equation	Solve for:	Given
(a) $PV = nRT$	P	$n = 1.50$ mol $R = \dfrac{0.0821 \text{ L·atm}}{\text{(mol·K)}}$ $T = 305$ K $V = 4.00$ L	(e) $\pi V = nRT$	π	$V = 2.50$ L $n = 0.0300$ mol $T = 298$ K $R = \dfrac{0.0821 \text{ L·atm}}{\text{(mol·K)}}$
(b) $M = n/V$	V	$M = 1.65$ mol/L $n = 0.725$ mol	(f) $P_1V_1 = P_2V_2$	P_2	$V_1 = 242$ mL $V_2 = 505$ mL $P_1 = 0.989$ atm
(c) $d = m/V$	V	$d = 2.23$ g/mL $m = 133$ g	(g) $E = h\nu$	ν	$E = 6.76 \times 10^{-19}$ J $h = 6.63 \times 10^{-34}$ J·s
(d) $\dfrac{P_1}{T_1} = \dfrac{P_2}{T_2}$	T_1	$T_2 = 273$ K $P_1 = 888$ torr $P_2 = 777$ torr	(h) $\Delta t = k_b m$	k_b	$\Delta t = 0.400°$C $m = 0.250$ m

Ans.

(a) $P = \dfrac{nRT}{V} = \dfrac{(1.50 \text{ mol})[0.0821 \text{ L·atm}/(\text{mol·K})](305 \text{ K})}{4.00 \text{ L}} = 9.39$ atm

(b) $V = \dfrac{n}{M} = 0.725 \text{ mol}\left(\dfrac{1 \text{ L}}{1.65 \text{ mol}}\right) = 0.439$ L

(c) $V = \dfrac{m}{d} = 133 \text{ g}\left(\dfrac{1 \text{ mL}}{2.23 \text{ g}}\right) = 59.6$ mL

(d) $T_1 = \dfrac{P_1 T_2}{P_2} = \dfrac{(888 \text{ torr})(273 \text{ K})}{777 \text{ torr}} = 312$ K

(e) $\pi = \dfrac{nRT}{V} = \dfrac{(0.0300 \text{ mol})[0.0821 \text{ L·atm}/(\text{mol·K})](298 \text{ K})}{2.50 \text{ L}} = 0.294$ atm

(f) $P_2 = \dfrac{P_1 V_1}{V_2} = \dfrac{(0.989 \text{ atm})(242 \text{ mL})}{505 \text{ mL}} = 0.474$ atm

(g) $\nu = E/h = (6.76 \times 10^{-19} \text{ J})/(6.63 \times 10^{-34} \text{ J·s}) = 1.02 \times 10^{15}/\text{s}$

(h) $k_b = \Delta t/m = (0.400°\text{C})/(0.250 \text{ m}) = 1.60°\text{C/m}$

CALCULATOR MATHEMATICS

A.18. Note the difference between $(5.71)^2$ and 5.71×10^2. Which calculator operation key should be used for entering each of these?

Ans. For $(5.71)^2$, the $\boxed{x^2}$ key should be pressed after entering 5.71, giving 32.6. For 5.71×10^2, the $\boxed{\text{EXP}}$ or $\boxed{\text{EE}}$ key followed by 2 should be pressed after entering 5.71, giving 571.

A.19. Calculate the reciprocal of 4.00×10^{-8} m.

Ans. $2.50 \times 10^7 \text{ m}^{-1} = 2.50 \times 10^7/\text{m}$

Note that the unit for the reciprocal is different from the unit for the given value.

A.20. If you have a value for the ratio x/y in the display of your calculator, what key(s) do you press to get the value y/x?

    Ans.   The reciprocal key, $\boxed{1/x}$

A.21. Perform the following calculations in your head. Check the results on your calculator:
    (*a*) $3+2\times7$ (*b*) $3\times2+7$ (*c*) $3\times(2-7)$ (*d*) $3\times2/(5+1)$ (*e*) $3\times2/(4\times1.5)$

    Ans.   (*a*) 17   (*b*) 13   (*c*) -15 (*d*) 1   (*e*) 1

A.22. Perform the following calculations in your head. Check the results on your calculator:
    (*a*) $2\times(-6)$ (*b*) $2\times2+(-6)$ (*c*) $-4\times(2-6)$ (*d*) $-3\times7/(-4-3)$

    Ans.   (*a*) -12 (*b*) -2 (*c*) 16   (*d*) 3

A.23. What value will be obtained if 4 is entered on a calculator and then the change-sign $\boxed{+/-}$ key is pressed *twice*?

    Ans.   The answer is 4. The sign is changed to -4 the first time $\boxed{+/-}$ is pressed, and -4 is changed back to 4 the second time it is pressed.

A.24. What value will be obtained if 3 is entered on a calculator and then the square $\boxed{x^2}$ key is pressed *twice*?

    Ans.   The answer is 81. The 3 is squared the first time $\boxed{x^2}$ is pressed, yielding 9 in the display. That 9 is squared when $\boxed{x^2}$ is pressed again.

A.25. Write the exponential number corresponding to each of the following displays on a scientific calculator:
    (*a*) 4.14 06   (*b*) 4.14−06   (*c*) −4.14 06   (*d*) −4.14−06   (*e*) 1 07

    Ans.   (*a*) 4.14×10^6 (*b*) 4.14×10^{-6} (*c*) -4.14×10^6 (*d*) -4.14×10^{-6}
           (*e*) 1×10^7

A.26. Write the display on a scientific calculator corresponding to each of the following exponential numbers:
    (*a*) 6.06×10^3 (*b*) 6.06×10^{-3} (*c*) -6.06×10^{-3} (*d*) -3.03×10^{-23}
    (*e*) 1.7×10^{-3}

    Ans.   (*a*) 6.06 03 (*b*) 6.06 − 03 (*c*) −6.06 − 03 (*d*) −3.03−23 (*e*) 1.7 − 03

A.27. Put parentheses around two successive variables in each expression so that the value of the expression is not changed. Use the precedence rules for algebra. For example, $a\times b+c$ is the same as $(a\times b)+c$.

    (*a*) $a/b+c$ (*c*) $a-b-c$

    (*b*) ab/c (*d*) a/bc

    Ans.   (*a*) $(a/b)+c$ (*c*) $(a-b)-c$
           (*b*) $(ab)/c$ or $a(b/c)$ (*d*) $a/(bc)$

A.28. Determine the value of each of the following expressions:

    (*a*) $\dfrac{(782)(1/760)(1.40)}{(0.0821)(273)}$ (*c*) $\dfrac{(3.00\times10^8)(6.63\times10^{-34})}{4.22\times10^{-10}}$

    (*b*) $\dfrac{(0.0821)(291)}{(0.909)(1.25)}$ (*d*) $\dfrac{(1.40)(7.82\times10^2)/(7.60\times10^2)}{(8.21\times10^{-2})(2.73\times10^2)}$

    Ans.   (*a*) 0.0643 (*b*) 21.0 (*c*) 4.71×10^{-16} (*d*) 0.0643
           Note that part *d* and part *a* are the same except that part *d* is presented in scientific notation.

A.29. Write the sequence of keystrokes required to do each of the following calculations on your calculator:

(*a*) 3×4^2 (*b*) $(3 \times 4)^2$ (*c*) -3^2 (*d*) 3^3 (*e*) 3^4

(*a*) $\boxed{3}\ \boxed{\times}\ \boxed{4}\ \boxed{x^2}\ \boxed{=}$

(*b*) $\boxed{3}\ \boxed{\times}\ \boxed{4}\ \boxed{=}\ \boxed{x^2}$

(*c*) $\boxed{3}\ \boxed{x^2}\ \boxed{+/-}$

(*d*) $\boxed{3}\ \boxed{x^2}\ \boxed{\times}\ \boxed{3}\ \boxed{=}$

(*e*) $\boxed{3}\ \boxed{x^2}\ \boxed{x^2}$

A.30. Report the result for each of the following expressions in scientific notation:

(*a*) $(2 \times 10^5)(3 \times 10^6)$

(*b*) $\dfrac{8 \times 10^3}{2 \times 10^4}$

(*c*) $\dfrac{5.5 \times 10^{-2}}{2.2 \times 10^2}$

(*d*) $(3.5 \times 10^7)(4.0 \times 10^{-3})$

(*e*) $\dfrac{-2.0 \times 10^7}{5.0 \times 10^{-2}}$

(*f*) $\dfrac{2.0 \times 10^{-5}}{-8.0 \times 10^{-2}}$

(*g*) $\dfrac{(1.0 \times 10^6)(8.0 \times 10^3)}{2.0 \times 10^8}$

Ans.   (*a*) 6×10^{11} (*d*) 1.4×10^5 (*g*) 4.0×10^1

(*b*) 4×10^{-1} (*e*) -4.0×10^8

(*c*) 2.5×10^{-4} (*f*) -2.5×10^{-4}

A.31. Which part of Problem A.27 has a different answer if the precedence rules for calculators are used?

Ans.   Part *d* is different. On the calculator, the quotient a/b is multiplied by c, so the expression $(a/b)c$ is the same as the value without the parentheses.

A.32. Solve each of the following equations for x.

(*a*) $\log x = 1.69$ (*b*) $\log (x/2.00) = 1.69$ (*c*) $\log (2.00/x) = 1.69$

Ans.   (*a*) 49 (*b*) 98 (*c*) 0.041

A.33. Report the result for each of the following expressions to the proper number of significant digits:

(*a*) $(103 - 152)/76.1$

(*b*) $(97 + 17)/15.0$

(*c*) $222/19.32 - 44.6$

(*d*) $(-13)^3$

(*e*) $(-3.59)(77.2)(66.3)$

(*f*) $-2.22 + (-3.33) \times 12.6$

(*g*) $-13.6 \times (-22.2)/66.6$

(*h*) $2.00^3 \times 3.00^2$

(*i*) $13.9/(-104 + 20.2)$

(*j*) $12.4/3.09 - 6.06$

Ans.   (*a*) -0.64 (*d*) -2.2×10^3 (*g*) 4.53   (*j*) -2.05

(*b*) 7.6   (*e*) -1.84×10^4 (*h*) 72.0

(*c*) -33.1 (*f*) -44.2 (*i*) -0.17

A.34. Report the result for each of the following expressions to the proper number of significant digits:

(*a*) $12.8 - 6.02$

(*b*) 12.8×6.02

(*c*) $-12.8 \div 6.02$

(*d*) $(1.28 \times 10^1) - (6.02 \times 10^0)$

(*e*) $(1.28 \times 10^{11}) - (6.02 \times 10^{10})$

Ans.   (*a*) 6.8   (*b*) 77.1   (*c*) -2.13 (*d*) 6.8   (*e*) 6.8×10^{10}

A.35. Report the result for each of the following expressions to the proper number of significant digits:

(a) $(8.93 \times 10^3) \times (7.17 \times 10^{-3})$ (d) $(1.11 \times 10^5)^4$

(b) $(1.70 \times 10^2)/(3.85 \times 10^{-7})$ (e) $(-3.47 \times 10^{-1})(71.07)$

(c) $(1.33 \times 10^6)/(1.69 \times 10^{-8})$ (f) $(2.00 \times 10^{27}) - (2.00 \times 10^{25})$

Ans.   (a) 64.0   (c) 7.87×10^{13} (e) -2.47×10^1

(b) 4.42×10^8 (d) 1.52×10^{20} (f) 1.98×10^{27}

A.36. Perform each of the following calculations, and report each answer to the proper number of significant digits:

(a) 6.24×2.000 (f) $31.5 + (-5.03)$

(b) $3.84/7.68$ (g) $-13 \times (-13.9)$

(c) $8.14/29.13$ (h) 4.00^4

(d) $(4.81)^2$ (i) $13.8/(-71.7)$

(e) $(-1.15)(33.8)$ (j) $112.7 - (-66.473)$

Ans.   (a) 12.5   (c) 0.279   (e) -38.9 (g) 1.8×10^2 (i) -0.192

(b) 0.500   (d) 23.1   (f) 26.5   (h) 256   (j) 179.2

A.37. To calculate 5.00×5.200, one student pressed the following keys:

$$\boxed{5}\,\boxed{.}\,\boxed{0}\,\boxed{0}\,\boxed{\times}\,\boxed{5}\,\boxed{.}\,\boxed{2}\,\boxed{0}\,\boxed{0}\,\boxed{=}$$

Another student pressed the following keys:

$$\boxed{5}\,\boxed{\times}\,\boxed{5}\,\boxed{.}\,\boxed{2}\,\boxed{=}$$

Both students got the same display: 26. Explain why.

Ans.   The calculator does not keep track of significant digits, so, for example, it does not differentiate between 5.00 and 5. The answer should be 26.0.

Supplementary Problems

A.38. Report the result for each of the following expressions to the proper number of significant digits:

(a) $1.98 \times 10^2 - 1.03 \times 10^2$ (c) $5.54 \times 10^7 + 6.76 \times 10^2$ (f) $-2.445/41.052$

(b) $\dfrac{1.59 \times 10^{16}}{1.98 \times 10^2 - 1.03 \times 10^2}$ (d) $5.54 \times 10^{-7} + 6.76 \times 10^{-2}$ (g) $17.0133 \div 17.0129$

(e) $4.184(125)(73.0 - 72.2)$

Ans.   (a) 95   (d) 6.76×10^{-2} (f) -0.05956

(b) 1.7×10^{14} (e) 4×10^2 (g) 1.00002

(c) 5.54×10^7

A.39. Solve for t:

$$t = \left(\frac{15.1 \text{ years}}{0.693}\right) \ln\left(\frac{17.4 \text{ g}}{12.1 \text{ g}}\right)$$

Ans. $t = 7.92$ years

A.40. State how you would obtain a value for the expression 2^{x^2}, given that $x = 3$.

Ans.   Exponentiation is done right to left. Thus, first square x, yielding 9, and then raise the 2 to the ninth power, yielding 512.

A.41. Solve for m:

$$\ln\left(\frac{10.0\text{ g}}{m}\right) = \frac{(0.693)(1.88\text{ years})}{2.79\text{ years}}$$

Ans. $m = 6.27$ g

A.42. Solve for t:

$$t = \left(\frac{13.1\text{ years}}{0.693}\right)\ln\left(\frac{72.3\text{ g}}{8.11\text{ g}}\right)$$

Ans. $t = 41.4$ years

A.43. Solve for m_0:

$$\ln\left(\frac{m_0}{1.50\text{ g}}\right) = \frac{(0.693)(4.22\text{ years})}{6.71\text{ years}}$$

Ans. $m_0 = 2.32$ g

A.44. Determine the values of a, b, and c in each of the following quadratic equations after you put the equation in the form $ax^2 + bx + c = 0$. Then calculate two values for each x.

(a) $x^2 + x - 6 = 0$ (d) $4x^2 - 21x + 20 = 0$ (g) $x^2 + 4 = 5x$ (j) $x^2 - 6x = -8$

(b) $x^2 - 5x + 4 = 0$ (e) $x^2 - 6x + 8 = 0$ (h) $x^2 = -(4x + 4)$

(c) $x^2 + 4x + 4 = 0$ (f) $x^2 = 6 - x$ (i) $4x^2 = 21x - 20$

Ans.

	a	b	c	x		a	b	c	x
(a)	1	1	−6	−3 and 2	(d)	4	−21	20	1.25 and 4
(b)	1	−5	4	1 and 4	(e)	1	−6	8	2 and 4
(c)	1	4	4	−2 and −2					

The equations of parts f to j may be rearranged to give the same equations as in parts a to e, respectively, and so the answers are the same.

A.45. What answer is displayed on the calculator after each of the following sequences of keystrokes?

(a) $\boxed{(}\ \boxed{2}\ \boxed{.}\ \boxed{3}\ \boxed{\text{EXP}}\ \boxed{+/-}\ \boxed{1}\ \boxed{1}\ \boxed{-}\ \boxed{7}\ \boxed{.}\ \boxed{2}\ \boxed{\text{EXP}}\ \boxed{+/-}\ \boxed{1}\ \boxed{0}\ \boxed{)}\ \boxed{\div}\ \boxed{3}$

(b) $\boxed{2}\ \boxed{.}\ \boxed{3}\ \boxed{\text{EXP}}\ \boxed{+/-}\ \boxed{1}\ \boxed{1}\ \boxed{-}\ \boxed{7}\ \boxed{.}\ \boxed{2}\ \boxed{\text{EXP}}\ \boxed{+/-}\ \boxed{1}\ \boxed{0}\ \boxed{\div}\ \boxed{3}$

(c) $\boxed{5}\ \boxed{.}\ \boxed{0}\ \boxed{\text{EXP}}\ \boxed{8}\ \boxed{\log}$

(d) $\boxed{2}\ \boxed{1/x}$

(e) $\boxed{2}\ \boxed{+}\ \boxed{6}\ \boxed{=}\ \boxed{1/x}\ \boxed{\times}\ \boxed{1}\ \boxed{6}\ \boxed{=}$

(f) $\boxed{1}\ \boxed{4}\ \boxed{\text{INV}}\ \boxed{\log}$

   (or $\boxed{\text{2ndF}}$)

(g) $\boxed{2}\ \boxed{0}\ \boxed{\ln}$

(h) $\boxed{3}\ \boxed{x^2}\ \boxed{\text{STO}}\ \boxed{x^2}\ \boxed{\times}\ \boxed{\text{RCL}}\ \boxed{=}$

Ans.   (a) −2.323 333 33 − 10   (e) 2

        (b) −2.17 − 10   (f) 1   14

        (c) 8.698 970 004   (g) 2.995 732 274

        (d) 0.5   (h) 729 (3^6 power)

A.46. Determine the value of x in each of the following:

(a) $x = \log 42.1$

(b) $\log x = 42.1$

(c) $x = \sqrt[3]{216}$

(d) $x = (100.0)^5$

(e) $x = \sqrt{4.48 \times 10^{18}}$

Ans. (a) 1.624 (b) 1×10^{42} (c) 6.00 (d) 1.000×10^{10} (e) 2.12×10^9

A.47. Determine the values of a, b, and c in each of the following quadratic equations after you put the equation in the form $ax^2 + bx + c = 0$. Then calculate two values for each x.

(a) $x^2 + (-4.5 \times 10^{-3})x - (3.8 \times 10^{-5}) = 0$

(b) $\dfrac{x^2}{0.100 - x} = 2.23 \times 10^{-2}$

Ans.

	a	b	c	x
(a)	1	-4.5×10^{-3}	-3.8×10^{-5}	8.8×10^{-3} and -4.3×10^{-3}
(b)	1	2.23×10^{-2}	-2.23×10^{-3}	3.74×10^{-2} and -0.0597

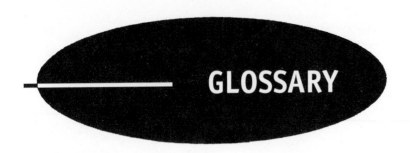

GLOSSARY

A   symbol for mass number.

absolute temperature   temperature on the Kelvin scale.

absolute zero $0\ K = -273.15°C$, the coldest temperature theoretically possible.

acetylene   ethyne; $CH \equiv CH$.

acid   (1) a compound containing ionizable hydrogen atoms. (2) a proton donor (Brønsted-Lowry theory).

acid, organic   a compound of the general type RCO_2H.

acid salt   a salt produced by partial neutralization of an acid containing more than one ionizable hydrogen atom, for example, $NaHSO_4$.

activated complex   the highest energy assembly of atoms during the change of reactant molecules to product molecules. This unstable species can decompose to form products or decompose in a different manner to re-form the reactants.

activation energy   the energy difference between the reactants and the activated complex.

activity   reactivity; tendency to react.

alcohol   an organic compound with molecules containing a covalently bonded —OH group on the radical: R—OH.

aldehyde   an organic compound of the general type RCHO.

alkali metal   a metal of periodic table group IA: Li, Na, K, Rb, Cs, or Fr.

alkaline earth metal   an element of periodic group IIA.

alkane   a hydrocarbon containing only single bonds.

alkene   a hydrocarbon each molecule of which contains one double bond.

alkyl radical   a hydrocarbon radical from the alkane series.

alkyne   a hydrocarbon each molecule of which contains one triple bond.

alpha particle   a ^4He nucleus ejected from a larger nucleus in a spontaneous radioactive reaction.

amide   an organic compound of the type RCONHR' or RCONR'R," formed by reaction of an organic acid with ammonia, or a primary or secondary amine.

amine   an organic compound of the general type RNH_2, R_2NH, or R_3N.

ammonia NH_3.

ammonium ion NH_4^+.

amphiprotic   the ability of a substance to react with itself to produce both a conjugate acid and base. Water is amphiprotic, producing H_3O^+ and OH^-.

amu atomic mass unit.

-ane name ending for the alkane series of hydrocarbons.

anhydrous without water. The waterless salt capable of forming a hydrate, such as $CuSO_4$ (which reacts with water to form $CuSO_4 \cdot 5H_2O$).

anion a negative ion.

anode electrode at which oxidation takes place in either a galvanic cell or an electrolysis cell.

aqueous solution a solution in water.

aromatic hydrocarbon a hydrocarbon containing at least one benzene ring, perhaps with hydrogen atoms replaced by other groups.

Arrhenius theory theory of acids and bases in which acids are defined as hydrogen-containing compounds that react with bases. Bases are defined as OH^- containing compounds.

aryl radical a hydrocarbon radical derived from the aromatic series.

atmosphere, standard a unit of pressure equal to 760 torr.

atmospheric pressure the pressure of the atmosphere; barometric pressure.

atom the smallest particle of an element that retains the composition of the element.

atomic energy energy from the nuclei of atoms; nuclear energy.

atomic mass the relative mass, compared to that of ^{12}C, of an average atom of an element.

atomic mass unit one-twelfth the mass of a ^{12}C atom.

atomic number the number of protons in the nucleus of the atom.

atomic theory Dalton's postulates, based on experimental evidence, which proposed that all matter is composed of atoms.

autoionization reaction of a substance with itself to produce ions.

auto-oxidation reaction of a substance with itself to produce a product with a lower oxidation number and another with a higher oxidation number; disproportionation.

Avogadro's number a mole; 6.02×10^{23} units; the number of atomic mass units per gram.

balancing an equation adding coefficients to make the numbers of atoms of each element the same on both sides of an equation.

balancing a nuclear equation ensuring that the total of the charges and the total of the mass numbers are the same on both sides of a nuclear equation.

barometer a device for measuring gas pressure.

barometric pressure the pressure of the atmosphere.

base (1) a compound containing OH^- ions or which reacts with water to form OH^- ions. (2) a proton acceptor (Brønsted-Lowry theory).

base of an exponential number the number (or unit) that is multiplied by itself. For example, the base in 1.5×10^3 is 10, and in 5 cm^3 it is cm.

battery a combination of two or more galvanic cells.

benzene a cyclic compound with molecular formula C_6H_6; the base of the aromatic hydrocarbon series.

beta particle a high-energy electron ejected from a nucleus in a nuclear reaction.

binary   composed of two elements.

binding energy   the energy equivalent of the difference between the mass of a nucleus and the sum of the masses of the (uncombined) protons and neutrons that make it up.

Bohr theory   the first theory of atomic structure which involved definite internal energy levels for electrons.

Boltzmann constant, k   a constant equal to the gas law constant divided by Avogadro's number: $k = 1.38 \times 10^{-23}$ J/molecule·K.

bonding   the chemical attraction of atoms for each other within chemical compounds.

Boyle's law   the volume of a given sample of gas at constant temperature is inversely proportional to the pressure of the gas: $V = k/P$.

Brønsted-Lowry theory   a theory of acids and bases that defines acids as proton donors and bases as proton acceptors.

buffer solution   a solution of a weak acid and its conjugate base or a weak base and its conjugate acid. The solution resists change in pH even on addition of small quantities of strong acid or base.

buret   a calibrated tube used to deliver exact volumes of liquid.

c (1) symbol for the velocity of light in a vacuum: 3.00×10^8 m/s. (2) symbol for specific heat.

carbonyl group   the $\diagdown\!\!\!\underset{\diagup}{C} = O$ group in an organic compound.

catalyst   a substance that alters the rate of a chemical reaction without undergoing permanent change in its own composition.

cathode   electrode at which reduction takes place in either a galvanic cell or an electrolysis cell.

cation   a positive ion.

Celsius temperature scale   a temperature scale with $0°$ defined as the freezing point of pure water and $100°$ defined as the normal boiling point of pure water.

centi-   prefix meaning 0.01.

chain reaction   a self-sustaining series of reactions in which the products of one reaction, such as neutrons, initiate more of the same reaction. One such particle can therefore start a whole series of reactions.

change-sign key   the key on a calculator that changes the sign of the number on the display. $\boxed{+/-}$

Charles' law   for a given sample of gas at constant pressure, the volume is directly proportional to the *absolute* temperature: $V = kT$.

circuit   a complete path necessary for an electric current.

classification of matter   the grouping of matter into elements, compounds, and mixtures.

coefficient   (1) the number placed before the formula of a reactant or product to balance an equation. (2) the decimal value in a number in standard exponential notation.

coinage metal   an element of periodic group IB: Cu, Ag, or Au.

collision theory   a theory that explains the effects of the various factors on the rates of chemical reactions in terms of the particles (molecules, ions, atoms) involved.

combination reaction   a reaction in which two elements, an element and a compound, or two compounds combine to form a single compound.

combined gas law   for a given sample of gas, the volume is inversely proportional to the pressure and directly proportional to the absolute temperature: $P_1 V_1 / T_1 = P_2 V_2 / T_2$

combustion   reaction with oxygen gas.

completion   complete consumption of at least one of the reactants in a chemical reaction. A reaction goes to completion if its limiting quantity is used up.

compound   a chemical combination of elements.

concentration   the number of parts of solute in a given quantity of solution (molarity or normality) or of solvent (molality).

condensed formula   a formula for an organic compound that shows all bonds between atoms except those to hydrogen atoms. The hydrogen atoms are placed to the right of the larger atoms to which they are attached.

conduct   allow passage of electricity by movement of electrons in a wire or ions in a liquid.

conjugates   an acid or base plus the product of reaction that differs by H^+.

control of electrons   assignment for oxidation number purposes of the electrons in a covalent bond to the more electronegative atom sharing them.

controlled experiment   a series of individual experiments in which all factors except one are held constant so that the effect of that factor on the outcome can be determined. Another series of experiments can be performed for each additional factor to be tested.

covalent bonding   bonding by shared electron pairs.

critical mass   the smallest mass of a sample that will sustain a chain reaction. Smaller masses will lose neutrons or other projectile particles from their bulk, and there will not be sufficient projectile particles to keep the chain going.

current   the flow of electrons or ions.

cycloalkane   a hydrocarbon containing a ring of carbon atoms and only single bonds.

cycloalkene   a hydrocarbon containing a ring of carbon atoms and one double bond per molecule.

Dalton's law of partial pressures   the total pressure of a mixture of gases is equal to the sum of the partial pressures of the components.

Daniell cell   a galvanic cell composed of copper/copper(II) ion and zinc/zinc ion half-cells.

daughter nucleus   the large nucleus (as opposed to a small particle such as an alpha particle) that results from the spontaneous disintegration of a (parent) nucleus.

decay   radioactive disintegration.

decomposition reaction   a reaction in which a compound decomposes to yield two substances.

definite proportions   having the same ratio of masses of individual elements in every sample of the compound.

delocalized double bonds   double bonds that are not permanently located between two specific atoms. The electron pairs can be written equally well between one pair of atoms or another, as in benzene.

Δ   Greek letter delta, meaning "change of."

density   mass divided by volume. A body of lower density will float in a liquid of higher density.

derivative   a compound of a hydrocarbon with at least one hydrogen atom replaced by a functional group.

deuterium   the isotope of hydrogen with a mass number of 2. Also called heavy hydrogen.

deuteron   the nucleus of a deuterium atom.

diatomic   composed of two atoms.

diffusion   the passage of one gas through another.

dimensional analysis   factor-label method.

direct proportion   as the value of one variable rises, the value of the other rises by the same factor. Directly proportional variables have a constant quotient, for example, $V/T = k$ for a given sample of a gas at constant pressure.

disintegration   spontaneous emission from a radioactive nucleus of an alpha, a beta, or a gamma particle.

disproportionation   reaction of a reactant with itself to produce a product with a lower oxidation number and another with a higher oxidation number.

double bond   a covalent bond with two shared pairs of electrons.

double decomposition   double substitution.

double replacement   double substitution.

double-substitution reaction   a reaction of (ionic) compounds in which the reactant cations swap anions. Also called double replacement, double decomposition, or metathesis.

EE key   the key on a calculator meaning "times 10 to the power," used to enter exponential numbers.

effusion   the escape of gas molecules through tiny openings in the container holding the gas.

Einstein's equation $E = mc^2$.

elastic collision   a collision in which the total kinetic energy of the colliding particles does not change.

electric current   the concerted (nonrandom) movement of charged particles such as electrons in a wire or ions in a solution.

electrode   a solid conductor of electricity used to connect a current-carrying wire to a solution in an electrolysis cell or a galvanic cell.

electrolysis   a process in which an electric current produces a chemical reaction.

electron   a negatively charged particle that occurs outside the nucleus of the atom and is chiefly responsible for the bonding between atoms.

electron dot diagram   a scheme for representing valence electrons in an atom with dots.

electronegativity   the relative attraction for electrons of atoms involved in covalent bonding. The higher the attraction, the higher the electronegativity.

electronic charge   the magnitude of the charge on an electron; 1.60×10^{-19} C.

electroplating   depositing a thin layer of a metal on the surface of another metal by means of an electrolysis reaction.

element   a substance that cannot be broken down into simpler substances by ordinary chemical means.

empirical formula   formula for a compound that contains the simplest whole number ratio of atoms of the elements.

endpoint   the point in a titration at which the indicator changes color permanently and the titration is stopped.

-ene   name ending for the alkene series of hydrocarbons.

energy   the capacity to produce change or the ability to do work.

equation   notation for a chemical reaction containing formulas of each reactant and product, with coefficients to make the numbers of atoms of each element equal on both sides.

equilibrium   state in which two opposing processes occur at equal rates, causing no apparent change.

equilibrium constant a constant equal to the ratio of concentrations of products to reactants, each raised to a suitable power, which is dependent for a given reaction on temperature only.

equivalent the quantity of a substance, (1) in a redox reaction, that reacts with or produces 1 mol of electrons; (2) in an acid-base reaction, that reacts with or produces 1 mol of H^+ or OH^- ions.

equivalent mass the number of grams per equivalent of a substance.

ester an organic compound of the general type RCO_2R'.

ether an organic compound of the general type ROR'.

ethylene ethene; $CH_2 = CH_2$.

event the reaction of one nucleus (plus a projectile, if any) in a nuclear reaction.

excess quantity more than sufficient of one reagent to ensure that another reagent reacts completely.

excited state the state of an atom with the electron(s) in higher energy levels than the lowest possible.

EXP key the key on a calculator meaning "times ten to the power," used to enter exponential numbers.

exponent a superscript telling how many times the coefficient is multiplied by the base. For example, the exponent in 2.0×10^3 is 3; the 2.0 is multiplied by 10 three times: $2.0 \times 10^3 = 2.0 \times 10 \times 10 \times 10$.

exponential number a number expressed with a coefficient times a power of 10, for example, 1.0×10^3.

extrapolation reading a graph beyond the experimental points.

factor-label method a method of problem solving that uses the units to indicate which algebraic operation to do for quantities that are directly proportional; dimensional analysis.

feeble acid or base an acid or base that has practically no tendency to react with water.

fission the process in which a nucleus is split into two more or less equal parts by bombardment with a projectile particle such as a neutron or a proton.

fluid a gas or liquid.

formula a combination of symbols with proper subscripts that identifies a compound or molecule.

formula mass the sum of the atomic masses of all the atoms in a formula.

formula unit the material represented (1) by the simplest formula of an ionic compound, (2) by a molecule, or (3) by an uncombined atom.

functional group the reactive part of an organic molecule.

fusion (1) melting. (2) combining two nuclei in a nuclear reaction.

galvanic cell a cell in which a chemical reaction produces an electric potential.

gamma ray a stream of gamma particles, essentially photons of high-energy light with zero rest mass and zero charge.

Graham's law the rate of effusion or diffusion of a gas is inversely proportional to the square root of its molecular mass.

gram the fundamental unit of mass in the metric system.

graphical formula a formula in which all atoms are shown with their bonding electron pairs represented by lines; structural formula.

ground state the lowest energy state of an atom.

half-life   the time it takes for one-half of the nuclei of any given sample of a particular radioactive isotope to disintegrate spontaneously.

half-reaction   the oxidation or reduction half of a redox reaction.

halogen   an element of periodic group VIIA: F, Cl, Br, I (or At).

heat   a form of energy; the only form of energy that cannot be completely converted into another form.

heat capacity   the energy required to change a certain quantity of a substance by a certain temperature.

heavy hydrogen   deuterium.

Heisenberg uncertainty principle   the location and the energy of a small particle such as an electron cannot both be known precisely at any given time.

heterogeneous   having distinguishable parts.

homogeneous   alike throughout; having parts that are indistinguishable even with an optical microscope.

Hund's rule of maximum multiplicity   when electrons partially fill a subshell, they remain as unpaired as possible.

hydrate   a compound composed of a stable salt plus some number of molecules of water, for example, $CuSO_4 \cdot 5H_2O$.

hydrocarbon   a compound of carbon and hydrogen only.

hydronium ion H_3O^+; the combination of a proton and water.

hypo   prefix meaning "still fewer oxygen atoms."

hypothesis   a proposed explanation of observable results.

ideal gas law $PV = nRT$.

indicator   a substance that has an intense color that changes depending on the acidity of a solution. Indicators are used to determine when the endpoint of a titration has been reached.

initial concentration   the concentration in an equilibrium system before the equilibrium reaction has proceeded at all.

inner transition series   the two series of elements at the bottom of the periodic table. The series containing elements 58 to 71 and 90 to 103, arising from the filling of $4f$ and $5f$ subshells.

interpolation   reading a graph between the experimental points.

inverse proportion   as the value of one variable rises, the value of the other goes down by the same factor. Inversely proportional variables have a constant product, for example, $PV = k$ for a given sample of gas at constant temperature.

ion   a charged atom or group of atoms.

ion-electron method   a method of balancing redox equations using ions in half-reactions.

ionic bond   the attraction between oppositely charged ions.

ionization constant   the equilibrium constant for the reaction of a weak acid or base with water.

isomerism   existence of isomers.

isomers   different compounds having the same molecular formula.

isotopes   two or more atoms of the same element with different numbers of neutrons in their nuclei.

joule   a unit of energy. 4.184 J raises the temperature of 1.00 g of water 1.00°C.

k Boltzmann constant: $k = R/N = 1.38 \times 10^{-23}$ J/(molecule·K).

K_a acid ionization constant.

K_b base ionization constant.

K_i acid or base ionization constant.

K_w ionization constant for the autoionization of water.

kelvin the unit of the Kelvin temperature scale.

Kelvin temperature scale a temperature scale with its 0 at the lowest theoretically possible temperature (0 K = $-273°$C) and with temperature differences the same as those on the Celsius scale.

ketone an organic compound of the type $RCOR'$.

kilo prefix meaning 1000.

kilogram the legal standard of mass in the United States; 1000 g.

kinetic energy energy of motion, as in a 10-ton truck going 50 mi/h: KE $= \frac{1}{2}mv^2 = \frac{1}{2}(10 \text{ ton})(50 \text{ mi/h})^2$ (where m is mass and v is velocity).

kinetic molecular theory a theory that explains the properties of gases in terms of the actions of their molecules.

law an accepted generalization of observable facts and experiments.

law of conservation of energy energy cannot be created or destroyed.

law of conservation of mass mass is neither created nor destroyed during any process.

law of conservation of matter matter is neither created nor destroyed during any ordinary chemical reaction. This law differs from the law of conservation of mass because rest mass can be converted into energy (which has its own mass associated with it). Some matter is converted to energy in nuclear reactions.

law of definite proportions all samples of a given compound, no matter what their source, have the same percentage of each of the elements.

law of multiple proportions in two compounds of the same elements, for a given mass of one element, the two masses of each other element are in a ratio of small whole numbers.

lead storage cell a cell composed of $Pb/PbSO_4$ and $PbO_2/PbSO_4$ electrodes in an H_2SO_4 electrolyte.

Le Châtelier's principle if a stress is applied to a system at equilibrium, the equilibrium will shift in a tendency to reduce the stress.

light a form of energy. Light has both wave and particle properties, with a speed in vacuum of 3.00×10^8 m/s.

limiting quantity the reagent that will be used up before all the other reagent(s); the substance that will be used up first in a given chemical reaction, causing the reaction to stop.

line formula a formula for an organic compound written on one line, in which bonded groups of atoms are written together, for example, $CH_3CHClCH_3$.

liter the fundamental unit of volume of the classical metric system (about 6% greater than a U.S. quart). (The cubic meter is the fundamental unit of volume in SI.)

ln key the key on the calculator that takes the natural logarithm of the number in the display.

log key the key on the calculator that takes the base 10 logarithm of the number in the display.

m (1) symbol for mass. (2) symbol for meter. (3) symbol for milli. (4) symbol for molal (unit of molality).

M symbol for molar, the unit of molarity.

main group   one of the eight groups at the left and right of the periodic table that extend up to the first or second period.

mass   a quantitative measure of the quantity of matter and energy in a body. Mass is measured by its direct proportionality to weight and/or to inertia (the resistance to change in the body's motion).

matter   anything that has mass and occupies space.

metathesis   double substitution.

meter   the basic unit of length in the metric system (about 10% greater than a yard).

metric system   a system of measurement based on the decimal system and designed for ease of use, in which prefixes are used that have the same meanings no matter what unit they are used with.

metric ton 10^6 g $= 1000$ kg $= 2200$ lb.

milli   prefix meaning 0.001.

millimole   0.001 mol.

mixture   a physical combination of elements and/or compounds.

mol   abbreviation for mole.

molal   unit of molality.

molality   number of moles of solute per kilogram of solvent.

molar   unit of molarity.

molar mass   the mass of 1 mol of a substance. The molar mass is numerically equal to the formula mass, but the dimensions are grams per mole.

molar volume   the volume of 1 mol of gas at STP; 22.4 L.

molarity   number of moles of solute per liter of solution.

mole 6.02×10^{23} units: Avogadro's number.

mole fraction   the ratio of moles of a component to total number of moles in a solution: $X_A = \dfrac{\text{moles A}}{\text{total moles}}$.

mole ratio   a ratio of moles, such as given by a chemical formula or a balanced chemical equation.

molecular formula   the formula of a compound that gives the ratio of number of atoms of each element to number of molecules of compound. The molecular formula is an integral multiple of the empirical formula.

molecular mass   the formula mass of a molecular substance. The sum of the atomic masses of the atoms in a molecule.

molecular weight   molecular mass.

molecule   a combination of atoms held together by covalent bonds.

monatomic   consisting of one atom. (A monatomic ion contains only one atom.)

net ionic equation   an equation in which spectator ions (ions that begin in solution and wind up unchanged as the same ion in solution) are omitted.

neutralization   the reaction of an acid with a base.

neutron   a neutral particle in the nucleus of the atom.

noble gas configuration   an octet of electrons in the outermost shell, such as the noble gases possess when they are uncombined. A pair of electrons in the first shell if that is the only shell, such as in helium.

normal   unit of normality.

normality   the number of equivalents of solute per liter of solution.

nuclear energy   energy from reactions of nuclei.

nuclear reaction   a reaction in which at least one nucleus undergoes change.

nucleus   the tiny center of an atom containing the protons and neutrons.

octet   a set of eight electrons in the outermost shell.

octet rule   a generalization that atoms tend to form chemical bonds to get eight electrons in their outermost shells or that eight electrons in the outermost shell of an atom is a stable state.

orbit   the circular path of an electron about the nucleus in the Bohr theory.

orbital   a subdivision of an energy level in which an electron will have a given value for each of n, l, and m_l.

organic chemistry   chemistry of compounds with C — C and/or C — H bonds.

outermost shell   the largest shell containing electrons in an atom or ion.

overall equation   an equation with complete compounds present, as opposed to a net ionic equation.

oxidation   raising of oxidation number, by loss of (control of) electrons.

oxidation number   the number of outermost electrons of a free atom minus the number of electrons the atom "controls" in a compound, ion, or molecule.

oxidation number change method   a method for balancing redox equations by balancing changes in oxidation numbers first.

oxidation state   oxidation number.

oxidizing agent   reactant that causes an increase in the oxidation number of another reactant.

oxy-   prefix meaning combined oxygen, as in oxyvanadium(IV) ion: VO^{2+}.

oxyanion   a negative ion containing oxygen as well as another element, for example, NO_{3-}.

paraffin   a saturated hydrocarbon.

parent nucleus   a nucleus that disintegrates spontaneously yielding a small particle plus another nucleus of size approximately equal to itself.

partial pressure   the pressure of a component of a gas mixture.

pascal   the SI unit of pressure. (1.000 atm $= 1.013 \times 10^5$ Pa $= 101.3$ kPa)

Pauli exclusion principle   no two electrons in a given atom can have the same set of four quantum numbers.

per   (1) divided by. For example, the number of miles per hour is calculated by dividing the total number of miles by the total number of hours. (2) prefix meaning "more oxygen."

percent composition   the number of grams of each element in a compound per 100 g of the compound.

percent yield   100 times the actual yield divided by the calculated yield: $100 \times \dfrac{\text{actual yield}}{\text{calculated yield}}$.

percentage   the number of units of an item present per 100 units total. For example, 73 g of sand in 100 g of a mixture (or 7.3 g in 10.0 g mixture, or 730 g in 1000 g mixture) is 73% sand.

period   a horizontal array of elements in the periodic table.

periodic table   a tabulation of the elements according to atomic number with elements having similar properties in the same (vertical) group.

peroxide ion $O_2{}^{2-}$ ion.

pH $-\log[H_3O^+]$.

phase change   change of state, as for example change of solid to liquid or solid to gas.

photon   a particle of light.

physical change   a process in which no change in composition occurs.

pipet   a piece of glassware calibrated to deliver an exact volume of liquid.

polyatomic   composed of more than one atom.

positron   a subatomic particle that may be ejected from a nucleus, with all the properties of an electron except for the sign of its charge, which is positive.

potential   the driving force for electric current.

potential energy   energy of position, as for example in a rock on top of a mountain.

precedence   the required order of operations in algebra when more than one operation is to be done.

precipitate   (1) *verb*: form a solid from solute in solution. (Form solid or liquid water from the water vapor in air.) (2) *noun*: the solid so formed.

pressure   force per unit area.

primary amine   an amine RNH_2 with one and only one R group.

product   element or compound produced in a chemical reaction.

property   a characteristic of a substance by which the substance can be identified.

propylene   propene; $CH_3CH{=}CH_2$.

proton   (1) a positive particle in the nucleus of the atom. (2) the H^+ ion (Brønsted-Lowry theory).

quadratic formula   a formula for solving for x in an equation of the general form $ax^2 + bx + c = 0$.

$$x = \frac{-b \pm \sqrt{b^2 - 4ac}}{2a}$$

quantum (plural, quanta)   a particle of energy.

quantum number   one of four values that control the properties of the electron in the atom.

R   ideal gas law constant: 0.0821 L·atm/(mol·K) or 8.31 J/(mol·K).

R   Rydberg constant (see a general chemistry text).

R—   symbol for a radical in organic chemistry.

radical   the hydrocarbon-like portion of an organic compound with one hydrogen atom of the hydrocarbon replaced by another group.

radioactive decay   the disintegration of a sample of a naturally radioactive isotope.

radioactive series   a series of isotopes produced one from the other in a sequence of spontaneous radioactive disintegrations.

rate law   an equation that relates the rate of a chemical reaction to the concentrations of its reactants: Rate $= k[A]^x[B]^y$

rate of reaction   the number of moles per liter of product that is produced by a chemical reaction per unit time.

reactant   element or compound used up in a chemical reaction.

reacting ratio   mole ratio in which substances react and are produced, given by the coefficients in the balanced chemical equation.

reaction   a chemical change; a process in which compositions of substances are changed.

reactivity   activity; tendency to react.

reagent   substance used up in a chemical reaction.

reciprocal key   the key on a calculator used to take the reciprocal of a number, that is, to divide the number into 1. $\boxed{1/x}$

redox   oxidation reduction.

reducing agent   reactant that causes a lowering of the oxidation number of another reactant.

reduction   lowering of oxidation number, by gain of (control of) electrons.

rest mass   the mass of a body at rest. (According to Einstein's theory, a body's mass will increase as it is put into motion—as its energy increases—despite its having a constant quantity of matter and thus a constant rest mass.)

rounding   reducing the number of digits in a calculated result to indicate the precision of the measurement(s).

salt   a combination of a cation and an anion (except for H^+, H_3O^+, O^{2-}, or OH^-). Salts may be produced by the reaction of acids and bases.

salt bridge   a connector between the two halves of a galvanic cell necessary to complete the circuit and prevent a charge buildup that would cause stoppage of the current.

saturated hydrocarbon   a hydrocarbon containing only single bonds.

saturated solution   a solution holding as much solute as it can stably hold at a particular temperature. Solutions in equilibrium with excess solute are saturated.

scientific notation   standard exponential notation.

secondary amine   an amine R_2NH with two and only two R groups.

shell   (1) a "layer" of electrons in an atom; (2) a set of electrons all having the same principal quantum number.

shift   a change in the position of an equilibrium system. Shift to the left produces reactants from products; shift to the right produces products from reactants.

SI (Système International d'Unités)   a modern version of the metric system.

significant digits   significant figures.

significant figures   the digits in a measurement or in the calculations resulting from a measurement that indicate how precisely the measurement was made; significant digits.

single bond   a covalent bond with one shared pair of electrons.

solubility   the concentration of a saturated solution at a given temperature.

solute   substance dissolved in a solvent, such as salt dissolved in water.

solution   a homogeneous mixture; a combination of a solvent and solute(s).

solvent   substance that dissolves the solute(s).

specific heat   the heat capacity per gram of substance; the number of joules required to raise 1 g of substance 1°C.

spectator ions   ions that are in solution at the start of a reaction and wind up unchanged (as the same ion in solution).

stable   resistant to reaction.

standard conditions for gases   0°C and 1 atm pressure; STP.

standard exponential notation   exponential notation with the coefficient having a value of 1 or more but less than 10.

standard state   the state of a substance in which it usually occurs at 25°C and 1 atm pressure. For example, the standard state of elementary oxygen is O_2 molecules in the gaseous state.

standard temperature and pressure   0°C and 1 atm pressure.

Stock system   the nomenclature system using oxidation numbers to differentiate between compounds or ions of a given element.

stoichiometry   the science of measuring how much of a substance can be produced from given quantities of others.

STP   standard temperature and pressure for gases; 0°C and 1 atm pressure.

stress   a change in conditions imposed on a system at equilibrium.

strong acid   an acid that reacts completely with water to form ions.

structural formula   a formula in which all atoms are shown with their covalently bonded electron pairs represented by lines; graphical formula.

subdivision   particle size of solids.

sublimation   changing directly from a solid to a gas.

subshell   the set of electrons in an atom all having the same value of n and the same value of l.

substance (pure substance)   an element or a compound.

substitution reaction   the reaction of a free element plus a compound in which the free element replaces one of the elements in the compound and enters the compound itself. For example: $2\,Na + ZnCl_2 \longrightarrow 2\,NaCl + Zn$

superoxide ion O_2^- ion.

supersaturated solution   a solution holding more solute than the solution can hold stably at the particular temperature.

symbol   one- or two-letter representation of an element or an atom of an element. The first letter of a symbol is capitalized; the second letter, if any, is lowercase.

system   portion of matter under investigation, such as the contents of a particular beaker.

t   symbol for Celsius temperature.

T   symbol for absolute temperature.

temperature   the intensity of heat in an object. The property that determines the direction of spontaneous heat flow when bodies are connected thermally.

ternary compound   a compound of three elements.

tertiary amine   an amine with three R groups; R_3N.

theory   an accepted explanation of experimental results.

thermochemistry   the science that investigates the interaction of heat and chemical reactions.

titration measured neutralization reaction (or other type) to determine the concentration or number of moles of one reactant from data about another.

total bond order the number of pairs of electrons on an atom that are shared with other atoms.

transition groups the groups that extend only as high as the fourth period of the periodic table. That is, the groups containing the elements with atomic numbers 21 to 30, 39 to 48, 57, 72 to 80, 89, and 104–109, arising from the filling of the $3d$, $4d$, $5d$, and $6d$ subshells with electrons.

transmutation the change of one element into another (by a nuclear reaction).

triple bond a covalent bond with three shared pairs of electrons.

tritium the isotope of hydrogen with mass number 3.

unary minus the minus sign attached to a number, which makes it negative, as in –2.

unsaturated solution a solution holding less solute than a saturated solution would hold at that temperature.

valence shell the outermost electron shell of a neutral atom.

vaporization changing from a liquid to a gas; evaporation.

vapor pressure the pressure of the vapor of a liquid in equilibrium with that same substance in the liquid phase. The vapor pressure for a given substance is determined by the temperature only.

voltage electric potential.

volumetric flask a flask calibrated to hold an exact volume of liquid.

w symbol for weight.

weak acid a hydrogen-containing compound that reacts somewhat but not completely with water to form ions. (Weak acids react almost completely with bases, however.)

weight the measure of the attraction of a body to the earth.

work force times the distance through which the force is applied.

X_A mole fraction of A.

-yne name ending for the alkyne series of hydrocarbons.

Z symbol for atomic number.

PRACTICE QUIZZES

For each quiz, you may use a periodic table and/or an electronic calculator, as necessary. The estimated time that might be given to you is stated before each quiz.

Chapter 1 (5 minutes)

1. Name the element represented by each of the following symbols:

    (a) Na   (b) S   (c) Sn   (d) Ca

2. Write the symbol for each of the following elements:

    (a) Potassium   (b) Aluminum   (c) Chlorine   (d) Lead

3. Identify the element in group IB of the fifth period of the periodic table.

Chapter 2 (15 minutes)

1. Calculate each answer to the proper number of significant digits:

    (a) $\dfrac{0.0450 \text{ kg}}{17.40 \text{ mL}} =$

    (c) $72.03 \text{ cm} - 67.5 \text{ cm} =$

    (b) $45.0 \text{ cm} + 67.5 \text{ cm} =$

2. Calculate the volume of 125 g of mercury. (density = 13.6 g/mL)

3. For the following problem, calculate the answer in kilograms to the proper number of significant digits, and give the answer in standard exponential notation:

$$5.25 \times 10^{-3} \text{ kg} + 1.17 \times 10^{2} \text{ g} =$$

Chapter 3 (15 minutes)

1. (a) Calculate the mass of sodium chloride formed from 20.0 g of sodium and 30.9 g of chlorine. Sodium chloride is the only compound formed by this pair of elements and each sample of the elements reacted completely. (b) State the law that allows this calculation. (c) Calculate the mass of sodium chloride formed from 20.0 g of sodium and 40.0 g of chlorine. (d) State the law that allows this calculation.

2. State the number of neutrons in each of the following isotopes:

    (a) ^{235}U (b) ^{3}H

3. Determine the atomic mass of bromine if 50.54% of naturally occurring bromine is ^{79}Br with mass of each atom 78.92 amu, and the rest is ^{81}Br, with mass of each atom 80.92 amu.

Chapter 4 (15 minutes)

1. Give the detailed electronic configuration of each of the following elements:

    (a) Na   (b) Fe   (c) Hf (element 72)

2. Draw the shapes of the (a) $1s$ orbital, (b) $2p_x$ orbital, and (c) $3d_{xz}$ orbital.

Chapter 5 (15 minutes)

1. State the formula of the compound of each of the following pairs:

   (*a*) Al and O   (*b*) Mg and S   (*c*) Pb^{4+} and O^{2-} (*d*) NH_4^+ and S^{2-}

2. Write an electron dot diagram for (*a*) PCl_3, (*b*) H_2O, and (*c*) NH_4^+.

Chapter 6 (15 minutes)

1. Name of each of the following ions:

   (*a*) Mn^{2+} (*b*) Mg^{2+} (*c*) NH_4^+ (*d*) Pb^{4+}

2. Write the formula of each of the following compounds:

   (*a*) sodium sulfite   (*b*) aluminum nitrate   (*c*) nickel(II) sulfate   (*d*) copper(I) sulfide

3. Name each of the following compounds:

   (*a*) $CoCl_3$ (*b*) $K_2Cr_2O_7$ (*c*) $Sn_3(PO_4)_2$ (*d*) $AgNO_3$

Chapter 7 (15 minutes)

1. Calculate the empirical formula of a compound containing 71.05% cobalt and 28.95% oxygen.

2. Determine the number of moles of O atoms in 175 g $CaCO_3$.

Chapter 8 (10 minutes)

1. Balance the following equations:

   (*a*) $C_3H_6O_2 + O_2 \longrightarrow CO_2 + H_2O$
   (*b*) $Al + O_2 \longrightarrow Al_2O_3$

2. Complete and balance the following equations. If no reaction occurs, write "NR."

   (*a*) $Zn + Cu(NO_3)_2 \longrightarrow$
   (*b*) $BaO + SO_3 \longrightarrow$

Chapter 9 (15 minutes)

1. Write a balanced net ionic equation for each of the following reactions:

   (*a*) $HClO_3(aq) + Ba(OH)_2(aq) \longrightarrow$
   (*b*) $HClO_2(aq) + Ba(OH)_2(aq) \longrightarrow$
   (*c*) $HClO_3(aq) + Ba(OH)_2(s) \longrightarrow$

2. Balance the following net ionic equation:

$$Ce^{4+} + CuI \longrightarrow I_2 + Ce^{3+} + Cu^{2+}$$

Chapter 10 (20 minutes)

1. Calculate the mass of PbI_2 that can be prepared by the reaction of 105 g NaI with $Pb(NO_3)_2$.

2. Calculate the number of moles of each substance present after 24.0 g of $CaCl_2$ is treated with 52.0 g of $AgNO_3$.

Chapter 11   (15 minutes)

1. Calculate the concentration of each ion in solution after 225 mL of 0.100 M $BaCl_2$ is treated with 208 mL of 0.250 M NaCl and diluted to 0.500 L.

2. Calculate the concentration of each ion in solution after 225 mL of 0.100 M $Ba(OH)_2$ is treated with 208 mL of 0.250 M HCl and diluted to 0.500 L.

Chapter 12   (15 minutes)

1. Calculate the volume occupied by 0.0250 mol of chlorine at STP.

2. Calculate the volume of oxygen gas, collected over water at 836 torr barometric pressure and 298 K, that can be prepared by the thermal decomposition of 2.40 g of $KClO_3$. ($P_{water} = 24$ torr at $25°C$)

Chapter 13   (10 minutes)

1. Calculate the ratio of the average velocity of neon molecules to argon molecules. Both gases are at the same temperature.

2. If two gases are at the same temperature, which of the following *must* be true?

   (a) The average speeds of their molecules must be the same.

   (b) Their pressures must be the same.

   (c) The average kinetic energies of their molecules must be the same.

   (d) Their rates of effusion must be the same.

   (e) Their numbers of molecules must be the same.

Chapter 14   (10 minutes)

1. Identify the oxidation state of the underlined element in each of the following:

   (a) H$\underline{N}O_3$ (b) Na$_2\underline{O}_2$ (c) Na$_2\underline{Cr}_2O_7$ (d) C$\underline{F}_4$

2. Complete and balance the following equation in acid solution:

$$Cr_2O_7{}^{2-} + Cl_2 \longrightarrow ClO_3{}^- + Cr^{3+}$$

Chapter 15   (15 minutes)

1. Calculate the molality of alcohol in an aqueous solution with mole fraction alcohol 0.150.

2. Calculate the mole fraction of a solution 20.0% by mass methanol (CH_4O) in water.

Chapter 16   (15 minutes)

1. Write an equilibrium constant expression for the following reaction:

$$2\,C(s) + O_2(g) \rightleftharpoons 2\,CO(g)$$

2. Calculate the concentration of each substance at equilibrium if 0.0700 mol N_2 and 0.100 mol O_2 are placed in a 1.00-L vessel and allowed to come to equilibrium according to the following equation:

$$N_2(g) + 2\,O_2(g) \rightleftharpoons 2\,NO_2(g) \qquad K = 6.0 \times 10^{-6}$$

Chapter 17   (20 minutes)

1. Calculate the pH after 0.100 mol of NH_3 and 0.200 mol of NH_4Cl are dissolved in 1.00 L of solution and allowed to come to equilibrium. $K_b = 1.8 \times 10^{-5}$.

2. Calculate the hydronium ion concentration after 0.250 mol $HC_2H_3O_2$ and 0.100 mol NaOH are dissolved in 1.00 L of solution and allowed to come to equilibrium. $K_a = 1.8 \times 10^{-5}$.

Chapter 18   (10 minutes)

1. State the class of compound represented by each of the following compounds:

    (a) CH_4 (b) CH_3OH (c) CH_3COCH_3 (d) $CH_3CONHCH_3$ (e) $CH_3CO_2CH_3$

2. Identify all isomers of $CH_3CH_2CH_2OH$.

3. Identify all isomers of $CH_3CH_2CH_2NH_2$.

Chapter 19   (15 minutes)

1. Complete the following nuclear equations:

    (a) $^{22}Na \longrightarrow {}^{22}Ne + ?$
    (b) $^{14}N + ? \longrightarrow {}^{17}O + {}^{1}H$
    (c) $^{235}U + {}^{1}n \longrightarrow ? + {}^{94}Kr + 2\ {}^{1}n$

2. Calculate the time required for an isotope to disintegrate to 40.0% of its original mass if its half-life is 50.0 years.

ANSWERS TO QUIZZES

Chapter 1

1. (a) Sodium (b) Sulfur (c) Tin (d) Calcium

2. (a) K (b) Al (c) Cl (d) Pb (Note that the first letter **must** be capitalized and the second, if any, lower case. If not, you are wrong.)

3. Silver

Chapter 2

1. (a) 0.00259 kg/mL (Three significant digits. Do not forget the units!)
   (b) 112.5 cm
   (c) 4.5 cm

2. $125 \text{ g}\left(\dfrac{1 \text{ mL}}{13.6 \text{ g}}\right) = 9.19 \text{ mL}$

3. $5.25 \times 10^{-3} \text{ kg} + 1.17 \times 10^{-1} \text{ kg} = 1.22 \times 10^{-1} \text{ kg}$

Chapter 3

1. (a) $20.0 \text{ g} + 30.9 \text{ g} = 50.9 \text{ g}$ (c) 50.9 g (9.1 g of chlorine does not react.)
   (b) The law of conservation of mass (d) The law of definite proportions

2. (a) 143 (b) 2

3. $\left(\dfrac{50.54}{100}\right)78.92 \text{ amu} + \left(\dfrac{49.46}{100}\right)80.92 \text{ amu} = 79.91 \text{ amu}$

Chapter 4

1. (a) $1s^2 2s^2 2p^6 3s^1$ (b) $1s^2 2s^2 2p^6 3s^2 3p^6 4s^2 3d^6$
   (c) $1s^2 2s^2 2p^6 3s^2 3p^6 4s^2 3d^{10} 4p^6 5s^2 4d^{10} 5p^6 6s^2 5d^2 4f^{14}$

2. See Figure 4-4. (However, you do not have to be an artist. For example, draw only the x and z axes and the outline of the four lobes of the $3d_{xz}$ orbital.)

Chapter 5

1. (a) Al_2O_3 (b) MgS (c) PbO_2 (d) $(NH_4)_2S$

2. (a) :C̈l:P̈:C̈l: (b) H:Ö:H (c) $\left[\begin{array}{c} \text{H} \\ \text{H:N:H} \\ \text{H} \end{array} \right]^{+}$
       :C̈l:

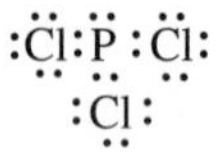

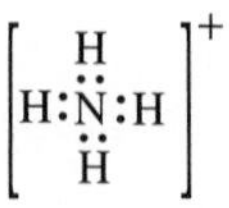

Chapter 6

1. (a) Manganese(II) ion (b) Magnesium ion (c) Ammonium ion (d) Lead(IV) ion

2. (a) Na_2SO_3 (b) $Al(NO_3)_3$ (c) $NiSO_4$ (d) Cu_2S

3. (a) Cobalt(III) chloride (b) Potassium dichromate (c) Tin(II) phosphate (d) Silver nitrate

Chapter 7

1. $71.05 \text{ g Co} \left(\dfrac{1 \text{ mol Co}}{58.9 \text{ g Co}} \right) = 1.21 \text{ mol Co} \qquad 28.95 \text{ g O} \left(\dfrac{1 \text{ mol O}}{16.0 \text{ g O}} \right) = 1.81 \text{ mol O}$

$\dfrac{1.81 \text{ mol O}}{1.21 \text{ mol Co}} = \dfrac{1.50 \text{ mol O}}{1.00 \text{ mol Co}} = \dfrac{3 \text{ mol O}}{2 \text{ mol Co}}$

The formula is Co_2O_3.

2. $175 \text{ g CaCO}_3 \left(\dfrac{1 \text{ mol CaCO}_3}{100 \text{ g CaCO}_3} \right) \left(\dfrac{3 \text{ mol O}}{1 \text{ mol CaCO}_3} \right) = 5.25 \text{ mol O}$

Chapter 8

1. (a) $2\,C_3H_6O_2 + 7\,O_2 \longrightarrow 6\,CO_2 + 6\,H_2O$

   (b) $4\,Al + 3\,O_2 \longrightarrow 2\,Al_2O_3$

2. (a) $Zn + Cu(NO_3)_2 \longrightarrow Cu + Zn(NO_3)_2$

   (b) $BaO + SO_3 \longrightarrow BaSO_4$

Chapter 9

1. (a) $H^+ + OH^- \longrightarrow H_2O$

   (b) $HClO_2 + OH^- \longrightarrow H_2O + ClO_2^-$ (The acid is weak.)

   (c) $2\,H^+ + Ba(OH)_2(s) \longrightarrow Ba^{2+} + 2\,H_2O$ [The $Ba(OH)_2$ is not in solution.]

2. $4\,Ce^{4+} + 2\,CuI \longrightarrow I_2 + 4\,Ce^{3+} + 2\,Cu^{2+}$
   (First balance the iodine, then the copper. The cerium has to provide for four positive charges, so there are four cerium ions on each side.)

Chapter 10

1. $2\,NaI + Pb(NO_3)_2 \longrightarrow PbI_2 + 2\,NaNO_3$

$105 \text{ g NaI} \left(\dfrac{1 \text{ mol NaI}}{150 \text{ g NaI}} \right) \left(\dfrac{1 \text{ mol PbI}_2}{2 \text{ mol NaI}} \right) \left(\dfrac{461 \text{ g PbI}_2}{1 \text{ mol PbI}_2} \right) = 161 \text{ g PbI}_2$

2. $24.0 \text{ g CaCl}_2 \left(\dfrac{1 \text{ mol CaCl}_2}{111 \text{ g CaCl}_2} \right) = 0.216 \text{ mol CaCl}_2$

$52.0 \text{ g AgNO}_3 \left(\dfrac{1 \text{ mol AgNO}_3}{170 \text{ g AgNO}_3} \right) = 0.306 \text{ mol AgNO}_3$

$2\,AgNO_3 + CaCl_2 \longrightarrow Ca(NO_3)_2 + 2\,AgCl$

$\dfrac{0.306 \text{ mol AgNO}_3}{2} = 0.153 \text{ mol AgNO}_3$

$\dfrac{0.216 \text{ mol CaCl}_2}{1} = 0.216 \text{ mol CaCl}_2$

$AgNO_3$ is in limiting quantity.

All in moles:	$2\,AgNO_3$	$+\ CaCl_2$	$\longrightarrow Ca(NO_3)_2$	$+\ 2\,AgCl$
Initial:	0.306	0.216	0.000	0.000
Change:	0.306	0.153	0.153	0.306
Final:	0.000	0.063	0.153	0.306

Chapter 11

1. $225 \text{ mL} \left(\dfrac{0.100 \text{ mmol BaCl}_2}{1 \text{ mL}} \right) = 22.5 \text{ mmol BaCl}_2$

$208 \text{ mL} \left(\dfrac{0.250 \text{ mmol NaCl}}{1 \text{ mL}} \right) = 52.0 \text{ mmol NaCl}$

These substances do not react, so we merely add the numbers of millimoles of chloride ion, and divide the number of millimoles of each ion by the total volume.

22.5 mmol $BaCl_2$ consists of 22.5 mmol Ba^{2+} and 45.0 mol Cl^-.

52.0 mmol NaCl consists of 52.0 mmol Na^+ and 52.0 mmol Cl^-.

There is a total of 97.0 mmol of Cl^- present.

The concentration of barium ion is 22.5 mmol/500 mL = 0.0450 M.

The concentration of sodium ion is 52.0 mmol/500 mL = 0.104 M.

The concentration of chloride ion is 97.0 mmol/500 mL = 0.194 M.

2. $225 \text{ mL} \left(\dfrac{0.100 \text{ mmol Ba(OH)}_2}{1 \text{ mL}} \right) = 22.5 \text{ mmol Ba(OH)}_2$

$208 \text{ mL} \left(\dfrac{0.250 \text{ mmol HCl}}{1 \text{ mL}} \right) = 52.0 \text{ mmol HCl}$

These substances react, so we have a limiting-quantities problem. We use the net ionic equation, and we recognize that the number of millimoles of barium ion and the number of millimoles of chloride ion will not change.

22.5 mmol $Ba(OH)_2$ consists of 22.5 mmol Ba^{2+} and 45.0 mmol OH^-.

52.0 mmol HCl consists of 52.0 mmol H^+ and 52.0 mmol Cl^-.

The concentration of barium ion is 22.5 mmol/500 mL = 0.0450 M.

The concentration of chloride ion is 52.0 mmol/500 mL = 0.104 M.

The hydroxide ion is limiting.

All in millimoles:	H^+	$+$	OH^-	$\longrightarrow$	H_2O
Initial:	52.0		45.0		
Change:	45.0		45.0		
Final:	7.0		0.0		

The concentration of hydrogen ion is 7.0 mmol/500 mL = 0.014 M.

Chapter 12

1. $V = \dfrac{nRT}{P} = \dfrac{(0.0250 \text{ mol})(0.0821 \text{ L·atm/mol·K})(273 \text{ K})}{(1.00 \text{ atm})} = 0.560 \text{ L}$

2.   Note that it is the volume of oxygen gas, not solid $KClO_3$ that we are to find. The oxygen pressure is the barometric pressure minus the water vapor pressure:

$P_{\text{oxygen}} = 836 \text{ torr} - 24 \text{ torr} = 812 \text{ torr} \left(\dfrac{1 \text{ atm}}{760 \text{ torr}} \right) = 1.07 \text{ atm}$

$2.40 \text{ g KClO}_3 \left(\dfrac{1 \text{ mol KClO}_3}{122 \text{ g KClO}_3} \right) \left(\dfrac{3 \text{ mol O}_2}{2 \text{ mol KClO}_3} \right) = 0.0295 \text{ mol O}_2$

$V = \dfrac{nRT}{P} = \dfrac{(0.0295 \text{ mol})(0.0821 \text{ L·atm/mol·K})(298 \text{ K})}{(1.07 \text{ atm})} = 0.675 \text{ L}$

Chapter 13

1. According to Graham's law,

$$\frac{v_{Ne}}{v_{Ar}} = \sqrt{\frac{MM_{Ar}}{MM_{Ne}}} = \sqrt{\frac{39.9 \text{ amu}}{20.2 \text{ amu}}} = 1.41$$

The neon molecules are on average 1.41 times faster than those of argon.

2. Only (c) "The average kinetic energies of their molecules must be the same," must be true. The average speeds and their effusion rates depend on the molar masses (Graham's law). Their pressures and numbers of moles (and thus molecules) are governed by the ideal gas law.

Chapter 14

1. (a) +5 (b) −1 (a peroxide) (c) +6 (d) −1

2. $34 \text{ H}^+ + 5 \text{ Cr}_2\text{O}_7^{2-} + 3 \text{ Cl}_2 \longrightarrow 6 \text{ ClO}_3^- + 10 \text{ Cr}^{3+} + 17 \text{ H}_2\text{O}$

Chapter 15

1. Assume that there is 1.000 mol total of alcohol plus water. Then there is 0.150 mol of alcohol and 0.850 mol of water.

$$0.850 \text{ mol H}_2\text{O}\left(\frac{18.0 \text{ g H}_2\text{O}}{1 \text{ mol H}_2\text{O}}\right) = 15.3 \text{ g H}_2\text{O} = 0.0153 \text{ kg H}_2\text{O}$$

$$\frac{0.150 \text{ mol alcohol}}{0.0153 \text{ kg H}_2\text{O}} = 9.80 \text{ m alcohol}$$

2. Assume 100.0 g of solution total. Then there are 20.0 g CH_4O and 80.0 g H_2O.

$$20.0 \text{ g CH}_4\text{O}\left(\frac{1 \text{ mol CH}_4\text{O}}{32.0 \text{ g CH}_4\text{O}}\right) = 0.625 \text{ mol CH}_4\text{O}$$

$$80.0 \text{ g H}_2\text{O}\left(\frac{1 \text{ mol H}_2\text{O}}{18.0 \text{ g H}_2\text{O}}\right) = 4.44 \text{ mol H}_2\text{O}$$

$$X_{methane} = \frac{0.625 \text{ mol methanol}}{5.07 \text{ mol total}} = 0.123 \qquad X_{water} = 0.877$$

Chapter 16

1. $K = \dfrac{[CO]^2}{[O_2]}$ (Note that solids are not included in the expression.)

2.

All in mol/L:	N_2	+	$2\,O_2$	$\rightleftharpoons$	$2\,NO_2$
Initial:	0.0700		0.100		0.000
Change:	$x/2$		x		x
Final:	$0.0700 - x/2$		$0.100 - x$		x

$$K = \frac{[NO_2]^2}{[N_2][O_2]^2} = \frac{x^2}{(0.0700)(0.100)^2} = 6.0 \times 10^{-6}$$

$$x^2 = 4.2 \times 10^{-9}$$

$$x = 6.5 \times 10^{-5} \text{ mol/L} = [NO_2]$$

$$[N_2] = 0.0700 \text{ mol/L}$$

$$[O_2] = 0.100 \text{ mol/L}$$

Chapter 17

1.

$$NH_3 + H_2O \rightleftharpoons NH_4^+ + OH^-$$

Initial:	0.100	0.200	0.000
Change:	x	x	x
Final:	$0.100 - x$	$0.200 + x$	x

$$K_b = \frac{[NH_4^+][OH^-]}{[NH_3]} = 1.8 \times 10^{-5}$$

$$= \frac{(0.200)x}{0.100} = 1.8 \times 10^{-5}$$

$$x = 9.0 \times 10^{-6} = [OH^-]$$

$$[H_3O^+] = K_w/[OH^-] = (1.0 \times 10^{-14})/(9.0 \times 10^{-6}) = 1.1 \times 10^{-9}$$

$$pH = 8.96$$

2. $HC_2H_3O_2 + NaOH \longrightarrow NaC_2H_3O_2 + H_2O$

Assume that this acid-base reaction proceeds 100%, yielding 0.100 mol $NaC_2H_3O_2$ and leaving 0.150 mol $HC_2H_3O_2$ in excess. These are the initial concentrations for the equilibrium calculation.

$$HC_2H_3O_2 + H_2O \longrightarrow C_2H_3O_2^- + H_3O^+$$

Initial:	0.150	0.100	0.000
Change:	x	x	x
Final:	$0.150 - x$	$0.100 + x$	x

$$K_a = \frac{[C_2H_3O_2^-][H_3O^+]}{[HC_2H_3O_2]} = 1.8 \times 10^{-5}$$

$$= \frac{(0.100)x}{0.150} = 1.8 \times 10^{-5}$$

$$x = [H_3O^+] = 2.7 \times 10^{-5}$$

Chapter 18

1. (*a*) Alkane (hydrocarbon)   (*b*) Alcohol   (*c*) Ketone   (*d*) Amide   (*e*) Ester

2. $CH_3CHOHCH_3$ and $CH_3CH_2OCH_3$

3. $CH_3CH(NH_2)CH_3$, $CH_3CH_2NHCH_3$, and $(CH_3)_3N$

Chapter 19

1. (*a*) $^{22}Na \longrightarrow {}^{22}Ne + {}_{+}\beta$ (a positron)

(*b*) $^{14}N + {}^4He \longrightarrow {}^{17}O + {}^1H$

(*c*) $^{235}U + {}^1n \longrightarrow {}^{140}Ba + {}^{94}Kr + 2\,{}^1n$

2. $\ln \dfrac{m_0}{m} = \dfrac{0.693t}{t_{1/2}} = \dfrac{0.693t}{50.0 \text{ years}} = \ln \dfrac{100.0}{40.0} = 0.9163$

$t = \dfrac{0.9163(50.0 \text{ years})}{0.693} = 66.1 \text{ years}$

INDEX

Letters with page numbers have the following meanings:
f = figure
p = problem
t = table
See also listings in the Glossary.

CONVERSIONS

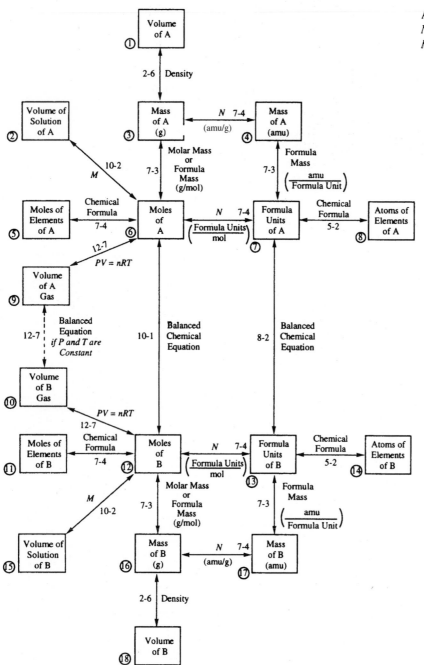

A and B are different substances.
N is Avogadro's number.
M is molarity.

Numbers on each arrow indicate the section in which the conversion was first introduced.

The bottom half of the figure is essentially a reflection of the top half.

348

TABLE OF THE ELEMENTS

Element	Symbol	Atomic Number	Atomic Mass	Element	Symbol	Atomic Number	Atomic Mass
Actinium	Ac	89	(227)	Meitnerium	Mt	109	(266)
Aluminum	Al	13	26.9815	Mendelevium	Md	101	(256)
Americium	Am	95	(243)	Mercury	Hg	80	200.59
Antimony	Sb	51	121.75	Molybdenum	Mo	42	95.94
Argon	Ar	18	39.948	Neodymium	Nd	60	144.24
Arsenic	As	33	74.9216	Neon	Ne	10	20.179
Astatine	At	85	(210)	Neptunium	Np	93	237.0482
Silver	Ag	47	107.868	Nickel	Ni	28	58.71
Gold	Au	79	196.9665	Niobium	Nb	41	92.9064
Barium	Ba	56	137.34	Nitrogen	N	7	14.0067
Berkelium	Bk	97	(249)	Nobelium	No	102	(254)
Beryllium	Be	4	9.01218	Sodium	Na	11	22.9898
Bismuth	Bi	83	208.9806	Osmium	Os	76	190.2
Bohrium	Bh	107	(262)	Oxygen	O	8	15.9994
Boron	B	5	10.81	Palladium	Pd	46	106.4
Bromine	Br	35	79.904	Phosphorus	P	15	30.9738
Cadmium	Cd	48	112.40	Platinum	Pt	78	195.09
Calcium	Ca	20	40.08	Plutonium	Pu	94	(242)
Californium	Cf	98	(251)	Polonium	Po	84	(210)
Carbon	C	6	12.011	Potassium	K	19	39.102
Cerium	Ce	58	140.12	Praseodymium	Pr	59	140.9077
Cesium	Cs	55	132.9055	Promethium	Pm	61	(145)
Chlorine	Cl	17	35.453	Protactinium	Pa	91	231.0359
Chromium	Cr	24	51.996	Lead	Pb	82	207.2
Cobalt	Co	27	58.9332	Radium	Ra	88	226.0254
Copper	Cu	29	63.546	Radon	Rn	86	(222)
Curium	Cm	96	(247)	Rhenium	Re	75	186.2
Dubnium	Db	105	(260)	Rhodium	Rh	45	102.9055
Dysprosium	Dy	66	162.50	Rubidium	Rb	37	85.4678
Einsteinium	Es	99	(254)	Rutherfordium	Rf	104	(257)
Erbium	Er	68	167.26	Ruthenium	Ru	44	101.07
Europium	Eu	63	151.96	Samarium	Sm	62	150.4
Fermium	Fm	100	(253)	Scandium	Sc	21	44.9559
Fluorine	F	9	18.9984	Seaborgium	Sg	106	(263)
Francium	Fr	87	(223)	Selenium	Se	34	78.96
Iron	Fe	26	55.847	Silicon	Si	14	28.086
Gadolinium	Gd	64	157.25	Silver	Ag	47	107.868
Gallium	Ga	31	69.72	Sodium	Na	11	22.9898
Germanium	Ge	32	72.59	Strontium	Sr	38	87.62
Gold	Au	79	196.9665	Sulfur	S	16	32.06
Hafnium	Hf	72	178.49	Antimony	Sb	51	121.75
Hassium	Hs	108	(265)	Tin	Sn	50	118.69
Helium	He	2	4.00260	Tantalum	Ta	73	180.9479
Holmium	Ho	67	164.9303	Technetium	Tc	43	98.9062
Hydrogen	H	1	1.0080	Tellurium	Te	52	127.60
Mercury	Hg	80	200.59	Terbium	Tb	65	158.9254
Indium	In	49	114.82	Thallium	Tl	81	204.37
Iodine	I	53	126.9045	Thorium	Th	90	232.0381
Iridium	Ir	77	192.22	Thulium	Tm	69	168.9342
Iron	Fe	26	55.847	Tin	Sn	50	118.69
Krypton	Kr	36	83.80	Titanium	Ti	22	47.90
Potassium	K	19	39.102	Tungsten	W	74	183.85
Lanthanum	La	57	138.9055	Uranium	U	92	238.029
Lawrencium	Lr	103	(257)	Vanadium	V	23	50.9414
Lead	Pb	82	207.2	Xenon	Xe	54	131.30
Lithium	Li	3	6.941	Ytterbium	Yb	70	173.04
Lutetium	Lu	71	174.97	Yttrium	Y	39	88.9059
Magnesium	Mg	12	24.305	Zinc	Zn	30	65.37
Manganese	Mn	25	54.9380	Zirconium	Zr	40	91.22

PERIODIC TABLE

| Classical | IA | IIA | IIIB | IVB | VB | VIB | VIIB | VIII | VIII | | IB | IIB | IIIA | IVA | VA | VIA | VIIA | 0 |
| Amended | IA | IIA | IIIA | IVA | VA | VIA | VIIA | VIII | VIII | | IB | IIB | IIIB | IVB | VB | VIB | VIIB | 0 |
Modern	1	2	3	4	5	6	7	8	9	10	11	12	13	14	15	16	17	18
	1 H 1.0080																	2 He 4.00260
	3 Li 6.941	4 Be 9.01218											5 B 10.81	6 C 12.011	7 N 14.0067	8 O 15.9994	9 F 18.9984	10 Ne 20.179
	11 Na 22.9898	12 Mg 24.305											13 Al 26.9815	14 Si 28.086	15 P 30.9738	16 S 32.06	17 Cl 35.453	18 Ar 39.948
	19 K 39.102	20 Ca 40.08	21 Sc 44.9559	22 Ti 47.90	23 V 50.9414	24 Cr 51.996	25 Mn 54.9380	26 Fe 55.847	27 Co 58.9332	28 Ni 58.71	29 Cu 63.546	30 Zn 65.37	31 Ga 69.72	32 Ge 72.59	33 As 74.9216	34 Se 78.96	35 Br 79.904	36 Kr 83.80
	37 Rb 85.4678	38 Sr 87.62	39 Y 88.9059	40 Zr 91.22	41 Nb 92.9064	42 Mo 95.94	43 Tc 98.9062	44 Ru 101.07	45 Rh 102.9055	46 Pd 106.4	47 Ag 107.868	48 Cd 112.40	49 In 114.82	50 Sn 118.69	51 Sb 121.75	52 Te 127.60	53 I 126.9045	54 Xe 131.30
	55 Cs 132.9055	56 Ba 137.34	57 La 138.9055	72 Hf 178.49	73 Ta 180.9479	74 W 183.85	75 Re 186.2	76 Os 190.2	77 Ir 192.22	78 Pt 195.09	79 Au 196.9665	80 Hg 200.59	81 Tl 204.37	82 Pb 207.2	83 Bi 208.9806	84 Po (210)	85 At (210)	86 Rn (222)
	87 Fr (223)	88 Ra 226.0254	89 Ac (227)	104 Rf (257)	105 Db (260)	106 Sg (263)	107 Bh (262)	108 Hs (265)	109 Mt (266)									

*

58 Ce 140.12	59 Pr 140.9077	60 Nd 144.24	61 Pm (145)	62 Sm 150.4	63 Eu 151.96	64 Gd 157.25	65 Tb 158.9254	66 Dy 162.50	67 Ho 164.9303	68 Er 167.26	69 Tm 168.9342	70 Yb 173.04	71 Lu 174.97

†

90 Th 232.0381	91 Pa 231.0359	92 U 238.029	93 Np 237.0482	94 Pu (242)	95 Am (243)	96 Cm (247)	97 Bk (249)	98 Cf (251)	99 Es (254)	100 Fm (253)	101 Md (256)	102 No (254)	103 Lr (257)